METHODS IN VIROLOGY

VOLUME VI

METHODS IN VIROLOGY
Advisory Board

METHODS IN VIROLOGY

EDITED BY

KARL MARAMOROSCH

WAKSMAN INSTITUTE OF MICROBIOLOGY
RUTGERS UNIVERSITY
NEW BRUNSWICK, NEW JERSEY

AND

HILARY KOPROWSKI

THE WISTAR INSTITUTE OF ANATOMY AND BIOLOGY
PHILADELPHIA, PENNSYLVANIA

Volume VI

ACADEMIC PRESS   New York   San Francisco   London   1977
A Subsidiary of Harcourt Brace Jovanovich, Publishers

ACADEMIC PRESS, INC.
111 Fifth Avenue, New York, New York 10003

United Kingdom Edition published by
ACADEMIC PRESS, INC. (LONDON) LTD.
24/28 Oval Road, London NW1

LIBRARY OF CONGRESS CATALOG CARD NUMBER: 66-30091

ISBN 0–12–470206–6

PRINTED IN THE UNITED STATES OF AMERICA

Table of Contents

Chapter 1—Application of Immunofluorescence to Diagnosis of Viral Infections
RICHARD W. EMMONS AND JOHN L. RIGGS

Chapter 2—Experimental Analysis of Effector Thymus-Derived Lymphocyte Function in Viral Infections
PETER C. DOHERTY, JON C. PALMER, AND
ROLF M. ZINKERNAGEL

Chapter 3—Methods for Assays of RNA Tumor Viruses
V. KLEMENT AND M. O. NICOLSON

Chapter 4—Assay Methods for Viral Pseudotypes

J. ZÁVADA

Chapter 5—Methods for Assaying Reverse Transcriptase

DANIEL L. KACIAN

Chapter 6—Methods for Studying Viroids

T. O. DIENER, A. HADIDI, AND R. A. OWENS

Chapter 7—Electron Microscopy of Viral Nucleic Acids

DONALD P. EVENSON

Chapter 8—Rapid Immune Electron Microscopy of Virus Preparations

ROBERT G. MILNE AND ENRICO LUISONI

Chapter 9—Transfection Methods

V. M. ZHDANOV

Chapter 10—Invertebrate Cell Culture Methods for the Study of Invertebrate-Associated Animal Viruses

D. L. KNUDSON AND S. M. BUCKLEY

Chapter 11—Techniques of Invertebrate Tissue Culture for the Study of Plant Viruses

D. V. R. REDDY

Chapter 12—Use of Protoplasts for Plant Virus Studies

S. SARKAR

Chapter 13—Nucleic Acid Hybridization Technology and Detection of Proviral Genomes

ENG-SHANG HUANG AND JOSEPH S. PAGANO

List of Contributors

Numbers in parentheses indicate the pages on which the authors' contributions begin.

S. M. BUCKLEY, Yale Arbovirus Research Unit, Department of Epidemiology and Public Health, Yale University School of Medicine, New Haven, Connecticut (323).

T. O. DIENER, Plant Virology Laboratory, Plant Protection Institute, ARS, U.S. Department of Agriculture, Beltsville, Maryland (185).

PETER C. DOHERTY, The Wistar Institute of Anatomy and Biology, Philadelphia, Pennsylvania (29).

RICHARD W. EMMONS, Viral and Rickettsial Disease Laboratory, California State Department of Health, Berkeley, California (1).

DONALD P. EVENSON, Memorial Sloan-Kettering Cancer Center, New York, New York (219).

A. HADIDI, Plant Virology Laboratory, Plant Protection Institute, ARS, U.S. Department of Agriculture, Beltsville, Maryland (185).

ENG-SHANG HUANG, The Cancer Research Center and Department of Medicine, The University of North Carolina at Chapel Hill, Chapel Hill, North Carolina (457).

DANIEL L. KACIAN, Institute of Cancer Research, and Department of Human Genetics and Development, College of Physicians and Surgeons, Columbia University, New York, New York (143).

V. KLEMENT, Departments of Pediatrics, Microbiology, and Biochemistry, University of Southern California School of Medicine, and Childrens' Hospital of Los Angeles, Los Angeles, California (59).

D. L. KNUDSON, Yale Arbovirus Research Unit, Department of Epidemiology and Public Health, Yale University School of Medicine, New Haven, Connecticut (323).

ENRICO LUISONI, Laboratorio di Fitovirologia applicata del C.N.R., Turin, Italy (265).

ROBERT G. MILNE, Laboratorio di Fitovirologia applicata del C.N.R., Turin, Italy (265).

M. O. NICOLSON, Departments of Pediatrics, Microbiology, and Biochemistry, University of Southern California School of Medicine, and Childrens Hospital of Los Angeles, Los Angeles, California (59).

R. A. OWENS, Plant Virology Laboratory, Plant Protection Institute, ARS, U.S. Department of Agriculture, Beltsville, Maryland (185).

ix

JOSEPH S. PAGANO, The Cancer Research Center and Department of Medicine, The University of North Carolina at Chapel Hill, Chapel Hill, North Carolina (457).

JON C. PALMER, The Wistar Institute of Anatomy and Biology, Philadelphia, Pennsylvania (29).

D. V. R. REDDY,* Department of Genetics and Development, University of Illinois, Urbana, Illinois (393).

JOHN L. RIGGS, Viral and Rickettsial Disease Laboratory, California State Department of Health, Berkeley, California (1).

S. SARKAR,† Max-Planck-Institut für Biologie, Abteilung-Melchers, Tübingen, Germany (435).

J. ZÁVADA, Institute of Virology, Slovak Academy of Sciences, Bratislava, Czechoslovakia (109).

V. M. ZHDANOV, The D. I. Ivanovsky Institute of Virology, Moscow, USSR (283).

ROLF M. ZINKERNAGEL, Scripps Clinic and Research Foundation, La Jolla, California (29).

*Present address: International Crops Research Institute for the Semi-Arid Tropics, 1-11-256 Begumpet, Hyderabad—500 016, A.P., India.

†Present address: Institut für Phytomedizin, Universität Hohenheim, Otto-Sander St. 5, D-7000 Stuttgart 70, West Germany.

Preface

Six years after the publication of the fifth volume of *Methods in Virology*, the Editors have decided that the time has come to acquaint old and new readers with the new tools available for the study of viruses. The Editors have followed closely the development of improved techniques in virology and have been faced with the difficult task of selecting the methods applicable not only to the research problems of today but also to those envisioned for the future. Although the contributors have been asked to explain the pertinence of their methods to current research as well as to project possible future uses, the Editors anticipate that readers will see possibilities for a broader application or a narrower refinement of a technique to meet their unique research needs—possibilities beyond those envisaged by the author.

Volume VI presents detailed discussions of the methods used to study interactions between viruses and host cells and between viral nucleic acids and proteins. The volume also deals with methodology in the fields of immunochemistry and morphology and contains specialized information on virus infection of higher and lower animals and of plants and bacteria. As were the earlier volumes, this volume is intended to be a general reference work for laboratory investigators as well as an informational source for graduate students and teachers. The Editors consider that it is the prerogative of each author to choose the form of presentation for his chapter; all the volumes, therefore, include chapters that deal exclusively with methodology and chapters that provide a more elaborate interpretation of the results obtained by the method described.

Although this volume and its predecessors cover many techniques in virology, the Editors realize that the task of keeping the reader informed about developments in the technology of science is a continuous one; it is their intention, therefore, to meet this need through the publication of additional volumes.

The Editors wish to thank their Board of Advisors for invaluable assistance in the conception of this volume and to thank all the contributors for the promptness and excellence of their collaboration. For their continual encouragement and for meeting the many challenges of book production, we are greatly indebted to Academic Press.

KARL MARAMOROSCH
HILARY KOPROWSKI

Contents of Other Volumes

1 Application of Immunofluorescence to Diagnosis of Viral Infections*

Richard W. Emmons and John L. Riggs

I. Introduction

The immunofluorescence (IF) technique, first developed and applied clinically by Coons et al. (1942), was recognized early as having great potential for research and diagnostic applications in virology. One of the first examples of its diagnostic use was that of Liu (1956), who examined nasal washings from suspected cases of influenza and found that positive diagnosis could often be made within 2 hours, while the conventional techniques took many days. Another early use of the method was that of Goldwasser and Kissling (1958) for detecting rabies virus antigen in brains of inoculated mice. McQueen et al. (1960) extended this use to the direct detection of rabies in animals, and its simplicity and sensitivity has made it the diagnostic method of choice. The technique was also utilized diagnostically for measles (Cohen et al., 1955), herpes simplex (Biegeleisen et al., 1959), varicella-zoster (Weller and Coons, 1954), poliomyelitis (Kalter et al., 1959), ECHO and

*A part of the work upon which this review is based was supported by Grant AI-01475 from the National Institute of Allergy and Infectious Diseases, National Institutes of Health, U.S. Public Health Service, Department of Health, Education and Welfare.

1

Coxsackie viruses (Shaw *et al.*, 1961), smallpox (Murray, 1963), respiratory syncytial virus (Schieble *et al.*, 1965), and rubella virus (Schmidt *et al.*, 1966).

After more than three decades, extensive research experience and a flood of literature have accumulated. Well over 500 English-language references to IF methods in diagnostic virology alone could be listed, but examples of their routine use remain surprisingly rare. Among the reasons for this are: lack of commercially produced, reliable reagents; inadequate quality control of IF procedures; shortage of experienced technical personnel; and the medical community's lack of confidence in the reliability of diagnostic IF tests.

In an earlier volume of this series (Volume III), the basic principles and methods of IF staining were presented in detail (Casals, 1967; Spendlove, 1967). After nearly a decade, it is appropriate to review the recent technical advances and practical applications that have been made in this field. A comprehensive restatement of IF theory and methods for preparation of IF reagents is beyond the scope of this chapter and is unnecessary. The reader is referred to many excellent reviews of these subjects (Beutner, 1971; Blundell, 1970; Cherry, 1974; Gardner and McQuillin, 1974; Goldman, 1968; Hebert *et al.*, 1972; Hijmans and Schaeffer, 1975; Kawamura, 1969; Liu, 1969; Marks, 1974; Nairn, 1969; Riggs, 1965; Rose and Bigazzi, 1973). Instead, emphasis is placed on the techniques needed to effectively transfer IF methods from their traditional research setting to practical clinical use.

II. Principles of Immunofluorescence Staining

The IF technique provides a means of observing an antigen–antibody reaction by chemically linking a fluorescent dye, such as fluorescein isothiocyanate (FITC), to specific antibody molecules. Such labeled antibodies retain the ability to react specifically with their respective antigens and, when viewed in a fluorescence microscope, the reaction site is visually detected by its fluorescence. The specificity of antigen–antibody reactions, coupled with the relatively short time required to prepare and observe the specimen, makes the IF procedure an ideal diagnostic tool for many infectious diseases, especially those of viral origin.

As with the case of any immunological procedure, the successful use of the IF technique requires careful preparation and standardization of reagents. A high-titered immune serum that is specific for the antigen in question must be labeled with the proper amount of dye to be useful in the direct technique. The reaction should be properly controlled (negatively and positively) and the conjugate properly characterized before being routinely used in a diagnostic laboratory.

III. Methods for Immunofluorescence Staining

A. EQUIPMENT

The standard fluorescence microscope supplied by most manufacturers uses a mercury lamp light source, such as the Osram HBO 200, and a glass filter system usually consisting of a heat-absorbing filter (e.g., Zeiss KG-1), excitation filters (Schott BG-12, UG-2, UG-5), and a barrier filter excluding wavelengths below certain levels. These filter combinations are utilized in conjunction with a dark-field condenser in a binocular microscope with 10 × oculars. Some workers prefer the more intense illumination provided by a monocular microscope, but generally the ease and convenience of a binocular microscope more than offsets this advantage.

The recent introduction of interference filters for fluorescence microscopy (Rygaard and Olsen, 1969, 1971) has made possible the use of ordinary tungsten bulbs as light source. These 2-band interference filters have been developed to produce a transmission band of up to 85% in the 400–500 nm (blue) range, and a smaller band (1% transmission) at approximately 630 nm (red). The maximum absorption peak of fluorescein isothiocyanate is at approximately 495 nm and thus the interference filter effectively excites typical green fluorescence at this wavelength, with the smaller band providing a contrasting red background which helps in interpreting morphological detail. The red contrast band can be reduced or eliminated by using a blue, red-excluding filter (BG-18) of appropriate thickness as part of the barrier filter system (Rygaard and Olsen, 1971).

The usual barrier filters are used with the interference filter, allowing the microscopist to observe the green fluorescence of fluorescein at approximately 520 nm. The Osram 12-V 100-W halogen lamp can be substituted for the ordinary tungsten bulb as a more intense light source for this system of filters.

Another recent advance in the instrumentation of fluorescence microscopy is the use of incident light, or epi-illumination (Ploem, 1967). In this system the illumination is directed to the specimen from above, eliminating the need for a substage condenser. Light from the source is directed vertically to a dichroic mirror, and then downward to the specimen. In this manner the objective acts as a condenser. Light from the excited fluorescence is then directed through the dichroic mirror and a system of barrier filters to the observer.

The choice of instrumentation and types of illumination is of course influenced by the intended use (routine diagnostic or research), the fluorochromes to be used (fluorescein, rhodamine, etc.), and whether counterstaining (such as with Evans blue or rhodamine conjugated to bovine serum albumin) is to be employed.

B. PREPARATION OF REAGENTS

1. *Preparation of Immune Serum*

An important aspect of the preparation of immune serum for IF reagents is that the procedure should be tailored to suit each host–virus system. If possible, one should prepare the viral antigen in animal cells of the same species as the animal to be immunized, thus avoiding production of unwanted antihost cell components in the immune serum. If viral antigen must be used to immunize heterologous animal species, it must be purified in some manner, preferably by density-gradient or differential centrifugation. However, even with such purified antigens, the antibody produced may reveal contamination with antihost cell components and growth medium components when used for IF staining.

The use of serum from animals convalescent from a natural infection may also avoid unwanted antitissue component staining, since the antibody in such animals is theoretically directed against the viral antigens alone. However, the presence of other specific antiviral antibodies must be considered (see Section III, C, 8).

The immunization schedule should be designed to elicit a high-titered, specific antibody response without producing antibodies cross-reactive with other viruses, as prolonged schedules may do. Various laboratory animals can be used for immune serum production, including rabbits, mice, guinea pigs, hamsters, monkeys, or goats. Mice can be used to produce large volumes of ascitic fluid containing high-titered antibodies. Newly purchased animals should be conditioned for a period of time and any animals showing signs of disease should not be used for immune serum production. Young adult animals should be used, since older animals may have had experience with a variety of antigens and may therefore possess unwanted antibodies. The animal colony should be monitored regularly to detect and eliminate unwanted viral or bacterial infections.

A procedure similar to the standard one used for producing rabies-immune hamster serum in our laboratory (Lennette *et al.*, 1965) can be implemented as a general model for production of IF reagents, and is briefly summarized below.

a. Antigen Preparation. 1. Inoculate 4-week-old hamsters intracerebrally with 0.03–0.05 ml of a 2% suspension of fixed rabies virus (CVS strain), containing 500 units each of penicillin and streptomycin per milliliter.

2. Harvest infected brain tissue when the hamsters become ataxic and moribund (usually 4 to 5 days postinfection) and prepare a 20% brain suspension in 0.85% sterile saline.

3. Inactivate the virus suspension by the addition of β-propiolactone (BPL): prepare a 10% solution of BPL in cold distilled water; add 10% by

volume of cold 8.8% sodium bicarbonate solution to the virus-infected brain suspension; then add 2% by volume of the freshly prepared BPL solution to produce a final concentration of approximately 0.2% BPL. The added bicarbonate neutralizes the acid produced when the BPL decomposes. Incubate the mixture at 37°C for 2 hours, safety test to assure complete inactivation, then store at −60°C until needed. Additional virus suspension is prepared without BPL inactivation for part of the immunization schedule.

b. *Antiserum Production.* 1. Young adult hamsters are inoculated intraperitoneally at weekly intervals (see below). A 1:1 mixture of the inactivated virus suspension and incomplete Freund's adjuvant is used for the first and third injections: (a) 1 ml of killed antigen plus adjuvant, (b) 1 ml of killed antigen, (c) 1 ml of killed antigen plus adjuvant, (d) 1 ml of killed antigen, (e) 1 ml of live antigen, and (f) 1 ml of live antigen.

2. Fourteen days following the final injection of live virus, the animals are bled out and the sera are tested for rabies antibodies by the indirect IF procedure. The high-titered sera are pooled, stored at −20°C, then conjugated with FITC as needed for use in the direct IF procedure for rabies.

Other examples of useful systems for producing specific FITC-conjugated sera are shown in Table I. Where cell cultures are used as antigen, the cells and the serum used in the medium should be from the same animal species to be immunized. Details of the immunization schedule and choice of antigen will vary, but the basic principles remain the same: avoid including any protein antigens in the immunizing material except homologous host tissue or serum and the desired viral antigen, and use the minimum number of

TABLE I

SYSTEMS FOR PREPARATION OF IMMUNOFLUORESCENCE REAGENTS

Immunofluorescence conjugate	Animal used to prepare immune serum	Tissue used for immunizing antigen
Herpes simplex, types 1 and 2	Hamster	Hamster brain
Rabies	Hamster	Hamster brain
Measles	Hamster	Hamster brain
Colorado tick fever	Hamster	Hamster brain
Varicella-zoster	Rhesus monkey	Primary rhesus monkey kidney cell culture (rhesus monkey serum)
Vaccinia	Rabbit	Rabbit kidney cell line (RK13)
Rubella	Rabbit	(rabbit serum)
Influenza A and B	Rabbit	Allantoic fluid ("purified")
Coronavirus	Rabbit	Human fetal diploid cell culture ("purified")

inoculations sufficient to produce a good antibody titer. Inactivation of the antigen is necessary only when the virus is lethal for the animal species.

2. *Labeling with Fluorescein Isothiocyanate*

It has been shown that overlabeled antibodies combine nonspecifically with many normal cellular components, thus inducing nonspecific staining, while underlabeled antibodies combine with the antigen and block specific staining. The fluorescein-to-protein (F/P) ratio of a conjugate has been used by many investigators as an indication of the amount of nonspecific staining that will be obtained, a high F/P ratio giving the greatest amount of nonspecific reactivity. One must consider the protein content of the antibody to be labeled, the procedures followed in the conjugation reaction, the "purification" of the antibody after conjugation, and the use of the conjugate in the staining procedure itself (Wells *et al.*, 1966; Hebert *et al.*, 1967; Cherry, 1974).

Although at present there is no universally accepted procedure for producing suitable conjugates of the desired specificity and sensitivity for all IF work, the following conjugation procedure has given us consistently good results in our laboratory:

a. A crude globulin fraction of the immune serum is obtained by two consecutive precipitations with 35–45% ammonium sulfate (obtained by adding an equal volume of 70–90% saturated ammonium sulfate to the serum).

b. The precipitates are recovered by centrifugation and the final precipitate is dissolved in 0.01 M phosphate-buffered saline (PBS), pH 7.2, in a smaller volume than that of the starting serum (e.g., precipitate from 20 ml of serum is dissolved in 10–15 ml of PBS).

c. The solution of crude globulins is dialyzed against PBS to remove the ammonium sulfate. This can be accomplished by dialyzing for 24 hours against three 1-liter changes of PBS at 4°C.

d. The protein content of the crude globulin solution is determined by either the method of Lowry *et al.* (1951) or by the biuret procedure (Gornall *et al.*, 1949).

e. The globulin solution (0.5–1.5% protein) is buffered by the addition of 0.5 M carbonate–bicarbonate buffer, pH 9.0, to 10–15% of the volume of the globulin solution.

f. Fluorescein isothiocyanate dissolved in a small volume ($\lesssim 0.5$ ml) of acetone is added drop by drop to the stirred, buffered solution in the ratio of 1 mg of dye to 100 mg of protein. The acetone–dye mixture can be heated on a water bath at 37°C to help the dye dissolve, or if it fails to do so, the acetone–dye mixture can be added as a slurry.

g. The mixture is transferred to the cold room (4°C) and is stirred slowly overnight with a magnetic stirrer.

3. Purification Procedure

In order to obtain the antibodies that are optimally labeled and to remove the antibodies that are over- and underlabeled, DEAE fractionation of the conjugate is performed.

a. The DEAE-cellulose is prepared by first washing with a 0.5 M NaOH solution, followed by copious washing with distilled water. The washed DEAE-cellulose is then equilibrated by washing with 0.0175 M phosphate buffer, pH 6.3, and is stored as a slurry in the buffer at 4°C.

b. The globulin solution conjugated to the FITC is dialyzed against the 0.0175 M buffer. A precipitate forms at this pH and is discarded.

c. A chromatography column is prepared with the washed, buffered DEAE-cellulose (a 2 cm diameter column of 15 cm height is sufficient for the globulins obtained from 20 ml of whole immune serum).

d. The column is washed and packed with the 0.0175 M buffer, with approximately 3–5 lb/in² air pressure, and the dialyzed conjugate is then applied to the column. Once the conjugate is on the DEAE-cellulose bed, the column is washed with 100 ml of the starting buffer. The eluate from this wash is discarded, since it contains the unwanted, unlabeled, and under-labeled globulins.

e. The column is then eluted with a solution of 0.075 M NaCl in the pH 6.3 buffer. A colored fraction can be seen moving down the column, and this fraction is retained as it contains the optimally labeled globulins. The eluate is caught in small fractions (3–5 ml) and as the desired fraction moves down the column and is collected, the intensity of the color of the fractions "tails off." The most highly colored fractions are then combined (usually 1.5 to 2 times the volume of the starting conjugate solution).

f. The combined fractions are then dialyzed against PBS, pH 7.2, and are ready for use in the IF procedure.

An alternative to the DEAE fractionation procedure is to carry out the conjugation procedure as described, and then either dialyze the conjugate exhaustively against PBS, pH 7.2, or pass it through a Sephadex column to remove unwanted FITC and bring the pH and ionic concentration back to the physiologic range. If the conjugate shows an excessive amount of non-specific staining in subsequent testing, it can be diluted in 20% normal mouse or beef brain suspension in PBS to help reduce such staining (see Section III, C, 4).

4. Specificity Testing and Titration of Conjugates

Before use as a diagnostic reagent, each conjugate must be thoroughly characterized as to its specificity and sensitivity. In the case of the direct technique the conjugate from the DEAE column is serially diluted in PBS while the conjugate prepared without DEAE fractionation is serially diluted in the 20% mouse or beef brain suspension. The dilutions of the conjugates are applied to preparations of the antigens (infected cell cultures, slip smears prepared from known virus-positive tissues, etc.) and are allowed to react for 30–45 minutes in a humid atmosphere at 37 °C. The preparations are then washed in 3 changes of PBS, 5-minute washing per change, are mounted in 25% glycerol in PBS or Elvanol mounting medium, and are examined in the fluorescence microscope. The last dilution showing a 1–2 + reaction determines the staining titer of the conjugate. In practice, the conjugate is usually used at a dilution of 1 to 2 tubes below the last dilution showing a 3–4 + reaction (usually a 1:5 or 1:10 dilution of the conjugate obtained from the DEAE column or a 1:40–1:80 dilution of the conjugate not purified by DEAE chromatography). Nonspecificity staining titers are also determined for each conjugate by using the same dilutions on uninfected cell cultures or tissue specimens. Cross-reaction staining titers are determined in the same manner, utilizing specimens prepared from cell cultures or tissues infected with heterologous viral antigens. A conjugate showing much nonspecific staining or a cross-reaction with a heterologous viral antigen, except where the cross-reaction is expected (e.g., V-Z and herpes simplex), should not be used as a diagnostic reagent.

For the indirect procedure a chessboard titration of dilutions of the intermediate serum against dilutions of the antispecies conjugate is carried out in the same manner, utilizing an antiserum of known antiviral reactivity as the intermediate serum. Nonspecific staining is also determined, utilizing uninfected specimens in the staining technique, and using normal serum on virus-infected specimens from the same species of animal against which the antiglobulin conjugate was prepared.

5. Preservation of Reagents

The conjugate is divided into 2 ml portions, suitably labeled, and is either lyophilized or stored frozen at $-20°C$. Upon reconstitution or thawing, the conjugate is diluted to the working dilution, filtered through an 0.45 μm Millipore filter to remove any precipitate which may have formed, and is stored at 4 °C. Conjugates stored frozen in this manner have retained titer in our laboratory as long as 8 years.

6. Photography

For a permanent record of IF stained specimens, the best photographs are transparencies obtained by using a 35 mm camera back attached to the fluorescence microscope. Either Ektachrome (ASA 125) or Super Ansco-chrome (ASA 200–500) color film can be utilized, and can be processed in the laboratory using the development kits available for such purposes. For black and white photography, Kodak Plus-X film (ASA 125) is a relatively fast film which can be processed using Kodak Microdiol-X developer at a 1:3 dilution to give a finer grain. Printing of such film on high-contrast paper results in an accurate record of what is actually seen by the observer.

C. Immunofluorescence Staining Procedures

1. *Collection and Transport of Clinical Specimens*

Specimens for IF tests should be collected aseptically using separate sterile instruments and separate vials for each specimen to avoid cross-contamination, since they may also be needed for virus isolation tests. Each specimen should be labeled with the patient's identity, date, and type of specimen. Glass slides should be labeled by etching the glass with a diamond marking pencil, or by marking the frosted section of the slide with a pencil that will not smudge or dissolve in acetone. Marks should always be on the same side as the tissue smear. A virus examination request form should accompany each specimen, giving lesion site, patient identification, and clinical details that will help the virologist decide appropriate tests to perform.

Ideally, tissue or body fluid samples should be tested in the fresh state, as promptly as possible. In practice, some delay is usually unavoidable. Samples should then be held in the refrigerator and hand-carried to the laboratory on wet ice. If longer storage and mailing are necessary, the specimen should be kept frozen (preferably at $-50\,^\circ$C or colder) and transported on Dry Ice in a flame-sealed glass vial or screw-capped vial properly sealed in a plastic bag to prevent absorption of the CO_2 and a detrimental lowering of the pH. However, freezing and thawing tissue may damage cell and tissue structure, making interpretation of the IF staining more difficult. For certain viruses (e.g., cytomegalovirus) freezing may be detrimental to the virus: fluids should be mixed with an equal volume of 70% sorbitol, and tissue should be immersed in 70% sorbitol before freezing and shipping on Dry Ice. For the rabies IF test on animal brain samples, preservation and mailing of the tissue in 50% neutral glycerol saline in screw-capped jars without refrigeration has proved satisfactory. The specimen should be washed in 0.85% saline three times for 20-minute periods to remove the glycerol, which can inhibit fluorescence. This economic method may be applicable to the routine diagnosis of other viral diseases if many specimens must be mailed to a distant laboratory.

Clotted blood samples for the diagnosis of Colorado tick fever are also mailed unfrozen to the laboratory. The serum is used for antibody tests, and the blood clot is used for attempts at virus isolation and to prepare smears for IF staining. The relatively heat-stable cell-associated virus and viral antigen easily survive the few days' transport time (Emmons and Lennette, 1966). This method might be applicable to the routine diagnosis of other diseases with a significant viremia period (e.g., measles, Rocky Mountain spotted fever, or cytomegalovirus); however, the brief viremic period in most viral diseases and the lability of the viruses require either immediate inoculation of blood samples or preservation by freezing until virus isolation can be attempted.

Other body fluids [urine, cerebrospinal fluid (CSF), saliva, and nasopharyngeal or bronchial washings] may be submitted for virus isolation attempts and IF staining of cell sediment. These should be examined fresh when possible, since freezing and thawing might damage the cells for IF staining or expose intracellular virus to antibody or enzyme action.

Tissue smears or cellular material from skin lesions or body fluids are prepared on glass slides (see Section III,C,2,a), which are air dried, then mailed to the laboratory without refrigeration. A protective plastic box or mailing tube, rather than a cardboard slide container, should be used to avoid breakage or contamination of the outer packing material.

Serum or whole blood samples sent for serologic tests, such as the indirect IF test, usually are sent unrefrigerated; in contrast to the problem of frozen whole blood being unsatisfactory for complement-fixation (CF) or hemagglutination-inhibition (HI) tests, frozen blood can be used for the indirect IF test, if serum is not available. Antibody titration can also be performed on CSF and other body fluids, or on tissue suspensions prepared for virus isolation attempts.

Cell cultures submitted for identification of isolated viruses are mailed as an intact cell monolayer in bottle or tube culture, with the container filled nearly full of maintenance medium, so that shaking in transit will not damage the cell sheet. Further passage of the isolate or harvesting and preparation of cell smears for the IF test can then be done (see Section III,C,2,a).

The latest requirements for packaging and labeling potentially infectious material ("etiologic agents"), particularly when interstate shipment is involved, should be adhered to (refer to Federal Register Code of Federal Regulations, Section 72.25 of Part 72, Title 42, Revised June 6, 1972, and National Institutes of Health Guide, February 10, 1975). All materials for viral study should be considered potentially hazardous, and appropriate precautions should be taken in handling packages in the mailroom and in opening them in the laboratory. Basically, diagnostic specimens must be packaged to withstand leakage of contents, shock, pressure changes, or other conditions of shipping and handling; use a securely closed, watertight primary container surrounded by absorbent material in case of leakage, then enclosed in a durable watertight secondary container, then in an outer shipping container, appropriately labeled with an approved "Biomedical Material Red Label." On receipt in the laboratory, each specimen and its accompanying submittal request form is assigned an accession number and properly stored until testing. A portion of each specimen is kept frozen (−65°C or colder, if possible) for future reference, research, or for medical–legal purposes (at least 1 year in this laboratory). Selected positive specimens are kept in permanent storage.

2. Preparation of Slides

a. *Tissue Impression Smears or Slip Smears.* The standard procedure as used for rabies tests (Lennette *et al.*, 1965) is suitable for any tissue that can be easily crushed and spread in a thin smear. Frozen sections may be better for tougher tissues, or where preservation of tissue architecture is desired. The knife blade and cabinet should be considered contaminated after each procedure, and should be appropriately disinfected. Impression smears may be used for autopsy or biopsy tissues from the patient, or for identification of viral isolates in an animal host tissue such as mouse brain (rabies, herpes, or arboviruses).

Three or more clean, standard glass microscope slides 1.1 mm thick or less, with one frosted end, are prepared for each specimen, and are labeled with the laboratory accession number, type of specimen, and date slide is prepared. Adequate sampling to include various portions of the involved tissue or organ is important, since viral antigen may be unevenly distributed. The special precautions relating to rabies diagnosis in this regard have been emphasized to avoid a serious false-negative result (Lennette and Emmons, 1971; Dean and Abelseth, 1973).

Place a small section of tissue (1.0–2.0 mm in diameter) in the center of the slide. Place another slide on top of the first slide and press to crush the tissue (use of disposable plastic gloves is advisable); then move the slides back and forth a short distance a few times and draw the top slide across the length of the bottom slide. Prepare at least two more slides in the same manner. The top slides can be discarded, since they often will have too thick a tissue smear for satisfactory staining; but they may also be kept to increase the total amount of tissue examined (particularly if the sample submitted is small, or if fluorescent antigen is expected to be sparse). Place the slides back to back in a glass slide carrier and allow to air dry at room temperature. Then immerse the slides entirely in acetone at −20°C overnight for best fixation. If emergency testing is needed, slides may be fixed in acetone at room temperature for 10 minutes, but more care in handling, staining, and washing them may be necessary to prevent washing off of the tissue smear. Some of the slides can be either held in acetone at −20°C, or dried and held frozen at −65°C for later examination.

For tissue that cannot be easily crushed or where the tissue surfaces are to be examined (lung, spleen, pericardial or liver surfaces, excised skin from vesicular lesions, etc.), grasp tissue with forceps, blot excess fluid on sterile filter paper, and gently press, or smear in small circular motion over an area of the slide 30–40 mm in length. For frozen sections, 3 or 4 sections can be placed on each slide. Prepare three or more slides for each specimen, dry, and fix as above. The recommendations given for labeling, fixing, and storing specimens apply to all the procedures described subsequently.

b. Vesicular Lesions or Epithelial Tissue. Gently sponge several early-stage cutaneous vesicles with 70% alcohol, then allow to dry. Gently blot mucous membrane lesions with sterile gauze to remove excess fluid and mucus (do not cause bleeding). Avoid older or pustular lesions, as the antigen and virus may be sparse, and leukocytes may fluoresce nonspecifically. Carefully remove the caps from vesicles and blot vesicular fluid and cell debris; take up cellular material from the base of the lesion onto a sterile swab slightly premoistened with holding medium. Place the swab into 2–3 ml of holding medium (tryptose–phosphate broth with 0.5% gelatin and phenol red indicator, pH 7.2–7.4) for later virus isolation attempt. Fluid may also be collected into glass capillary pipettes for virus isolation attempts or electron microscopy. Scrape cellular material from the base of lesions with a sterile scalpel blade, tightly rolled swab, or the broken end of a swab stick, and prepare 2 or 3 thin smears 10–15 mm in diameter on each of 3 to 4 labeled slides. Similarly, swabs can be used to obtain nasal or conjunctival epithelial cells. Prepare smears by rolling the swab over the slide. Allow slides to air dry, and submit unfixed to the laboratory along with virus isolation specimens and blood samples for serologic tests.

c. Blood Smears. Blood smears for the rapid diagnosis of Colorado tick fever (Emmons and Lennette, 1966) may be prepared using portions of blood clot for slip smears, as described above for brain tissue, or by smearing a drop of blood or a small piece of blood clot in a small circle on the slide (as for a malaria thick smear).

d. Cellular Sediment from Body Fluids. Although it is not a routine procedure in this laboratory, testing of cellular sediment from CSF, pleural fluid, urine, saliva, bronchial or nasal washings, etc., may be diagnostically helpful or of research interest. Examination of CSF has been promoted for the rapid diagnosis of herpetic encephalitis, although we have not been able to confirm the validity or routine usefulness of this procedure. The non-specific staining of polymorphonuclear leukocytes and other cell debris, particularly if an indirect IF staining method has been used, is troublesome and may give false-positive results. In this type of study the IF method should not be used alone, but in conjunction with virus isolation attempts and serologic tests.

e. Cell Cultures for Identification of Virus Isolates. Cell cultures suspected of virus infection (ideally, cultures showing 2–3 + CPE, without appreciable loss of cells from the glass) can be used for identification by IF staining (e.g., measles, rubella, herpes simplex, vaccinia, and varicella-zoster viruses).

A single virus-infected cell culture tube usually has sufficient antigen for IF identification; however, the usual practice is to mix homologous normal cells with infected cells to give sufficient cell volume for easier manipulation and better contrast when viewed in the fluorescence microscope.

Certain cell cultures require trypsin or trypsin-versene treatment to aid in dispersing cells [e.g., human fetal diploid (lung), BSC-1, VERO cells], while other cell types can be sufficiently dispersed by scraping the cell sheet into the maintenance medium with a 1 ml pipette and then re-peatedly bulb-pipetting the mixture (e.g., primary monkey kidney, RK13, BHK-21 cells).

For each virus-infected culture to be identified, select 2 to 4 normal, uninoculated homologous cell culture tubes and pool the virus-infected and normal cells. If trypsin or Versene is required, remove the maintenance medium and add 1 ml of prewarmed 0.25% trypsin or Versene in PBS, pH 7.5, to each tube. Rinse the cell sheets by rotating the tubes by hand for 10–15 seconds. Remove the trypsin completely and allow the tubes to stand 2–5 minutes until the cells begin to loosen from the glass. Wash down, collect, and pool the virus-infected and normal cells with 3 ml of PBS containing 2% FBS. A normal cell culture suspension is also prepared as a negative control for the IF staining. Centrifuge each cell suspension at 2000 rpm for 5 minutes, then completely remove and discard the supernatant fluid without disturbing the cell pellet. Resuspend the cells uniformly in 0.05 ml PBS with 2% FBS.

Take up 0.05 ml (control) normal cell suspension with a 0.2 ml pipette or Pasteur pipette equipped with bulb. Hold the pipette vertically, touch the tip to a control slide, and quickly prepare 2 or 3 small spot drops (approximately 5 mm in diameter, representing approximately 0.005 ml) on each of the control slides. Take up 0.05 ml of the virus-infected suspension and similarly prepare 2 or 3 small spot drops on each of the 4 test slides. Spot drops should appear as hazy suspensions, not granular (too dilute) or milky (too thick). Allow the slides to air dry, then fix in acetone and proceed with IF staining procedures or store the slides at −65°C until needed.

If desired for the study of *in vitro* growth characteristics, or for better determination of cytoarchitecture, cell cultures can be prepared on glass cover slips. Somewhat different and more troublesome methods are then needed to carry these cultures through the fixing, staining, and examination procedures, and this technique offers little advantage for the routine diagnostic laboratory. The reader should consult the literature for references to cover slip methods or microculture procedures designed to handle large numbers of specimens rapidly and conveniently.

3. *Preparation of Control Slides*

Positive and negative control slides, used for serum antibody assays by indirect IF and as checks on the potency and specificity of direct IF staining reactions, are prepared ahead of time and stored at −65 °C. In general, the same type of tissue (mouse brain, blood smear, or specific cell culture) to be used in the test slides is used for control slides. Rabies-positive mouse brain slides usually maintain antigenicity for at least a month when held at −20 °C in acetone, and virus-infected tissue culture smears maintain antigenicity a year or more when held at −65 °C (electric freezer, not Dry Ice box). In routine processing of large numbers of specimens (as in a rabies laboratory) sufficient negative brain specimens are usually submitted to act as the negative control in each test run. Positive and negative control slides for each virus are prepared essentially as described in Section III,C,2,e, using bottle instead of tube cultures. Positive control slides should be prepared with a mixture of about 10–30% virus-infected cells, and 70–90% normal cells, to provide a good contrast and ease of interpretation of the IF staining. The pooled cell harvest from one 8-oz. bottle of virus-infected human fetal diploid kidney cells, and two normal 8-oz. bottles, resuspended in 0.9 ml of **PBS** containing 2% FBS, can be used to prepare 60–80 positive control slides, with 2 or 3 spot drops per slide. Dilute the suspension as necessary so it is hazy, but not granular (too dilute) or milky (too concentrated). Slides may be stored conveniently at −65 °C in a 100-slot slide box (2 slides back to back per slot). The box is labeled with cell strain, passage level of cells, virus identity and passage number, date cells were inoculated, and date cells were harvested and slides made. Antigenic reactivity is preserved up to a year or more without appreciable loss.

4. *Direct IF Staining Procedures*

The direct staining method is usually preferable to the indirect method for detecting virus antigen, since it is simpler, more rapid, and less subject to misinterpretation of nonspecific background staining. The necessity to prepare and maintain many different conjugates may seem impractical for the occasional user, but the more reliable results justify the effort. Properly prepared antisera conjugated with FITC are remarkably stable reagents and maintain their titer indefinitely when stored frozen and for a year or more when stored at 4 °C if kept from bacterial contamination. The direct IF method is required for rabies, and is strongly recommended for herpes, varicella-zoster, vaccinia, Colorado tick fever, measles, influenza, and respiratory syncytial viruses. Many of the problems and disappointments experienced by laboratories attempting to initiate IF methods may have been due to the use of the indirect IF method with improperly prepared immune serum for the intermediate step, along with inadequate specificity controls. The following is a step-by-step procedure for using the direct IF method for the examination of tissue preparations, as previously described.

a. Acetone-fixed test slides and positive and negative control slides are prepared, as described, and are air dried.

b. Appropriate areas of the slides are then ringed with a Tri Chem (Tri Chem, Inc., Belleville, New Jersey) or Artex (Artex Hobby Products, Inc., Lima, Ohio) liquid embroidery pen.

c. Place the slides on a tray lined with wet paper towels in order to form a humid environment and prevent drying out of the conjugate.

d. The appropriate working dilution (based on previous titrations) of the conjugate is prepared fresh for each day's use, allowing about 0.05 ml of diluted conjugate for each ringed area to be stained. (A general-purpose diluent for IF procedures utilizing conjugates which have not been DEAE fractionated consists of the 20% normal mouse brain or beef brain suspension described previously.)

e. Apply the diluted conjugate to the entire ringed area by pipette tip without touching the smear surface, as this may dislodge cells; overlap approximately one-half of the embroidery ink line, so that drying at the smear edge does not occur.

f. Cover the tray with a second inverted tray to form a humid chamber and place at 35–37 °C for 20 minutes.

g. Remove the slides to a slide carrier, rinse by immersing completely 5–10 times in 0.01 M PBS, pH 7.2–7.5, then transfer to a fresh dish of PBS for a 10-minute period of washing with occasional agitation. Transfer to fresh FBS for a second 10-minute washing interval with occasional agitation.

h. Rinse in distilled water, transfer the slides to a dry paper-lined tray, and allow to dry completely.

i. Add a small drop of 25% glycerol in PBS to each ringed area and mount with 22 × 40 or 22 × 50 mm glass cover slip, without producing air bubbles.

j. Examine the preparations with the fluorescence microscope, interpret, record, and report the results.

5. *Indirect Immunofluorescence Staining Method for the Detection and Identification of Viral Antigen*

Indirect IF staining may also be used to detect and identify viral antigens in cell cultures, tissue smears, or frozen tissue sections, but the method is more cumbersome than the direct IF method and there may be a greater problem with reaction specificity. The indirect IF method is most easily applied to detecting viral antigen in cell cultures. Detection of antigens in tissue smears is more difficult, primarily because of the serum globulins present intracellularly (such as in lymphocytes or blood vessel endothelial cells) and throughout the intercellular spaces, which react with the fluorescein-conjugated antispecies globulin. Attempts to remove these globulins (such as by treatment at pH 3.0) may also degrade the specific viral antigens. The specific immune serum must be especially free from all other antibodies (such as to heterologous tissue or serum proteins) because of the greater sensitivity of the indirect IF procedure. For example, specific antibody to rubella virus can be prepared in rabbits, using rubella-infected RK13 cell cultures grown and maintained in medium containing only rabbit serum as the protein. A brief description of the indirect IF staining procedure follows:

a. Three or more sets of uninoculated normal and the mixture of test virus-infected/normal cell culture smears are prepared as described previously (Section III,C,2,e), and are outlined with embroidery paint.

b. The appropriate working dilutions of specific immune serum and homologous species normal serum are prepared in 20% normal beef brain suspension, and are applied to two of the sets of normal and virus-infected/normal cell culture smears, respectively. The 20% beef brain diluent is applied to the third set of normal and virus-infected/normal cell smears. This latter step is especially important in the final definition of reaction specificity.

c. The slides are incubated at 35 °C in a humid atmosphere for 20 minutes, then washed twice (10 minutes for each washing) in PBS, pH 7.2–7.5, and rinsed in distilled water.

d. The working dilution of the FITC-labeled antispecies globulin (previously determined as the lowest dilution giving maximal specific staining and minimal background staining) is then applied to each smear.

e. After a second incubation for 20 minutes, the slides are washed twice (5 minutes each washing), rinsed in distilled water, and mounted in buffered glycerol saline and examined (appropriate positive and negative control slides are included in each test run).

6. *Indirect Immunofluorescence Staining Method for Titration of Antibody*

The methods used in our laboratory for rubella (Lennette *et al.*, 1967) and Colorado tick fever (Emmons *et al.*, 1969) are applicable to any virus that can be suitably grown in cell culture. Preparation of slides is as described in Section III,C,3.

a. Inactivate the sera to be tested by incubating at 56 °C for 30 minutes. It may be necessary to remove lipids and debris (which might contribute to nonspecific staining) by centrifugation or filtration through a 0.45 μm Millipore filter.

b. Prepare serial 2-fold dilutions of test sera (usually 1:4 to 1:128) in the 20% beef brain suspension. Add each serum dilution to cover the ringed area of a slide with virus-infected cells, and for controls, add the 1:4 and 1:8 dilutions to a slide with normal cells. Known positive control sera are included in each run.

c. Incubation, washing, and staining with antihuman γ-globulin conjugate are then done, as outlined in Section III,C,5. The antibody titer is the highest dilution of serum giving at least a 1 + specific staining reaction with virus-infected cells (1–4 + scale).

Antihuman IgM or IgG conjugated γ-globulin may similarly be used to determine differential titers of these two classes of immunoglobulins. However, preparation or purchase of reagents from commercial sources may be difficult.

7. *Fluorescent Focus Inhibition Test for Neutralizing Antibody*

Determination of neutralizing antibody end point titers by detection of the breakthrough virus with the use of IF staining, rather than by animal death, CPE, or plaque formation, was described for rabies by King *et al.* (1965). An adaptation of this method for routine rabies antibody testing has been used in our laboratory since 1968, and has proved to be practical and reliable for large-scale use (Lennette and Emmons, 1971). A similar but more rapid test is used routinely by the Center for Disease Control (Smith *et al.*, 1973). This type of test has advantages in economy, ease, and adaptation to routine large-scale use. It is especially helpful for viruses that do not produce easily recognizable CPE or plaques.

Bottle cultures of baby hamster kidney cells (BHK), line 0853, are maintained in continuous passage, or stock cells are held frozen in liquid nitrogen for use as needed. Outgrowth medium consists of Eagle's MEM in Hanks' BSS plus 10% fetal bovine serum. Trypsin-dispersed cells from confluent monolayers are used to prepare slide microcultures as follows.

a. Arrange slides (100 can be prepared conveniently at a time) on paper-covered work desks with two 15 mm diameter circular cover slips on each slide, then frost the slides with Fluoroglide spray (Chemplast Co., Inc., Wayne, New Jersey), allow to dry, then remove cover slips and save them for repeat use. Autoclave the slides (5 slides per 150 mm, filter paper-lined petri dish).

b. The BHK cell suspension, diluted 1:5 to 1:8 in Eagle's MEM in Earle's BSS plus 10% FBS, is then added to the unfrosted areas of the slides (approximately 0.1 ml per area). The filter paper is soaked with sterile distilled water to maintain humidity, and the slide microcultures are incubated in a CO_2 incubator at 35°C for about 48 hours, until confluent monolayers have formed.

c. Sera to be tested are heat-inactivated (56 °C, 30 minutes) and 2-fold dilutions from 1:4 to 1:128 in Eagle's MEM and Earle's BSS plus 3% FBS are incubated 90 minutes at 35 °C with equal-volumes (0.2 ml) of mouse passage stock LEP Flury rabies virus, diluted to contain 100–300 rabies fluorescent focus/50 (RFF/50) doses per 0.1 ml of the final serum–virus mixtures. In each test run, controls are included consisting of uninoculated cells, a virus titration series, and antibody-positive and antibody-negative sera.

d. The medium is removed and each control and serum–virus mixture is inoculated (0.1 ml) onto two microculture monolayers; then the cultures are incubated for about 4 days more.

e. The maintenance medium is then removed, each entire slide is fixed in acetone, and the microculture areas are ringed with embroidery paint, stained by the standard direct IF method for rabies, and examined with the fluorescence microscope.

The antibody titer is the highest dilution of serum that limits the development of fluorescent foci to only 1–2 in the microculture (examine 20–40 microscope fields with 25 × objective). Good agreement with the standard neutralization test in mice is found, and the titers are reproducible on repeated testing. The method can be adapted to use commercial microculture slide systems (Smith *et al.*, 1973).

8. *Specificity Tests*

In addition to the usual safeguards against false-negative or false-positive results (positive and negative control tests, accurate use of test procedures, repeat testing when questionable results are obtained, avoidance of errors in specimen and slide identification), certain additional tests may be helpful to interpreting atypical staining. These have had most use in rabies IF staining, where the tolerance for error must be near zero. In the absorption test, the usual working dilution of the conjugate in normal mouse brain suspension is used in parallel with the same dilution of conjugate in a 20% suspension of virus-infected mouse brain: the latter should fail to show fluorescence with the test specimen because specific reactivity has been absorbed out of the conjugate. In the "blocking test," the tissue smear is first reacted with unconjugated immune γ-globulin, which attaches to the specific antigenic sites and prevents the subsequent uptake of the conjugated γ-globulin. In practice, this is less satisfactory than the absorption test, as complete blocking is difficult to achieve. Heterologous staining tests may also be done in which the test smears are stained by the specific conjugate, but should not be stained at all by other conjugates, except those with expected cross-reactions (e.g., herpes simplex and V–Z viruses, or distemper and measles viruses). Of course, the possibility of a dual infection in the test tissue must be considered. Each conjugated γ-globulin should be produced free of antibody to any other viruses, which might cause false-positive results. The animal colony should be monitored serologically at regular intervals for such viruses as lymphocytic choriomeningitis, ectromelia, reovirus, Sendai, mouse adenovirus, and others. Antibody to these agents, labeled unwittingly during the conjugation procedure with the intended virus, could react with homologous or related viral antigens in human tissues or in cell cultures and animal host tissues used for virus isolation, and result in erroneous interpretations. For example, when using conjugated immune monkey serum to stain viral isolates in monkey kidney cell culture, beware of false-positive staining due to the presence of simian viruses (such as SV40). Normal control monkey kidney cells should be stained along with the infected cells to detect such latent viruses. Each conjugated immune serum can also be screened for heterologous antibodies likely to give trouble in test situations. We may also make a strong argument against using an "undefined" immune serum (such as convalescent human serum) in an indirect IF test to identify viruses.

Finally, isolation attempts for the suspect virus and confirmatory serologic tests for specific viral antibody titer rise should be used routinely along with IF staining until accumulated experience by the laboratory shows the accuracy and reliability of the IF method. It can thereafter be used as the primary diagnostic tool. Submission of problem specimens to a reference laboratory for an independent assessment, and a regular system of performance evaluation, via "unknown" test specimens, are helpful. There is no substitute for good "bench-level" training and experience, since diagnostic skills are hard to obtain through written descriptions or photographs alone. The IF tests should be performed frequently, preferably daily, to maintain technical and interpretive expertise.

9. Interpreting, Recording, and Reporting Results

Sufficient numbers of cells (neurons in brain smears, basal epithelial cells from skin lesions, etc.) must be present in most microscope fields examined to conclude that a definitive examination has been done. If slides are not satisfactory and no typical fluorescence is seen, then the test should be reported as "unsatisfactory," *not* as "negative." The absence of typical staining when sufficient acceptable slides have been thoroughly examined is considered a negative result. For positive slides, an estimate of the *amount* of antigen present ($\pm$ to 4 + scale) should be recorded. The distribution of antigen in relation to cell morphology, as well as its mere presence, is important in interpreting the result (e.g., the neuronal cytoplasmic distribution of rabies virus antigen, or the neuronal and glial cell nuclear and cytoplasmic staining in herpes encephalitis). Control slides must show the expected reactions (3–4 + fluorescence in positive control slide and no fluorescence in negative control slide).

In certain cases, cross-reactions because of shared antigens are to be expected. Thus, herpes simplex virus and varicella–zoster virus antigens in skin lesions may show low-level (1–2 +) cross-reaction by IF staining. Since herpes, V–Z, and vaccinia/variola may sometimes be difficult to distinguish clinically, we routinely stain vesicular lesion smears separately with the three conjugates. We have not encountered the same problem of cross-reactions in brain, lung, liver, or other tissue smears infected with either V–Z or herpes viruses, but the three conjugates are used routinely as a part of the specificity control methods. Typical staining of cytoplasmic and nuclear antigen with the herpes simplex conjugate, no reaction with the vaccinia/variola conjugate, and no reaction or only low-level cross-reaction with the V–Z conjugate are interpreted as a positive test for herpes simplex.

Herpes simplex types 1 and 2 conjugates may also be used. Although cross-reactions occur, homologous staining is readily differentiable from heterologous staining and thus IF can be used to type isolates in cell culture. However, determination of type specificity by direct staining of lesion material is often unsatisfactory because of unevenness in antigen distribution on the smears, and therefore is not routinely attempted in our laboratory. Similarly, type-specific herpes antibody determination by indirect IF staining is not sufficiently reliable or of benefit clinically to incorporate the procedure into our routine diagnostic service.

Smears with epithelial cells showing 3–4 + staining by V–Z conjugate of both cytoplasmic and nuclear antigen, no reaction with the vaccinia/variola conjugate, and no or only low cross-reaction with the herpes conjugate, are positive for V–Z. Smears with epithelial cells showing 3–4 + staining by the vaccinia/variola conjugate of strictly cytoplasmic-associated antigen, and no reaction with either the herpes or V–Z conjugates, are positive for vaccinia. The IF method cannot distinguish between vaccinia and variola antigens. Electron microscopy is used in conjunction with IF staining for the rapid diagnosis of skin lesions, but it also cannot distinguish between vaccinia and variola viruses. Standard biological tests are necessary for this distinction. However, the likelihood now of encountering variola virus is extremely small, and the history of travel or exposure, and clinical and epidemiologic considerations are diagnostically helpful.

The same interpretations as described above are applied to identification of herpes, V–Z, or vaccinia viruses in cell culture. Results are most easily interpreted when about 10–30% of the cells are infected, providing more uniformity of fluorescence against a contrasting background of uninfected cells.

Nonspecific reactions are encountered infrequently, usually because of unsatisfactory tissue or improperly prepared smears. Polymorphonuclear leukocytes can show nonspecific flourescence and cell debris and tissue fragments may trap conjugate and prevent it from being completely removed in the washing process.

Similarly, stained nasal smears, other biopsy or autopsy tissue smears, and blood smears should be examined and interpreted with consideration for the expected cytopathologic effects of the specific virus, not just for the presence of fluorescent particles. Examples include the presence of influenza antigen in the ciliated epithelial cells of nasal tissue; typical cytoplasmic elementary bodies of the psittacosis agent in tissues of infected birds; intraerythrocytic location of Colorado tick fever virus antigen in peripheral blood smears; syncytial formation with intracytoplasmic and intranuclear antigen in measles; and the diffuse, fine, homogeneous cytoplasmic staining in various arbovirus infections of cell cultures or suckling mouse brain.

Results of each examination are recorded in a permanent daily log workbook and also on each laboratory specimen transmittal form. The original form is kept in a central office file, and copies are returned to the submitting physician or laboratory and also to the appropriate local public health jurisdiction for epidemiologic and disease control purposes. Regular compilations and analysis of laboratory results are essential for evaluating the reliability and practicality of the IF technique, as compared with other diagnostic methods.

10. Safety Aspects of IF Staining

The usual precautions for safety in a virology laboratory should be adhered to, and need not be reiterated here, but some points peculiar to IF procedures might be emphasized. All procedures should be done on a workbench or tray lined with absorbent plastic-backed paper (plastic side down), which is folded up and autoclaved or incinerated after use; thus the table top itself is not contaminated with the specimen or with contaminated equipment. All contaminated materials are placed in stainless steel pans for autoclaving. Avoid contaminating eyes, mouth, fingers, and skin; handle infectious material with forceps or disposable plastic gloves only. Use bulb or propipette procedures only, not mouth-pipetting. The frosted end of slides ordinarily is not contaminated, but even acetone fixation may not inactivate virus or assure adherence of tissue to the slide, so handle slides with care. Acetone used for fixing slides and PBS used for washing procedures may be autoclaved or disinfected by adding hypochlorite bleach. The microscope stage can be wiped clean with disinfectant regularly, and kept free of glass chips. Specific immunization (rabies, rubella) may be used for the microbiologists as indicated.

IV. Clinical Applications of Immunofluorescence

Numerous references have already been made to diagnostic IF procedures for specific diseases, but a brief summary review may be helpful. In theory, IF staining could be used to identify any viral, rickettsial, or chlamydial agent for which suitable immune serum can be prepared, and to detect and titrate antibody for any agent that can be propagated in a suitable cell culture or animal host tissue. Table II gives a partial listing of diseases of clinical or public health importance where rapid diagnosis may be a substantial benefit, either by allowing specific antibiotic or antiviral drug treatment or passive immunization (Group A) or for other considerations. The efficacy of therapy in such diseases as herpes simplex infections (antiviral drugs) or Lassa fever (convalescent human plasma) is still uncertain but may be established in the future, and other diseases will doubtless be added to this list. Certain diseases (Group B) can be better managed if the diagnosis is confirmed early, even though specific treatment may not be available. For

TABLE II

DISEASES OF CLINICAL AND PUBLIC HEALTH IMPORTANCE FOR WHICH RAPID
IMMUNOFLUORESCENCE TESTS MAY BE HELPFUL

A. Specific therapeutic or preventive treatment of the patient facilitated

Herpes (eye)	Rabies (animal source)
Herpes (genital, skin, brain)?	Rocky Mountain spotted fever
Herpes B (animal source)	Trachoma
Lassa fever?	Typhus
Mycoplasma	Vaccinia?
Q fever	Varicella-zoster?

B. Better management of the patient facilitated

Colorado tick fever	Lymphocytic choriomeningitis
Contagious ecthyma (orf)	Parainfluenza
Cytomegalovirus	Rabies
Herpes (genital lesions in pregnancy)	Respiratory syncytial
Influenza A and B	Rubeola

C. Public health action, disease prevention in contacts facilitated

California encephalitis	Rubella
Herpes (venereal disease)	Rubeola
Herpes B (animal source)	St. Louis encephalitis
Influenza A and B	Typhus
Lassa fever	Vaccinia
Lymphocytic choriomeningitis	Varicella
Psittacosis	Variola
Q fever	Venezuelan encephalitis
Rabies	Western encephalitis
Rocky Mountain spotted fever	

example, genital herpes in late pregnancy may be an indication for cesarean section to prevent neonatal herpes. The diagnosis of rubeola or of Colorado tick fever eliminates the concern and need for treatment of Rocky Mountain spotted fever, with which they may be confused. In lymphocytic choriomeningitis (LCM), the specific indirect IF antibody titer rise is so rapid, in contrast to the CF antibody, that the diagnosis can be presumptively established within a few days of onset of symptoms, thus differentiating it from other viral meningitides or encephalitides. Early rabies diagnosis by corneal test, skin biopsy, or the indirect IF antibody test (usually the earliest antibody response detectable) helps in management of the case and in determining whether heroic therapeutic measures are justified.

For several other diseases besides rabies and LCM, the indirect IF antibody test has advantages over other antibody tests in rapidity, ease, and sensitivity. It is often useful as a confirmatory diagnostic test when the CF or HI tests are equivocal. Thus, it can be used for herpes simplex, vaccinia, varicella, rubella, rubeola, Colorado tick fever, psittacosis, Rocky Mountain spotted fever, St. Louis encephalitis, Western encephalitis, Venezuelan encephalitis, California encephalitis, dengue, and other arboviruses, or for unusual diseases such as orf or milker's nodules for which CF antigens may not be available.

In addition to direct benefit to the individual patient, the rapid diagnosis of certain diseases (Table II, Part C) has special significance to public health actions and prevention of further spread of the disease from person to person or from animal or vector source in the environment to people. The diagnosis of vaccinia or varicella-zoster permits rational use of Vaccinia Immune Globulin or Zoster Immune Globulin to prevent spread to contacts, such as in a cancer chemotherapy ward. Identification of rubella or rubeola assists in managing pregnancies or planning immunization programs. Immunofluorescence staining of nasal smears can give the earliest proof of epidemic influenza activity in a community. The prompt diagnosis of Lassa fever leads to surveillance of contacts and possibly to early and more effective passive immunization. Other examples will probably be added to this list in the future. Identification of rabies, herpes B, LCM, psittacosis, Q fever, typhus, Rocky Mountain spotted fever, and various arboviruses in animal or arthropod vector sources, guides in taking preventive steps for exposed individuals or the community at risk.

The scope of IF diagnostic activities in our laboratory is demonstrated in Table III, which lists the number of test procedures performed for various viruses (direct identification in the patient, antibody assay, or identification of viral isolates). When incorporated into the laboratory diagnostic routine, IF staining to identify viral isolates provides great savings in time, money, and resources. Identification can be completed the same day the presence

TABLE III

SELECTED LIST OF IMMUNOFLUORESCENCE TESTS BY THE VIRAL AND RICKETTSIAL
DISEASE LABORATORY, 1965–1974

Antigen tested	Direct detection of antigen in tissue	Antibody titration	Identification of isolate cell culture or laboratory animal
Colorado tick fever	694	87	233
Herpes simplex	2132	522	1768
Influenza A and B	339	—[a]	171
Lymphocytic choriomeningitis	—	861	—
Parainfluenza	—	—	90
Psittacosis	52	22	90
Rabies	13,643	1161	353
Respiratory syncytial	—	—	48
Rubella	—	2727	1796
Rubeola	194	360	113
St. Louis encephalitis	—	223	570
Vaccinia	593	288	466
Varicella	1397	276	466
Venezuelan encephalitis	—	159	—
Western encephalitis	—	205	587

[a] Data incomplete, not included for this review

FIG. 1. Human blood clot smear showing intraerythrocyte staining of the antigen of Colorado tick fever after reacting with anti-Colorado tick fever FITC conjugate. × 470.

FIG. 2. Rabies-infected skunk brain smear stained with FITC-conjugated hamster immune serum. × 370.

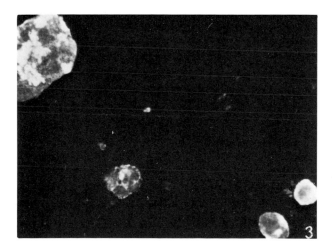

FIG. 3. A human genital vesical smear stained with FITC-conjugated hamster immune serum to herpes simplex virus type 2. × 370.

of virus is recognized. For some viruses (such as rubeola, rubella, or LCM), CPE may be difficult to recognize in cell culture and IF staining is the best method for recognition of virus presence and for identification. Detection of latent viruses in animal stocks or cell cultures (such as LCM, SV40, or ectromelia) is simplified by having specific IF reagents and methods available.

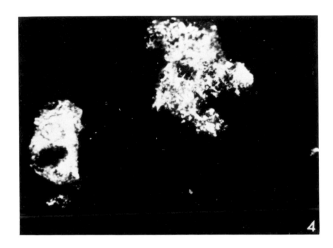

FIG. 4. An isolate of measles virus infecting primary rhesus monkey kidney cell culture stained with an antimeasles virus conjugate. × 370.

A few examples (rabies, herpes, CTF, measles) of IF staining, as used routinely in the diagnostic laboratory, are shown in Fig. 1–4. The numerous other examples described above attest to the valuable role IF methods can play. As new, effective antiviral therapeutic modalities or preventive measures are discovered, the need and demand for rapid, early, and accurate diagnostic methods will increase accordingly.

ACKNOWLEDGMENTS

The valuable contributions of Mr. James D. Woodie in preparation of reagents and standardization of immunofluorescence methods, and of Mrs. Dana Gallo in assisting Mr. Woodie in the routine diagnostic application of these methods, are gratefully acknowledged.

REFERENCES

Beutner, E. H., ed. (1971). *Ann. N.Y. Acad. Sci.* **77**, 1–529.
Biegeleisen, J. Z., Jr., Scott, L. V., and Lewis, V., Jr. (1959). *Science* **129**, 640.
Blundell, G. P. (1970). *Prog. Clin. Pathol.* **3**, 211.
Casals, J. (1967). *In* "Methods in Virology" (K. Maramorosch and H. Koprowski, eds.), Vol. 3, Chapter 4, pp. 182–185. Academic Press, New York.
Cherry, W. B. (1974). *In* "Manual of Clinical Microbiology" (E. H. Lennette, E. H. Spaulding, and J. P. Truant, eds.), 2nd ed. pp. 29–44. Am. Soc. Microbiol. Washington, D.C.
Cohen, S. M., Gordon, I., Rapp, F., Macaulay, J. C., and Buckley, S. M. (1955). *Proc. Soc. Exp. Biol. Med.* **90**, 118.

Coons, A. H., Creech, H. J., Jones, R. N., and Berliner, E. (1942). *J. Immunol.* **45**, 159.

Dean, D. J., and Abelseth, M. K. (1973). *In* "Laboratory Techniques in Rabies" (M. M. Kaplan and H. Koprowski, eds.), 3rd. ed. pp. 73–84. World Health Organ., Geneva.

Emmons, R. W., and Lennette, E. H. (1966). *J. Lab. Clin. Med.* **68**, 923.

Emmons, R. W., Dondero, D. V., Devlin, V., and Lennette, E. H. (1969). *Am. J. Trop. Med. Hyg.* **18**, 796.

Gardner, P. S., and McQuillin, J. (1974). "Rapid Virus Diagnosis. Application of Immuno-fluorescence." Butterworth, London.

Goldman, M. (1968). "Fluorescent Antibody Methods." Academic Press, New York.

Goldwasser, R. A., and Kissling, R. E. (1958). *Proc. Soc. Exp. Biol. Med.* **98**, 219.

Gornall, A. G., Bardawill, C. J., and David, M. M. (1949). *J. Biol. Chem.* **177**, 751.

Hebert, G. A., Pittman, B., and Cherry, W. B. (1967). *J. Immunol.* **98**, 1204.

Hebert, G. A., Pittman, B., McKinney, R. M., and Cherry, W. B. (1972). "The Preparation and Physicochemical Characterization of Fluorescent Antibody Reagents." U. S. Dept. of Health, Education and Welfare, Public Health Service, H.S. and Mental Health Administration, Center for Disease Control, Atlanta, Georgia.

Hijmans, W., and Schaeffer, M., eds. (1975). *Ann. N.Y. Acad Sci.* **254**, 1–627.

Kalter, S. S., Hatch, M. H., and Ajello, G. W. (1959). *Bacteriol. Proc.* pp. 89–90.

Kawamura, A., Jr. (1969). "Fluorescent Antibody Techniques and Their Applications." University Park Press, Baltimore, Maryland.

King, D. A., Croghan, D. L., and Shaw, E. L. (1965). *Can. Vet. J.* **6**, 187.

Lennette, E. H., and Emmons, R. W. (1971). *In* "Rabies" (Y. Nagano and F. M. Davenport, eds), pp. 77–90. University Park Press, Baltimore, Maryland.

Lennette, E. H., Woodie, J. D., Nakamura, K., and Magoffin, R. L. (1965). *Health Lab. Sci.* **2**, 24.

Lennette, E. H., Woodie, J. D., and Schmidt, N. J. (1967). *J. Lab. Clin. Med.* **69**, 689.

Liu, C. (1956). *Proc. Soc. Exp. Biol. Med.* **92**, 883.

Liu, C. (1969). *In* "Diagnostic Procedures for Viral and Rickettsial Diseases" (E. H. Lennette and N. J. Schmidt, eds.), 4th ed., Chapter 4, pp. 179–204. Am. Public Health Assoc., New York.

Lowry, O. H., Rosebrough, N. J., Farr, A. L., and Randall, R. J. (1951). *J. Biol. Chem.* **193**, 265.

McQueen, J. L., Lewis, A. L., and Schneider, N. J. (1960). *Am. J. Public Health Nation's Health* **50**, 1743.

Marks, M. I. (1974). *In* "Viral Immunodiagnosis" (E. Kurstak and R. Morisset, eds.), pp. 173–179. Academic Press, New York.

Murray, H. G. (1963). *Lancet* **1**, 847.

Nairn, R. C. (1969). "Fluorescent Protein Tracing". Williams & Wilkins, Baltimore, Maryland.

Ploem, J. S. (1967). *Z. Wiss. Mikrosk. Mikrosk. Tech.* **68**, 129.

Riggs, J. L. (1965). *In* "Applied Virology" (M. Sanders and E. H. Lennette, eds.), pp. 43–57. Olympic Press, Sheboygan, Wisconsin.

Rose, N. R., and Bigazzi, P. E., eds. (1973). "Methods in Immunodiagnosis." Wiley, New York.

Rygaard, J. and Olsen, W. (1969). *Acta Pathol. Microbiol. Scand.* **76**, 146.

Rygaard, J., and Olsen, W. (1971). *Ann. N.Y. Acad. Sci.* **177**, 430.

Schieble, J. H., Lennette, E. H., and Kase, A. (1965). *Proc. Soc. Exp. Biol. Med.* **120**, 203.

Schmidt, N. J., Lennette, E. H., Woodie, J. D., and Ho, H. H. (1966). *J. Lab. Clin. Med.* **68**, 502.

Shaw, E. D., Newton, A., Powell, A. W., and Friday, C. J. (1961). *Virology* **15**, 208.

Smith, J. S., Yager, P. A., and Baer, G. M. (1973). *In* "Laboratory Techniques in Rabies" (M. M. Kaplan and H. Koprowski, eds.), 3rd ed., pp. 354–357. World Health Organ., .Geneva.

Spendlove, R. S. (1967). *In* "Methods in Virology" (K. Maramorosch and H. Koprowski, eds.), Vol. 3, pp. 475–520. Academic Press, New York.

Weller, T. H., and Coons, A. H. (1954). *Proc. Soc. Exp. Biol. Med.* **86**, 789.

Wells, A. F., Miller, C. E., and Nadel, M. K. (1966). *Appl. Microbiol.* **14**, 271.

2 Experimental Analysis of Effector Thymus-Derived Lymphocyte Function in Viral Infections*

Peter C. Doherty, Jon C. Palmer, and
Rolf M. Zinkernagel

I. Introduction

Our intention here is to briefly summarize an experimental approach to the study of cell-mediated immunity (CMI) in virus infections and to indicate a few of the problems that exist and require attention. Some of the points made are based more on intuition than on scientific evidence, but we wish to present an interpretive and technical basis for experimentation rather than a rigorous account of current knowledge. This latter function has been adequately served by several extensive, recent reviews (Wheelock and

*This is publication No. 1110 from the Department of Immunopathology at the Scripps Clinic and Research Foundation, La Jolla, California.

Toy, 1973; Allison, 1974; Blanden, 1974; Burns and Allison, 1974; Doherty and Zinkernagel, 1974; Doherty et al., 1976a). Furthermore, we do not apologize for the fact that the discussion is concerned essentially with CMI in the mouse: this is our area of expertise and murine systems are the only virus-pathogenesis models that have been thoroughly investigated from the immunological aspect.

Cell-mediated immunity is, for our present purpose, defined as an adaptive response depending on participation of specifically sensitized thymus-derived lymphocytes (T cells). Little consideration is given here to antibody-dependent cell-mediated effector function (MacLennan, 1972; Perlmann et al., 1972). The reason for this is that, as yet, such phenomena have only been defined in vitro and their possible role in vivo is quite unclear. Our prejudice is that no primary role should be attributed to any immune function unless the phenomenon can be adoptively induced in immunologically naive recipients infected with the pathogen. This constraint, i.e., the necessity to partially fulfill Koch's postulates by transfer of effector cell populations, is obviously only applicable to experimental situations and should not be taken as a criticism of clinical studies. Also, even if such evidence is not forthcoming, it is possible that antibody-dependent cell-mediated mechanisms may play some ancillary role secondary to invasion of immune T cells.

The present contribution is divided into two broad parts. Initially the systems used are considered in a fairly general way which should, hopefully, be moderately readable. The second part comprises a more detailed description of techniques.

II. Measurement of T-Cell Function in Vitro

The reason that CMI is worth considering now (so many years after the role of antibody in most virus infections was reasonably well understood) is that, until very recently, there have been no well-defined in vitro assays for analyzing T-cell function. Macrophage-migration inhibition tests (Bloom, 1971) and microcytotoxicity assays (based on reduction of target cell numbers) may be confounded by antibody-dependent mechanisms (Lamon et al., 1973), though there can undoubtedly be a major T-cell component. Both systems are, however, technically tedious and interpretation may be rather subjective.

The breakthrough came with the development of cytotoxic assays (reviewed by Cerottini and Brunner, 1974) that depend on release of isotope from virus-infected cells which are radiolabeled, and then specifically lysed by virus-immune T cells (Oldstone and Dixon, 1970; Cole et al., 1973; Marker and Volkert, 1973; Doherty et al., 1974; Gardner et al., 1974a,b; Zinkernagel and Doherty, 1974a). Most investigators have used ^{51}Cr, which is both convenient and does not seem to be reincorporated into cells to any significant extent. Availability of these quantitative *in vitro* assays has, in turn, allowed development of rational protocols for *in vivo* adoptive transfer experiments. Effector function demonstrated by these two basic mechanisms has, to date, shown a good correlation.

A. The Assay System

The principle underlying the cytotoxic T-cell assay is simple. Target cells are infected with virus and radiolabeled by exposure to $Na_2{}^{51}CrO_4$ (see Section IV,C). This sequence may be reversed for rapidly growing lytic viruses (e.g., influenza), or persistently infected cell lines may also be used [e.g., with lymphocytic choriomeningitis (LCM) virus]. Target cells are then incubated in contact with lymphocytes from virus-immune mice, generally for intervals ranging from 6 to 16 hours. Timing may be very critical for lytic viruses [though this does not seem to be a problem with the ectromelia (mouse pox) model], but most assays used so far can be run overnight. This is convenient, as setting up a reasonably large experiment is generally a full day's work for two people.

Assays are generally done in 6 mm, flat-bottomed, 96-hole tissue culture plates (IS-FB-96TC, Linbro Chemical Co., New Haven, Connecticut), which hold from 300 to 400 μl of medium. Comparable plates with V-bottomed wells (Linbro IS-MVC-96) may also be used if only small numbers of lymphocytes and targets are available, so that maximal cell-to-cell contact is achieved at the apex of the container. Lymphocytes may either be allowed to settle onto the targets or, if a more rapid test is desired, centrifuged lightly (200 g for 1 minute) prior to commencing incubation.

The characteristics of the interaction are that the T cell and the target must make contact, and that each specifically sensitized lymphocyte may kill more than one target cell. Activity of individual T cells is thus maximized (for adherent targets) by overlaying confluent virus-infected cells with immune populations at somewhat less than monolayer density—with L929 fibroblasts (L cells), this represents a ratio of less than 30 mouse spleen cells to one target. The T cells are thus not physically inhibited from moving around the dish. Proportionally more lymphocytes may be mixed with non-adherent targets (Fig. 1) to achieve high levels of ^{51}Cr release.

Availability of. targets that can be manipulated in suspension is, if achievable, of considerable advantage. Nonadherent cells, e.g., the P815 mastocytoma which has been used successfully with a number of different viruses (LCM, ectromelia, Sendai, influenza), can be infected with virus and labeled with ^{51}Cr in suspension (see Section IV,C). Equal numbers of targets (generally 2 × 10^4) are then dispensed into each assay well (in 50 μl

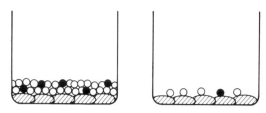

ADHERENT TARGETS

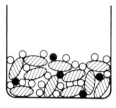

SUSPENSION TARGETS

FIG. 1. Maximal activity of individual sensitized T cells is achieved by mixing T cells (●) and targets (○) in suspension. Use of excessive lymphocytes with adherent target cells results in inhibition of T-cell motility.

or 100 μl) and lymphocytes are added, either in a set volume or in sufficient medium to fill the well (the latter may be a quicker procedure). Completion of the assay then requires only that 100 μl of supernatant be removed for counting, as the number of ^{51}Cr-labeled cells in each well is identical.

Processing of adherent cells may be much more time-consuming. Unequal cell loss will generally occur if monolayers have to be infected, washed, ^{51}Cr-labeled, and washed when growing in the well (see Section IV,C). It may then become necessary (though not invariably) to determine the amount of input ^{51}Cr in each well. At the completion of the assay the whole of the supernatant medium is carefully removed (with a short Pasteur pipette) and the well is filled with water. The cells are then stood in water for at least 2–3 hours at room temperature, and the water lysate is removed by vigorous pipetting. Release of ^{51}Cr is assessed by comparing counts in supernatant with total counts in supernatant and water lysate. A further water lysis correction factor, i.e., the percentage of ^{51}Cr release by incubating targets overnight in water, is usually applied.

Obviously the more procedures that need to be performed on cells growing in wells, the greater the variability in target numbers and the more tedious the operation overall. Considerable time may be saved if cells can be infected and labeled prior to plating. Overnight labeling of adherent targets, the cells being dispersed in medium containing $Na_2{}^{51}CrO_4$, has proved useful in this regard (see Section IV,C).

Even so, experiments with targets that are adherent may ultimately be more practical in some instances. This is particularly true for mouse peritoneal macrophages (see Section IV,C) which have been used extensively for genetic studies. Also some cells show high-background ^{51}Cr release if they are trypsinized and maintained in suspension prior to use. Furthermore, if only small numbers of targets are available, or required, it may be more economical to explant initially into wells.

B. IMMUNIZATION OF MICE

In attempting to develop a cytotoxic T-cell assay at least as much attention must be given to procedures used to generate immune T cells as to the *in vitro* assay system itself. Many virologists are much more familiar with cell cultures than with mice, so it may be worthwhile to consider this aspect in some detail. Some of the points to be made may seem naive, but we know of instances where lack of awareness of such simple concepts has led to considerable wasted effort.

1. *Necessity for Infectious Virus*

All successful experiments to date have, so far as we are aware, been done using lymphocytes from mice injected with infectious virus or virus-infected cells, generally given in a single dose. This apparent requirement for infectious virus may simply reflect that the extent of antigen expression is insufficient when an inactivated preparation is used. Obviously overall exposure of the immune system is greatly enhanced by the process of dissemination and replication of virus in a variety of cell types throughout the organism.

Furthermore, a killed agent is probably processed quite differently, presumably by macrophages, and may not induce alterations in cell membranes that are recognized by cytotoxic T cells. Some inactivated virus vaccines stimulate quite satisfactory levels of antibody formation, which could indicate that the CMI response is biased more toward T-cell helper function rather than to generation of surveillance (cytotoxic) T cells. There are, however, indications that not all viruses need go through a complete growth cycle in order to modify cell membranes in an immunogenic way. Ectromelia virus irradiated with ultraviolet light has been used successfully to prepare target cells for the *in vitro* assay (Ada *et al.*, 1976). Even so, irradiated poxviruses induce considerable modification of both cell morphology and normal metabolism.

The requirement for infectious virus may not be absolute, but success would seem more likely if this constraint is initially observed.

2. *Virus Dose and Route of Inoculation*

The dose of virus given and the route of inoculation used depend on an understanding of both the rate and site of virus multiplication. Generally, it is convenient if the spleen can be taken as a source of immune T cells. The reasons for this are obvious. Normally at least 30% of nucleated cells in the spleen are T cells, so recovery of relatively large numbers of lymphocytes is usually possible. Furthermore, the organ is easily located, readily removed, and good cell viability is routinely found.

Total numbers of nucleated cells per spleen range from about 2.5×10^7 for mice injected intracerebrally with viscerotropic LCM virus, to 2.0×10^8 for mice given ectromelia virus. The normal value is generally between 5.0×10^7 and 1.0×10^8. Total cell counts may also vary considerably among the different inbred strains and in animals suffering from intercurrent infections.

Another point that is extremely important in all experiments involving spleen cells is that total numbers and viability of lymphocytes should be determined. This can easily be done using white blood cell diluting fluid, for total cell counts, and trypan blue exclusion, for viability. Use of "spleen equivalents" may be extremely misleading. Splenomegaly is also a poor indicator of the degree of immune response, e.g., spleens of mice dying of LCM may be enlarged, but contain relatively few leukocytes and a lot of debris.

Generation of an immune T-cell response in spleen requires that virus be inoculated by an appropriate route. Virus injected intravenously (into the lateral tail vein) must circulate only once to reach the spleen, whereas antigen given peripherally (e.g., into the footpad) may be trapped in the draining (popliteal) lymph node. This means that if the agent is not one that either causes viremia or grows extensively in many tissues, the T-cell response may be essentially localized to the regional lymph node. For instance, in mice given influenza virus intranasally, more potent cytotoxic lymphocyte populations are found in the mediastinal lymph nodes than in the spleen (Cambridge *et al.*, 1976). The same is true for Sendai virus where the cytotoxic T-cell response detected in spleen is stronger following intraperitoneal rather than intracerebral exposure (Doherty and Zinkernagel, 1976).

The dose of virus given may also considerably influence the characteristics of the CMI response. In experiments with Sendai virus optimal immunization (in spleen) was achieved by injecting a maximal concentration intraperitoneally. Viremia is not an obvious feature of this infection, and any replication of virus may be localized to the peritoneal cavity. Similar findings were made for influenza, with the additional observation that presence of excess inactivated virus tended to inhibit generation of CMI.

With viscerotropic LCM virus, however, it is important that a low infectious dose be used. Administration of high titers of virus may result in early appearance of immune T-cell activity, but the maximum level of cytotoxicity achieved is relatively low. This virus multiplies extensively in most tissues, and the apparent high dose immune paralysis (Hotchin, 1971; Doherty *et al.*, 1974) may reflect either recruitment of T cells to these sites, or some form of central tolerance. It should be emphasized that LCM is the only virus for which this phenomenon has yet been observed.

Another contraindication for giving high titers of virus would be, of course, if the mouse dies from the infectious process before an adequate CMI response is generated. For this reason a relatively avirulent strain of virus may, as in the ectromelia system (Blanden, 1974), prove an optimal immunogen.

In all of the virus systems examined so far the maximum cytotoxic T-cell response is attained between 6 and 9 days after immunization with the infectious agent. This varies with virus dose, route of inoculation, and mouse strain. Thus, in attempting to establish a T-cell assay it is best to concentrate on this interval, perhaps with the addition of time points at day 4 and day 12.

C. Sensitization of T Cells *in Vitro*

Attempts at generating cytotoxic T-cell responses to infectious virus *in vitro* are, as yet, at a preliminary stage, though early results are encouraging (R. V. Blanden and M. B. C. Dunlop, personal communication). Extremely potent secondary responses (see Section IV,B) can, however, be routinely produced by culturing memory T cells (from mice that had been immunized with infectious virus at least 3 or 4 weeks previously) with virus-infected macrophages for 4 or 5 days (Dunlop and Blanden, 1976; Dunlop *et al.*, 1976; Gardner and Blanden, 1976); Availability of these highly selective immune lymphocyte populations enables the use of small numbers of T cells and short assay times, thus facilitating more rigorous analysis of the T cell–target cell interaction.

D. Requirement for a Histocompatible Interaction

Cytotoxic T-cell activity is, at least in the mouse, demonstrable only when lymphocyte and virus-infected target share genes coding for strong transplantation (H-2) antigens (Fig. 2). Mutuality of one allele at either of the two relevant loci (K and D) in the H-2 gene complex is sufficient, and there is no requirement for identity of immune response (*Ir*) genes. Outbred mouse strains generally behave in this regard as if they are inbred, and cross reacti-

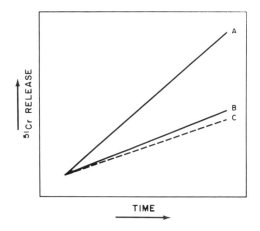

FIG. 2 Typical graph of release of ^{51}Cr with time from virus-infected target cells cultured with (A) syngeneic (H-2 compatible), (B) allogeneic (H-2 different) virus-immune T cells, or (C) normal syngeneic or allogeneic lymphocytes. Release of ^{51}Cr would, in the LCM-L cell system, remain linear for about 14 to 16 hours, or until (with syngeneic immune T cells) > 70% lysis had been achieved.

vity is not seen between different inbred and outbred mouse strains (Zinkernagel and Doherty, 1975b).

Two basic mechanisms have been invoked to explain this phenomenon. One possibility is that H-2 genes are involved in some form of physiological recognition between T cell and target, the lymphocyte also possessing an immunologically specific receptor for virus. The alternative, which at present seems more likely, is that the T-cell receptor recognizes an antigen which is partly coded for by H-2 genes and partly by the viral genome, an operational concept loosely defined by the term "altered self." These models have recently been discussed at length elsewhere (Blanden *et al.*, 1976; Doherty *et al.*, 1976a; Zinkernagel and Doherty, 1976), and consideration here will only be given to practical aspects arising from the H-2 compatibility requirement.

First, it is obviously essential to use an H-2 compatible system. The H-2 gene complex has recently been discussed in great detail by Klein (1975), and a useful table of H-2 types is presented by Shreffler and David (1975) which we have abstracted here for the common mouse strains (Table I). Furthermore, phenotypic expression of H-2 genes is essential in the target cells: virus-infected teratocarcinoma cells that do not express any detectable H-2 antigen are not killed by virus-immune T cells. Also, the extent of virus-specific immune lysis may, in a pluripotent teratoma line, be directly correlated with the proportion of cells with H-2 antigen on their surface (P. C. Doherty, unpublished data). The second major point is that this phe-

TABLE I

H-2 HAPLOTYPES OF SOME COMMON MOUSE STRAINS AND TARGET CELLS

H-2 Type			
K	D	Mouse strains	Target cells
k	k	CBA/H, CBA/J, C3H/HeJ, AKR	L cells
d	d	BALB/c, DBA/2	P815
b	b	C57BL/6J, BI0, 129	EL4
k	d	A/J, B10.A	
s	s	SJL	
q	q	DBA/1	

nomenon provides a rigorous and extremely rapid means of assessing whether lysis is mediated by immune T cells or by antibody-dependent mechanisms. Antibody-dependent cell-mediated cytotoxicity operates both across H-2 barriers, and between lymphocytes and targets from different species. Such lysis is independent of added complement and superficially resembles T cell-mediated cytotoxicity. However, it can be shown definitively, as was done by Ramshaw (1975) for a mouse-herpesvirus model, that ^{51}Cr release from allogeneic or xenogeneic virus-infected targets is mediated by non-thymus-derived cells (K cells, null cells) that interact with small amounts of antibody secreted by plasma cells in the assay well.

It now seems safe to assert that all CMI phenomena requiring H-2K or H-2D compatibility between lymphocyte and target reflect activity of specifically sensitized T cells. This constraint has been shown for arenaviruses (Zinkernagel and Doherty, 1974b), poxviruses (Blanden et al., 1975; Koszinowski and Ertl, 1975), paramyxoviruses (Doherty and Zinkernagel, 1976; Lewandowski et al., 1976), SV40 (Trinchieri et al., 1976), and Coxsackie virus (C. Y. Wong, J. J. Woodruff, and J. F. Woodruff, unpublished data).

E. CONTROLS AND INTERPRETATION OF RESULTS

1. *Normal Lymphocytes and Targets*

When establishing a cytotoxic T-cell assay two obvious controls should be incorporated at the outset. First, target cells that are not infected with virus must also be exposed to the various lymphocyte preparations. Second, labeled cells should be incubated with comparable lymphoid populations from uninfected mice. These comparisons are particularly important when experiments are done to evaluate cytotoxic activity across H-2 barriers (see Section II,D), or for situations where *in vitro* sensitization methods are used to generate immune T cells (see Section IV,B). Cytolytic capacity of normal spleen cells may vary considerably between inbred mouse strains, e.g., spleen preparations from uninfected BALB/c mice may cause considerable ^{51}Cr release from H-2 incompatible L cells expressing LCM viral antigens.

Also, there is some evidence that spleen preparations taken early (day 2–day 4) from mice immunized with virus may induce high levels of ^{51}Cr release from normal, syngeneic target cells (Pfizenmaier et al., 1975). This has led to the interpretation that the process of virus infection has resulted in derepression, or induction, of self-reactive T-cell clones. However, such phenomena may, at least in some instances, simply reflect that virus present in spleen cells is secondarily infecting the normal targets, which are then susceptible to lysis by virus-immune lymphocytes. Cell surface changes recognized by such cytotoxic T cells may, at least in the ectromelia system, be expressed within 1 to 2 hours of exposure to virus (Ada et al., 1976).

2. Necessity for a Quantitative Approach

The limitations of the cytotoxic T-cell assay must be recognized, and the numerical data obtained should not be regarded with undue reverence. Rigorous quantitative interpretation is only valid if thorough dose-response and kinetic experiments are done, using different numbers of lymphocytes at several time points. Background levels of ^{51}Cr release vary considerably from target to target (1–5% per hour), but are generally, at least until some essential medium constituent is exhausted, substantially linear (Fig. 2). Furthermore, in some experiments with the more rapidly lytic viruses, presence of overlaying lymphocytes may tend to stabilize infected targets, resulting in lower spontaneous release when high numbers of normal spleen cells are used. In such situations, therefore, immune and normal spleen cells should be compared at all dose levels tested. A useful initial protocol is to make 2-fold dilutions of lymphocyte populations, starting at a ratio of 100 spleen cells to one target cell.

Any assessment of T-cell potency is best made in situations where less than 50% specific ^{51}Cr release is observed; otherwise, the numbers of targets available will be limiting. In this respect it may be convenient to express cytotoxic capacity as lytic units (Cerottini and Brunner, 1974), calculated from the ratio of lymphoid cells to target cells necessary to cause 30% specific ^{51}Cr release. Total spleen cell counts must also be considered if such criteria are used to compare the magnitude of the immune response in different mouse strains.

All such assay systems obviously lack the precision of techniques available for studying antibody-producing cells (Cunningham, 1974). Any estimate of the numbers of effector T cells must be inaccurate, as we have no rigorous means of assessing how many targets are lysed by each individual lymphocyte. For the H-antigen systems, the problem is being approached by using minimal numbers of targets in tissue culture "rafts" with extremely small wells (L. M. Pilarski, personal communication). Development of a single cell assay for virus-immune T cells may prove impractical. However, several groups are attempting to establish clones of thymus-derived effector lymphocytes.

3. *Establishing That Cytotoxic T Cells Are Specific Effectors*

Virus-specific ^{51}Cr release recognized only for H-2K or H-2D compatible interactions may reasonably be considered to reflect activity of specifically sensitized T cells (see Section II,D). Where possible, cross-specificity experiments should also be done with other viruses (Doherty *et al.*, 1976a). Other criteria can also be used to demonstrate that cytotoxicity is mediated by thymus-derived lymphocytes rather than by antibody-dependent cell-mediated mechanisms (Cerottini and Brunner, 1974; Zinkernagel and Doherty, 1974a). Any activity due to nonspecific lymphotoxins secreted generally into the tissue culture fluid may be tested for by incubating immune spleen cells and virus-infected targets overnight, and then by determining if the supernatant medium is cytotoxic for uninfected cells either in the presence or absence of normal spleen cells (Perlmann and Holm, 1969).

Immune T-cell function may be abolished by treatment with anti-θ (Thy-1,2) ascitic fluid and complement (Haughton and McGhee, 1969; Blanden and Langman, 1972). Macrophages may be removed with carbonyl iron (Golstein and Blomgren, 1973) and an electromagnet, by incubating spleen cells at low density (1×10^6/ml) on plastic for 4 hours at 37°C, or by active adherence to glass beads (Shortman *et al.*, 1971). Nonadherent cells are then gently washed from the surface of the dish with warm medium and presence of residual phagocytes is tested for by microscopic examination of a cell aliquot that has been exposed to colloidal carbon. Depletion of antibody-forming cells and macrophages may also be achieved by using nylon wool columns (Julius *et al.*, 1973).

The times at which T-cell effector function and antibody-dependent cell-mediated cytotoxicity can be detected tend to be rather different. Cytotoxic T-cell activity generally peaks at between 6 and 12 days after exposure to virus, and then decreases either to background or to much lower levels. Antibody-dependent mechanisms may, however, be reasonably considered to persist for as long as specific Ig is circulating—perhaps for years with many viruses. Another distinguishing feature is that cytotoxicity mediated by non-thymus-derived cells may be greatly inhibited by performing the assay in the presence of excess antiviral antibody whereas, in most instances, T-cell effector function is not diminished by this procedure. On the other hand, lysis mediated directly by T cells is considerably depressed by treating target cells, but not the lymphocytes themselves, with antisera specific to structures coded for by genes mapping at H-2K or H-2D.

At least in the alloantigen systems, T cells can be readily removed by adsorption onto monolayers (Golstein et al., 1971). So far this has not been possible for the virus models, except with the chicken Rous sarcoma system (Wainberg et al., 1974). Development of such techniques would obviously be advantageous for comparing specificities of different T-cell subsets. At present this is most readily done by use of "cold target" competitive inhibition protocols, details of which have been described elsewhere (Ortiz de Landazuri and Herberman, 1972; Zinkernagel and Doherty, 1975a).

III. Assessment of T Cell-Mediated Immunity in Vivo

The principle of transferring immunity with cells was first systematically implemented in an infectious disease model (listeriosis in the mouse) by Mackaness and his colleagues (Mackaness, 1969). This approach was not applied rigorously to a virus system until Blanden, who had worked with Mackaness, performed a series of elegant experiments using the ectromelia (mouse pox) system, a condition in which the classic work of Fenner had indicated a central role for CMI (reviewed by Blanden, 1974). Application of similar cell-transfer protocols established a central role for the T-cell response in another virus disease of mice, LCM, which had long been thought to represent a cell-mediated immunopathological process (Hotchin, 1971; Cole et al., 1972; Doherty and Zinkernagel, 1974). Experiments with some other viruses have, however, failed to demonstrate a clear role for immune T cells. This may reflect that either phenomena mediated by thymus-derived lymphocytes are unimportant, or that the protocols used were inadequate for this purpose.

A. Basic Approach

Experiments with both the LCM and ectromelia systems have clearly shown that optimal transfer of specific CMI is closely correlated with highest levels of cytotoxic T-cell activity as measured in vitro. Furthermore, in both systems, significant effector function is seen only for H-2K or H-2D compatible interactions (Doherty et al., 1976a). Rational selection of a donor system thus depends on development of an in vitro cytotoxic T-cell assay, and on maximizing activity in the cell population (generally spleen) to be transferred.

A typical experimental protocol is that 5.0×10^7 immune or normal spleen cells are inoculated intravenously (via the lateral tail vein) into each of 4 or 5 recipient mice. Effector function is then assessed (by one of the methods described below) from 24 to 72 hours later. There seems little point in transferring cells to an extravascular site, e.g., the peritoneal cavity, unless one wishes to investigate local phenomena. Even then it is much more physiological to have T cells "home" to the target organ from blood.

The status of the recipient must also be carefully considered. Inoculation of immune cells before exposure to the infectious agent is unlikely to give an interpretable result. Any inhibition of virus growth could be due to neutralization of the inoculum by antibody secreted subsequent to cell transfer. Thus, it is reasonable to delay dosing with cells for at least 24 hours after administration of the virus.

The time that cells are transferred depends very much on the rate of virus growth in the recipient. Cell-mediated immunity apparently reflects, at least initially, direct membrane-to-membrane contact between lymphocyte and target. Such interactions are, of course, very quantitative–a 2-fold decrease in T-cell numbers may also result in effective halving of the overall inflammatory process (Doherty et al., 1976a). Thus, if very many cells throughout the organism are supporting virus growth, it may be quite difficult to show evidence of CMI subsequent to adoptive transfer: the sensitized T-cell population may be so "diluted" that significant effects are not seen in a particular organ.

The experiment must, therefore, be biased so as to produce a maximal effect in the relevant tissue site. For instance, in the LCM model, recipient mice are injected intracerebrally with the "neurotropic" Armstrong strain of virus (which grows mainly in the brain) rather than with the "viscerotropic" WE3 strain, as the target organ of interest is the central nervous system. On the other hand, there is obviously no point in transferring immune populations until sufficient antigen is expressed to promote T-cell recruitment from blood.

The point must be made that adoptive transfer is a rather drastic procedure, and should be considered essentially as an assay system. For instance, we cannot hope to accurately duplicate events that may be occurring in a localized immune response. The model used must therefore be manipulated to produce a maximal effect that can be observed and analyzed, rather than to attempt exact reproduction of the physiological situation. For instance, it would seem valid to immunize donors intraperitoneally with influenza virus in order to produce an effect with spleen which may be manifest in recipient lung, rather than to adoptively transfer mediastinal lymph nodes from intranasally inoculated mice. Once a phenomenon is produced subsequent protocols may be biased in this direction.

B. ASSAY PROCEDURES

1. *Survival*

The simplest and most dramatic demonstration of CMI is to modify survival patterns in recipient mice. This may be manifest as protection, in the ectromelia model, or induction of neurological disease, in the LCM system. The protocol is important to establish a case for a particular role of CMI, and may also be useful as a simple screening procedure. However, a maximal effect must be achieved to allow use of such a gross end point, and rigorous analysis of mechanism requires more quantitative techniques.

2. *Deletion of Recipient T-Cell Function*

Demonstration that effector function is a property of transferred T cells depends on there being no specific CMI generated by the recipient. This can be achieved by completing the experiment before the response in the recipient develops, say, within 3 days of infection with virus, or by using an immunosuppression schedule.

Some means of immunosuppression are probably contraindicated, e.g., rabbit antimouse thymocyte serum may have a half-life of up to 6 days (Jeejeebhoy and Singla, 1972), and could thus tend to eliminate donor T cells. Congenitally hypothymic nude (nu/nu) mice, or ATxBM mice (adult, thymectomized, lethally irradiated, and reconstituted with syngeneic bone marrow) may be used. However, if the assay system reflects lymphocyte-mediated activation of macrophages (Blanden, 1974), the background in such mice may be abnormally high. Mice chronically depleted of T cells apparently develop, presumably as a general protective mechanism, nonspecifically activated macrophages that tend to be more effective than those found in normal mice with intact thymic function (Zinkernagel and Blanden, 1975).

One means of immunosuppression that has been extremely useful is that developed for virus models by G. A. Cole and his colleagues (Gilden *et al.*, 1972). Recipients are treated with cyclophosphamide (150–200 mg/kg, Asta-Werke A. G., Brakwede, Germany) at 1 to 2 days prior to inoculation of donor lymphocytes. The drug is rapidly metabolized and has no effect on cells transferred 24 hours later. Treatment with cyclophosphamide completely eliminates the virus-immune cytotoxic T-cell response (Doherty and Zinkernagel, 1974). However, macrophage function is also greatly depressed. Thus, if the assay system depends on participation of recipient macrophages (Blanden, 1974), a new source of potential monocytes must be provided. This may be achieved by administering syngeneic bone marrow cells at the time of, or prior to, immune lymphocyte transfer (McFarland *et al.*, 1972; Hapel, 1975).

3. *Suppression of Virus Titer*

Measurement of virus titer in target organs of mice given immune or normal (injected with diluent) lymphocyte populations is probably the most sensitive *in vivo* assay available for any T-cell function (Blanden, 1971). The technique has proved particularly successful when spleen and liver are used as sites for estimating reduction of virus growth in both the ectromelia and LCM models. Adoptive transfer of LCM virus-immune spleen cells largely eliminated virus from visceral sites, but did not cause any decrease in titers in brain (Mims and Blanden, 1972). McFarland *et al.* (1972) and Hapel (1975), however, successfully used adoptive transfer protocols to depress virus titers in brains of mice infected with arboviruses. In McFarland's experiment, where cyclophosphamide was given to suppress the recipient response, addition of normal bone marrow and spleen cells proved essential to provide a source of macrophages.

Specificity of the CMI response apparently resides in direct interaction between receptors on donor T cells and membrane changes on virus-infected targets in solid tissues. However, virus clearance is completely dependent on recruitment and activation of recipient macrophages. Phagocytic cells in the donor spleen population apparently play no significant role in the second, nonspecific component of CMI.

4. *Localization of Inflammatory Cells*

A somewhat less functional assay is to assess homing of transferred cell populations to sites of virus growth. One means of doing this is to radiolabel the donor lymphocytes and then remove the target organ for direct counting and/or autoradiography. Such techniques have been used extensively by Woodruff and her colleagues (Woodruff and Gesner, 1969) for monitoring lymphocyte circulation in virus-infected rats. A problem with this type of experiment may be that the liver and lung tend to serve as "graveyards" for irrelevant transferred cells, so the model may not be suitable for studies involving these organs. The method of Miller *et al.* (1975), where antigen is localized to the ear and radiolabeled cells home to this site (which is readily taken for counting), may also be useful for virus systems.

An alternative technique is to inject virus, or virally modified cells, into the mouse footpad and then to measure the extent of footpad swelling at 6, 24, and 48 hours after adoptive transfer. Control preparations are either injected into the contralateral footpad, or control material is given to separate groups of mice. This method has also been successfully used for monitoring the development of CMI following primary infection (Tosolini and Mims, 1971). The extent of distension can be accurately assessed using dial-gauge calipers (Pocket Thickness Gauge, Catalog No. 52545001, Schlesinger's, Brooklyn, New York).

Also, the extent of T-cell activity in at least one anatomical site may be directly measured by counting the number of cells present in the inflammatory exudate (Doherty, 1973). Mouse cerebrospinal fluid (CSF) generally contains less than 100 cells per milliliter whereas, in acute meningitis, cell counts of more than 50,000 per milliliter may be achieved. Small amounts of clear CSF (5 μl) may be readily obtained from the cisterna magna of all the mouse strains tested so far. However, in moribund animals, acute brain swelling may tend to obliterate the subarachnoid cistern.

The mouse-CSF model has the added advantage that lymphocytes localizing in a site of virus-induced damage can be assayed functionally *in vitro*, though large numbers of mice (10 or more) may need to be sampled to obtain sufficient numbers. Both in LCM and in ectromelia (when the virus is injected intracerebrally) the inflammatory exudate contains many cytotoxic T cells (Zinkernagel and Doherty, 1973; Hapel and Gardner, 1974) which, at least in LCM, may comprise the majority of invading mononuclear cells. In arbovirus infections, however, recruitment of monocytes is apparently essential to magnify the inflammatory process (Doherty, 1973; Hapel, 1975). Furthermore, the use of the central nervous system as a target organ has the added advantage that there is no resident lymphoid tissue, so lymphocytes localizing to this site are likely to be involved in the disease.

C. Controls and Interpretation

Most adoptive transfer systems have been controlled by inoculating groups of virus-infected mice with either normal spleen cells or lymphocytes from mice immunized with other infectious agents. Further analysis has then depended on elimination of T cells by treatment with anti-θ(Thy-1,2) ascitic fluid (see Section IV,A), or by separating lymphocytes using a rosetting technique (Parish *et al.*, 1974; Blanden *et al.*, 1975; Parish, 1975). Additional procedures that have been used are removal of macrophages with carbonyl iron (Golstein and Blomgren, 1973) and depletion of B cells (antibody-forming cell precursors), either on nylon wool columns (Julius *et al.*, 1973) or by incubation with various anti-Ig sera (Blanden, 1971).

Such experiments may establish a requirement for immune T cells, but need not necessarily demonstrate how effector function is manifest. Recent results with LCM, ectromelia, and a delayed-type hypersensitivity (DTH) model indicate that genetic studies may allow a more precise definition of the roles of different lymphocyte populations (Doherty and Zinkernagel, 1975; Miller *et al.*, 1975; Doherty *et al.*, 1976b; Kees and Blanden, 1976). Adoptive induction of maximal inflammatory process, or protection, in the virus systems requires (as in the *in vitro* cytotoxic assay, Section II,D) that donor and recipient must share genes mapping at either H-2K or H-2D. Little or no effector function is manifest in the absence of such compatibility. Conversely, transfer of DTH to a soluble antigen (Fowl γG) is only successful in combinations sharing genes in the Ir region of the H-2 gene complex (Miller *et al.*, 1975). Identity at H-2K or H-2D is irrelevant. Expression of helper function, involved in differentiation of B-cell (antibody-forming cell precursor) clones to synthesize IgG, also requires (at least in some studies) that T cells and B cells share *Ir* genes (Katz and Benacerraf, 1975).

Thus, by selecting suitable H-2 recombinant mouse strains (Table II), it is possible to map effector T-cell function within the H-2 gene complex. Any immunologically mediated recovery (or pathological) process reflecting exposure to soluble antigens, e.g., bacterial toxins, or operating via IgG secreted as a result of T-cell helper function, may be recognized in I-region compatible interactions. Events requiring direct membrane-to-membrane contact between T cell and virus-infected target may, alternatively, occur only when H-2K or H-2D genes are shared.

Furthermore, by doing reciprocal experiments using allogeneic (H-2 incompatible) immune populations, it is possible to control for effects following transfer of cell suspensions containing virus, activated macrophages or antibody-producing cells. Thus, if allogeneic immune spleen cells are functionally inert, it would seem unlikely that antibody-mediated (or antibody-dependent cell-mediated) mechanisms are of primary importance in the immune process. This seems to be the case for all studies to date with the virus systems. In such experiments it is essential to compare the potency

TABLE II

MOUSE STRAINS SUITABLE FOR DEFINING T-CELL EFFECTOR FUNCTION ON THE BASIS OF THE H-2 COMPATIBILITY REQUIREMENT[a]

Mouse strains	H-2 Gene complex					
	K	*Ir*-A	*Ir*-B	*Ir*-C	S	D
A/J	k	k	k	d	d	d
A.TH	s	s	s	s	s	d
A.TL	s	k	k	k	k	d
A.QR	q	k	k	d	d	d
CBA/H	k	k	k	k	k	k
BALB/c	d	d	d	d	d	d
SJL	s	s	s	s	s	s
DBA/1	q	q	q	q	q	q

[a]From Shreffler and David (1975). The B10 series of recombinants, as described herein, are also extremely useful in this regard.

of different immune populations in F1 recipients (or suitable H-2 recombinants), and to complete the assay before commencement of allograft rejection (within 3 days).

IV. Techniques

Many of the methods described here are, with various minor differences, followed widely in cellular immunology laboratories. We may thus be guilty of describing techniques without due acknowledgment, as they have been in general use for some time. The presentation of such established procedures here is in accord with the stated purpose of this volume, to provide a source of readily available information for virologists. Only essential, straightforward methods with which we are familiar are considered; otherwise reference is made to original papers.

A. Preparation of Effector Lymphocytes

1. *Making a Single Cell Suspension*

If the lymphocytes are to be used for overnight cytotoxicity assays or for cell transfer, these procedures need not be done with scrupulous attention to sterile technique. Obviously sterile medium and equipment should be used, but the same instruments may serve to remove a number of spleens and all operations may be performed on the bench. The converse is true for establishing 4- to 6-day mixed lymphocyte cultures for secondary stimulation *in vitro*, where precautions must be taken to avoid bacterial contamination.

a. Spleens are removed into sterile balanced salt solution, taking care to trim off excess adherent fat. Mesenteric lymph nodes are, in the mouse, also quite prominent (forming a single large conglomerate in the mesentery) and constitute a good source of T cells. Location of smaller lymph nodes may be facilitated by injecting India ink into the draining site prior to attempted removal.

b. Single cell suspensions are prepared either by gentle homogenization in Tenbroeck tissue grinders or by chopping with scissors and pressing through stainless steel grids using the plunger of a 5.0 ml disposable syringe.

c. The cell suspension is then allowed to stand for 4 minutes (to allow debris to settle) and the supernatant is decanted, centrifuged lightly (200 g for 5 minutes), and the cells are resuspended in tissue culture medium containing 10% FCS at 4°C. Alternatively, the decanting step may be dispensed with and the whole spleen suspension pelleted and quickly resuspended in alkaline (pH 8.0) Eagle's minimal essential medium or Puck's A saline (Blanden and Langman, 1972).* This is followed by rapid centrifugation (maximum speed on a bench centrifuge for 5 seconds) to deposit capsular material (which tends to clump under alkaline conditions). The suspension containing the lymphocytes is then pipetted off (remaining capsular debris tends to adhere to the sides of the pipette), and the cells are pelleted (200 g for 5 minutes) and resuspended in medium. It is advisable to follow the second procedure if lymphocytes are to be injected intravenously into mice. Also, preparations to be given intravenously are best suspended in acid medium that does not contain serum, as this tends to minimize clumping.

All procedures may be done in RPMI (RPMI 1640, Flow Laboratories, Rockville, Maryland) or F15 (GIBCO, Grand Island, New York) containing 10% heat-inactivated (56 °C for 30 minutes) fetal calf serum. However, this is more expensive and it is generally sufficient to use Puck's A saline throughout, except for the final suspension in tissue culture medium. Lymphocytes tend to prefer rather acid conditions (pH 7.0–7.4) and medium should, if necessary, be gassed with CO_2. Also, cells should be held at 4 °C subsequent to removal of capsular material or red blood cells. However, it is important not to pipette cells vigorously when in the cold, as the lymphocyte membrane tends to be stabilized and high mortality may result.

2. Lysis of Erythrocytes

Removal of red blood cells is advisable in experiments with viruses that haemadsorb, but may make surprisingly little difference in other systems, e.g., the LCM assay.

a. The lysing solution (Boyle, 1968) is stored at 4 °C as a 10 × concentrate, and diluted 1:10 in distilled water prior to use. The formula is as follows: 4.12 gm Tris base, 150 ml double deionized H_2O, 30 ml IN HCl, 14.94 gm NH_4Cl. Dissolve 4.12 gm Tris base in 150 ml DDH_2O, adjust pH to 7.5 with 30 ml IN HCl. Add 14.94 gm NH_4Cl and adjust pH to 7.5 with 1.1 ml IN HCl. The solution must be filtered, not autoclaved.

b. Centrifuged cells are suspended in the solution by pipetting and held at room temperature for 10 minutes. Lymphocytes are then washed 3 times in balanced salt solution before suspension in medium containing serum. The cell pellet changes from a red to a cream color, though spleen preparations from black mice, or from old mice, may appear rather gray.

---

*Puck's A saline:

Phenol red (water soluble)	0.4 gm	Glucose	20 gm
Sodium chloride NaCl	160 gm	Sodium bicarbonate $NaHCO_3$	7 gm
Potassium chloride KCl	8 gm	Distilled water (made up to)	20 liters

3. *Removal of Dead Cells*

Viable cell counts are assessed by trypan blue (0.05%) or erythrosin B (0.07%) dye exclusion, and are generally greater than 80%. This is usually acceptable for both *in vitro* and *in vivo* assays. Dead cells and erythrocytes may be removed by centrifugation through Isopaque/Ficoll, following the procedure described by Davidson and Parish (1975). Alternatively, cell populations of low viability may be resuspended in 5% glucose in distilled water at 4 °C and filtered through cotton wool (von Boehmer and Shortman, 1973). This yields about 80% of the original live cells with 90 to 95% viability, but the experience of one of us (JCP) is that some activity may be lost from this treatment.

4. *Intravenous Inoculation of Mice*

Mice are generally injected with immune cells suspended in from 0.2 ml to 0.4 ml of acid medium not containing serum. Physiological cooling in the mouse is, at least in part, achieved by dilation of the lateral tail veins. Holding mice under a heat lamp (GEC R-30 reflector photoflood) for a minute or two prior to inoculation results in considerable enlargement of the lateral tail veins, which can then be entered with a 25 or 26 gauge needle. It is important that the needle point be kept horizontal (with the bevel up), otherwise the needle will pass straight through the vein. It is also essential that the mouse be adequately restrained (E-C mouse-vise, Arthur H. Thomas Co., Philadelphia, Pennsylvania). Correct positioning of the needle is immediately apparent, as the inoculum flows extremely smoothly. Depending on the speed of the operator, 5 or 10 mice can be safely heated together in the one cage. Profuse sweating around the nostrils is a warning sign presaging death from heat exhaustion.

Most mice readily tolerate inoculation of from 5.0 to 10.0×10^7 immune spleen cells, providing that the dose is given slowly, the cells are not clumped, and capsular material has been adequately removed. At least initially, it is advisable to give one half of the inoculum and then pause for 4 or 5 seconds before slowly injecting the remainder. Problems with cell emboli may be encountered if the mice have suffered previously from pneumonia. In this case it is necessary to give a smaller dose, or to infuse cells extremely slowly. Generally the same tail vein can be used again later the same day if required.

5. *Anti-θ (Thy-1) Treatment for Characterization of Effector Cells*

An established way to verify that the cytotoxic cell being studied is a thymus-derived cytotoxic lymphocyte (killer cell) is to determine whether treatment with anti-Thy-1 (θ) serum complement abolishes cytotoxic activity. The easiest way to accomplish this is to obtain, either commercially or from an established immunology laboratory, some well characterized anti-Thy-l serum and use it exactly as was done to abolish known alloreactive cytotoxic T-cell activity. Nonspecific toxicity of antiserum or of complement by absorption with an equal volume of normal spleen cells or with agarose (80 mg/ml for 30 minutes at 4°C). It is essential to include both a negative and a positive control in order to properly interpret the results of this characterization. A convenient positive control is to treat, in parallel with the population to be characterized, a population of cytotoxic cells generated in response to allogeneic cells, for example, by injecting a mouse intraperitoneally with 30×10^6 allogenic spleen cells and assaying the spleen on day 10. This allogeneic killer activity is known to be T-cell mediated and cytotoxicity should be completely abolished by the anti-Thy-l treatment. A general anti-Thy-l treatment protocol is incubation of cells with anti-Thy-l serum for 30 minutes at 0°C, centrifugation, addition of complement to the pellet, resuspension and incubation for 45–60 minutes at 37°C, followed by washing. A convenient negative control is to test in parallel a population of antisheep erythrocyte plaque-forming cells generated by injecting a mouse intraperitoneally 5 days previously with 0.2 ml of a 10% suspension of sheep red blood cells. Treatment with complement only, or normal mouse serum plus complement, is necessary here to control for possible nonspecific loss of plaque-forming cells as a result of manipulation and incubation. See Kiessling et al. (1976b) for a good discussion of characterization of different kinds of effectors.

B. SECONDARY STIMULATION in Vitro

Extremely active effector T-cell populations can be generated by taking memory spleen cells from mice immunized at least 3 weeks previously and exposing them to virus-infected stimulator cells in vitro for 4 or 5 days (Dunlop and Blanden, 1976; Dunlop et al., 1976; Gardner and Blanden, 1976). The kinetics of this response, optimal stimulator-responder ratios, and time of maximal lytic capacity may vary considerably from virus to virus, and must be determined for each experimental system. The protocols used are based on those developed by Wagner et al. (1973) for secondary stimulation of thymus-derived lymphocytes reactive to histocompatibility antigens. High levels of effector function are achieved when peritoneal macrophages or spleen cells are used as stimulators [perhaps such cells deliver an inductive signal (Bretscher, 1974)]: fibroblasts or tumor cells (except those of lymphoid or reticuloendothelial origin) are generally not suitable for this purpose.

Potent cytotoxic T-cell preparations were generated, in the ectromelia system, by incubating 1×10^6 virus-infected spleen cells in contact with 2×10^6 responders (from mice given virus 4 or 5 weeks previously) for 4 days (Gardner and Blanden, 1976). Cultures were established in 2.0 ml of F15 (GIBCO, Grand Island, New York), incorporating 10% heat-inactivated (56°C for 30 minutes) fetal calf serum and 10^{-4} M 2-mercaptoethanol (2 ME) (Carbiochem, La Jolla, California. A 100 × stock solution is made by dissolving 0.35 ml 2-ME in 100 ml phosphate-buffered saline) and maintained in 16 mm tissue culture wells (Linbro FB-16-24TC, New Haven, Connecticut) in an atmosphere of 10% CO_2, 7% O_2 in N_2. No difference was observed when irradiated (850 rad) or unirradiated stimulators were used, but it was found essential to hold cultures at 39°C throughout (in a water bath) to prevent the virus from infecting and killing responder T cells.

This was not a problem in the LCM model (Dunlop and Blanden, 1976). Cultures were maintained at 37°C and maximal activity was seen on day 5. However, the optimal stimulator-responder ratio was greatly different (1:10), and virus-infected primary peritoneal macrophages (Section III, C) were used as stimulators. Also, experiments were done with 2.5×10^6 stimulators and 2.5×10^7 responders in 10.0 ml of medium (see above), and were maintained in 25 cm^2 tissue culture flasks (Falcon No. 2012, Becton Dickinson, Oxnard, California). Other potentially useful aspects of these experiments are that formaldehyde (2% w/v formaldehyde in phosphate-buffered saline for 10 minutes) or glutaraldehyde (0.25% w/v glutaraldehyde in phosphate-buffered saline for 15 seconds) fixed virus-infected macrophages could, following five medium washes, be used for sensitization. Also, presence of a phagocytic cell (removed by carbonyl-iron depletion) in the responder population was essential for maximal stimulation.

C. PREPARATION OF TARGET CELLS

A variety of cell types have been used successfully in these assays. The only absolute constraints are: (a) the cell must be susceptible to infection with the virus in question, but not die (i.e., release >40% of incorporated label) before lymphocyte effector function is manifest; (b) the majority (at least 90%) of targets must express the virus-induced antigenic change; this may be checked by immunofluoresence; (c) normal lymphocytes must not cause high levels of nonspecific cytotoxicity, a problem with some tumor lines (Kiessling et al., 1976a); and (d) the target must phenotypically express at least one H-2K or H-2D gene shared with the effector T cell. Otherwise, no rigid generalizations can be made, e.g., primary mouse embryo fibroblasts are not particularly useful for the ectromelia and LCM systems, but have proved the target of choice in the Coxsackie virus model (C. Y. Wong, J. J. Woodruff, and J. F. Woodruff, unpublished data).

1. *Peritoneal Macrophages*

Primary mouse peritoneal macrophages have proved extremely useful for genetic studies, being a readily available source of targets from different strains of mice (Zinkernagel and Doherty, 1975c; Zinkernagel, 1976).

Mice are either killed by cervical dislocation or anesthetized with ether. Cold phosphate-buffered saline (10.0–14.0 ml) is then injected vigorously through the body wall (using a 20.0 ml syringe and a 20 gauge needle), the abdomen is massaged gently, the majority of the suspending phosphate-buffered saline is withdrawn and then reinjected, and the medium is finally removed into centrifuge tube held on melting ice. Generally, with anesthetized mice, between 50 and 80% of the inoculum is recovered, containing from 3.0 to 8.0 $\times$ 10^6 large viable nucleated cells.

More macrophages can usually be obtained if the mice are first killed, as the abdominal wall may then be exposed and the position of the needle point (which is likely to clog with mesentery) can be readily observed. It is best to keep the bevel of the needle downward, and to make a "tent" of the body wall in the right, anterior section of the peritoneal cavity. The cells are then centrifuged (200 g for 5 minutes), resuspended in medium (RPMI + 10% fetal calf serum, incorporating 10 μg/ml of penicillin and streptomycin), and dispensed at a concentration of between 5.0 $\times$ 10^5 and 1.0 $\times$ 10^6 per milliliter. Samples heavily contaminated with red blood cells should either be discarded or first treated with lysing solution (see Section IV,A,2).

Many more cells can be recovered if mice are first stimulated with thioglycollate (Argyris, 1967). Donors are injected intraperitoneally with 1.0 ml or 2.0 ml of thioglycollate medium (Baltimore Biological Laboratories, Cockeysville, Maryland) and sampled 3 days later. Cells obtained in this way are suitable for use with LCM virus. The procedure is much more practical than using unstimulated mice, as some strains (e.g., wild and BALB/c mice) may normally be poor sources of peritoneal macrophages.

2. *Labeling*

In order to assay cytotoxicity as opposed to growth inhibition or loss of adherence to surfaces, some observation must be made which is dependent on cell death or lysis. The most convenient and widely adopted measure of cell death is release of radioisotope from prelabeled target cells: ^{14}C- or ^{3}H-labeled amino acids (Bean *et al.*, 1973), thymidine-^{3}H or ^{125}IUdR (Le Mevel *et al.*, 1973), and ^{51}CrO$_4$ have been used as labeling reagents for preparing target cells for assay. Chromium as ^{51}CrO$_4$ is the isotope of choice in all cases where it can be used, since it does not require dividing cells for efficient labeling, is inexpensive, and can be counted in a gamma counter without sample preparation or counting cocktails. The chromate ion is taken up by viable cells, bound to intracellular small molecular weight proteins, and leaks out of the cell when its plasma membrane is damaged. However, all cells release incorporated chromate spontaneously, usually at the rate of about 1–5% per hour, so chromium release is most suitable for assays of short duration. Possible mechanisms of cytolysis have been revviewed by Henney (1973).

a. Cells in Suspension. Harvest at least 5×10^6 cells, or 1.5 times the required number of target cells, from an exponentially growing culture which has a viability of at least 90%. Wash the cells and resuspend in assay medium, optimally medium containing serum to which 1% v/v, pH 7.2, 1 M morpholenopropane sulfonic acid (MOPS) (Calbiochem, La Jolla, California) or 1 M N-2-hydroxyethyl-piperazine-N'-2-propane-sulfonic acid, (HEPES) (Calbiochem, La Jolla, California) has been added to stabilize pH during manipulation outside of a CO_2 incubator. Add ^{51}Cr as sodium chromate to achieve a concentration of 100–500 $\mu Ci/ml$ depending on how high a labeling level is needed and how readily the target cells being used take up the label. Incubate at 37°C with agitation every 15 minutes, or constantly to prevent settling of the cells. Cells should be labeled for at least 30 minutes and for up to 2 hours, again depending on the ease of labeling and the required level. Mouse lymphoblasts or mastocytoma cells label in 90 minutes at 250 $\mu Ci/ml$ to a level of about 0.3–0.5 cpm/per cell.

In order to minimize dilution of the assay medium with chromium solution, purchase the sodium chromate at 10 mCi/ml so that less than 10% of the final mixture is chromium stock solution. For example, we use New England Nuclear, NEZ030, delivered at 10 mCi/ml. At the end of the labeling period the target suspension is diluted 2- to 3-fold with assay medium, centrifuged, and the radioactive supernatant is removed to a radioactive waste container to be stored for decay or processed by radiation safety personnel. The cell pellet is then washed 2–3 times in assay medium with centrifugation through a serum underlay introduced slowly into the bottom of the tube by Pasteur pipette. This serum underlay is advisable as it greatly increases the effectiveness of the washes, but is not essential. The washed targets are again counted, viability is measured, and they are diluted to the desired concentration for addition to the assay, usually 150–200 $\times 10^3/ml$ dispensed in 100 μl of medium. Total label incorporated is assessed by incubating one group (4 wells) of targets in water or detergent (1% Triton X-100 in water, Fisher Scientific Co., Pittsburgh, Pennsylvania) for the length of the assay (See Section II,A).

b. Monolayers. Adherent cells can be effectively labeled by exposure to ^{51}Cr at a level of 2–4 μCi per assay well. The isotope can be diluted, so as to be available in 400 μl of growth medium to which the targets are exposed overnight. Alternatively, cells are incubated with 2–4 μCi in 50 μl for 1 to 2 hours at 37°C. Whichever method is used, the monolayers are washed twice with warm medium containing serum prior to exposure to lymphocytes. The former may be preferable, as it reduces the number of manipulations performed (and, perhaps, spontaneous ^{51}Cr release) on the day targets are to be used. Total label is estimated by removing all the supernatant medium from 3 or 4 wells and incubating in water or detergent throughout the assay (See Section II,A).

D. APPARATUS FOR MICROASSAY

For small experiments totaling 100–200 assay wells, or for complicated combinations of effectors and targets and ratios of the two, it is most convenient to add target cells and lymphocytes to the assay plate with a Biopette (Schwarz/Mann, Orangeburg, New York). However, for experiments involving larger numbers of replicates (or total assay wells) requiring removal of one-half the supernatant medium for radioactive counting at the end of the assay, the use of a multiple dispensing apparatus greatly reduces the amount of work involved. The Terasaki Dispenser (Hamilton, Reno, Nevada) can be modified by bending the needles to make the separation correspond to the well-to-well distance of a standard 96-hole microtiter plate. This dispenser will then permit addition of, or supernatant removal from, 6 wells simultaneously. Aliquots removed for radioactive counting can be dispensed into 6 × 50 mm tubes (Fisher) held in a 96-well microtiter plate, or directly into a flexible microtiter plate (u. 1–220–24B, Cooke Lab, Alexandria, Virginia), allowed to evaporate to dryness, and the wells can be separated by cutting with scissors and counted.

For the greatest possible counting efficiency, in situations where miniaturization and maximal sensitivity are required, the released ^{51}Cr can be counted in liquid scintillation. This modification is more costly and laborious but should permit the use of as few as 100 target cells per well (Thorn *et al.*, 1974).

E. ESTIMATION OF CYTOTOXICITY

Quantitative approaches to cytotoxic T-cell effector function in the allograft system have been described by Henney (1971), Miller and Dunkley (1974), and Cerottini and Brunner (1974). Analysis of the virus models in this regard has not received the same degree of attention, the methods used (Zinkernagel and Doherty, 1974a, 1975a) being based on concepts developed for alloreactivity. Consideration here is confined to simple presentation of results.

Results for monolayer targets have generally been expressed as:

$$\frac{\text{counts in supernatant} \times 100}{\text{counts in supernatant} + \text{counts in cells}} \times \frac{100}{\% \text{ water lysis}}$$

the final value being given as the mean $\pm$ S.E. for 3 or 4 wells. In most experiments, water lysis releases 70–80% of total counts, i.e., incubation of one group of targets in water throughout the assay results in this percentage of counts being present in the supernatant, the remainder being recovered by adding more water to the well and pipetting vigorously.

With some viruses (e.g., Sendai) and some batches of $Na^{51}Cr$ the water lysis value may be very low. In such cases it may be necessary to use detergent, but if this happens, release mediated by immune lymphocytes may also be minimal.

When the above method is used, separate sets of data are obtained for immune lymphocytes (I) incubated with virus-infected (i) or normal (n) targets and likewise for normal lymphocytes (N). Once the system is established and it has been adequately shown that there are no major problems with nonspecific cytotoxicity, it may be more convenient to give results as specific ^{51}Cr release. Mean values are obtained as described above, except that no correction is made for the percentage of water lysis (W), and cytotoxicity is expressed as:

$$\frac{Ii - Ni}{Wi - Ni} \times 100$$

With suspended targets, however, water lysis is not generally calculated for each well, as the number of cells is constant (See Section II,A). Readings are made on 100 μl of supernatant, and this is compared with ^{51}Cr release by detergent (D). Providing there is no significant cytotoxicity for normal targets, mean values are obtained and release is calculated as:

$$\frac{Ii - Ni}{Di - Ni} \times 100$$

Activity of immune lymphocyte populations may also be expressed as lytic units (Cerottini and Brunner, 1974). Determining the plot of numbers of immune lymphocytes versus the percent of ^{51}Cr release reveals a substantially linear relationship over, say, the 25–65% interval (depending on the system used). The number of lytic units in a population is then calculated by dividing the total number of cells in the population by the number of cells per well necessary to cause 33% specific lysis.

ACKNOWLEDGMENTS

We thank Drs. R. V. Blanden, M. B. C. Dunlop, B. B. Knowles, G. Trinchieri, and J. Woodruff for advice and permission to cite unpublished material.

REFERENCES

Ada, G. L., Jackson, D. C., Blanden, R. V., Tha Hla, R., and Bowern, N. A. (1976). *Scand J. Immunol.* **5**, 23.

Allison, A. C. (1974). *Transplant. Rev.* **19**, 3.

Argyris, F. (1967). *J. Immunol.* **99**, 744.

Bean, M. A., Pees, H., Rosen, G., and Oettgen, H. F. (1973). *Natl. Cancer Inst., Monogr.* **37**, 41.

Blanden, R. V. (1971). *J. Exp. Med.* **133**, 1074.

Blanden, R. V. (1974). *Transplant. Rev.* **19**, 56.

Blanden, R. V., and Langman, R. E. (1972). *Scand. J. Immunol.* **1**, 379.

Blanden, R. V., Bowern, N. A., Pang, T. E., Gardner, I. D., and Parish, C. R. (1975). *Aust. J. Exp. Biol. Med. Sci.* **53**, 187.

Blanden, R. V., Hapel, A. J., and Jackson, D. C. (1976). *Immunochemistry* **13**, 179.

Bloom, B. R. (1971). *Adv. Immunol.* **13**, 101.

Boyle, W. (1968). *Transplantation* **6**, 761.

Bretscher, P. A. (1974). *Cell. Immunol.* **13**, 171.

Burns, W. H., and Allison, A. C. (1974). *In* "The Antigens" (M. Sela, ed.), Vol. 3, p. 479 Academic Press, New York.

Cambridge, G., MacKenzie, J. S., and Keast, D. (1976). *Infect. Immun.* **13**, 36.

Cerottini, J. -C., and Brunner, K. T. (1974). *Adv. Immunol.* **19**, 67.

Cole, G. A., Nathanson, N., and Prendergast, R. A. (1972). *Nature (London)* **238**, 335.

Cole, G. A., Prendergast, R. A., and Henney, C. S. (1973). *In* "Lymphocytic Choriomeningitis and Other Arenaviruses" (F. Lehmann-Grube, ed.), pp. 61–71. Springer-Verlag, Berlin and New York.

Cunningham, A. J. (1974). *Contemp. Top. Mol. Immunol.* **3**, 1.

Davidson, W. F., and Parish, C. R. (1975). *J. Immunol. Methods* **7**, 387.

Doherty, P. C. (1973). *Am. J. Pathol.* **73**, 607.

Doherty, P. C., and Zinkernagel, R. M. (1974). *Transplant. Rev.* **19**, 89.

Doherty, P. C., and Zinkernagel, R. M. (1975). *J. Immunol.* **114**, 30.

Doherty, P. C., and Zinkernagel, R. M. (1976). *Immunology.* **31**, 27.

Doherty, P. C., Zinkernagel, R. M., and Ramshaw, I. A. (1974). *J. Immunol.* **112**, 1548.

Doherty, P. C., Blanden, R. V., and Zinkernagel, R. M. (1976a). *Transplant. Rev.* **29**, 89.

Doherty, P. C., Dunlop, M. B. C., Parish, C. R., and Zinkernagel, R. M. (1976b). *J. Immunol.* **117**, 187.

Dunlop, M. B. C., and Blanden, R. V. (1976). *Immunology* **31**, 171.

Dunlop, M. B. C., Doherty, P. C., Zinkernagel, R. M., and Blanden, R. V. (1976). *Immunology* **31**, 181.

Gardner, I. D., and Blanden, R. V. (1976). *Cell. Immunol.* (in press).

Gardner, I. D., Bowern, N. A., and Blanden, R. V. (1974a). *Eur. J. Immunol.* **4**, 63.

Gardner, I. D., Bowern, N. A., and Blanden, R. V. (1974b). *Eur. J. Immunol.* **4**, 68.

Gilden, D. H., Cole, G. A., Monjan, A. A., and Nathanson, N. (1972). *J. Exp. Med.* **135**, 860.

Golstein, P., and Blomgren, H. (1973). *Cell. Immunol.* **9**, 127.

Golstein, P., Svedmyr, E. J., and Wigzell, H. (1971). *J. Exp. Med.* **134**, 1385.

Hapel, A. J. (1975). *Scand. J. Immunol.* **4**, 267.

Hapel, A. J., and Gardner, I. D. (1974). *Scand. J. Immunol.* **3**, 311.

Haughton, G., and McGhee, M. P. (1969). *Immunology* 16, 447.
Henney, C. S. (1971). *J. Immunol.* 107, 1558.
Henney, C. S. (1973). *Transplant. Rev.* 17, 37.
Hotchin, J. (1971). *Monogr. Virol.* 3, 1.
Jeejeebhoy, H. F., and Singla, O. (1972). *Immunology* 22, 789.
Julius, M. H., Simpson, E., and Herzenberg, L. A. (1973). *Eur. J. Immunol.* 3, 645.
Katz, D. H., and Benacerraf, B. (1975). *Transplant. Rev.* 22, 175.
Kees, U., and Blanden, R. V. (1976). *J. Exp. Med.* 143, 450.
Kiessling, R., Petranyi, G., Klein, G., and Wigzell, H. (1976a). *Int. J. Cancer* 17, 275.
Kiessling, R., Petranyi, G., Karre, K., Jondel, M., Tracey, D., and Wigzell, H. (1976b). *J. Exp. Med.* 143, 772.
Klein, J. (1975). "Biology of the Mouse Histocompatibility-2 Complex." Springer-Verlag, Berlin and New York.
Koszinowski, U., and Ertl, H. (1975). *Nature (London)* 255, 552.
Lamon, E. W., Wigzell, H., Klein, E., Anderson, B., and Skurzak, H. M. (1973). *J. Exp. Med.* 137, 1472.
LeMevel, B. P., Oldham, R. K., Wells, S. A., and Herberman, R. B. (1973). *J. Natl. Cancer Inst.* 51, 1551.
Lewandowski, L. J., Gerhard, W. U., and Palmer, J. C. (1976). *Infect. Immun.* 13, 712.
McFarland, H. F., Griffin, D. E. and Johnson, R. T. (1972). *J. Exp. Med.* 136, 216.
Mackaness, G. B. (1969). *J. Exp. Med.* 129, 973.
MacLennan, I. C. M. (1972). *Transplant. Rev.* 13, 67.
Marker, O., and Volkert, M. (1973). *J. Exp. Med.* 137, 1511.
Miller, J. F. A. P., Vadas, M. A., Whitelaw, A., and Gamble, J. (1975). *Proc. Natl. Acad. Sci. U.S.A.* 72, 5095.
Miller, R. G., and Dunkley, M. (1974). *Cell. Immunol.* 14, 284.
Mims, C. A., and Blanden, R. V. (1972). *Infect. Immun.* 6, 695.
Oldstone, M. B. A., and Dixon, F. J. (1970). *Virology* 42, 805.
Ortiz de Landazuri, M., and Herberman, R. B. (1972). *Nature (London), New Biol.* 238, 18.
Parish, C. R. (1975). *Transplant. Rev.* 25, 98.
Parish, C. R., Kirov, S. M., Bowern, N., and Blanden, R. V. (1974). *Eur. J. Immunol.* 4, 808.
Perlmann, P., and Holm, G. (1969). *Adv. Immunol.* 11, 117.
Perlmann, P. H., and Wigzell, H. (1972). *Transplant. Rev.* 13, 91.
Pfizenmaier, K., Trostmann, H., Röllinghoff, M., and Wagner, H. (1975). *Nature (London)* 258, 238.
Ramshaw, I. A. (1975). *Infect. Immun.* 11, 767.
Shortman, K., Williams, N., Jackson, H., Russell, P., Byrt, P., and Diener, E. (1971). *J. Cell Biol.* 48, 556.
Shreffler, D. C., and David, C. S. (1975). *Adv. Immunol.* 20, 125.
Thorn, R. M., Palmer, J. C., and Manson, L. A. (1974). *J. Immunol. Methods* 4, 301.
Tosolini, F. A., and Mims, C. A. (1971). *J. Infect. Dis.* 123, 134.
Trinchieri, G., Aden, D., and Knowles, B. B. (1976). *Nature (London)* 261, 312.
von Boehmer, H., and Shortman, K. (1973). *J. Immunol. Methods* 2, 293.
Wagner, H., Röllinghoff, M., and Nossal, G. J. V. (1973). *Transplant. Rev.* 17, 3.
Wainberg, M. A., Markson, Y., Weiss, D. W., and Doljanski, F. (1974). *Proc. Natl. Acad. Sci. U.S.A.* 71, 3565.
Wheelock, E. F., and Toy, S. (1973). *Adv. Immunol.* 16, 124.
Woodruff, J. J., and Gesner, B. M. (1969). *J. Exp. Med.* 129, 551.
Zinkernagel, R. M. (1976). *J. Exp. Med.* 143, 437.
Zinkernagel, R. M., and Blanden, R. V. (1975). *Experientia* 31, 591.

Zinkernagel, R. M., and Doherty, P. C. (1973). *J. Exp. Med.* **138**, 1266.
Zinkernagel, R. M., and Doherty, P. C. (1974a). *Scand J. Immunol.* **3**, 287.
Zinkernagel, R. M., and Doherty, P. C. (1974b). *Nature (London)* **248**, 701.
Zinkernagel, R. M., and Doherty, P. C. (1975a). *J. Exp. Med.* **141**, 1427.
Zinkernagel, R. M., and Doherty, P. C. (1975b). *J. Immunol.* **115**, 1613.
Zinkernagel, R. M., and Doherty, P. C. (1975c). *J. Immunol. Methods* **8**, 263.
Zinkernagel, R. M., and Doherty, P. C. (1976). *Contemp. Top. Immunobiol.* (in press).

3 Methods for Assays of RNA Tumor Viruses*

V. Klement and M. O. Nicolson

*This work was supported by a contract from the National Cancer Institute, NIH, no. NOI CP 5-3500.

I. Introduction

RNA tumor viruses are members of a larger group of enveloped viruses whose particles contain an RNA genome and an enzyme, RNA-directed DNA polymerase. The nomenclature and complete classification of these viruses has not been definitely established. Since the first well-studied members of this group were oncogenic, inducing mostly leukemias and to a lesser extent solid tumors, the whole group was originally called RNA tumor viruses, oncornaviruses, or leukoviruses. The subdivision of RNA tumor viruses into A-, B-, and C-type was based on morphological criteria (Bernhard, 1960): A particles were intracytoplasmic and donut-shaped, while B and C particles were extracellular, the former with eccentric, the latter with centrally positioned nucleoid. More recently, when it appeared that the group also included nononcogenic members, such as visna, maedi, progressive pneumonia virus (Lin and Thormar, 1970; Schlom et al., 1971; Stone et al., 1971; Brahic and Vigne, 1975), and foamy viruses (Parks and Todaro, 1972; Hruska and Takemoto, 1975), the terms RNA–DNA viruses, ribodeoxy-viruses, or retraviruses have been suggested (Dalton et al., 1974; Temin, 1974).

In this chapter, we limit our discussion to the assays for the C-type subgroup of oncogenic members of the family, i.e., for leukemia and sarcoma viruses. We still list the methods under the headings of sarcoma and leukemia viruses, but in doing so, we realize that this division might be obsolete. Clearly, RNA tumor viruses with the variety of genetically stable forms of sarcoma viruses (Aaronson et al., 1972, 1975) and with the "transforming" and "nontransforming" leukemia viruses (see Section V,C and D) represent a continuous spectrum of biological qualities, rather than two distinct categories. Since the recent work in our laboratories has dealt mainly with mammalian C-type viruses, the methods described below will cover mostly that area, and no attempt will be made to describe methodological details of work with avian tumor viruses. However, since the research on avian RNA tumor viruses in many respects preceded and directly influenced the work with mammalian C-type viruses, we also shall give the pertinent references from that field.

We will describe the procedures presently used for routine quantitative infectivity assays and biological characterization of mouse and mammalian RNA tumor viruses, namely, transformation focus assay, leukemia virus plaque assays, and immunological titrations based on the induction of synthesis of virus-specific antigens.

Other assay techniques have been developed that do not directly involve the measurement of infectivity. These assays utilize morphological, physical, and biochemical characteristics of RNA tumor viruses for which we do not have a suitable infectivity assay at present. In addition, due to the capacity of RNA tumor viruses for cellular integration of the viral genome, techniques were developed for the detection of viral genetic information and viral structural components. Of these techniques, not directly measuring infectivity, we will cover in detail only the assay for RNA-directed DNA polymerase. For others, the reader will be referred to pertinent literature.

A. BIOHAZARD

The RNA tumor viruses represent agents of unknown biological activity in man and should be treated with caution. Due to rapid expansion of this field, investigators from many specialized disciplines (immunology, biochemistry, molecular biology, etc.) have recently become involved in RNA tumor virus research. Both established laboratories and investigators who are newly entering the field are urged to study the guidelines for viral oncology research facilities specified by the National Cancer Institute (Office of Biohazard Safety, Viral Oncology, Division of Cancer Cause and Prevention, 1972, 1974, 1975), containing the general criteria for architectural design and operational practices for research involving oncogenic viruses. Technical information on laboratory safety equipment can be obtained directly from the manufacturers (Baker Co., Inc., Sanford, Maine; Bioquest, Cockeysville, Maryland).* In addition, standard virological procedures, sterility of vessels, tools, and materials, and aseptic operations are essential to prevent cross-contamination of virus stock and cell lines and to protect the laboratory personnel from exposure to oncogenic viruses.

---

* One or more manufacturing or distributing companies in the United States are listed for the convenience of the reader only and not as an endorsement of the product.

B. STABILITY AND STORAGE

RNA tumor viruses are thermolabile. The half-life of virus infectivity at 37 °C was estimated for avian viruses to be 2–6 hours with variations depending upon the virus strain and the composition of the incubation medium (Vogt, 1965). Viruses should be harvested into prechilled vessels and subsequent handling should be carried out at 0° to 5 °C.

Virus stocks are collected in tissue culture medium containing stabilizing protein, such as 5–10% calf or fetal bovine serum. The sera should be pretested and known to be free of nonspecific virus inhibitors. Twenty percent veal infusion broth (Difco Laboratories, Detroit, Michigan) in Eagle's minimum essential medium (EMEM) is used as a routine virus diluent in our laboratory.

Repeated assays of standard virus stocks should give identical titers. The virus stocks are stored in small aliquots (1.0 ml) in one dram vials (Wheaton Glass Company, Millville, New Jersey) in the Revco freezing cabinets at −80 °C. Standard laboratory strains of mouse leukemia or sarcoma viruses maintain titers under these conditions for years. However, fresh field virus isolates tend to lose their titers more rapidly (0.5–1.0 $\log_{10}$ within weeks or months) and their biological activity is better preserved in liquid nitrogen (−195.8 °C) in sealed vials (Wheaton gold band cryule, 1.2 ml).

C. TITRATIONS

The purpose of the quantitative virus titration assay is to determine the amount of infectious material in a biological sample. This is done by measuring a specific response in a susceptible host. In the case of RNA tumor viruses, the host has been an animal or embryo (usually the species of origin of the virus) or, more recently, tissue culture. To measure the infectivity, individual groups of animals or tissue cultures are inoculated with dilutions of the original material and the incidence of, or incubation periods for, specific pathologic changes are recorded for each group. Monodisperse virion suspensions are essential for accurate titrations. The RNA tumor viruses are produced at the cell surface and detached often in aggregates of surface mucopolysaccharides (Morgan, 1968). For aggregate elimination, the viruses can be subjected to treatment by ultrasound and/or filtration through 0.45 μm membrane filters or preferably through 0.22 μm filters (if the virus titer is high enough to allow some loss) before inoculation. Tissue culture assay systems have the advantage over *in vivo* titrations of being faster, less expensive, and more suitable for determination of titration patterns since they offer the possibility of observing focal areas of cytopathology (foci or plaques) induced by single infectious particles. The main disadvantage of tissue culture is the necessity for correlation of *in vitro* cytopathology with tumorigenesis or leukemogenesis *in vivo*.

II. In Vivo Infectivity Assays

Early quantitative biological studies of RNA tumor viruses were done *in vivo* and were based on the occurrence of disease in the individual groups of virus-inoculated animals. The criteria of virus activity were either the over-all incidence of tumors or leukemias as determined by death rate, or the latent period of development of disease. The latent period varied widely from several days or weeks for avian or murine sarcomas and myeloblastic leukemias, to months or years for lymphoblastic leukemias. Although these *in vivo* quantitative assays (Bryan, 1946a,b) are highly accurate, they are currently little used because of their cost, space, and time require-ments.

The use of early signs of disease (host immune response, palpable tumor, enlarged spleen) significantly shortened the time requirements for the *in vivo* tests. After the introduction of more quantitative parameters in these tests, such as spleen weight, antigen concentration, or spleen tumor count, a variety of rapid and sensitive assays were developed, some of which are pre-sently in use.

A. Antibody Production Assay

This test is based on the inverse relationship between the dose of inocu-lated mouse leukemia virus (MLV) (Moloney) into mice and the time re-quired for the appearance of detectable antibodies. These are measured by complement-dependent cytotoxicity of the sera against Moloney lymphoma cells. Although much faster than *in vivo* leukemia induction assay, the test still requires up to 9 weeks when used as an end point assay (Grunder *et al.*, 1972).

B. Spleen Weight Assay

Splenomegaly after inoculation of MLV or mouse sarcoma (MSV) viruses is measured by the spleen weight of experimental mice. This technique was used for quantitation of Friend (Rowe and Brodsky, 1959) and Rauscher (Chirigos, 1964) leukemia viruses and for MSV (Hirsch and Harvey, 1969). The test can be evaluated within 2 to 3 weeks after inoculation and gives reproducible titers within 2- to 5-fold difference.

C. HOST RESISTANCE ASSAY

The spleen weight response to infection with Friend leukemia virus was used as an indicator of induction of resistance to leukemogenesis by previous inoculation of Moloney leukemia virus, which has a long latent period. Weanling mice were inoculated with Moloney virus and challenged 22 days later with Friend virus. Animals were sacrificed 14 days after the challenge and the absence or diminution of spleen weight response to Friend virus was recorded. The test was evaluated as an end point assay (Rowe, 1963).

In a similar approach, a virus-induced myeloid leukemia (chloroleukemia) was assayed in various inbred strains of mice by its capacity to induce resistance to a challenge with syngeneic leukemia transplant (Prigogina and Stavrovskaja, 1964). Leukemia cells were transplanted 20 days after virus infection and the animals were sacrificed 2 months after the challenge. The results were evaluated as an end point assay by finding a 50% resistance dose.

D. SPLEEN FOCUS ASSAY

This method was introduced independently by Axelrad and Steeves (1964) as an assay for Friend leukemia virus and by Pluznik and Sachs (1964) for Rauscher leukemia virus. The spleen colonies can be read macroscopically as small, yellow-white focal lesions on the spleen surface of mice, sacrificed 9 days after virus inoculation. The average colony count per spleen is directly proportional to the dose of virus injected.

E. SPLEEN ANTIGEN TEST

Groups of newborn mice are inoculated intraperitoneally with serial dilutions of the virus. Fourteen days after inoculation, a 10% tissue extract is prepared from each spleen and the amount of MLV–group-specific (gs) antigen is determined in each sample, using 4 units of antisera reactive against MLV–gs antigen (see Section VI,A). Both the accuracy and the sensitivity of the spleen antigen test (SPAT) are similar to the *in vitro* assays for MLV (XC plaque assay and COMUL test; see Sections V,A and VI,A). In assays of fresh virus isolates, which have not yet been adapted *in vitro*, SPAT correlates better with the isolates' leukemogenic potential than do the *in vitro* assays (Peters *et al.*, 1974).

F. END POINT DILUTION IMMUNOFLUORESCENT ASSAY

This test was developed specifically for work with radiation leukemia virus (Declève *et al.*, 1974a). The test virus preparation is inoculated directly into the thymic tissue of 4- to 5-week-old mice. The test is evaluated at various intervals after inoculation by reading the percentage of cells in a thymus cell suspension that are positive for mouse gs antigen, as detected by immunofluorescence. The end point titer is detectable with high accuracy as early as 7 days after inoculation. Like SPAT, the *in vivo* immunofluorescent end point assay correlates well with the lymphoma-inducing capacity of the given virus stock.

G. ECTODERMAL LESIONS ON CHICK EMBRYO CHORIOALLANTOIC MEMBRANE

This unique quantitative assay for avian sarcoma viruses developed by Keogh (1938) according to the original Burnet's method (Burnet, 1936) employs titration of viruses on chorioallantoic membranes of chick embryos, and is the direct predecessor of the *in vitro* transformation focus assay. The essential step in the method is to establish an artificial air space between the inner shell membrane and the chorioallantoic membrane of a 12-day-old chick embryo, which will allow deposition of inoculum on the flat surface of chorioallantoic membrane. The inoculated eggs are incubated at 37.5 °C for 7 days, the chorioallantoic membranes are harvested, and the focal lesions are counted against black background with side illumination. The number of lesions is directly proportional to the concentration of the inoculum (Keogh, 1938). This method was widely used (Rubin, 1955; Prince, 1958a; Vigier, 1959; Vrba, 1963) but later replaced by a tissue culture assay (Manaker and Groupe, 1956; Temin and Rubin, 1958). It was complicated by the resistance of individual chick embryos, which was found to be caused by a true genetic resistance to Rous sarcoma virus (RSV) infection (Prince, 1958b) or by an infection of the embryo by a lymphoid leukosis virus (Payne and Biggs, 1965).

III. In Vitro Infectivity Assays: Theoretical Considerations

A. PROBABILITY OF INFECTION

Common stocks of RNA tumor viruses contain heterogeneous populations of physical particles, consisting of infectious and noninfectious virions. While only the first score in the infectivity assay, noninfectious particles have their importance in tests measuring amounts of antigens or biochemically active products (RNA-directed DNA polymerase, RDDP). The proportion of these subpopulations is found by comparing the physical particle count on electron microscopy (Sharp, 1965) with the infectivity titer.

The infectious unit [plaque-forming unit (PFU); focus-forming unit (FFU)] is the amount of infectivity (number of infectious particles) which on the average records in the tests as one focal change. It could be one infectious particle, but usually not all potentially infectious particles score as infectious units. This scoring depends on design and technical performance of individual tests, and the physiological conditions of the assay cells are a major determining factor. For the quantitative deliberation below, let us consider a situation in which each infectious particle represents one infectious unit. Let us consider having an aliquot of virus stock that contains v number of infectious virus particles. When a population of c cells in tissue culture is exposed to v virus particles, the ratio $m = v/c$ is designated the multiplicity of infection (m.o.i.). After the exposure, some cells will be infected and some might remain uninfected. Some of the infected cells will be infected originally by a single infectious particle, some with 2 particles, and some with even 3, 4, etc.

The probability P_x of each such event (P_0 = probability of noninfection, P_1 = probability of infection by a single virus particle, P_2 = probability of simultaneous infection with 2 virus particles, etc.) can be predicted according to the equation for Poisson distribution

$$P_x = \frac{e^{-m} \times m^x}{x!} \tag{1}$$

where

$x = 0, 1, 2, 3, \ldots$, indicating noninfection, single infection, double infection, etc.

e = base of natural logarithms $= 2.718 = 10^{0.434}$

m = multiplicity of infection $= v/c$

v = number of infectious particles

c = number of cells

$x! = x \times (x - 1) \times (x - 2) \ldots 3 \times 2 \times 1$

$1! = 1$

$0! = 1$

For example, if we were to find how many cells of 10^5 will remain uninfected when exposed to 10^5 infectious virus particles, the multiplicity of infection m in this case is $m = 10^5/10^5 = 1.0$, and the probability of noninfection according to Eq. (1) is

$$P_0 = \frac{e^{-1} \times 1^0}{0!} = 0.368 \tag{2}$$

Therefore 36.8%, or approximately 37%, of cells will remain initially noninfected at $m = 1.0$. Accordingly, the probability that a cell will be infected with one or more infectious particles $(P_{\geq 1})$ when $m = 1.0$ is

$$P_{\geq 1} = 1 - P_0 = 1 - 0.368 = 0.632 \tag{3}$$

Therefore, at $m = 1.0$, 63% of the cells will be initially infected, but not all of them with only a single virus particle. If each infected cell results in one focal lesion, some lesions will be induced by double or multiple infections.

Since, practically, it is not possible to distinguish lesions induced by a single virion from those induced by two or more, to find the titer, we have to determine P_1 values or design the experimental conditions such that $P_{\geq 2}$ will be insignificant as compared with P_1. What are these conditions? As demonstrated in Table I, the determining factor is m.o.i. Table I, which was computed according to Eq. (1), shows that at high m.o.i. the majority of cells are infected by 2 or more particles. At m.o.i. equal to 1.0, approximately 42% of the lesions could be considered to be induced by 2 or more particles. At m.o.i. equal to 0.1, 5% of countable foci or plaques might be expected to have developed from double or multiple infection. Finally, at m.o.i. equal to 0.01, only 0.5% of observed focal lesions could be considered to have originated from 2 or more particles (Table I).

In practice, 1–200 foci or plaques per dish can be comfortably counted. Therefore, to keep the probable incidence of foci induced by double or multiple infection low (below 0.5%), one should always use a large excess of cells over the number of infectious particles (100-fold excess is needed for the above-mentioned 0.5% level). The Poisson distribution does not have to be taken under consideration for evaluation of titration kinetics, providing m.o.i. is of the order of 0.01 or smaller.

TABLE I

PROPORTIONS OF CELLS INITIALLY INFECTED WITH A SINGLE VIRUS PARTICLE AS A FUNCTION OF MULTIPLICITY OF INFECTION

Multiplicity of infection	P_0	P_1	$P_{\geq 2}$ [a]	$P_{\geq 2}$ as percentage of $P_{\geq 1}$ [b]
10	0.000045	0.000454	0.999501	99.9546
1	0.367879	0.367879	0.264241	41.8023
0.1	0.904837	0.090484	0.004679	4.9168
0.01	0.990050	0.009900	0.000050	0.5025
0.001	0.999000	0.000999	<0.000001	0.04995

[a] $P_{\geq 2}$ = probability of infection with 2 or more particles.
[b] $P_{\geq 1}$ = probability of infection with 1 or more particle.

B. TITRATION PATTERNS

The titer of those viruses that can induce morphologically distinguishable and countable focal areas (foci) in cell monolayers can be expressed in FFUs per unit of volume.

Let us consider preparing 2-fold dilutions (1:1, 1:2, 1:4, 1:8, etc.) from a given virus stock. A certain volume aliquot (e.g., 0.1 ml) of each dilution is inoculated into a minimum of 2 tissue culture plates containing a known initial number of adherent cells (e.g., in our further calculations, let this number be 6.4×10^5 per plate). When focal changes develop, they are counted in each dish and the average number of foci is determined for each dilution. If the numbers of foci per dilution are directly proportional to the amount of virus inoculated, the titration is said to be of 1-hit type, meaning that one infectious particle induced one focal change (Table II; Fig. 1A and B). Each 2-fold dilution of the virus would result in a halving of the number of foci (Table II). The slope of the straight line connecting the values of the mean number of foci per dilution is 45° (Fig. 1A). If the theoretical titer is determined for each virus dilution by multiplying the average number of foci per dilution by the dilution factor, a straight line parallel to the abscissa will result (Fig. 1B). These titration patterns indicate that each focal change was induced by a single infectious particle.

Let us now consider the situation in which focal change could develop only if it was originated by 2 particles. The virus pool, then, for our consideration and for simple computation, could be visualized as consisting of 2 equal subpopulations of defective particles, v_1 and v_2. The focal change could be initiated only by simultaneous infection of one cell by both v_1 and v_2 particles which complement each other in a certain important biological function(s) necessary for development of the lesion. A single infection by

TABLE II

EXAMPLE OF 1-HIT AND 2-HIT TITRATION PATTERNS

Virus dilution	Average number of foci per plate	
	1-Hit pattern	2-Hit pattern
1:1	640	640
1:2	320	160
1:4	160	40
1:8	80	10
1:16	40	2.5
1:32	20	—
1:64	10	—
1:128	5	—

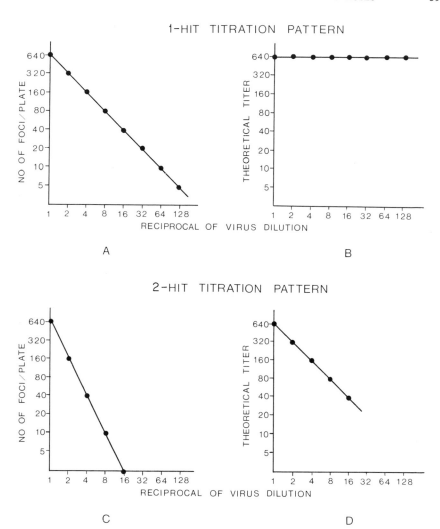

FIG. 1. One-hit and two-hit titration patterns. See text, Section III,B.

either v_1 or v_2 does not lead to a focal change. Let each of the cultures be plated again with 6.4×10^5 cells and let us assume that the stock of this defective virus induced 640 foci when used undiluted. These 640 foci were induced by initial infection of 640 cells by both v_1 and v_2 particles. The probability of such an event is the combined probability $P_{(v_1 + v_2)}$ representing multiplication of individual probabilities P_{v_1} and P_{v_2}, each of which determines the probability of infection by a single virus particle P_{v_1} and P_{v_2}, respectively.

If v_1 and v_2 are present in equal amounts, $P_{v_1} = P_{v_2}$, and

$$P_{v_1}^2 = P_{(v_1 + v_2)} = \frac{640}{640000} = 0.001 \tag{4}$$

$$P_{v_1} = \left(\frac{1}{1000}\right)^{1/2} = \frac{1}{31.6} = 0.032 \tag{5}$$

In this case, therefore, 3.2% of all cells [Eq. (5)] were infected with v_1 alone, another 3.2% of cells were infected with v_2 alone, and only 0.1% [(Eq. (4)], or 640 of 6.4×10^5 cells, were infected by both v_1 and v_2. After 1:2 dilution of virus, P_{v_1} and P_{v_2} will be reduced to 1.5% each. This approximation can be done since, at the level of m.o.i. $= 0.032$, the figure will not change significantly whether the Poisson statistics are included or not (see Table I). The probability $P_{(v_1 + v_2)}$ will then become 0.025%, or 160 foci. After 1:2 dilution of inoculum, therefore, the number of foci decreases from 640 to 160, or in 1:4 ratio. Similarly, it can be shown that each further step of 1:2 dilution of inoculum will further reduce the number of observed foci by a factor of 4 (Table II, Fig. 1C and D).

By analogy we can conclude that when simultaneous infection by 3 particles is necessary for induction of one focal change, each 2-fold dilution step will result in an 8-fold (or 2^3) decrease in the actual number of foci observed, although this conclusion is of theoretical interest only.

In general, for each dilution step k, the actual number of foci observed per plate will decrease by the factor k^n, where n indicates the number of infectious particles necessary for induction of one focal change, i.e., the hit number of the titration pattern.

For the above considerations, we have made several simplifying assumptions. First, we have considered the ratio of two types of particles participating in a 2-hit titration pattern to be 1:1. In practice, where the two types of viruses are represented by defective sarcoma virus and competent leukemia virus, one component might be in excess. The large excess of one component might result in a 1-hit type titration pattern of the other component, which normally would give 2-hit kinetics. The theoretical analysis of complex titration curves for virus stocks consisting of defective and competent particles in various proportions has been reported elsewhere (O'Connor and Fischinger, 1968; Hahn et al., 1970).

C. PRECISION OF THE FOCUS (PLAQUE) ASSAY

In titrations, it is practical to evaluate focus (plaque) counts in orders of 10 to 100. Virus dilutions are inoculated in replicate plates, preferably in more than two plates. When the test is evaluated, the number of foci for each plate is determined, and the average number of foci (f) for each group of plates inoculated with the same virus dilution is determined. The variability of the number of foci per plate is approximately proportional to $f^{1/2}$ (this is the approximate standard deviation). The variations of individual readings in replicate plates are expected to be within the interval of $f \pm 2f^{1/2}$ for a test with reasonably high accuracy. Since, in a test with a low average number of foci, $f^{1/2}$ represents the higher percentage of f, the accuracy of titrations generally increases with the average number of focal changes. The practical upper limit is set, however, around 100–200 foci per 60 mm petri plate (depending on the size of the lesions), since at that level some individual focal areas might become confluent. When small areas, such as 15 mm plastic cups, are used for focus or plaque assay, the relatively broad variability range can be offset by an increased number of replicate assays per dilution.

D. End Point Assay

In some virus assays it is difficult to recognize focal lesions, since the cytopathology has a rather diffuse character. In such cases, each culture is scored as either positive or negative, and each replicate group of cultures inoculated with identical virus dilutions is evaluated for the percentage of positive and negative cultures. The titer is expressed in statistical units, such as 50% tissue culture infectious dose ($TCID_{50}$). The $TCID_{50}$ is defined as the virus dose capable of infecting just one half of the inoculated cultures.

Since the actual biological infectious unit cannot be smaller than one infectious particle, it is necessary to establish the relationship between one infectious unit, such as FFU or PFU, and $TCID_{50}$ as a statistical term.

According to the Poisson distribution, the requirement that one-half of the cultures be infected (with one or more infectious units each) and the other half remain uninfected, means that

$$P_{\geq 1} = 1 - P_0 = 0.5 = e^{-\nu} \tag{6}$$

To determine the average number of infectious units per culture that will leave one-half of the cultures uninfected requires computation of the value of ν from Eq. (6): $0.5 = e^{-\nu}$; $\nu = 0.69$.

In order to infect one-half of the cultures, we have to use 0.69 infectious units per culture. One $TCID_{50}$ is equivalent to 0.7 FFU (or PFU).

IV. In Vitro Assays for Sarcoma Viruses

In vitro virus-induced oncogenic transformation is a loosely defined phenomenon. The "transformed" cell phenotype has been characterized by changed morphology, increased population density, loss of contact inhibition and multilayered growth, a capacity to grow in suspension in liquid or semisolid medium (agar), relative independence of proliferative capacity on the serum component of growth media, and a capacity to establish immortal tissue culture lines. None of these characteristics, however, is uniformly associated with the neoplastic behavior of the cells following *in vivo* implantation (Hayflick, 1967; Sanford, 1967; Dulbecco, 1969; Sanford *et al.*, 1972).

In vitro transformation by RNA tumor viruses conforms to most of the above-listed criteria; however, in many cases, cells are not capable of inducing immortal tissue culture lines. Thus, human and bovine cells transformed by RSV (Stenkvist, 1966) and human cells transformed by a Kirsten variant of MSV (V. Klement, unpublished) fail to attain unlimited capacity for *in vitro* propagation, although they exhibit increased growth capacity at low serum level, higher saturation density, and multilayered growth.

Cytopathic changes induced *in vitro* by RNA tumor viruses were first observed by Halberstaedter *et al.* (1941), and by Lo *et al.* (1955), who reported changes in morphology of chick embryo cells inoculated with material from Rous sarcoma.

In the first quantitative transformation experiment by Manaker and Groupe (1956) chick embryo fibroblasts were inoculated with various doses of RSV. In higher virus dilutions the focal transformation changes were easily recognized in the cell monolayer. The number of foci was directly related to virus dilution. This assay was further improved and developed into a highly accurate titration method by Temin and Rubin (1958). They showed conclusively that the titration pattern was of 1-hit type, that one virus particle was sufficient to initiate one focus, and that foci could be formed by either or both of the two basic mechanisms: division of sarcoma virus-transformed cells, and infection of surrounding cells by virus released from initially infected cells. The major factor in focus development, however, was attributed to division of transformed cells, since there was no difference in the size of foci observed under conditions permitting or excluding reinfection. The procedures of Temin and Rubin, with some modifications, are essentially used at the present time. The methodology of the updated RSV focus assay was more recently reviewed by Vogt (1969a).

A. EFFICIENCY OF TRANSFORMATION

In the original studies on RSV transformation focus assay (Temin and Rubin, 1958; Rubin, 1960a) the highest proportion of transformed cells observed was only 10%. A total of 90% of the cells remained noninfected even with high multiplicity of infection. Since the transformation frequency in cellular clones was approximately the same as that in the uncloned population, it was concluded that the low efficiency did not result from genetic heterogeneity of the target cell population but most likely from the physiological state of the cells. Proliferation of the transformed cells appeared to be the main mechanism leading to the formation of foci.

Careful studies of factors influencing the efficiency of transformation revealed that the α-globulin "fetuin" (found in calf and especially fetal calf serum) and certain polyanions inhibited the expression of transformation (Rubin, 1960b; Parshad and Sanford, 1968; Sanford et al., 1972).

Polycations, on the other hand, increased the efficiency of transformation by the viruses of several avian antigenic subgroups (Vogt, 1967; Toyoshima and Vogt, 1969). The increase of infectivity of avian sarcoma virus was up to 80-fold (Toyoshima and Vogt, 1969), probably through increased adsorption of the virus. When the polycation DEAE-dextran (DEAE-D) was used in combination with synchronization of target cells, virtually 100% infectivity efficiency was achieved (Hanafusa, 1969).

B. TRANSFORMATION FOCUS ASSAY: MOUSE SARCOMA VIRUS

A quantitative in vitro assay for MSV (Harvey, Moloney), similar to that used in an avian system, was developed by Hartley and Rowe (1966) using secondary mouse embryo cells. Unlike RSV, the MSV titration pattern followed 2-hit kinetics, which was possible to correct to a 1-hit pattern by coinfection with a mouse leukemia helper virus. This finding indicated that MSV was defective in transforming function and therefore different from some strains of RSV, which were known to be defective in replication but competent in transforming activity' (Hanafusa et al., 1963). Both RSV and MSV, however, had an excess of a nontransforming helper virus.

When MSV stocks were assayed on tissue culture cell lines, exhibiting flat, contact-inhibited growth pattern, the titration kinetics was of the 1-hit type (Jainchill et al., 1969; Aaronson et al., 1970; Aaronson and Rowe, 1970). Detailed analysis of foci of transformation in such lines revealed lesions of two types. Larger foci occurred early, and these were dependent upon the presence of a helper virus. The other type was represented by small foci appearing later and growing mainly by cell division. When small foci were isolated they maintained transformed morphology, but they did not produce infectious virus. After superinfection with leukemia virus, sarcoma virus genome was rescued, and both leukemia and sarcoma viruses were produced. The former (large) type of foci gave a 2-hit titration pattern. When combined counts were made, which included both large and small foci, the titration was that of the 1-hit type. Since the secondary mouse embryo cells do not form absolutely contact-inhibited monolayers, they revealed essentially only helper-dependent foci, thus giving 2-hit titration kinetics. Only the use of contact-inhibited lines uncovered the helper-independent foci.

1. Equipment and Supplies

a. Disposable plastic and glass tissue culture supplies are available from: Falcon, Division of Becton Dickinson and Co., Oxnard, California: Corning Glass Works, Corning, New York; or Nalge Co., Inc., Nalgene Labwear Division, Rochester, New York. Items: 150 cm² or 75 cm² tissue culture flasks, 60 × 15 mm tissue culture dishes, 10, 5, and 1.0 ml serological pipettes, Pasteur pipettes, 15 and 50 ml centrifuge tubes, 0.45 or 0.22 μm filters.

b. Chemicals and solutions: Tissue culture media, sera, and solutions are available from several major companies (Microbiological Associates, Inc., Bethesda, Maryland; Flow Laboratories, Rockville, Maryland; or Grand Island Biological Co., Grand Island, New York. Items: Eagle's minimum essential medium with Earle's Salts (EMEM); other media might be specified. Phosphate-buffered saline (PBS), calf serum (CS), fetal bovine serum (FBS), trypsin–EDTA solution (T–E), penicillin–streptomycin solutions (P–S). Other chemicals: DEAE-D (Pharmacia Fine Chemicals, Inc., Piscataway, New Jersey), polybrene (Aldrich, San Leandro, California), Bacto agar and veal infusion broth (Difco Laboratories, Detroit, Michigan), dimethylsulfoxide (DMSO), methanol, Giemsa blood stain.

c. Equipment: Incubators, inverted microscope, dissecting microscope, hemocytometer.

2. Procedure

a. Trypsinize monolayer cultures of cells for assay. Prepare cell suspension with 10^5 cells per milliliter in growth medium (EMEM with 10% FBS and P–S, 100 IU/ml and 100 μg/ml, respectively) and plate cell suspension in petri dishes (60 mm diameter), 3×10^5 cells per plate.

b. Incubate 24 hours at 37°C in CO_2 incubator (5% CO_2 in air) and check the quality of the monolayer.

c. Remove the fluid with Pasteur pipette, using vacuum line and a fluid collection flask; add 2 ml of growth medium with 25 μg of DEAE-D per milliliter and incubate 1 hour in the incubator.

d. Remove the fluid, wash once with EMEM or with PBS, and add EMEM with antibiotics, supplemented with 5% postnatal calf serum and 1% DMSO, 4 ml per dish.

e. Inoculate each dish with 0.4 ml of respective virus dilution. Inoculate two dishes with 0.4 ml of virus diluent only (20% veal infusion broth in EMEM) as uninfected controls.

f. Change the fluid (EMEM, P–S, 5% CS, 1% DMSO) on second or third day after infection. Observe for foci under the dissecting microscope on the sixth or seventh day after infection. Observe either living cultures or fixed with methanol and stained with Giemsa.

3. Comments and Variations

Readers seeking more information on general tissue culture methodology are referred to the tissue culture manual by Kruse and Patterson (1973). The number of cells plated per dish might vary depending upon the type of cells used for assay. When secondary mouse embryo cells are used, the number is 3.5×10^5. The value 3×10^5 is commonly used for normal rat kidney (NRK) cell line (Duc-Nguyen et al., 1966) whereas faster growing mouse cell lines such as NIH 3T3, BALB/3T3, and SC-1 (Jainchill et al., 1969; Aaronson et al., 1970; Hartley and Rowe, 1975) are plated at 2×10^5 cells per plate. While this does not significantly change the initial m.o.i. in the range of 10–100 countable foci per plate and, therefore, the sensitivity of the assay, it assures comparable rates of growth in the individual assays. It also provides for active proliferation of cells in the critical postinfection period. Confluency and, therefore, marked decrease in sensitivity for induction of new satellite foci is reached, on the average, on the third day.

The treatment of cells with DEAE-D (Vogt, 1967; Toyoshima and Vogt, 1969) could be substituted by adding polybrene to the growth medium from the time of plating to the time of infection (24 hours). The optimal nontoxic concentration for mouse cells which increased the titer comparably to DEAE-D was 10 μg/ml. Addition of DMSO (Vogt et al., 1970) in tissue culture medium did not change the titer, but improved the distinction of foci in continuous tissue culture lines.

In this method, which is used mostly for routine comparative titrations, the inoculation procedure was simplified by adding the virus inoculum directly to the tissue culture fluid. If high sensitivity of the test is desired, the experimenter should consider inoculating the dish with a small volume of virus material (0.4 ml) for an hour (Temin and Rubin, 1958) followed by addition of growth medium. In our experience, this increased the sensitivity of the assay, although not 10-fold, as might be expected from the theoretical dilution of the inoculum.

The FBS is replaced by postnatal CS in transformation focus assay in view of the inhibitory effect of FBS on transformation. When a cell line is used which grows in other than EMEM base, the special medium base is used in place of EMEM.

In the described procedure (Hartley and Rowe, 1966), liquid medium was used instead of the agar medium used originally by Temin and Rubin (1958). Therefore, it is important to read the foci as soon as they fully develop, usually the sixth or seventh day (Figs. 2 and 3). Secondary (satellite) foci are virtually not seen in the contact-inhibited lines, since by the time of first virus release the cultures are already close to confluency and most of the cells are not synthesizing DNA, the chance of establishing second generation foci is minimal.

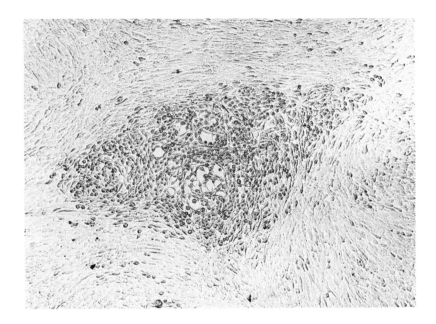

FIG. 2. A focus of transformed rat (NRK) cells infected with MSV (Kirsten) unstained. ×
63.

4. *Evaluation*

In Table III an example is given of computation of a focus-forming titer.
The titration pattern in this example indicates 1-hit titration kinetics. Only
under these circumstances is it possible to compute the average titer from
the focus counts in individual virus dilutions.

C. SELECTIVE ASSAY SYSTEMS FOR MAMMALIAN SARCOMA VIRUSES AND THEIR PSEUDOTYPES

Similarly to RSV (Hanafusa *et al.*, 1964), MSV has a defective transforming genome. The infectivity of the progeny depends upon the presence of a
competent, nontransforming helper virus, which provides the virus coat for
the production of a respective MSV pseudotype (Hartley and Rowe, 1966;
Aaronson *et al.*, 1970; Parkman *et al.*, 1970). Mouse sarcoma viruses (Harvey, Moloney, Kirsten) and their mouse-tropic pseudotypes can be characterized as N-tropic, B-tropic, or NB-tropic, according to their capacity to
grow in mouse cells of NIH Swiss (N-type) or BALB/c (B-type) origin
(Hartley *et al.*, 1970).

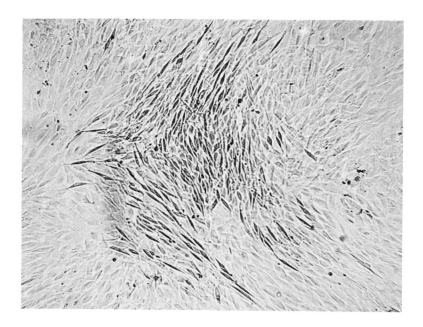

FIG. 3. A focus of transformed mouse (NIH 3T3) cells infected with MSV (Moloney) unstained. × 200.

The mouse xenotropic MSV pseudotypes cannot transform either B- or N-type mouse cells, but transform efficiently a wide variety of mammalian and avian cells (Levy, 1975). After the first transspecies rescue of MSV had been demonstrated (Sarma *et al.*, 1970) a wide variety of pseudotypes of murine, feline (Snyder and Theilen, 1969), and simian (Theilen *et al.*, 1971) sarcoma viruses became available. These included helper virus specificities of MLV (both mouse tropic and xenotropic), feline epigenetic (Jarrett *et al.*, 1964) and endogenous (McAllister *et al.*, 1972; Livingston and Todaro, 1973; Fischinger *et al.*, 1973; Sarma *et al.*, 1973) leukemia viruses, and primate epigenetic and endogenous C-type viruses (Theilen *et al.*, 1971; Kawakami *et al.*, 1972; Benveniste *et al.*, 1974).

The search for a suitable cell system which would allow comparative titrations of these viruses produced a subclone derived from simian virus 40 (SV40) transformed human skin fibroblasts which was susceptible to MSV pseudotypes produced with mouse, woolly monkey (Theilen *et al.*, 1971) and gibbon (Kawakami *et al.*, 1972) type-C viruses (Todaro and Meyer, 1974). Further, after screening of a large variety of mammalian tissue culture lines, a mink lung cell line CCL64 from the American Type Culture Collection (ATCC) (Shannon, 1972), was found to be selectively sensitive to endogenous

TABLE III

EXAMPLE OF COMPUTATION OF TRANSFORMATION FOCUS-FORMING TITER

Virus dilution	Number of foci per dish (0.4 ml inoculum)	A Number of foci per dilution (0.8 ml inoculum)	B Theoretical number of foci per 1.0 ml inoculum (A × 5/4)	Log_{10} of B	Correction for dilution factor	Average titer (FFU or PFU per 1.0 ml)
Uninfected control	0,0	0 (<1.0)	0 (<1.25)	<0.1	<0.1 [a]	
10^{-2}	88,72	160	200	2.30	2.30 + 2.00 = 4.30	
10^{-3}	6,10	16	20	1.30	1.30 + 3.00 = 4.30	$10^{4.33}$
10^{-4}	2,0	2	2.5	0.40	0.40 + 4.00 = 4.40	
10^{-5}	0,0	0 (<1.0)	0 (<1.25)	—	—	
10^{-6}	0,0	0 (<1.0)	0 (<1.25)	—	—	

[a] Limit of sensitivity of this method.

xenotropic mouse (AT 124), feline (RD114/CCC), and primate (M-7) viruses, while being resistant to N- and B-tropic endogenous murine viruses and to NB-tropic standard mouse laboratory virus strain (Rauscher). The CCL64 line has been very useful for comparison of various interference patterns among mammalian C-type viruses (Henderson et al., 1974).

D. Other Assays for Transforming C-Type Viruses

1. Metabolic Assay

The observation of Manaker and Groupe (1956) that the morphological alteration of chick embryo fibroblasts infected with RSV was accompanied by increased acidity of tissue culture fluid was the basis of a metabolic assay for RSV. The test (Calnek, 1964a) is based on measuring acid production associated with RSV transformation of chicken fibroblasts. It is performed in disposable microplates and a color indicator scale is used to determine the end-point titer. The method was reported to be 96.4% as accurate as the transformation focus assay.

2. Proteolytic Plaque Assay

Cells transformed with both DNA and RNA viruses produce a factor that can cause lysis of fibrin through serum plasminogen activation (Unkeless et al., 1973), as well as proteolysis of other proteins, such as casein (Goldberg, 1974). These observations led to the development of a plaque assay for avian sarcoma virus, based on formation of proteolytic plaques in agar nutrient medium with casein (Balduzzi and Murphy, 1975). The assay was reported to be as sensitive as the transformation focus assay. Nontransforming viruses or nontransforming mutants of RSV did not produce plaques in casein medium. Some variants of RSV such as the morph-f (Temin, 1960) were only weakly positive.

3. Search for Biochemical Correlates of Transformation

Limitation of the transformation focus assay, due to the necessity for development of a colony of morphologically transformed cells in order to recognize the changed phenotype, has led to a search for biochemical markers which might indicate the transformed phenotype at the one-cell level.

The known increase in hexose uptake (Hatanaka and Hanafusa, 1970) and synthesis of hyaluronic acid in transformed cells (Temin, 1965) was explored by Bader (1973) as a possible approach to early identification of transformation utilizing differential incorporation of 2-(^{3}H)-deoxyglucose in hyaluronic acid synthesis. The differential activity of transformed and nontransformed cells could be recorded as early as 16 hours after transformation.

V. Assays for Leukemia Viruses

The *in vitro* transformation focus assay (Temin and Rubin, 1958) was instrumental in the discovery of resistance-inducing factor (RIF) (Rubin, 1960c), an avian leukosis virus which induced resistance in chicken fibroblasts to subsequent transformation by RSV. On this principle was based the first accurate *in vitro* assay for leukemia viruses.

Another approach to *in vitro* titrations of leukemia viruses was detection of virus-specific antigens in infected cultures, either by immunofluorescence (Vogt and Rubin, 1963) or by complement fixation (Armstrong *et al.*, 1964; Huebner *et al.*, 1964; Sarma *et al.*, 1964; Hartley *et al.*, 1965). These methods have been reviewed in detail by Huebner (1967).

More recently, some leukemia viruses were found to be able to induce distinct cytopathic changes *in vitro* and several methods developed from this principle are outlined below.

A. XC Plaque Assay

Although nontransforming RNA tumor viruses (leukemia viruses) are generally thought not to produce any detectable cytopathology *in vitro*, large multinucleated cells were observed initially among RSV transformed cells by Manaker and Groupe (1956) and have been seen in various *in vitro* cell culture systems infected with RNA tumor viruses, as well as in some tumors (Ahlström and Forsby, 1962; Purchase and Okazaki, 1964; Stenkvist, 1966; Sinkovics *et al.*, 1969; Todaro and Meyer, 1974). Although this phenomenon might have been caused in some instances by the presence of syncytial "foamy" viruses (Cook, 1969; Hruska and Takemoto, 1975), the syncytium inducing capacity of C-type RNA viruses is not surprising considering their close relationship to "slow" and syncytial viruses (Dalton *et al.*, 1974).

When rat Rous sarcoma cell line XC (Svoboda, 1960) was cocultivated with MLV-infected mouse embryo cells, a reproducible induction of syncytia was observed. This phenomenon was utilized for an *in vitro* end point dilution assay for MLV (Klement *et al.*, 1969). Wistar rat embryo lung cell line (RSL) was reported to respond similarly (Koga, 1973). Based upon this principle, a plaque assay technique (Fig. 4) which had wide application in work with MLV was developed (Rowe *et al.*, 1970). Attempts were made to apply a similar technique to feline and simian C-type viruses (Klement and McAllister, 1972; Rand and Long, 1972; Rangan *et al.*, 1972a,b, 1973).

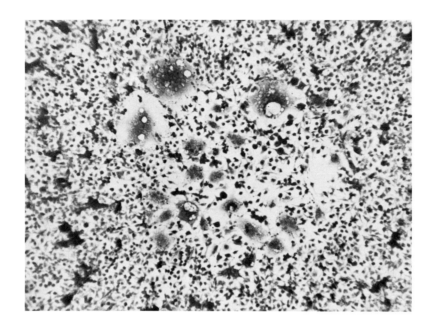

FIG. 4. XC plaque in NIH 3T3 cells infected with Gross leukemia virus, Giemsa. × 100.

1. Equipment and Supplies

a. All items listed as disposable plastic and glass tissue culture supplies, media, and chemicals under transformation focus assay; all pieces of equipment listed under transformation focus assay.

b. In addition, germicidal lamp as a source of ultraviolet (UV) radiation (germicidal lamp G15T8), and UV meter (black ray ultraviolet intensity meter: Ultra Violet Products, Inc., San Gabriel, California).

2. Procedure

a. Proceed as described in steps 1–3 in the procedure for transformation focus assay (Section IV,B,2).

b. Remove the fluid containing DEAE-D, wash once with EMEM or PBS. Add EMEM with antibiotics supplemented with 10% FBS, 4 ml per dish.

c. Inoculate each dish with 0.4 ml of virus dilution, inoculate two dishes with 0.4 ml of virus diluent only (20% veal infusion broth) as uninfected controls.

d. Change the fluid (EMEM, P–S, 10% FBS) on second or third day after infection.

e. On the sixth day after infection: (1) prepare XC cell suspension in EMEM 10% FBS containing 10^6 XC cells per 4 ml; and (2) remove the fluid from dishes and irradiate the cells with UV source, total dose of 1600 erg/mm^2.

f. Add 4 ml of EMEM 10% FBS containing 10^6 XC cells per dish.

g. On the second day after irradiation, change the fluid (EMEM 10% FBS, P–S). Add fresh medium very gently, drop by drop, as the cell sheet is very fragile.

h. On the third or fourth day after irradiation, fix with methanol for 10 minutes and stain with Giemsa.

3. *Comments and Variations*

Number of cells plated varies; when secondary embryo cells are used for assay, 3.5×10^5 cells are plated per dish. The growth medium is EMEM, P–S, 10% FBS, as indicated above. However, when faster growing tissue culture cell lines are used, such as NIH 3T3, BALB 3T3, or SC-1, the number of cells plated per dish is reduced to 2×10^5 and the concentration of FBS in steps b–d is reduced to 5%.

A modified procedure designated as reverse XC test was introduced by Niwa *et al.* (1973). In this test, which was used mainly for radiation leukemia virus assay, the plating sequence was reversed; namely, the cells (10^2–10^5) infected with MLV were plated onto dishes that had been previously seeded with 10^5 XC cells. This test was used as an end point dilution assay with regard to virus-infected cells plated on top of XC.

The dose of 1600 erg/mm^2 will decrease approximately 5-fold the ability of individual cells to induce syncytia in XC cells (Rowe, 1971) and approximately 10-fold the titer of free infectious virus (Latarjet, 1970). The area that develops into an XC plaque consists at the time of irradiation of relatively large numbers of virus-producing cells (of the order of 10 and 100; see fluorescent focus assay, Section VI,B). Each of these areas will be inactivated as a multiple target system and, therefore, the above dose should not significantly change the number of recorded XC plaques.

Other comments, variations, and exceptions listed under the transformation focus assay also apply here.

4. *Evaluation*

The number of XC plaques (Fig. 4) is determined and the titer in plaque-forming units per milliliter (PFU/ml) is computed as demonstrated for transformation focus assay in Table III.

The XC plaque titration patterns obtained in susceptible assay systems have been uniformly of a 1-hit type. In resistant cells, however, both 1-hit type and complex titration curves have been reported (Jolicoeur and Baltimore, 1975; Pincus et al., 1975).

B. NONSYNCYTIAL CYTOPATHIC EFFECTS

A nonspecific cytopathic effect, characterized by the appearance of rounded cells with increased cytoplasmic eosinophila and vacuolization, and marked variation in size, shape, and intensity of staining of cell nuclei was observed in a mouse cell line persistently infected with Rauscher leukemia virus (Tyndall et al., 1965). Several other investigators reported similar findings with avian leukosis viruses and even developed quantitative cytopathic plaque virus assays.

An avian leukosis virus associated with Schmidt–Ruppin Rous sarcoma virus (SR–RSV) induced atypical foci in chicken cells propagated at 41°C. The lesions appeared as slightly granular areas and could be visualized after 8 days by increased uptake of neutral red (Dougherty and Rasmussen, 1964). A similar observation was made by Graf (1972), who reported formation of plaques in chick embryo fibroblasts infected with avian leukosis virus subgroups B and D.

Chick embryo fibroblasts infected with a temperature-sensitive mutant of SR–RSV were transformed when maintained at permissive temperature (37°C), but had normal flat morphology when kept at nonpermissive temperature (41°C). Superinfection with certain avian leukosis viruses (e.g., RAV-2) and further incubation at 41°C produced severe cytopathic changes which could be recorded in certain virus dilutions as cytopathic plaques (Kawai and Hanafusa, 1972).

C. TRANSFORMATION INDUCED BY LEUKEMIA VIRUSES

Avian myeloblastosis virus (AMV) induced morphological transformation of chick hematopoietic cells in vitro (Beaudreau et al., 1960; Baluda and Goetz, 1961) and also transformed chicken fibroblasts, although the efficiency of transformation was 10,000-fold lower than that of hematopoietic tissue (Moscovici et al., 1969). In a recent and more accurate assay (Moscovici et al., 1975) AMV induced foci of myeloblasts in cultures of chick yolk sac macrophages. Avian myelocytomatosis strain MC29 also induced rapid and massive morphological changes in chick embryo fibroblasts (Langlois et al., 1967). Chick embryo cells infected with several strains of avian leukosis virus showed gradual changes in morphology during serial passage and colonies of morphologically transformed cells were observed, which were absent in noninfected cultures (Calnek, 1964b).

The Abelson MLV, a variant originally derived from Moloney leukemia virus, induced lymphoblastic leukemia of B-cell type and plasmacytomas (Abelson and Rabstein, 1970), but generally not MSV-type sarcomas. Yet, *in vitro*, this virus can be assayed by transformation focus assay on NIH 3T3 and BALB 3T3 cells (Scher and Siegler, 1975). The *in vitro* studies revealed that Abelson transforming leukemia virus is defective, contains excess Moloney leukemia helper virus, can establish nonproductive transformation of fibroblasts *in vitro*, and the defective genome of the transforming leukemia virus can be rescued by a competent mouse leukemia helper virus. Another example of a similar combination of defective transforming leukemia virus and competent nontransforming helper virus is Friend and Rowson–Parr leukemia virus complex (Dawson *et al.*, 1968; Fieldsteel *et al.*, 1969; Rowson and Parr, 1970). Recently an *in vitro* assay for Friend leukemia virus was described, based on induction of erythroid colonies in bone marrow cells (Clarke *et al.*, 1975).

In biological classification, these viruses might possibly bridge the gap between traditional sarcoma and leukemia viruses.

D. TRANSFORMATION FOCUS ASSAY IN SPECIAL CELL LINES

Hackett and Sylvester (1972) derived a subline of BALB 3T3 with changed growth characteristics (growth rate and saturation density) as compared to the original parental BALB 3T3 line. It retained flat morphology and infection with MLV (Moloney) resulted in induction of transformed foci.

Bassin *et al.* (1971) derived several MSV-transformed mouse cell lines by cloning of transformed cells in soft agar. Those lines that did not produce MSV or leukemia virus were designated S^+L^-. One S^+L^- line which had rather flat morphology appeared to be an excellent assay system for a variety of otherwise nontransforming viruses. A quantitative virus assay of MLV determined by induction of transformed foci on this S^+L^- line gave titers identical to those obtained by XC test. Similar assays were developed for leukemia viruses (including endogenous xenotropic viruses) of other species. A cat cell line with flat morphology, carrying cat endogenous RD-114 virus genome (McAllister *et al.*, 1972; Fischinger *et al.*, 1973; Sarma *et al.*, 1973; Todaro *et al.*, 1973b) superinfected by MSV (Moloney) produced replication-positive and transformation-negative (r^+t^-) RD-114 virus, as well as MSV (RD-114) pseudotype (Fischinger *et al.*, 1974). Upon infection with either feline leukemia virus or xenotropic MLV, focal lesions consisting of piled up and loosely attached cells were produced. The focus assay of both viruses gave 1-hit titration kinetics.

A similar assay for murine xenotropic viruses, feline leukemia viruses of group C, RD-114, and endogenous baboon type-C virus (M-7) was developed by Peebles (1975) using the S^+L^- mink lung cell line transformed by MSV (RD-114). The line has flat morphology, rescuable MSV, and mouse group-specific (gs) antigen. Foci were induced by superinfection with endogenous xenotropic mouse, cat, and baboon viruses. However, the line was resistant to Gross and Kirsten leukemia viruses, as well as to feline leukemia viruses types A and B.

VI. Immunological Assays

As the result of a virus-productive interaction of an RNA tumor virus with the host cell, all the virus structural proteins and new virus-specific cellular antigens are synthesized, and practically any of these proteins and antigens lends itself to qualitative and quantitative determination by an appropriate method.

The infection of a heterologous host generally does not result in virus-productive interaction. Partial expression of the virus genome, however, may be present and may elicit strong immunological reaction against the respective antigen. Observations of such phenomena led to the discovery of a gs antigen of avian C-type RNA tumor viruses by Huebner et al. (1964). The authors found that hamsters and guinea pigs carrying nonproductive tumors induced by SR–RSV formed complement-fixing antibodies against an antigen present both in their own tumors, as well as in chicken sarcomas. The relevant antigen was identified as avian gs antigen (Bauer and Schäfer, 1965).

The gs antigen, the major virion internal protein, p30 (August et al., 1974) was also referred to as the major species-specific protein. The latter designation, however, should be avoided, since no single gs antigen is representative for all RNA tumor viruses of a certain species. This was demonstrated by Cook (1969), who isolated chicken syncytial virus, as C-type RNA virus lacking the gs antigen of the avian leukosis type. Chicken syncytial virus appeared to be a member of a new class of avian C-type viruses later identified as the reticuloendotheliosis group (Halpern et al., 1973; Purchase et al., 1973). A similar situation also exists in cats (McAllister et al., 1972; Fischinger et al., 1973; Sarma et al., 1973; Todaro et al., 1973b).

Other structural viral antigens and a new virus-specific cellular antigen were also used for infectivity assays (Vogt and Rubin, 1961; Vogt, 1964; Nordenskjöld et al., 1970) and the respective methods are listed below.

A. COMPLEMENT-FIXATION ASSAY FOR MURINE LEUKEMIA VIRUSES

The virus infectivity assay based on quantitative detection of a gs antigen was developed by Sarma et al. (1964) for avian leukosis viruses. Known as

COFAL test (COmplement Fixation test for Avian Leukosis virus), it has been widely used and later adapted for murine (COMUL) and feline (COCAL) RNA tumor viruses (Hartley *et al.*, 1965, 1969; Sarma *et al.*, 1971).

The test is based on the induction of synthesis of gs antigen of the respective RNA tumor virus in susceptible cells. Serial virus dilutions are inoculated into groups of tissue culture plates previously seeded with susceptible cells. After a period of incubation, the cells from individual dishes are harvested, disrupted by ultrasonic disintegration, and the amount of a gs antigen in the homogenate is determined by complement fixation using an antiserum specifically reactive with the respective gs antigen.

1. *Equipment and Supplies*

a. Plastic and glass tissue culture supplies as listed in Section IV,B. In addition, disposable tubes, Falcon #2058, and disposable components of microtitration system, as specified below.

b. Reagents and biologicals: Bacto-antisheep hemolysin glycerinated (Difco Laboratories, Detroit, Michigan), complement-fixation test diluent tablets (Oxoid Limited, London; distributed by Flow Laboratories, Inc., Rockville, Maryland), guinea pig complement (Texas Biological Laboratories, Inc., Fort Worth, Texas), and sheep red blood cells.

c. Equipment: Sonifier Cell-Disrupter, Model W 185 with microprobe (Heat Systems Ultrasonics, Inc., Plainview, New York); and microtitration system, complete with accessories and supplies (Cooke Laboratory Products, Alexandria, Virginia).

2. *Antiserum against Mouse Leukemia gs Antigen*

An *in vivo* transplant passage line of mouse sarcoma induced by MSV carried in suckling inbred Fisher rats was inoculated into weanling Fisher rats. The rats were bled as late as possible after the appearance of tumor. Individual sera were tested by standard complement-fixation microtechnique (Palmer *et al.*, 1969) against 4–8 units of standard MLV–gs antigen and against the battery of normal antigens (normal NIH Swiss mouse tissues and NIH Swiss cultures). The sera reacting 1:20 or less with normal control antigens and 1:160 or more with MLV antigens were pooled and such pools used in the complement-fixation test (Hartley *et al.*, 1965).

More recently, rabbit antisera against mouse p30 antigen purified by guanidine agarose chromatography (Nowinski *et al.*, 1973) were prepared. Adult New Zealand white rabbits were immunized 3 times, with an interval of 3 weeks between immunizations, with 100 μg of protein in complete Freund's adjuvant (50% homogenate, total volume 1.0 ml per immunization, 0.1 ml injected into each of the 4 footpads and 0.6 ml injected by the dorsal intradermal route).

3. *Procedure*

a. Proceed as described in steps a–d in Section V,A,2. For higher accuracy of titration use four or more dishes per each virus dilution.

b. Keep the cultures up to 21 days after inoculation, changing the fluid twice a week. Use individual Pasteur pipette for each dish during fluid change.

c. Harvest the cells from each dish individually, 21 days after infection, as follows:

1. Remove the fluid and keep in tube on ice until used.

2. Scrape cell monolayer completely with rubber scraper.

3. Resuspend the scraped fragments of the cell monolayer in 0.6 ml of the culture's own fluid, and keep on ice. (Falcon tube #2058 is suitable for later sonication with a microprobe.)

4. Freeze all samples at −70°C.

d. Thaw and disintegrate each sample individually (2–3 seconds) with sonifier equipped with a microprobe and set aside aliquot (0.2 ml) for a blind passage (See Section VI,A,4).

e. Test each sample in a standard complement-fixation microtechnique (Palmer *et al.*, 1969) using 1.8 units of complement and 4 units of antiserum.

4. *Comments and Variations*

What was said about the number of cells plated per dish in Sections IV,B,3 and V,A,3 also applies to the COMUL test. In addition, since the cultures in COMUL test are maintained for 3 weeks, when fast-growing tissue culture lines are used, it is often necessary to passage the cells to prevent cell overgrowth. The cultures are trypsin passed at 1:2 or 1:4 ratio on the seventh and fourteenth days (Sarma *et al.*, 1971) and harvested on day 21.

For routine screening, it is sufficient to perform the COMUL test as described above. For exact determination of the titer, however, especially when low positivities were detected by CF in some samples, it is advisable to perform a blind passage. This is done by inoculation of freshly plated dishes (1 or 2 per sample) with cell homogenate from and beyond the titration end point.

5. *Evaluation*

The results of complement-fixation reaction in individual samples are scored as $+ + + +$, $+ + +$, $+ +$, $+$, and $\pm$. A sample with reading $+ + + +$ or $+ + +$ is considered positive, that with $+ +$, $+$, or $\pm$ is considered negative.

The cultures in individual dilutions are scored as positive or negative, and the titer in $TCID_{50}$ is determined from statistical tables on end-point dilution methods (Fisher and Yates, 1963) or by a Spearman–Körber method, as is demonstrated below (Table IV).

In Table IV are summarized data from a titration performed in five dishes per dilution. The sum of positive cultures, expressed as the percentage of the total for each dilution group, was $\Sigma = 320$.

The $TCID_{50}$ titer of a volume of an inoculum (in our case this was 0.4 ml) is computed from the equation:

$$\log_{10} TCID_{50} = \frac{\Sigma}{100} - 0.5 \tag{7}$$

According to Eq. (7) the titer of the virus stock in our example was $10^{2.7}/0.4$ ml. After correction to a 1.0 ml volume, the titer was $10^{3.1}$ $TCID_{50}$ per milliliter.

If the titration in Table IV had started from a dilution other than 10^0, e.g., 10^{-n}, and if 10^{-n} dilution still had 100% positive cultures, then the number $n \times 100$ would be added to Σ before entering Σ in Eq. (7).

B. Immunofluorescent Focus Assay

The fluorescent focus assay (Rapp et al., 1959) was first used for titration of RNA tumor viruses by Vogt and Rubin (1963). It allowed direct visualization of virus productive and nonproductive transformed foci as well as nontransformed but helper virus infected centers (Vogt, 1964). This method has also been used as an infectivity assay for murine leukemia viruses (Rowe et al., 1966; Woods et al., 1970; Declève et al., 1974b). The technique was described in detail by Vogt (1969b). Since many of the newly described mouse

TABLE IV

Example of Computation of $TCID_{50}$ End Point Titer in COMUL Test

Virus dilution	Total number of cultures	Total number of positive cultures	Percent positive cultures
Uninfected control	5	0	0
10^0	5	5	100
10^{-1}	5	5	100
10^{-2}	5	4	80
10^{-3}	5	2	40
10^{-4}	5	0	0
10^{-5}	5	0	0
			$\Sigma = 320$

C-type virus isolates (Levy, 1973; Todaro *et al.*, 1973a; Benveniste *et al.*, 1974; Hopkins and Jolicoeur, 1975; Fischinger *et al.*, 1975; Bryant and Klement, 1976; Hartley and Rowe, 1976; Vredevoe and Hays, 1976) are negative in the XC assay, the importance of a fluorescent focus assay as an alternative procedure has increased.

We describe below a simplified method for fluorescent focus assay, which can be performed directly in plastic cups of disposable tissue culture trays and which is presently being used in our laboratory as a routine titration procedure for MLVs (Fig. 5).

1. *Equipment and Supplies*

a. Standard plastic and glass disposable supplies for tissue culture, as described for previous methods. Twenty-four well tissue culture trays FB 16–24 TC (Linbro Chemical Co., Division of Flow Laboratories, Hamden, Connecticut) or tissue culture clusters with 24 wells, 16 mm diameter (Costar Division of Cooke Laboratory Products, Cambridge, Massachusetts).

b. Rabbit antiserum against mouse gs antigen (p30 protein) prepared by immunization of New Zealand rabbits, as described above. Fluorescein iso-

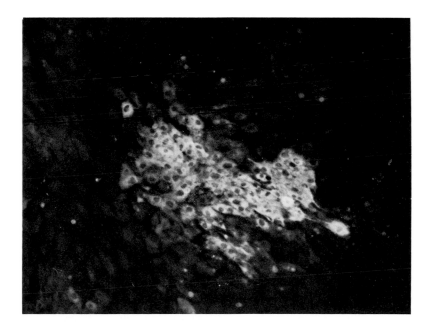

FIG. 5. Fluorescent focus in NIH 3T3 cells infected with wild mouse leukemia virus isolate #292 (Bryant and Klement, 1976) stained with rabbit antimouse p30 antiserum and goat antirabbit γ-globulin FITC conjugate. × 200.

thiocyanate (FITC) conjugated antisera against rabbit 7 S γ-globulin (Hyland Laboratories, Costa Mesa, California).

2. Procedure

a. Plate 2–3 $\times$ 10^4 cells per cup in medium consisting of EMEM, 10% FBS, P–S, and polybrene (see Section IV,B,1,3). Incubate 24 hours at 37 °C in CO$_2$ (5% in air) humidified incubator.

b. After 24 hours, remove the fluid; wash once with EMEM or PBS, add EMEM, 5% CS, P–S (1.0 ml per well) and inoculate with the virus (0.1 ml per well). Change the fluid (using multisuction device—Fig. 6) on second and third days.

c. On the sixth day after inoculation, remove the fluid and fix the monolayers with 95% methanol in saline for 10 minutes. Rinse with PBS, pH 7.2; add 50 μl of antiserum per well, and incubate in a wet chamber at 37 °C for 30 minutes on a horizontal shaker.

d. Remove the antiserum. Rinse once and wash twice (5 minutes each) with PBS on a horizontal shaker. Rinse again and incubate with respective FITC conjugate. Incubate for 30 minutes at 37 °C.

e. Follow again with one cycle of rinsing and washing and overlay each well with 3 drops of 90% glycerol in PBS.

f. Observe on fluorescent microscope. For best results use vertical illuminator and low magnification (10 $\times$ eyepieces and 4 $\times$ objective).

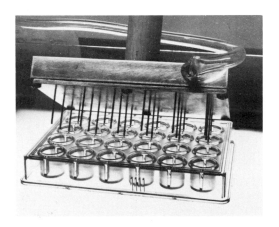

FIG. 6. Device for simultaneous fluid suction from the wells of the FB-16-24 TC tissue culture tray (Linbro Chemical Co., New Haven, Connecticut). The unit consists of aluminum box with embedded stainless steel needles, connected to a vacuum. The needles are sterilized by flame (see Section VI,B,2).

3. *Comments*

To improve the quality of the monolayers of some cell types, the wells of tissue culture trays are filled with 0.01% gelatin and held at 4 °C for two hours prior to the plating of the cells. Variation in the number of cells plated per well depends again on the type of cells used, as was previously explained (Sections IV, B and V, A)

When mouse cells are used in the assay, the antisera are absorbed, before they are used for immunofluorescent staining, on 20% NIH Swiss mouse tissue homogenate for 48 hours at 4 °C. The end point fluorescence titer of antisera is determined on fixed cells of rat cell line #58967 infected with Ki-MSV (Roy-Burman and Klement, 1975), but other chronically infected lines might possibly be used as well. Eight units of antisera are routinely used in the fluorescent focus assay. The FITC-conjugated antirabbit γ-globulin is titrated similarly, using 8 units of antisera. In most cases the conjugate is used at 1:10 or 1:20 dilution.

The method of titration in wells, as described, is useful also for transformation focus assay and XC plaque assay. The size of the well limits the number of countable lesions to approximately 30. Viruses could be cloned from wells with a single transformation focus, XC plaque, or fluorescent focus by two or more consecutive cycles of isolation. For that purpose, fluid is harvested from trays at the expected end points before the trays are fixed and stained.

The harvest of the fluid from individual wells was simplified by using silicone rubber or polyurethane foam gaskets, prepared in our laboratory, which matched exactly with the upper surface of the trays (Fig. 7). The gas-

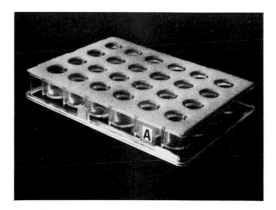

FIG. 7. Polyurethane foam gasket with openings matching the upper surface of the tissue culture tray (see Section VI,B,3).

kets were sterilized by autoclaving and placed on top of the trays from which the fluid was to be harvested. The recipient plate (plate B) in which the fluid was to be harvested was placed upside down on top of the assay plate A (Fig. 8), again matching the wells with the openings in the gaskets. Both plates were tightly pressed together and at the same time tilted over. The fluid passed from the wells in the master assay plate A to the corresponding wells in the recipient plate B without loss or cross-contamination. The recipient plate can be stored or frozen until used. The gaskets were used for the work with FB 16–24 TC Linbro trays. However, the #3524 Costar 24-tissue culture cluster trays do not require the use of a gasket for fluid harvests since the top surfaces of the wells fit tightly when two trays are placed together.

C. Cell Membrane Antigen Assay

Infection of a mouse bone marrow tissue culture line with Moloney leukemia virus was followed by production of a virus-specific membrane antigen, detected by immunofluorescence (Klein and Klein, 1964). This observation led to the development of an *in vitro* infectivity assay (Nordenskjöld *et al.*, 1970), which records the virus-specific membrane antigen either by membrane fluorescence or by immune adherence, and which has comparable sensitivity to the *in vivo* leukemia virus titration. A similar assay, utilizing complement-dependent cytotoxicity, was also described (Haughton, 1965).

VII. Interference

Viral interference (Rubin, 1960c, 1961) has been used as an end point infectivity assay for avian leukosis viruses (Vogt and Rubin, 1963). The mechanism of the interference was found to be the blocking of cell surface receptor sites (Steck and Rubin, 1966a,b). Therefore, interference occurs only between viruses that are related by host range and antigenicity (Vogt and Ishizaki, 1966). Although an analogous situation has been observed in feline leukemia sarcoma complex (Sarma and Log, 1971), the interference between murine leukemia viruses did not show a differential pattern (Sarma *et al.*, 1967). Thus, this type of assay has not been used often with murine or mammalian C-type viruses and no detailed procedure is given here. However, in view of the recent identification of a new class of mouse amphotropic viruses, which demonstrate differential interference patterns from ecotropic and xenotropic viruses (Hartley and Rowe, 1976), interest in the interference studies on mouse and mammalian viruses will be revived.

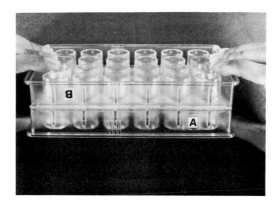

FIG. 8. Harvest of the fluid from master plate A to the recipient plate B, using the polyurethane gasket (Section VI,B,3).

VIII. *Assays for DNA Provirus*

Virus-specific sequences integrated into host cell DNA are termed *provirus* (Temin, 1964). These sequences can be detected by a variety of molecular hybridization techniques which employ either a complementary DNA (cDNA) synthesized from a known C-type virus (reviewed by Varmus *et al.*, 1974), or the viral genomic RNA (Melli *et al.*, 1971; Neiman, 1973; Quintrell *et al.*, 1974) hybridized under annealing conditions with provirus-containing cell DNA. These methods do not involve infectious virus, and are limited in their ability to assess the completeness of the integrated proviral genome. However, described herein are assays for DNA provirus which detect the presence of complete viral genomes inasmuch as the end point in each case is the production of infectious virus.

A. RESCUE

There are two basic types of nonproductive interactions of transforming RNA tumor viruses with the host cell: (1) nonproductive transformation of a heterologous cell by a competent transforming virus (Svoboda, 1960), and (2) transformation of a homologous or heterologous cell by a defective sarcoma virus (Hanafusa *et al.*, 1963; Huebner *et al.*, 1966; Sarma *et al.*, 1966). Both types of cells carry an integrated viral DNA genome or provirus. The first successful *in vitro* rescue of a sarcoma virus genome was done by Šimkovič *et al.* (1962) by cocultivation of the rat tumor XC cell line (Svoboda, 1960), which carried a competent RSV genome, with normal chick embryo fibroblasts. The induction of synthesis of infectious RSV required direct cell-to-cell contact of virogenic (XC) cells and indicator chick embryo cells. The process of rescue was significantly enhanced by using inactivated Sendai virus (Svoboda *et al.*, 1967).

To rescue defective sarcoma virus genome from permissive cells, one needs only to superinfect the cells with homologous helper leukemia virus to which the cells are susceptible. Study of the kinetics of virus release in a homologous system showed that each cell producing leukemia virus also produced sarcoma virus, and that both virus types were released at the same time (Rowe, 1971).

For rescue of defective sarcoma virus genome from heterologous host cells that might be resistant to the superinfecting leukemia helper virus, one needs to combine superinfection by a helper virus with cocultivation of the nonproducer cells and sensitive indicator cells (Huebner *et al.*, 1966; Sarma *et al.*, 1966). In this system the total yield of virus is increased by a factor of ten when β-propriolactone-inactivated Sendai virus is used. The process of sarcoma virus genome rescue, once started, is not significantly influenced by the presence of specific antiserum in the tissue culture medium prior to the harvest, thus indicating the primary importance of cell fusion and cell-to-cell transport of macromolecular components in the formation of permissive heterokaryons (Kelloff *et al.*, 1971).

B. TRANSFECTION

The DNA of transformed nonproductive cells carrying competent sarcoma virus genome can infect permissive cells and give rise to progeny virus identical to the transforming virus. This was first shown by Hill and Hillova (1972), who observed transformation and RSV production in chicken cells previously inoculated with DNA of the XC rat tumor, containing RSV provirus. Similar results were obtained with DNA of avian cells productively infected with avian viruses (Cooper and Temin, 1974; Hill *et al.*, 1974), and with DNA of mammalian cells, both productively and nonproductively infected with mammalian C-type viruses (Karpas and Milstein, 1973; Scolnick and Bumgarner, 1975; Nicolson *et al.*, 1976).

The method for detecting infectious proviral DNA, termed transfection, requires the complete proviral DNA genome, in the case of nontransforming viruses; defective sarcoma genomes can be rescued via transfection if the transfected cells are superinfected with the appropriate helper virus (Karpas and Milstein, 1973). The minimum molecular size of the provirus-containing DNA required for infectivity was reported for an avian virus (RSV) to be approximately equivalent to a double-stranded copy of the viral RNA subunit (Cooper and Temin, 1974). The isolated DNA is applied to the permissive cells after prior treatment of the cells with DEAE-D (Hill and Hillova,

1972) or by complexing the DNA with calcium phosphate (Graham and van der Eb, 1973). Events occurring in the process of transfection are not clear yet, particularly with respect to the question of integration of the exogenous DNA into the host cell genome prior to its transcription (Hill *et al.*, 1974; Nicolson *et al.*, 1976). Assay for virus production following DNA transfection can be carried out by previously described methods (Sections IV and V).

C. Virus Induction

Many *in vitro* methods have been employed for induction of latent C-type viruses. In general, these represent biological manipulation and exposure of cells to physical and chemical inducers.

Long-term *in vitro* propagation of mouse cell lines by itself often leads to C-type RNA virus induction (Aaronson *et al.*, 1969; Rowe *et al.*, 1971; Klement *et al.*, 1973; Lieber *et al.*, 1973). Other experimental manipulations leading to virus induction include transformation by DNA oncogenic virus (Rowe *et al.*, 1971; Rhim and Huebner, 1973), lymphocyte stimulation (Hirsch *et al.*, 1972; Moroni and Schumann, 1975), and inhibition of protein synthesis (Kotler *et al.*, 1972; Aaronson and Dunn, 1974). Physical and chemical mutagens and cancerogenic compounds were found to be good virus inducers (Igel *et al.*, 1969; Aaronson *et al.*, 1971; Klement *et al.*, 1971; Lowy *et al.*, 1971; Rowe *et al.*, 1971; Teich *et al.*, 1973; Weiss *et al.*, 1971).

IX. Detection of Viruses and Viral Components by Methods Not Utilizing Infectivity

The assays to be discussed in this section are not based on the property of viral infectivity but rely on physical or biochemical characteristics of the C-type viruses in general. As a result, the methods are less specific than the previously described infectivity assays and more difficult to quantitate. They are most useful for rapid, qualitative detection of C-type virus in infected (or induced) cultures. Further, they require full expression of the viral genome and virus replication inasmuch as the assays measure the production of extracellular virus.

A. Sucrose Density Gradients

Sucrose density gradients were used first by Robinson *et al.* (1965) and by Duesberg and Robinson (1966) for the separation and purification of avian and murine RNA tumor viruses from the culture medium of infected cells. With some modifications, this procedure is still the most widely utilized technique for C-type virus purification. The method takes advantage of the common density range of C-type RNA tumor viruses, i.e., 1.15–1.17 gm/ml in sucrose gradients. The method can also be utilized for the detection of C-type virus, which has been labeled with 5-[³H]uridine. This is a rapid, potentially nonspecific technique, which is qualitative in nature. The assay is based on 24-hour labeling of a virus-productive culture with 5-[³H]uridine, with the consequent incorporation of radioactivity into viral RNA, which is detected by isopycnic banding in sucrose density gradients. The presence and position of the virus are determined by measuring the radioactivity of the gradient fractions.

1. *Procedure*

All cultures to be assayed are grown to approximately 75% confluency in plastic flasks of 75 or 150 cm² area. The medium is replaced with fresh growth medium containing 20 μCi/ml of 5-[³H]uridine (10–20 Ci/mMole, New England Nuclear, Boston, Massachusetts) using 10 and 20 ml for 75 and 150 cm² flasks, respectively. Cultures are incubated for 24 hours at 37 °C. The culture medium is harvested into chilled centrifuge tubes and clarified at 17,000 g in a refrigerated centrifuge. All subsequent steps are carried out at 0°–5 °C. The clarified samples in 5/8 × 4 inch or 1 × 3 1/2 inch polyallomer tubes (Beckman Instruments, Inc. Fullerton, California) are underlayered first with 5.0 ml of 20% sucrose, then with 0.5 ml of 65% sucrose. All solutions are made in TNE buffer (NaCl, 0.1 M; EDTA, 0.001 M; Tris 0.01 M; pH 7.2). The discontinuous gradients are centrifuged in an SW 27 rotor in a Beckman preparative ultracentrifuge (Beckman Instruments) for 2 hours at 22,000 rpm. The top culture medium is carefully removed, and the interface area, between 20 and 65% sucrose, is recovered, diluted with TNE to give 15% sucrose or less (approximately 4–5 ml total volume) and layered onto a preformed linear 15–55% sucrose gradient (12 ml) in a 5/8 × 4 inch polyallomer tube. The gradients are centrifuged as above for a minimum of 4 hours, or a maximum of 16 hours, at 24,000 rpm. Fractions of 0.5 ml are collected either from the bottom of the gradient, using a peristaltic pump, or by displacement of the gradient upward by introduction of a dense solution. The refractive index of every other fraction is measured, and each fraction is precipitated in 5% TCA after the addition of 200 μg of carrier RNA, filtered with suction onto an individual glass fiber filter [GF/C, 2.4 cm (Whatman; W. and R. Balston, Ltd.; available from most laboratory supply companies)], washed twice with ethanol, once with ethyl ether, dried, and counted in toluene-based scintillation medium in a scintillation spectrometer.

2. *Comments*

C-Type viruses can be expected to form a sharp band of radioactivity at densities between 1.15 and 1.17 gm/ml (Robinson *et al.*, 1965; Duesberg and Robinson, 1966; Hardy *et al.*, 1969; Sarkar and Moore, 1974). The sensitivity of this method is less than that of assays utilizing virus infectivity.

Specificity of this assay relates to the whole group of leukemia and sarcoma viruses with similar densities in sucrose, but it is not possible to distinguish among individual members of this group. Furthermore, the method does not distinguish between infectious and noninfectious virus particles. Extracellular materials that can interfere with the interpretation of the gradient patterns would be other RNA-containing virions of similar density in sucrose (B-type, density 1.18 gm/ml; Sarkar and Moore, 1974). Mycoplasma, which bands at a density of 1.22–1.24 in sucrose (Todaro *et al.*, 1971), quite distinct from C-type viruses, could be confused with C-type viral cores, which band at a density of approximately 1.26 gm/cc (Fleissner and Tress, 1973).

Density gradients of solutes other than sucrose have been used successfully for banding C-type viruses. When Rauscher leukemia virus was sedimented in either polyglucose or Ficoll gradients, 3 sharply separated bands resulted, the middle (major) band containing the majority of complement-fixing antigens and virtually all of the infectivity (Oroszlan *et al.*, 1965). Gradient separations utilizing salts of heavy metals, such as CsCl or potassium tartrate, although allowing viral separation, have been shown to inactivate mammalian C-type viruses (Oroszlan *et al.*, 1965; McCrea *et al.*, 1961).

Complement-fixation tests for the direct titration of antigens in the viral bands can be performed by modification of conventional methods (see Section VI,A; Oroszlan *et al.*, 1965). The amount of radioactivity in bands can be correlated with infectious virus titer using a suitable infectivity assay mentioned above.

B. RNA-DIRECTED DNA POLYMERASE

RNA-directed DNA polymerase (RDDP) was first described in mid-1970 as a component of both Rauscher murine leukemia virus and RSV (Baltimore, 1970; Temin and Mizutani, 1970).

Since that time the enzyme has been found in virtually all RNA tumor, viruses, and it now appears that the enzyme is an integral part of at least 6 groups of viruses; C-, B-, and, recently, A-type RNA tumor viruses (Wong-Staal *et al.*, 1975), Mason–Pfizer monkey virus, slow viruses of sheep, and foamy or syncytium-forming viruses (see review by Green and Gerard, 1974). It is now well established that RDDP enzyme is viral coded and plays an essential role in virus replication and cell transformation through the formation of a DNA provirus. Although a few polymerase-defective viruses have been described, both avian (Hanafusa *et al.*, 1972; Friis *et al.*, 1975) and mammalian (Somers *et al.*, 1973; Peebles *et al.*, 1975; Tronick *et al.*, 1975), nearly ubiquitous presence of RDDP in C-type viruses has provided a useful and rapid means of assay for extracellular virons. The RDDP activity can be measured in two different ways. The endogenous reaction utilizes disrupted viruses with added substrates, and measures DNA synthesis using the viral RNA as template. It was early discovered that RDDP responds very well to exogenous natural, as well as to synthetic, polynucleotides (Spiegelman *et al.*, 1970). A number of studies have compared the template activities of various DNA's and RNA's, RNA · RNA, DNA · DNA, and RNA · DNA homopolymer duplexes, as well as homopolymer · oligomer duplexes for a variety of avian and mammalian RDDP activities (see reviews by Temin and Baltimore, 1972; Green and Gerard, 1974). The enzyme is primer dependent, copying only single-stranded regions of DNA or RNA template and proceeding in a $5' \to 3'$ direction. In general, the most widely used and active of the synthetic homopolymer templates have been the homopolymer · oligomer duplexes, poly(A) · oligo(dT), and poly(C) · oligo (dG) (Baltimore and Smoler, 1971). These templates, however, are not specific for viral RDDP, especially poly(A) · oligo(dT) which can be copied effectively by some cellular DNA polymerases (Fry and Weissbach, 1973). To date, apart from viral 70S RNA, the only synthetic template which appears to be specific for all viral RDDP enzymes is poly(2'-O-methyl-cytidylate) · oligo(deoxyguanylate) (Gerard *et al.*, 1974; Gerard, 1975). While this is obviously important in studying intracellular DNA polymerases, or in characterization of the viral RDDP, the specificity of the template becomes less critical when employed in a routine assay for the production of extracellular virus.

1. *Conditions for Assaying Mammalian RDDP in Tissue Culture Fluids*

a. Disruption of viral membrane by a nonionic surface-active agent is required since the intact membrane is impermeable to the reaction components and the virion RDDP is located in the viral core. The effectiveness of a variety of these surfactants has been evaluated by Stromberg (1972) for avian myeloblastosis virus. Much has been written about appropriate surfactant concentrations and the inhibitory effect of levels much above optimal for mammalian viruses (Temin and Baltimore, 1972; Green and Gerard, 1974). However, the information relates mainly to reactions utilizing endogenous viral RNA template, wherein the retention of close association between enzyme and template is critical. We and others have not found this to be the case using synthetic ribodeoxyhomopolymers as template (Baltimore and Smoler, 1971; M.O. Nicolson, unpublished data). In fact, optimal concentrations of NP40 in our routine assay are 5–10 times that recommended for the endogenous reaction.

b. There is an absolute requirement by RDDP enzyme for divalent cation, either Mg^{2+} or Mn^{2+}. The optimal concentration of Mg^{2+} (around 10 mM) is 5–10 times that of Mn^{2+} (1–2 mM).

c. The optimum pH for the reaction is approximately 8.0; the temperature optimum for the mammalian RDDP is 37 °C (Baltimore, 1970), and for the avian enzyme, 40 °C (Temin and Mizutani, 1970).

d. The addition of a reducing agent is required for full activity; dithiothreitol is usually the compound of choice.

e. Monovalent cations (Na^+ or K^+) are not essential, but generally somewhat increase the rate of reaction when added in low concentrations. For mammalian viral RDDP the commonly used range is from 20 to 120 mM.

f. In the case of synthetic homopolymer templates, only one deoxyribonucleoside triphosphate is required as substrate for synthesis of the complementary DNA homopolymer. References to these reaction conditions can be found in two reviews (Temin and Baltimore, 1972; Green and Gerard, 1974).

2. Reagents

Whatman #1 filter paper circles, 2.4 cm
Tris 0.4 M, KCL 0.5 M, pH 8.1
$MnCl_2$ 0.02 M
Poly(A)·oligo(dT) 0.2 mg/ml in Tris 0.01 M, NaCl 0.15 M, pH 8.1
  (Collaborative Research, Inc., Waltham, Massachusetts)
Dithiothreitol 0.10 M (make fresh) (Calbiochem, La Jolla, California)
Me-[³H]-TTP, 500 μCi/ml in 50% ETOH (sp. act. 10–20 Ci/mMole)
  (New England Nuclear, Boston, Massachusetts)
NP40 2% (Shell International Chemical Co., London, England)
Tris, 0.01 M, NaCl 0.15 M, pH 8.1
ATP 0.01 M (P. L. Biochemicals, Inc., Milwaukee, Wisconsin)
TCA 5% (trichloroacetic acid), sodium pyrophosphate 2%
TCA 5%, sodium pyrophosphate 1%
TCA 5%
Ethanol, 95 or 100%
Ethyl ether

3. Reaction Mixes (Made Fresh Daily)

Mix	Proportionate volumes
A	A
Tris 0.4 M, KCL 0.5 M, pH 8.1	0.1
DTT 0.10 M	0.2
NP40 2%	0.1
H_2O	0.6
	1.0

B and C	B	C
Tris 0.4 M, KCL 0.5 M, pH 8.1	0.1	0.1
MnCl$_2$ 0.02 M	0.1	0.1
Poly (A) · oligo (dT) 0.2 mg/ml	0.2	—
Tris 0.1 M, NaCl 0.15 M, pH 8.1	—	0.2
Me-[^{3}H]-TTP, 500 μCi/ml	0.2	0.2
H$_2$O	0.4	0.4
	1.0	1.0

4. *Procedure*

a. Harvest 24-hour culture medium, generally 10 ml from T-75 flask; smaller or larger volumes can also be used. Clarify at 17,000 g for 10 minutes at 3 °C in a refrigerated Sorvall centrifuge, rotor SS34.

b. Pellet virus either in rotor SW 27, 90 minutes, 25,000 rpm, or in rotor #40, 60 minutes, 40,000 rpm, Beckman preparative ultracentrifuge, 2 °C. Slow with brake and immediately aspirate off the supernatant fluid. Invert and drain in refrigerator. If screwcap polycarbonate tubes are used, to drain properly insert small piece of cotton in neck, then invert. If using the fixed angle rotor, orient tubes in known fashion in rotor in order to keep track of position of potential pellet.

c. To each tube, in ice, add 75 λ of reaction mix A. Stir in pellet region with closed off Pasteur pipette, then resuspend by pipetting up and down with a mechanical 25 λ micropipette. For each sample, pipette 25 λ into each of two tubes on ice (10 × 75 or 6 × 50 mm).

d. To one tube add 25 λ of reaction mix B, to the other add 25 λ of reaction mix C. Mix, cover, and incubate at 37 °C for 60 minutes.

e. Remove tubes to ice. Pipette 25 λ of each reaction mix onto a separate filter which has been pretreated (immediately before sample) with 10 μl of 0.01 M ATP. Allow liquid to adsorb and drop filters into beaker (in ice) containing 5% TCA and 2% sodium pyrophosphate (up to 100 filters in one beaker). Use approximately 10 ml solution per filter; wash 20 minutes, stirring gently every 5 to 10 minutes.

f. Transfer filters successively through the following cold solutions (10 ml × number of filters), 5 minutes each:

5% TCA, 1% sodium pyrophosphate; 5% TCA; 5% TCA; 95–100% ethanol; 95–100% ethanol; and ethyl ether.

g. Dry filters in air, oven, or under heat lamp for minimum of 30 minutes. Count in scintillation media according to normal procedure.

h. Counts per minute of sample without added template are subtracted as background from sample with template. Results are reported as counts per minute per milliliter of culture medium.

5. *Comments*

It is recommended that a known positive control be included in each RDDP assay. A 24-hour harvest of medium from a culture chronically infected and producing a high titer of C-type virus is clarified at 12,000 rpm for 10 minutes, diluted 1:5 with EMEM, containing P–S and 10% FBS, and frozen in 10 ml aliquots at −80°C. Virus titer by RDDP assay should remain stable for 6 months.

The cultures to be assayed for RDDP should not be heavily grown, preferably not yet confluent. This condition not only reduces contribution to the background cpm by cellular debris, but generally gives higher yields of virus relative to cells.

As a qualitative screening technique for extracellular C-type virus, the sensitivity of this method correlates well with the intracellular titer of p30 antigen determined by complement fixation (Filbert *et al.*, 1974). As a qualitative assay, values of 1000 cpm/ml above background are scored as positive, those lower are considered to be negative; however, these values cannot be considered to be absolute.

Quantitative data, comparing the RDDP assay with the COMUL test, have been recorded for five mammalian C-type viruses (Kelloff *et al.*, 1972). The authors show that in 6 days after inoculation, for instance, as little as $10^{0.4}$ $TCID_{50}$ of RLV (COMUL) can be detected by RDDP assay, thus providing a rapid and sensitive infectivity assay for C-type viruses. Results were similar for the other viruses tested.

The relationship of RDDP to total particle number can be determined by particle counts using electron microscopic methods (Sharp, 1965; Barbieri *et al.*, 1970). This relationship, however, will vary widely with the particular C-type virus and the cell culture from which it is produced. Not only does the enzyme specific activity vary among virus strains but the ability to copy various synthetic templates varies as well (Spiegelman *et al.*, 1970; Verma *et al.*, 1974).

C. Radioimmunoassay

The antigenic determinants of the major viral proteins of the RNA tumor viruses can be assayed by the technique of radioimmunoassay. The method is specific and sensitive. However, since it requires highly purified radiolabeled viral proteins, the preparation of which is both expensive and technically demanding, it is hardly a method to be used routinely in smaller laboratories. The reader is referred to the following references in which radioimmunoassay is adapted for assay of C-type viral proteins; Scolnick *et al.* (1972), Oroszlan *et al.* (1972), Parks *et al.* (1973), and Strand and August (1973).

ACKNOWLEDGMENTS

We thank Drs. Peter Vogt and Janet Hartley for reviewing the manuscript and for their suggestions, and Dr. Malcolm Pike for his kind help in computing Table I and for critical review of the theoretical part. We also thank Dr. Wallace Rowe, Dr. Robert McAllister, Dr. Susan Shimizu, Mary Dougherty, Martin Bryant, David Krempin, Belen Diaz, and Geraldine Trail for helpful suggestions during preparation of the manuscript.

REFERENCES

Aaronson, S. A., and Dunn, C. Y. (1974). *Science* **183**, 422.

Aaronson, S. A., and Rowe, W. P. (1970). *Virology* **42**, 9.

Aaronson, S. A., Hartley, J. W., and Todaro, G. J. (1969). *Proc. Natl. Acad. Sci. U.S.A.* **64**, 87.

Aaronson, S. A., Jainchill, J. L., and Todaro, G. J. (1970). *Proc. Natl. Acad. Sci. U.S.A.* **66**, 1236.

Aaronson, S. A., Todaro, G. J., and Scolnick, E. M. (1971). *Science* **174**, 157.

Aaronson, S. A., Bassin, R. H., and Weaver, C. (1972). *J. Virol.* **9**, 701.

Aaronson, S. A., Stephenson, J. R., Hino, S., and Tronick, S. R. (1975). *J. Virol.* **16**, 1117.

Abelson, H. T., and Rabstein, L. S. (1970), *Cancer Res.* **30**, 2213.

Ahlström, C. G., and Forsby, N. (1962). *J. Exp. Med.* **115**, 839.

Armstrong, D., Okuyan, M., and Huebner, R. J. (1964), *Science* **144**, 1534.

August, J. T., Bolognesi, D. P., Fleissner, E., Gilden, R. V., and Nowinski, R. C. (1974). *Virology* **60**, 595.

Axelrad, A. A., and Steeves, R. A. (1964), *Virology* **24**, 513.

Bader, J. P. (1973). *Science* **180**, 1069.

Balduzzi, P. C., and Murphy, H. (1975). *J. Virol.* **16**, 707.

Baltimore, D. (1970). *Nature (London)* **226**, 1209.

Baltimore, D., and Smoler, D. (1971). *Proc. Natl. Acad. Sci. U.S.A.* **68**, 1507.

Baluda, M. A., and Goetz, I. E. (1961). *Virology* **15**, 185.

Barbieri, D., Delain, E., Lazar, P., Hue, G., and Barski, G. (1970). *Virology* **42**, 544.

Bassin, R. H., Tuttle, N., and Fischinger, P. J. (1971). *Nature (London)* **229**, 564.

Bauer, H., and Schäfer, W. (1965). *Z. Naturforsch., Teil B* **20**, 815.

Beaudreau, G. S., Becker, C., Bonar, R. A., Wallbank, A. M., Beard, D., and Beard, J. W. (1960). *J. Natl. Cancer Inst.* **24**, 395.

Benveniste, R. E., Lieber, M. M., Livingston, D. M., Sherr, C. J., and Todaro, G. J. (1974), *Nature (London)* **248**, 17.

Bernhard, W. (1960). *Cancer Res.* **20**, 712.

Brahic, M., and Vigne, R. (1975). *J. Virol.* **15**, 1222.

Bryan, W. R. (1946a). *J. Natl. Cancer Inst.* **6**, 225.

Bryan, W. R. (1946b). *J. Natl. Cancer Inst.* **6**, 373.

Bryant, M. L., and Klement, V. (1976). *Virology* **73**, 532.

Burnet, F. M. (1936). *Med. Res. Counc. (G.B.), Spec. Rep. Ser.* **220**, 58.

Calnek, B. W. (1964a). *Avian Dis.* **8**, 163.

Calnek, B. W. (1964b). *Natl. Cancer Inst., Monogr.* **17**, 425.

Chirigos, M. A. (1964). *Cancer Res.* **24**, 1035.

Clarke, B. J., Axelrad, A. A., Shreeve, M. M., and McLeod, D. L. (1975). *Proc. Natl. Acad. Sci. U.S.A.* **72**, 3556.

Cook, K. (1969). *J. Natl. Cancer Inst.* **43**, 203.

Cooper, G. M., and Temin, H. M. (1974). *J. Virol.* **14**, 1132.

Dalton, A. J., Melnick, J. L., Bauer, H., Beaudreau, G., Bentvelzen, P., Bolognesi, D., Gallo,

R., Graffi, A., Haguenau, F., Heston, W., Huebner, R., Todaro, G., and Heine, U. I. (1974). *Intervirology* **4**, 201.

Dawson, P. J., Tacke, R. B., and Fieldsteel, A. H. (1968). *Br. J. Cancer* **22**, 569.

Declève, A., Lieberman, M., Niwa, O., and Kaplan, H. S. (1974a). *Nature (London)* **252**, 79.

Declève, A., Niwa, O., Hilgers, J., and Kaplan, H. S. (1974b). *Virology* **57**, 491.

Dougherty, R. M., and Rasmussen, R. (1964). *Natl. Cancer Inst., Monogr.* **17**, 337.

Duc-Nguyen, H., Rosenblum, E. N., and Zeigel, R. F. (1966). *J. Bacteriol.* **92**, 1133.

Duesberg, P. H., and Robinson, W. S. (1966). *Proc. Natl. Acad. Sci. U.S.A.* **55**, 219.

Dulbecco, R. (1969). *Science* **166**, 962.

Fieldsteel, A. H., Kurahara, C., and Dawson, P. J. (1969). *Nature (London)* **223**, 1274.

Filbert, J. E., McAllister, R. M., Nicolson, M. O., and Gilden, R. V. (1974). *Proc. Soc. Exp. Biol. Med.* **145**, 366.

Fischinger, P. J., Peebles, P. T., Nomura, S., and Haapala, D. K. (1973). *J. Virol.* **11**, 978.

Fischinger, P. J., Blevins, C. S., and Nomura, S. (1974). *J. Virol.* **14**, 177.

Fischinger, P. J., Nomura, S., and Bolognesi, D. P. (1975). *Proc. Natl. Acad. Sci. U.S.A.* **72**, 5150.

Fisher, R. A., and Yates, F. (1963). *In* "Statistical Tables for Biological, Agricultural and Medical Research," 6th ed., pp. 8 and 66. Hafner, New York.

Fleissner, E., and Tress, J. (1973). *J. Virol.* **12**, 1612.

Friis, R. R., Mason, W. S., Chen, Y. C., and Halpern, M. S. (1975). *Virology* **64**, 49.

Fry, M, and Weissbach, A. (1973). *J. Biol. Chem.* **248**, 2678.

Gerard, G., Rottman, F., and Green, M. (1974). *Biochemistry* **13**, 1632.

Gerard, G. F. (1975). *Biochem. Biophys. Res. Commun.* **63**, 706.

Goldberg, A. R. (1974). *Cell* **2**, 95.

Graf, T. (1972). *Virology* **50**, 567.

Graham, F. L., and van der Eb, A. J. (1973). *Virology* **52**, 456.

Green, M., and Gerard, G. F. (1974). *Prog. Nucleic Acid Res. Mol. Biol.* **14**, 187.

Grunder, G., Fenyö, E. M., Strouk, V., and Klein, E. (1972). *Proc. Soc. Exp. Biol. Med.* **140**, 378.

Hackett, A. J., and Sylvester, S. S. (1972). *Nature (London), New Biol.* **239**, 164.

Hahn, G. M., Declève, A., Lieberman, M., and Kaplan, H. S. (1970). *J. Virol.* **5**, 432.

Halberstaedter, L., Doljanski, L., and Tanenbaum, E. (1941). *Br. J. Exp. Pathol.* **22**, 179.

Halpern, M. S., Wade E., Rucker, E., Baxter-Gabbard, K. L., Levine, A. S., and Friis, R. R. (1973). *Virology* **53**, 287.

Hanafusa, H. (1969). *Proc. Natl. Acad. Sci. U.S.A.* **63**, 318.

Hanafusa, H., Hanafusa, T., and Rubin, H. (1963). *Proc. Natl. Acad. Sci. U.S.A.* **49**, 572.

Hanafusa, H., Hanafusa, T., and Rubin, H. (1964). *Proc. Natl. Acad. Sci. U.S.A.* **51**, 41.

Hanafusa, H., Baltimore, D., Smoler, D., Watson, K. F., Yaniv, A., and Spiegelman, S. (1972). *Science* **177**, 1188.

Hardy, W. D., Jr., Geering G., Old, L. J., de Harven, E., Brodey, R. S. and McDonough, S. (1969). *Science* **166**, 1019.

Hartley, J. W., and Rowe, W. P. (1966). *Proc. Natl. Acad. Sci. U.S.A.* **55**, 780.

Hartley, J. W., and Rowe, W. P. (1975). *Virology* **65**, 128.

Hartley, J. W., and Rowe, W. P. (1976). *J. Virol.* **19**, 19.

Hartley, J. W., Rowe, W. P., Capps, W. I., and Huebner, R. J. (1965). *Proc. Natl. Acad. Sci. U.S.A.* **53**, 931.

Hartley, J. W., Rowe, W. P., Capps, W. I., and Huebner, R. J. (1969). *J. Virol.* **3**, 126.

Hartley, J. W., Rowe, W. P., and Huebner, R. J. (1970). *J. Virol.* **5**, 221.

Hatanaka, M. and Hanafusa, H. (1970). *Virology* **41**, 647.

Haughton, G. (1965). *Science* **147**, 506.

Hayflick, L. (1967). *Natl. Cancer Inst., Monogr.* **26**, 355.

Henderson, I. C., Lieber, M. M., and Todaro, G. J. (1974). *Virology* **60**, 282.

Hill, M. and Hillova, J. (1972). *Nature (London), New Biol.* **237**, 35.

Hill, M., Hillova, J., Dantchev, D., Mariage, R., and Goubin, G. (1974). *Cold Spring Harbor Symp. Quant. Biol.* **39**, 1015.

Hirsch, M. S., and Harvey, J. J. (1969). *Int. J. Cancer* **4**, 440.

Hirsch, M. S., Phillips, S. M., Solnik, C., Black, P. H., Schwartz, R. S., and Carpenter, C. B. (1972). *Proc. Natl. Acad. Sci. U.S.A.* **69**, 1069.

Hopkins, N., and Jolicoeur, P. (1975). *J. Virol.* **16**, 991.

Hruska, J. F., and Takemoto, K. K. (1975). *J. Natl. Cancer Inst.* **54**, 601.

Huebner, R. J. (1967). *In* "Carcinogenesis: A Broad Critique," pp. 23–47. Williams & Willkins, Baltimore, Maryland.

Huebner, R. J., Armstrong, D., Okuyan, M., Sarma, P. S., and Turner, H. C. (1964). *Proc. Natl. Acad. Sci. U.S.A.* **51**, 742.

Huebner, R. J., Hartley, J. W., Rowe, W. P., Lane, W. T., and Capps, W. I. (1966). *Proc. Natl. Acad. Sci. U.S.A.* **56**, 1164.

Igel, H. J., Huebner, R. J., Turner, H. C., Kotin, P., and Falk, H. L. (1969). *Science* **166**, 1624.

Jainchill, J. L., Aaronson, S. A., and Todaro, G. J. (1969). *J. Virol.* **4**, 549.

Jarrett, W. F. H., Crawford, E. M., Martin, W. B., and Davie, F., (1964). *Nature (London)* **202**, 567.

Jolicoeur, P., and Baltimore, D. (1975). *J. Virol.* **16**, 1593.

Karpas, A. and Milstein, C. (1973). *Eur. J. Cancer* **9**, 295.

Kawai, S., and Hanafusa, H. (1972). *Virology* **48**, 126.

Kawakami, T. G., Huff, S. D., Buckley, P. M., Dungworth, D. L., Snyder, S. P., and Gilden, R. V. (1972). *Nature (London), New Biol.* **235**, 170.

Kelloff, G. J., Huebner, R. J., Long, C., and Gilden, R. V. (1971). *Virology* **46**, 965.

Kelloff, G. J., Hatanaka, M., and Gilden, R. V. (1972). *Virology* **48**, 266.

Keogh, E. V. (1938). *Br. J. Exp. Pathol.* **19**, 1.

Klein, E., and Klein, G. (1964). *J. Natl. Cancer Inst.* **32**, 547.

Klement, V. and McAllister, R. M. (1972). *Virology* **50**, 305.

Klement, V., Rowe, W. P., Hartley, J. W., and Pugh W. E. (1969). *Proc. Natl. Acad. Sci. U.S.A.* **63**, 753.

Klement, V., Nicolson, M. O., and Huebner, R. J. (1971). *Nature (London)* **234**, 12.

Klement, V., Nicolson, M. O., Nelson-Rees, W., Gilden, R. V., Oroszlan, S., Rongey, R. W., and Gardner, M. B. (1973). *Int. J. Cancer* **12**, 654.

Koga, M. (1973). *Gann* **64**, 321.

Kotler, M., Weinberg, E., Haspel, O., and Becker, Y. (1972). *J. Virol.* **10**, 439.

Kruse, P. F., Jr., and Patterson, M. K., Jr., eds. (1973). "Tissue Culture: Methods and Applications." Academic Press, New York.

Langlois, A. J., Sankaran, S., Hsiung, P. L., and Beard, J. W. (1967). *J. Virol.* **1**, 1082.

Latarjet, R. (1970). *Int. J. Cancer* **6**, 31.

Levy, J. A. (1973). *Science* **182**, 1151.

Levy, J. A. (1975). *Nature (London)* **253**, 140.

Lieber, M. M., Benveniste, R. E., Livingston, D. M., and Todaro, G. J. (1973). *Science* **182**, 56.

Lin, F. H., and Thormar, H. (1970). *J. Virol.* **6**, 702.

Livingston, D. M., and Todaro, G. J. (1973). *Virology* **53**, 142.

Lo, W. H. Y., Gey, G. O., and Shapras, P. (1955). *Bull. Johns Hopkins Hosp.* **97**, 248.

Lowy, D. R., Rowe, W. P., Teich, N., and Hartley, J. W. (1971). *Science* **174**, 155.

McAllister, R. M., Nicolson, M., Gardner, M. B., Rongey, R. W., Rasheed, S., Sarma, P. S.,

Huebner, R. J., Hatanaka, M., Oroszlan, S., Gilden R. V., Kabigting, A., and Vernon, L. (1972). *Nature (London), New Biol.* **235**, 3.

McCrea, J. F., Epstein, R. S., and Barry, W. H. (1961). *Nature (London)* **189**, 220.

Manaker, R. A., and Groupe, V. (1956). *Virology* **2**, 838.

Melli, M., Whitfield, C., Rao, K. V, Richardson, M., and Bishop, J. O. (1971). *Nature (London), New Biol.* **231**, 8.

Morgan, H. R. (1968). *J. Virol.* **2**, 1133.

Moroni, C., and Schumann, G. (1975). *Nature (London)* **254**, 60.

Moscovici, C., Moscovici, M. G., and Zanetti, M. (1969). *J. Cell. Physiol.* **73**, 105.

Moscovici, C., Gazzolo, L., and Moscovici, M. G. (1975). *Virology* **68**, 173.

Neiman, P. (1973). *Virology* **53**, 196.

Nicolson, M. O., Hariri, F., Krempin, H. M., McAllister, R. M., and Gilden, R. V. (1976). *Virology* **70**, 301.

Niwa, O., Declève, A., Lieberman, M., and Kaplan, H. S. (1973). *J. Virol.* **12**, 68.

Nordenskjöld, B. A., Klein, E., Tachibana, T., and Fenyö, E. M. (1970). *J. Natl. Cancer Inst.* **44**, 403.

Nowinski, R. C., Sarkar, N. H., and Fleissner, E. (1973). *Methods Cancer Res.* **8**, 237.

O'Connor, T. E., and Fischinger, P. J. (1968). *Science* **159**, 325.

Office of Biohazard Safety, Viral Oncology, Division of Cancer Cause and Prevention. (1972). "Revised 1972 Guide for the Case and Use of Laboratory Animals," DHEW Publ. No. (NIH) 73–23. *Natl. Cancer Inst.*, Bethesda, Maryland.

Office of Biohazard Safety, Viral Oncology, Division of Cancer Cause and Prevention. (1974). "Safety Standards for Research Involving Oncogenic Viruses," DHEW Publ. No. (NIH) 75–790. *Natl. Cancer Inst.*, Bethesda, Maryland.

Office of Biohazard Safety, Viral Oncology, Division of Cancer Cause and Prevention. (1975). "Design Criteria for Viral Oncology Research Facilities," DHEW Publ. No. (NIH) 76–891. *Natl. Cancer Inst.*, Bethesda, Maryland.

Oroszlan, S., Johns, L. W., Jr., and Rich, M. A. (1965). *Virology* **26**, 638.

Oroszlan, S., White, M. M. H., Gilden, R. V., and Charman, H. P. (1972). *Virology* **50**, 294.

Palmer, D. F., Casey, H. L., Olsen, J. R., Eller, V. H., and Fuller, J. M. (1969). "A Guide to the Performance of the Standardized Diagnostic Complement Fixation Method and Adaptation to Micro Test." U.S. Dept. of Health, Education and Welfare, Public Health Service.

Parkman, R., Levy, J. S., and Ting, R. C. (1970). *Science* **168**, 387.

Parks, W. P., and Todaro, G. J. (1972). *Virology* **47**, 673.

Parks, W. P., Scolnick, E. M., Noon, M. C., Watson, C. J., and Kawakami, T. G. (1973). *Int. J. Cancer* **12**, 129.

Parshad, R., and Sanford, K. K. (1968). *J. Natl. Cancer Inst.* **41**, 767.

Payne, L. N., and Biggs, P. M. (1965). *Virology* **27**, 621.

Peebles, P. T. (1975). *Virology* **67**, 288.

Peebles, P. T., Gerwin, B. I., Papageorge, A. G., and Smith, S. G. (1975). *Virology* **67**, 344.

Peters, R. L., Spahn, G. J., Rabstein, L. S. Huebner, R. J., and Kelloff, G. J. (1974). *Appl. Microbiol.* **28**, 614.

Pincus, T., Hartley, J. W., and Rowe, W. P. (1975). *Virology* **65**, 333.

Pluznik, D. H., and Sachs, L. (1964). *J. Natl. Cancer Inst.* **33**, 535.

Prigogina, E. L., and Stavrovskaja, A. A. (1964). *Nature (London)* **201**, 934.

Prince, A. M. (1958a). *J. Natl. Cancer Inst.* **20**, 147.

Prince, A. M. (1958b). *J. Natl. Cancer Inst.* **20**, 843.

Purchase, H. G., and Okazaki, W. (1964). *J. Natl. Cancer Inst.* **32**, 579.

Purchase, H. G., Ludford, C., Nazerian, K., and Cox, H. W. (1973). *J. Natl. Cancer Inst.* **51**, 489.

Quintrell, N., Varmas, H. E., Bishop, J. M., Nicolson, M. O., and McAllister, R. M. (1974). *Virology* **58**, 568.

Rand, K. H., and Long, C. (1972). *Nature (London), New Biol.* **240**, 187.

Rangan, S. R. S., Moyer, P. P., Cheong, M. P., and Jensen, E. M. (1972a). *Virology* **47**, 247.

Rangan, S. R. S., Wong, M. C., Ueberhorst, P. J., and Ablashi, D. V. (1972b). *J. Natl. Cancer Inst.* **49**, 571.

Rangan, S. R. S., Ueberhorst, P. J., and Wong, M. C. (1973). *Proc. Soc. Exp. Biol. Med.* **142**, 1077.

Rapp, F., Seligman, S. J., Jaross, L. B., and Gordon, I. (1959). *Proc. Soc. Exp. Biol. Med.* **101**, 289.

Rhim, J. G., and Huebner, R. J. (1973). *Proc. Soc. Exp. Biol. Med.* **144**, 210.

Robinson, W. S., Pitkanen, A., and Rubin, H. (1965). *Proc. Natl. Acad. Sci. U.S.A.* **54**, 137.

Rowe, W. P. (1963). *Science* **141**, 40.

Rowe, W. P. (1971). *Virology* **46**, 369.

Rowe, W. P., and Brodsky, I. (1959). *J. Natl. Cancer Inst.* **23**, 1239.

Rowe, W. P., Hartley, J. W., and Capps, W. I. (1966). *Natl. Cancer Inst., Monogr.* **22**, 15.

Rowe, W. P., Pugh, W. E., and Hartley, J. W. (1970). *Virology* **42**, 1136.

Rowe, W. P., Hartley, J. W., Lander, M. R., Pugh, W. E., and Teich, N. (1971). *Virology* **46**, 866.

Rowson, K. E., and Parr, I. B. (1970). *Int. J. Cancer* **5**, 96.

Roy-Burman, P., and Klement, V. (1975). *J. Gen. Virol.* **28**, 193.

Rubin, H. (1955). *Virology* **1**, 445.

Rubin, H. (1960a). *Virology* **10**, 29.

Rubin, H. (1960b). *Virology* **12**, 14.

Rubin, H. (1960c). *Proc. Natl. Acad. Sci. U.S.A.* **46**, 1105.

Rubin, H. (1961). *Virology* **13**, 200.

Sanford, K. K. (1967). *Natl. Cancer Inst., Monogr.* **26**, 387.

Sanford, K. K., Jackson, J. L., Parshad, R., and Gantt, R. R. (1972). *J. Natl. Cancer Inst.* **49**, 513.

Sarkar, N. H., and Moore, D. H. (1974). *J. Virol.* **13**, 1143.

Sarma, P. S., and Log, T. (1971). *Virology* **44**, 352.

Sarma, P. S., Turner, H. C., and Huebner, R. J. (1964). *Virology* **23**, 313.

Sarma, P. S., Vass, W., and Huebner, R. J. (1966). *Proc. Natl. Acad. Sci. U.S.A.* **55**, 1435.

Sarma, P. S., Cheong, M. P., Hartley, J. W., and Huebner, R. J. (1967). *Virology* **33**, 180.

Sarma, P. S., Log, T., and Huebner, R. J. (1970). *Proc. Natl. Acad. Sci. U.S.A.* **65**, 81.

Sarma, P. S., Gilden, R. V., and Huebner, R. J. (1971). *Virology* **44**, 137.

Sarma, P. S., Tseng, J., Lee, Y. K., and Gilden, R. V. (1973). *Nature (London), New Biol.* **244**, 56.

Scher, C. D., and Siegler, R. (1975). *Nature (London)* **253**, 729.

Schlom, J., Harter, D. H., Burny, A., and Spiegelman, S. (1971). *Proc. Natl. Acad. Sci. U.S.A.* **68**, 182.

Scolnick, E. M., and Bumgarner, S. J. (1975). *J. Virol.* **15**, 1293.

Scolnick, E. M., Parks, W. P., and Livingston, D. M. (1972). *J. Immunol.* **109**, 570.

Shannon, J. E., ed. (1972). "The American Type Culture Collection Registry of Animal Cell Lines," 2nd ed. ATCC, Rockville, Maryland.

Sharp, D. G. (1965). *Lab. Invest.* **14**, 831.

Šimkovič, D., Valentová, N., and Thurzo, V. (1962). *Neoplasma* 9, 104.

Sinkovics, J. G., Groves, G. F., Bertin, B. A., and Shullenberger, C. C. (1969). *J. Infect. Dis.* 119, 19.

Snyder, S. P., and Theilen, G. H. (1969). *Nature (London)* 221, 1074.

Somers, K. D., May, J. T., Kit, S., McCormick, K. J., Hatch, G. G., Stenback, W. A., and Trentin, J. J. (1973). *Intervirology* 1, 11.

Spiegelman, S., Burny, A., Das, M. R., Keydar, J., Schlom, J., Travnicek, M., and Watson, K. (1970). *Nature (London)* 227, 563.

Steck, F. T., and Rubin, H. (1966a). *Virology* 29, 628.

Steck, F. T., and Rubin, H. (1966b). *Virology* 29, 642.

Stenkvist, B. (1966). *Acta Pathol. Microbiol. Scand.* 67, 67.

Stone, L. B., Scolnick, E., Takemoto, K. K., and Aaronson, S. A. (1971). *Nature (London)* 229, 257.

Strand, M., and August, J. T. (1973). *J. Biol. Chem.* 248, 5627.

Stromberg, K. (1972). *J. Virol.* 9, 684.

Svoboda, J. (1960). *Nature (London)* 186, 980.

Svoboda, J., Machala, O., and Hlozanek, I. (1967). *Folia Biol. (Prague)* 13, 15.

Teich, N., Lowy, D. R., Hartley, J. W., and Rowe, W. P. (1973). *Virology* 51, 163.

Temin, H. M. (1960). *Virology* 10, 182.

Temin, H. M. (1964). *Virology* 23, 486.

Temin, H. M. (1965). *J. Natl. Cancer Inst.* 35, 679.

Temin, H. M. (1974). *J. Am. Med. Assoc.* 230, 1043.

Temin, H. M., and Baltimore, D. (1972). *Adv. Virus Res.* 17, 129.

Temin, H. M., and Mizutani, S. (1970). *Nature (London)* 226, 1211.

Temin, H. M., and Rubin, H. (1958). *Virology* 6, 669.

Theilen, G. H., Gould, D., Fowler, M., and Dungworth, D. L. (1971). *J. Natl. Cancer Inst.* 47, 881.

Todaro, G. J., and Meyer, C. A. (1974). *J. Natl. Cancer Inst.* 52, 167.

Todaro, G. J., Aaronson, S. A., and Rands, E. (1971). *Exp. Cell Res.* 65, 256.

Todaro, G. J., Arnstein, P., Parks, W. P., Lennette, E. H., and Huebner, R. J. (1973a). *Proc. Natl. Acad. Sci. U.S.A.* 70, 859.

Todaro, G. J., Benveniste, R. E., Lieber, M. M., and Livingston, D. M. (1973b). *Virology* 55, 506.

Toyoshima, K., and Vogt, P. K. (1969). *Virology* 38, 414.

Tronick, S. R., Stephenson, J. R., Verma, I. M., and Aaronson, S. A. (1975). *J. Virol.* 16, 1476.

Tyndall, R. L., Vidrine, J. G., Teeter, E., Upton, A. C., Harris, W. W., and Fink, M. A. (1965). *Proc. Soc. Exp. Biol. Med.* 119, 186.

Unkeless, J. C., Tobia, A., Ossowski, L., Quigley, J. D., Rifkin, D. B., and Reich, E. (1973). *J. Exp. Med.* 137, 85.

Varmas, H., Heasley, S., and Bishop, J. M. (1974). *J. Virol.* 14, 895.

Verma, I. M., Meuth, N. L., Fan, H., and Baltimore, D. (1974). *J. Virol.* 13, 1075.

Vigier, P. (1959). *Virology* 8, 41.

Vogt, P. K. (1964). *Natl. Cancer Inst., Monogr.* 17, 523.

Vogt, P. K. (1965). *Adv. Virus. Res.* 11, 293.

Vogt, P. K. (1967). *Virology* 33, 175.

Vogt, P. K. (1969a). *In* "Fundamental Techniques in Virology" (K. Habel and N. P. Salzman, eds.), Vol. 1, pp. 198–211. Academic Press, New York.

Vogt, P. K. (1969b). *In* "Fundamental Techniques in Virology" (K. Habel and N. P. Salzman, eds.), Vol. 1, pp. 316–326. Academic Press, New York.

Vogt, P. K., and Rubin, H. (1961). *Virology* **13**, 528.
Vogt, P. K., and Rubin, H. (1963). *Virology* **19**, 92.
Vogt, P. K., and Ishizaki, R. (1966). *Virology* **30**, 368.
Vogt, P. K., Toyoshima, K., and Yoshii, S. (1970). *In* "Defectivité Demasquaqe et Stimulations des Virus Oncogènes," p. 229. CNRS, Paris.
Vrba, M. (1963). *Acta Virol. (Engl. Ed.)* **7**, 525.
Vredevoe, D. L., and Hays, E. F. (1976). *Cancer Res.* **36**, 370.
Weiss, R. A., Friis, R. R., Katz, E., and Vogt, P. K. (1971). *Virology* **46**, 920.
Wong-Staal, F., Reitz, M. S., Jr., Trainor, C. D., and Gallo, R. C. (1975). *J. Virol.* **16**, 887.
Woods, W. A., Massicot, J., and Chirigos, M. A. (1970). *Proc. Soc. Exp. Biol. Med.* **135**, 772.

4 *Assay Methods for Viral Pseudotypes*

J. Závada

I. Introduction

A. SCOPE

Phenotypic mixing is the mixed assembly of structural components of two distinguishable viruses into one or both species of particle. It is known to occur frequently in dual infections of cells with two related and sometimes even with unrelated viruses. Phenotypic mixing was first described for the mixed assembly of tail fiber proteins following dual infection of bacterial cells with two closely related phages, T2 and T4, by Novick and Szilard (1951). Later it was observed in many pairwise infections of cells with various animal viruses (reviewed by Závada, 1976), and it is known also in various

109

phages and in plant viruses. An excellent review recently appeared covering phenotypic mixing in plant, bacterial, and animal viruses (Dodds and Hamilton, 1976).

For years phenotypic mixing was valued merely as an interesting laboratory artifact, until it was found to be not only common, but possibly of crucial importance in the transmission of several RNA tumor viruses. Many highly oncogenic members of this group of viruses are defective, lacking information for surface antigens, and often also for reverse transcriptase and for other structural proteins. These defective viruses can exist as infectious particles only as *pseudotypes*—particles with surface antigens, defective genomes, and, in many cases, other proteins, provided by a helper virus. The helper virus represents another, usually less oncogenic or even nononcogenic, member of this viral group. Since it was first described by Hanafusa *et al.* (1963), the problem of defectiveness and pseudotype formation between various RNA tumor viruses has received much attention and there are several reviews on this topic (Hložánek 1975; Sarma and Gazdar, 1974; Tooze, 1973; Weiss, 1976). These pseudotypes of defective RNA viruses with envelope antigens of related competent viruses of the same group will therefore be mentioned here only briefly.

During recent years it became apparent that phenotypic mixing, including the extreme form of genomic masking or formation of pseudotype particles, takes place between viruses which belong to different families of enveloped viruses. Even a combination of viruses as distinct as a rhabdovirus and a herpes virus, whose only common feature is the fact that both are enveloped, will form pseudotypes on mixed infection.

This chapter will concentrate on pseudotype formation between various families of enveloped animal viruses because we believe that it might help, together with other methods, to solve certain important problems. For example, it may: (1) contribute to the elucidation of mechanisms of host restriction in some viruses; (2) provide simple, precise, and rapid plaque assays and neutralization tests for viruses that replicate slowly and are not cytopathogenic; (3) lead to the detection and analysis of surface viral antigens of viruses that are defective or only partially expressed; and (4) be useful for the introduction of genomes of viruses with restricted host range into cells of otherwise resistant species.

It should be borne in mind, however, that when phenotypic mixing changes the biological properties of virus particles and may confer an extension of host range, such mixed viruses may constitute a medical or veterinary hazard. Furthermore, mixed infections might in certain cases result not only in phenotypic modification, but also in genotypic modification or recombination, giving rise to or promoting the selection of virus populations with permanently altered properties. In view of these possibilities, it is important to consider potential hazards to health before embarking upon experiments involving mixed infections, and to take suitable precautions to contain any new viral combinations that may arise.

B. Phenotypic Mixtures and Pseudotypes

Phenotypically mixed viral particles usually arise in cells mixedly infected with two viruses, but in some instances they have been produced even in cell-free systems. They contain the genome of one virus (for convenience, designated here as A), and the proteins of the other virus, B, or of both A and B viruses. If the viruses used are complex enough to contain several species of structural antigens, either all (in the case of closely related viruses), or only some of these (in combinations of distantly related viruses) can be exchanged. The easiest way of detecting phenotypic mixing is specific neutralization, or other properties of viral surface antigens. With the exception of reverse transcriptase provided to alpha-Rous sarcoma virus (RSVα) by the helper virus (Hanafusa and Hanafusa, 1971), virtually all papers on phenotypic mixing so far concern only the surface antigens. Therefore the terminology also refers to the surface specificities. A phenotypically mixed virus is designated as A(B), or A(A + B), and reciprocally, B(A) and B(A + B). The first letters denote the genome of such a particle (and often also internal viral proteins), and the bracketed letters denote the specificity of surface antigens (Rubin, 1965). The A(B) and B(A) particles are only neutralizable by anti-B or anti-A sera, respectively, and these will be referred to as pseudotype particles. They represent the extreme case of phenotypic mixing—the genomic masking. The A(A + B) and B(A + B) particles may be doubly neutralizable with both anti-A and anti-B sera and will be referred to as DN particles. At present, however, we do not know if the particles behaving as A(B) pseudotypes contain no A surface antigen at all, or contain antigen only in such small amounts or distribution that do not render the particles neutralizable with anti-A serum.

From what is known so far (see Section II, C and D) on phenotypic mixing between members of various families of enveloped viruses, we can propose a provisional simplified model (Fig. 1), showing that only the surface viral glycoproteins, and not internal components, participate in phenotypic mixing.

Pseudotype formation has been tested in several pairwise combinations of unrelated enveloped viruses, and successful reports are listed in Table I. At first sight, it is striking that in all of these instances one partner has always been vesicular stomatitis virus (VSV). We do not know why VSV holds such a privileged position. In combinations of other families of enveloped viruses DN particles have been detected at a high rate, but no pure pseudotypes

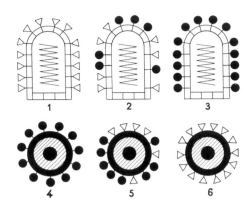

FIG. 1. Scheme of phenotypic mixing between VSV and ATV. (1–3): Particles with genome and internal proteins corresponding to VSV. (4–6): Particles corresponding to ATV. (1, 4): Parent-type particles; (2, 5): doubly neutralizable particles; (3): pseudotype particles VSV(ATV); and (6): pseudotype particles ATV(VSV).

TABLE I

PSEUDOTYPES OBTAINED WITH COMBINATIONS OF VIRUSES BELONGING TO DIFFERENT FAMILIES

| Genome | Donor of surface antigens | | Reference |
	Family	Virus	
VSV-Indiana	Paramyxoviridae	SV5	Choppin and Compans (1970), McSharry et al. (1971)
		HVJ/Sendai	Kimura (1973)
		Measles virus/SSPE	Wild et al. (1976)
VSV-Indiana	Orthomyxoviridae	Fowl plague virus	Závada and Rosenbergová (1972)
VSV-Indiana	Retroviridae	Avian tumor virus	Závada (1972a,b), Závada and Závodská (1973–1974), Love and Weiss (1974), Weiss et al. (1974), Boettiger et al. (1975)
		Mouse leukemia virus	Závada (1972a), Huang et al. (1973), Krontiris et al. (1973), Besmer and Baltimore (1975)
VSV-Indiana	Herpetoviridae	Herpes simplex virus type 1	Huang et al. (1974)
VSV-Indiana	Togaviridae	Sindbis virus	Z. Závadová and J. Závada (unpublished)
VSV-Indiana	Unknown	Unknown—permanent cell lines from human tumors	Závada et al. (1972, 1974a,b)
RSV	Rhabdoviridae	VSV-Indiana	Weiss et al. (1974, 1976)

could be found (Závada and Rosenbergová, 1968—fowl plague virus + avain myeloblastosis; Dyadkova and Kuznetsov, 1970—RSV + parotitis virus). Moreover, on purely statistical grounds the appearance of a pure pseudotype in mixed infection with two competent viruses is extremely improbable, but the occurrence of DN particles has a very high probability (Závada, 1976). But recent electron microscopic observations suggest that the assembly of VSV cores with oncornavirus spikes is not a random process, and that groups of spikes are acquired in patches (Ogura and Bauer, 1976).

II. Methods

A. Selective Assay for Viral Genomes

The first presumption for analyzing all classes of viral particles that could result from a mixed infection (see Fig. 1) is a selective assay for all particles with genome A, and, if possible, another parallel assay for all particles with genome B, irrespective of the envelope in which they are enclosed.

There is usually no problem about virus assay for at least one of the two viruses. Vesicular stomatitis virus is a rapid plaque producer in most species of vertebrate cells, while many surface antigen donors do not produce any plaques (e.g., leukosis viruses), or they produce foci of transformed cells that only appear much later (RSV). In some combinations (VSV–SV5, VSV–herpes simplex virus, VSV–measles virus; for references see Table I), VSV produces plaques much more rapidly, or produces plaques morphologically different from those of the other virus, which cannot be mistaken with each other.

Another possibility for the assay of each virus species harvested from mixed infection is to exploit different assay conditions, such as the temperature of incubation. Wild-type VSV plaques at temperatures between 31° and 39°C, and fowl plague virus (FPV) between 34° and 41.5°C; thus a selective assay for VSV plaques was to incubate at 31.5°C, and for FPV plaques at 40.5°C (Závada and Rosenbergová, 1972). The problem of assaying particles with RSV genomes in the presence of excess cytopathogenic VSV was similarly solved, in this instance by employing temperature-sensitive (ts) mutants of VSV (Weiss et al., 1974, 1976). The cultures for the focus assay of particles with RSV genomes were incubated at 40.5°C so that the excess VSV in the mixed stocks would not replicate; then pseudotype RSV particles bearing the envelope specificity of VSV were selectively revealed by focus assays on cells resistant to infection with the parental RSV strain used, but susceptible to infection with VSV. Thus conditional assay temperatures and restrictive host range assays are useful adjuncts to the use of selective neutralizing antisera (discussed below) for the identification of specific pseudotypes.

B. NEUTRALIZATION BY SPECIFIC ANTIBODIES

Once a selective assay of virus A is available, the three types of envelopes (A, A + B, B) enfolding A genomes can most simply be determined by using

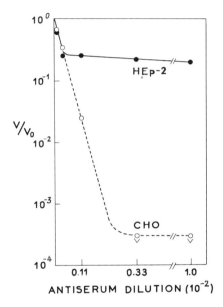

FIG. 2. Neutralization by anti-VSV serum of VSV(HSV) pseudotypes formed after mixed infection in the presence of cytosine arabinoside. HEp-2 cells were infected with herpes simplex virus at a multiplicity of 2 in the presence of 20 μg/ml of cytosine arabinoside, and incubated at 37°C for 4.5 hours. Then the cells were superinfected with VSV at a multiplicity of 0.01. After each attachment period the cells were washed 2 times with 10 ml of saline. The cells were covered with MEM containing 2% FCS, 5 μg/ml gentamicin, and 20 μg/ml cytosine arabinoside. After a total incubation time of 30 hours at 37°C, the extracellular fluid was decanted, cells and cell debris precipitated, and the supernatant frozen at −100°C. For neutralization mixtures the supernatant containing VSV and VSV(HSV) progeny was thawed and mixed with equal volumes of rabbit anti-VSV antiserum at the indicated dilutions. Neutralization was for 1 hour at 22°C. Then the samples were diluted and plaque assayed on either HEp-2 or CHO cells. Virus in duplicate aliquots of 0.1 or 0.2 ml of the appropriate dilutions were attached to monolayer of HEp-2 or Chinese hamster ovary (CHO) cells on 60 mm plastic petri plates for 40 minutes at 37°C. The cells were then rinsed with medium and overlaid with MEM containing 2% FCS, 5 μg/ml of gentamicin, and 0.9% agar. After incubation at 37°C for 48 hours in a 5% CO_2 humidified incubator, the cells were stained with 0.01% neutral red and the plaques counted after another 18 hours of incubation. The log of the percentage of nonneutralized virus V/V_0 is plotted versus the antibody concentration. [From Huang *et al.* (1974), with permission.]

specific anti-A and anti-B sera. The A(B) pseudotypes are those particles that remain infectious after neutralization with anti-A serum, and that are neutralized by adding anti-B scrum. In several papers the demonstration of pseudotypes used just this simple procedure, using one concentration of each serum, sufficient to neutralize all corresponding nonpseudotype virus (Choppin and Compans, 1970; Závada and Rosenbergová, 1972; Závada, 1972a).

As a refinement of this method, several dilutions of antiserum may be used to reveal the dependence of virus survival upon the concentration of serum, as shown in Fig. 2 (Huang *et al.*, 1973; 1974). This reduces the possibility of mistaken or insufficient concentration of serum, and at the same time shows the reproducibility of the result.

When using highly potent antiserum directed against a virus titrated by the plaque assay, we have often found it necessary to remove the inoculum after adsorption to the assay plates and to rinse the cell sheets. Excess of anti-VSV serum remaining in the assay plate would not neutralize the pseudotype, but would inhibit the development of VSV plaques after the progeny of the pseudotype particles reappear in the original VSV envelope.

Two more points should be mentioned here: (1) Some viruses always contain an antiserum-resistant fraction that is not pseudotype. Certain viruses (e.g., togaviruses) readily produce infectious virus-antibody complexes, which can be subsequently neutralized with serum directed against the IgG of the species immunized with the virus. Other viruses contain serum-resistant aggregates, which can be removed by filtration through Millipore filters. However, there has never been any difficulty with VSV, for all pure-grown virus was neutralized without any residual fraction. (2) It is advisable to carry out the neutralization in the absence of complement (i.e., sera heated at 56 °C for 30 minutes). Complement may enable nonspecific neutralization of enveloped viruses by antibodies reacting with normal cellular components (Aupoix and Vigier, 1975). There are also various reports from early studies in virological serology showing a widespread occurrence of heat-labile, nonspecific viral inhibitors in various sera.

C. GENETIC METHODS

1. *Complementation*

Complementation has not commonly been used as a way of constructing pseudotypes, but has been used as a method to determine which viral components participate in phenotypic mixing of VSV with an RNA tumor virus, avian myeloblastosis virus (AMV). Vesicular stomatitis virus provides an excellent model for such experiments because its genetics are well understood and mutants of several groups are available. Most of complementation groups have also been physiologically characterized (Pringle, 1975).

In mixed infection of AMV and VSV unilateral complementation of conditional lethal mutants of VSV by competent, wild-type AMV was examined (Závada and Závodská, 1973–1974). Temperature-sensitive (*ts*) mutants of VSV, representing five groups of complementation, were used in parallel for the infection of normal chick embryo cells and of chick cells preinfected with AMV. It was apparent that only mutants of group V were complemented by AMV at 39 °C, and that representatives of the other four groups were not complemented. Since mutants of group V affect viral glycoprotein, it was concluded that this is the only structural component which could be exchanged between particles. The complemented VSV particles possessed properties characteristic of the pseudotype, i.e., they exhibited the specific host range as well as neutralization and interference specificities of AMV (Závada, 1972a; Závada and Závodská, 1973–1974). A partial result of such complementation is presented in Fig. 3. Vesicular stomatitis virus

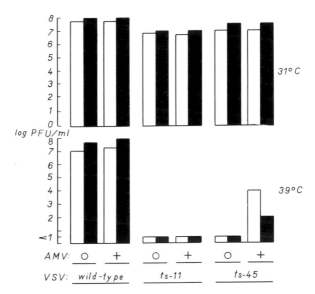

FIG. 3. Complementation of VSV *ts* mutants by AMV. Confluent secondary cultures of C/O chick embryo cells in 50 ml Müller flasks, uninfected or infected as primaries with AMV, were infected with VSV wild-type, *ts*(*G*)11 (group I), or with *ts*(O) 45 mutant (group V) at m.o.i. 2–5, and incubated for 14 hours at 31 °C or at 39 °C, immersed in a water bath. Virus yield was plaque assayed at 31 °C in C/O chick cells and in mouse L cells. Open bars: plaque assay in CEC. Solid bars: plaque assay in L cells.

mutants of the recently described group VI (Rettenmier *et al.*, 1975) are currently under test.

Complementation of *ts* mutants of VSV by other enveloped viruses proved to be of little use as a method of constructing the pseudotypes. The main obstacle was that the available VSV mutants which affect viral glyco-protein synthesis or assembly are often quite leaky and they revert at a relatively high rate. In addition, the mutants of group V show multiplicity-dependent leakiness, due to reutilization of the glycoprotein of infecting particles (Deutsch, 1975).

The glycoprotein of VSV is a relatively large molecule (65,000 daltons) and almost certainly it is a multifunctional protein with different active sites involved in maturation, adsorption, neutralization, penetration, uncoating, etc. So far, mutants of this group appear to be extremely rare compared with mutants of group I (RNA polymerase), and in the mutants available it is not known which functions of the glycoprotein are affected. The appli-cation of complementation for constructing the pseudotypes will in future depend on selecting specific mutants that would fail either to synthesize or to assemble viral spikes and thus would depend entirely on the foreign glycoprotein available.

Complementation of virus mutants that are nonconditionally defective in envelope glycoproteins can yield useful pseudotypes in situations where mixed infections of nondefective viruses might result in doubly neutralizable particles only. Thus VSV will complement glycoprotein-defective RSV to produce infectious RSV(VSV) particles (Weiss *et al.*, 1976) analogous to the rescue of RSV pseudotypes by avian leukemia "helper" viruses. Simi-larly, a MSV(VSV) pseudotype could be obtained (Livingston *et al.*, 1976). Xenotropic murine leukemia virus (X-MuLV) will also rescue defective RSV as infectious RSV (X-MuLV) pseudotypes (Levy and Vogt, 1976; Weiss and Wong, 1976), although the avian and mammalian C-type viruses are not genetically or antigenically related. When mixed infections are difficult to accomplish because of mutually exclusive host ranges (e.g., the helper virus cannot penetrate or replicate in the host cell harboring the defective virus), complemented pseudotypes may be produced by fusion of the two cell types infected with the different virus species (Weiss and Wong, 1976). A *ts* mutant of MLV defective in an envelope function was complemented by VSV (Breitman and Prevec, 1976).

2. *Phenotypic Stabilization*

Phenotypic stabilization has features both common to and different from complementation. In complementation of *ts* mutants, the interaction of the two viruses under test is examined during the replication cycle inside the cells at the restrictive temperature (39°–40°C). In contrast to this, phenotypic stabilization employs differences of thermal stability between the protein components of the viruses used, which are detected in the mature virions resulting from mixed infection at the permissive temperature. In this way phenotypic mixing between various togaviruses was demonstrated (Burge and Pfefferkorn, 1966).

The VSV mutant $ts(O)45$, belonging to group V, which was complemented by AMV in cells incubated at 39°C, is also thermolabile (tl). It is rapidly inactivated by heating at 45°C, whereas this temperature does not inactivate wild-type and most other mutants of VSV. In addition to complementation by AMV this mutant also showed phenotypic stabilization by AMV or RSV (Závada, 1972b; Závada and Závodská, 1973–1974; Weiss et al., 1976). The VSV mutant $ts(O)45$, reproduced at 31°C in normal chick cells and subsequently heated at 45°C, loses 10^4 or more units of infectivity for avian or for mammalian cells. The same mutant, reproduced in parallel cultures of chicken cells preinfected with AMV or RSV, showed different thermal inactivation curves when tested in chicken or in mammalian cells. For mammalian cells, thermal inactivation of this virus was the same as with control virus, but in chicken cells only 10^2 units of infectivity were inactivated. The residual heat-stable infectivity of VSV $ts(O)45$, grown at 31°C in AMV-infected cells, heated at 45°C and plaque assayed in normal chicken cells, was neutralizable with anti-AMV serum; this shows that the heat-stable fraction was predominantly the VSV(AMV) pseudotype. The "tails" of thermostable fractions (survival about 10^{-4}) of the $ts(O)45$ mutant grown in normal chick cells and heated at 45°C (assayed either in chick cells or in BHK cells), or of the same mutant, reproduced in AMV-infected cells, heated and assayed in BHK cells, represent the revertants of this ts mutant toward the wild type, as shown by retesting of the virus picked from plaques of heated virus. On the other hand, reisolation of virus from plaques, produced in chick cells by $ts(O)45$ reproduced in mixed infection with AMV and heated at 45°C for 60 minutes, always yielded heat-labile viral progeny with VSV neutralization specificity. This showed that acquisition of thermal stability was phenotypic and not due to recombination of VSV with AMV.

This way of producing VSV pseudotypes was further improved by selecting a new mutant of VSV, termed the tl 17 mutant (Závada, 1972b). To reduce the rate of reversion toward wild type, several thermolabile mutations all affecting the glycoprotein were accumulated. This was achieved in the following way: VSV was grown for several passages at 31°C in such a concentration of neutralizing anti-VSV serum that its infectivity was considerably reduced but not eliminated. The idea was that under such a selection pressure mutants of VSV with an altered neutralization specificity would prevail, and that some of these mutants might also possess increased heat lability. From this selected virus population 120 plaques were picked, and virus from each plaque was tested in parallel plates for infectivity when unheated, and when heated at 45 °C for 60 minutes. The most heat-labile clone was again grown with anti-VSV serum, and the isolation and testing of plaques for heat lability was repeated, this time at 42 °C, at which temperature the first-step mutant was not inactivated. Again a plaque isolate was selected, showing further increased heat lability, and the procedure was repeated once again, this time by heating at 39 °C. Consequently, the resulting clone should carry at least three thermolabile lesions affecting the viral glycoprotein known to be the neutralization antigen, since the selection pressure (anti-VSV serum) was always directed toward this. The resulting mutant, *tl* 17, can with advantage be used for the preparation of the VSV pseudotypes by phenotypic stabilization. Its major advantage is that it has a very low reversion rate (less than 10^{-6}). An experiment demonstrating phenotypic stabilization by AMV is presented in Fig. 4.

This way of preparing the VSV pseudotypes using the *tl* 17 mutant has also been employed for other avian tumor viruses (Love and Weiss, 1974; Weiss *et al.*, 1974; Boettiger *et al.*, 1975) and with measles virus (Wild *et al.*, 1976). The *ts*(O)45 mutant of VSV has also proved useful in obtaining

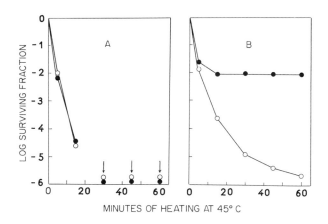

MINUTES OF HEATING AT 45° C

FIG. 4. Thermal inactivation of the *tl* 17 mutant of VSV grown in control and in AMV-preinfected chick cells. Secondary cultures of control (A) and AMV-infected chick cells (B), prepared as in Fig. 3, were infected with *tl* 17 mutant and incubated for 24 hours at 31 °C. Infectious tissue culture fluid (Eagle's medium with 10% calf serum) was diluted 1:10 with saline, and buffered with M/150 phosphate buffer, pH 7.2. One milliliter of each dilution was heated for intervals (abscissa) at 45.0° ± 0.1 °C. Surviving virus infectivity was determined by plaque assay in CEC (●) and in BHK-21 cells (○). [From Závada (1972b).]

VSV pseudotypes with murine leukemia virus (MuLV) by heating (Kron-tiris *et al.*, 1973). In all these instances it was shown that the heat-stabilized VSV *tl* mutants produced in mixed infection with other enveloped viruses also carried other surface properties corresponding to the pseudotype, the neutralization specificity, and, with the tumor viruses, the host-range and interference specificities.

The method of making VSV pseudotypes by employing thermal lability of its mutated glycoprotein has some technical advantages over the first method (use of anti-VSV serum). In addition, the *tl* 17 mutant may be more efficient in acquiring foreign neutralization antigens than wild-type VSV, for this mutant possibly synthesizes or assembles its glycoprotein less efficiently than the wild-type even at the low permissive temperature (Weiss *et al.*, 1974; J. Závada, unpublished).

There is one more point that should be stressed. Thermal inactivation of the *tl* 17 mutant is strongly dependent on pH, and to some extent also on the proteins (serum) in heated suspension. At a pH behow 7.0 inactivation is very slow and incomplete. To achieve reproducible results, infectious tissue culture fluids were either diluted 1:10 with phosphate-buffered saline at pH 7.2 (Závada, 1972b), or the pH was adjusted to 8.0 (Boettiger *et al.*, 1975) before thermal inactivation.

We do not know exactly what happens to the glycoprotein of *tl* 17 mutant on heating. One possibility was that the spikes became detached from vi-rions, resulting in "bald" particles. But this was not proved to be the case. Purified *tl* 17 was divided in two parts: one was heated for 60 minutes at 45 °C, and the other was not heated. Wild-type VSV was treated in the same way. Subsequently, all four virus samples were pelleted through a cushion of sucrose solution. Virus pellets were analyzed by acrylamide electro-phoresis, and all appeared to contain in approximately the same proportion all major viral polypeptides, including the glycoprotein (G. Russ, personal communication).

A limitation of the heat inactivation method is that the glycoprotein of the other virus must resist heating at 45 °C. A heat-labile antigen would of course escape detection. Therefore it is advisable to try both ways of pre-paring pseudotypes, surviving fractions following neutralization and follow-ing heat inactivation.

3. *Other Aspects*

Interference may occur between two viruses infecting a cell simultaneously or in sequence. This holds true mainly for viruses that either induce inter-feron, or that shut off macromolecular synthesis in the infected cell, often simultaneously shutting off the synthesis of components of other viruses present.

In experiments of phenotypic mixing of VSV with SV5, McSharry *et al.* (1971) first infected cultures with SV5, and after an interval of incubation superinfected these with VSV. In all experiments on mixed infections of VSV and RNA tumor viruses, the cultures were always infected first with the tumor viruses, which grow relatively slowly and do not interfere with VSV, and only after these had multiplied enough (usually after several days of incubation), was VSV added.

It has also been pointed out (Huang *et al.*, 1974) that no VSV (herpes virus) pseudotypes could be detected in cultures infected with more than one PFU per cell of VSV, while at lower multiplicities the proportion of pseudotype was very high. In experiments designed to detect a reciprocal pseudotype, RSV(VSV), it was shown that, using a mutant of VSV (*ts* 1026) with defective shutoff host cell synthesis, the pseudotypes were produced more efficiently than with other VSV mutants (Weiss *et al.*, 1976). Again, infection with a high multiplicity of VSV (>10 PFU/cell) reduced the titers both of RSV and of the pseudotype, RSV(VSV).

D. BIOCHEMICAL ANALYSIS OF PSEUDOTYPES

The most straightforward way of analyzing phenotypically mixed particles was employed by McSharry *et al.* (1971) in viral progeny resulting from mixed infection of cells with VSV and the paramyxovirus, SV5. The variety of virions in this population was separated in two steps. First, two distinct bands separated during isopycnic centrifugation in a density gradient of tartarate. The band sedimenting at density 1.16 gm/ml contained only particles with VSV genomes, while the other (density 1.23) contained predominantly particles with SV5 genomes. The second stage of separation, following dialysis of the fraction banding at 1.16 gm/ml, employed adsorption and elution of particles with paramyxovirus surface antigens to chicken erythrocytes. Thus it was possible to separate rhabdovirus particles with the VSV genome and surface antigens of SV5, though this technique does not distinguish pure pseudotypes from particles composed of antigenic mosaics. Phenotypically mixed VSV labeled with radioactive amino acids or glucosamine was analyzed by acrylamide gel electrophoresis. As shown in Figs. 5 and 6, phenotypically mixed VSV particles acquire surface glycoproteins from SV5, but no other structural polypeptides.

Antimetabolites provide a method for obtaining phenotypically mixed particles, with only one genomic component in mixed infections of RNA and DNA viruses. Cytosine arabinoside (20 μg/ml) has no effect on the reproduction of VSV, but it prevents not only replication of herpes simplex virus DNA, but also the synthesis of most of its polypeptides, except for the glycoprotein. In cells mixedly infected with VSV and with herpes simplex

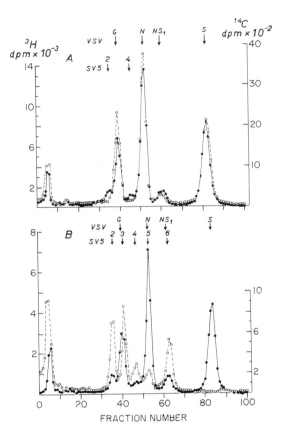

FIG. 5. Polypeptides of phenotypically mixed VSV-SV5 virions. (A) Coelectrophoresis on a polyacrylamide gel of ³H-amino acid-labeled VSV-SV5 hybrid virions (●—●) with ¹⁴C-amino acid-labeled VSV (O-- O) as a marker. (B) Coelectrophoresis of ³H-amino acid-labeled VSV-SV5 hybrid virions (●—●) with ¹⁴C-amino acid-labeled SV5 (O-- O) as a marker. [From McSharry *et al.* (1971), with permission.]

virus and incubated in medium containing the drug (Huang *et al.*, 1974), the VSV (herpes) pseudotype was produced efficiently, as seen in Fig. 2. This shows that for the production of this pseudotype only the glycoprotein of herpes virus is required.

E. Pseudotypes Constructed *In Vitro*

There are several ways of constructing viral pseudotypes or at least DN particles *in vitro*. The first, but nevertheless, the most clear-cut example was demonstrated with tobacco mosaic virus by Fraenkel-Conrat (1956). RNA

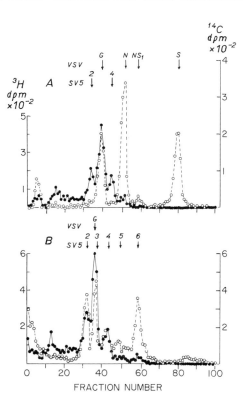

FIG. 6. Glycoproteins of phenotypically mixed VSV–SV5 virons. (A) Coelectrophoresis on a polyacrylamide gel of ^{3}H-glucosamine-labeled VSV–SV5 hybrid virions (●—●) with ^{14}C-amino acid-labeled VSV (○--○) as a marker. (B) Coelectrophoresis of ^{3}H-glucosamine-labeled VSV–SV5 hybrid virions (●—●) with ^{14}C-amino acid-labeled SV5 (○--○) as a marker. [From McSharry et al. (1971), with permission.]

isolated from various viral strains was reconstituted *in vitro* into complete virions with proteins prepared from genetically different strains of this virus. The progeny of reconstituted viruses corresponded entirely to the RNA used. This experiment was of great importance for providing empirical proof that hereditary traits of viruses are carried exclusively by nucleic acids, and that the protein moiety of virions is not genetically important.

Curious pseudotypes were constructed in a similar way *in vitro* by reconstitution of virions with RNA of animal viruses and proteins of a plant virus (Black et al., 1973). The RNAs extracted from animal picornaviruses (polio and Mengo viruses) were assembled *in vitro* with protein subunits isolated from cowpea chlorotic mottle virus. These artificial pseudotypes had a low infectivity, but their RNA was partially protected against the action of ribonuclease.

Such a complete reconstitution is at present not possible with more complicated animal viruses. Nevertheless, with VSV it was possible to produce an *in vitro* pseudotype (Bishop *et al.*, 1975). The VSV virions were denuded of surface spikes by treatment with proteases and thus rendered noninfectious. If these were mixed with glycoprotein isolated from the same (Indiana) or a different (New Jersey) serotype of VSV, infectivity was partially recovered. Such particles exerted the specificity of neutralization of the glycoprotein added. Where they were of different serological type than the internal components of the virus, such reconstructed particles can be considered as pseudotypes. Their exact shape and structure is not known but they show a very low buoyant density.

Another way of *in vitro* phenotypic mixing was demonstrated with mutant phages. An *amber* mutant of T7 phage, affecting gene 17, produces in restrictive hosts only noninfectious virions. These are rendered infectious by adding a cell-free extract from restrictive cells infected with T3 phage, with an amber mutation affecting some other gene. Such activated particles showed serological specificity of "coat chimeras" (Issinger *et al.*, 1973).

There are many instances of complexes being produced *in vitro* by interactions between various viruses (O'Connor and Rauscher, 1964; Compans, 1974). These complexes have not been studied from the viewpoint of their neutralization properties, but in the case of mixed, cosedimented murine and feline oncornaviruses, exhibited an extended host range (Fischinger and O'Connor, 1969). In some instances the complex possibly might behave as doubly neutralizable particles. DN particles indeed appeared when VSV was mixed *in vitro* with FPV or even with ether-disrupted FPV, although in this way no complete pseudotypes were evident (Závada and Rosenbergová, 1972).

In a sense, the observation that treatment of resistant cells with inactivated Sendai virus renders them susceptible to infection with polio virus (Neff and Enders, 1968) or RSV (Weiss, 1969a) may be regarded as a form of *in vitro* interaction. This technique also enables glycoprotein-defective particles of RSV or VSV to penetrate and therefore function in the host cell (Weiss, 1976). However, care should be taken to ensure that the Sendai virus stocks (propagated in chick eggs) are quite free of contaminating leukosis viruses in order to avoid unwanted phenotypic mixing with the leukosis virus (Weiss, 1969a). The phenomenon of Sendai virus-mediated infection is believed to be akin to the cell fusion promoted by the same agent.

III. Applications

A. Study of Host Restriction and of Specific Interference

The study of host restriction and of specific interference has proved to be the most fruitful field of application of VSV pseudotypes. The advantages are obvious: VSV is a virus of extremely broad host range; it grows and produces plaques in all species of vertebrate cells that have been tested; it can even grow in insect cells. Many viruses, on the other hand, and oncogenic RNA viruses in particular, show characteristic patterns of host restriction which often is not only species specific, but in some instances there even exist resistant and susceptible strains of one species. In these, restriction might occur either at the level of initial steps of cell infection (adsorption, penetration, and uncoating), or in later steps (regulation of expression of viral genome). In parallel to host restriction, the VSV pseudotypes are also suitable for the investigation of specific interference. In several instances VSV pseudotypes, with surface antigens provided by RNA tumor viruses have been used to determine whether early or late steps (or both) play a role in host restriction or viral interference.

1. Avian Tumor Viruses

The group of avian tumor viruses (ATV) contains viruses with various properties. Some can transform fibroblasts *in vitro* (mainly sarcoma viruses), and for these the focus assay of infectivity (Temin and Rubin, 1958), based on the appearance of transformed "foci" of cells, is available. Other viruses of this group, viruses that cause leukosis or are nonpathogenic viruses, give rise to inapparent infections in susceptible cultures of fowl fibroblasts— they grow without inducing a cytopathic effect or cell transformation. The sarcoma viruses are either competent or defective. Competent sarcoma viruses contain complete genetic information and they can multiply alone, like the Schmidt–Ruppin strain of RSV or the B77 virus. The genome of defective sarcoma viruses (Bryan high titer strain) appear to lack the gene for the surface structural antigen, gp85 (Scheele and Hanafusa, 1971; De Giuli et al., 1975; Ogura and Friis, 1975). Some RSV strains are also defective in reverse transcriptase (Hanafusa and Hanafusa, 1971; H. Hanafusa et al., 1972). These genes were apparently lost in exchange for determinants of high oncogenicity. Some of the defective strains produce virions which, however, are not infectious, but in many other respects resemble competent virions of this group.

These defective sarcoma viruses depend on other competent viruses belonging to the same group to provide them with the missing functions or structural components (Temin, 1962). Particles of defective RSV, produced in the presence of helper viruses, have surface antigenic specificities (or even the reverse transcriptase) of the corresponding helper virus. The term "pseudotypes" was first used for these particles (Rubin, 1965; Hanafusa *et al.*, 1964; Hanafusa, 1965).

The group of avian tumor viruses so far comprises seven subgroups. Subgroups A–E were isolated from chicken (Vogt and Ishizaki, 1965; Duff and Vogt, 1969; Weiss, 1969a,b); subgroups F and G are from pheasants (Fujita *et al.*, 1974). These subgroups share common internal antigens [group-specific (gs) antigens], but they differ from each other by the surface specificities of the virion, by neutralization, and host-range and interference patterns.

a. Viruses belonging to one subgroup show a strong cross-neutralization, but there is none, or only a low cross-neutralization, between viruses of different subgroups (Ishizaki and Vogt, 1966).

b. Subgroup-specific host range is manifested at two levels: (i) intraspecies specificity, determined in chicken and (ii) interspecies specificity, determined in mammalian cells and in cells of other avian species.

Chicken strains showing a specific resistance to individual subgroups are available: cells of C/O chicken are susceptible to A–E subgroups, chicken cells C/A are resistant to viruses belonging to the A subgroup, but not to the B subgroup, and the opposite holds true for C/B chicken cells (Vogt and Ishizaki, 1965). Genetic analysis of the host revealed a simple Mendelian dominance of susceptibility, governed by resistance and susceptibility alleles termed *tv-a*, *tv-b*, and *tv-c*, at different genetic loci (Payne and Biggs, 1966; Crittenden *et al.*, 1967). Viruses belonging to the E subgroup are endogenous chicken C-type viruses (Weiss, 1975b). They fail to infect cells of most chicken strains (which thus have the C/E phenotype), but their permissive hosts are Japanese quail, turkey, and various species of pheasant (Weiss, 1969a). Only sarcoma viruses belonging to C and D subgroups can transform mammalian as well as avian cells (Hanafusa and Hanafusa, 1966; Duff and Vogt, 1969). In most instances, however, this infection does not result in the production of infectious virions, although complete viral genomes are present and can be rescued by cocultivation or fusion with permissive chicken cells (Svoboda, 1960; Svoboda and Dourmashkin, 1969), and infectious proviral DNA has been isolated from such cultures (Hill and Hillová, 1972).

Specific resistance to individual subgroups in chicken cells appears to be due to a failure of these viruses to penetrate or uncoat in resistant cells; the rate of adsorption to susceptible and resistant cells is very similar (Piraino, 1967; Crittenden, 1968; Weiss, 1976).

c. The third subgroup-specific feature of avian tumor viruses is specific

interference; preinfection of susceptible chick fibroblasts with a nontransforming ATV renders them resistant to transformation by competent sarcoma viruses or by pseudotypes of defective sarcoma viruses belonging
to the same subgroup, but not to viruses of other subgroups of ATV or to
unrelated cytopathogenic viruses (Rubin, 1961; Vogt and Ishizaki, 1966).

Thus these three specificities, defining the subgroups of ATV, appear to
be determined by surface viral antigens and involve the process of penetration and/or uncoating. It is not yet clear if the resistance of mammalian
cells to A and B subgroups is due to lack of penetration or of adsorption.
Nevertheless, in addition to these very early steps of infection there are
also some instances of postpenetration blocks, which influence the genome
expression of these viruses in nonpermissive cells. This clearly applies at
least in two instances: for C and D subgroup viruses which infect and transform mammalian cells but do not produce complete virions, and for E subgroup viruses, which are as DNA proviruses integrated into cell genome of
perhaps all chickens, but are usually only partially expressed.

A more detailed determination of how far these two restrictions contribute to ATV resistance of cells was provided by the use of VSV(ATV)pseudo-
types. No host restriction applies to the replication of VSV (and if it were, it
could easily be determined). Thus any restriction for VSV(ATV) pseudotypes may occur at the level of adsorption or penetration. An additional
advantage is that the VSV(ATV) pseudotypes can be easily assayed in cells
in which transformation by RSV cannot be determined, e.g., cells already
transformed (HeLa), or even in cells transformed by RSV.

Coincidence of the three envelope specificities was demonstrated in the
first papers on VSV(AMV) pseudotypes (Závada, 1972a,b). In addition to
its neutralization specificity, the pseudotype also showed the host-range
and interference specificity corresponding to AMV (which is a mixture of
subgroups A and B). While control VSV with its own surface antigen showed
a very similar plating efficiency when tested in chick, human, mouse, or
hamster cells, the pseudotype was infectious only for chicken cells. Preinfection of chicken cells with AMV did not decrease plating efficiency
of normal VSV, but it rendered them resistant to the pseudotype VSV(AMV).
More detailed studies (Love and Weiss, 1974; Boettiger et al., 1975) examined the host range of VSV(ATV) pseudotypes not only for species specificity, but also employed cells of chick strains specifically resistant to various
groups of chick leukosis viruses. For example, pseudotypes of VSV with
antigens of RAV-1 (subgroup A) failed to plaque in chicken C/A cells,
but were infectious for other chick cells; correspondingly, C/E chick cells
were selectively resistant to VSV pseudotypes with surface antigens provided
by members of the E subgroup; these pseudotypes were infectious for golden
pheasant cells. Table II, adapted from Boettiger et al. (1975), summarizes

TABLE II

SPECIFICITY OF ENVELOPE PSEUDOTYPES OF VSV PREPARED WITH AVIAN LEUKEMIA AND SARCOMA VIRUSES: ASSAY ON RESISTANT CHICKEN CELLS[a]

Subgroup	Pseudotype	Assay cells[b]	t_0[c]	t_{60}[c]	Penetration ratio t_{60}/t_0 ($\log_{10}$)
A	RAV-1	C/E	8×10^6	2×10^4	-2.6
		C/A	7.2×10^6	1.0×10^1	-5.8
	PR-A	C/E	3.4×10^6	1.1×10^4	-2.5
		C/A	2.4×10^6	$<3 \times 10^1$	<-5.9
B	RAV-2	C/E	3.9×10^7	8.5×10^4	-2.7
		C/E-RAV-2	4.8×10^7	4.0×10^1	-6.1
		C/B	4.7×10^7	3.3×10^3	-4.2
	MAV-2	C/E	1.8×10^7	4.0×10^3	-3.7
		C/E-RAV-2	2.5×10^7	3.0×10^0	-6.9
		C/B	3.3×10^7	9.0×10^1	-5.6
C	RAV-49	C/E	6.7×10^7	4.2×10^5	-2.2
		C/E-CZAV	5.0×10^7	5.0×10^4	-3.0
		C/B	1.0×10^8	3.0×10^5	-2.5
		C/E-RAV-49	4.0×10^7	2.1×10^2	-5.3
	B77	C/E	1.2×10^8	7.0×10^5	-2.2
		C/E-RAV-49	5.4×10^7	8.0×10^2	-4.8
		C/E-CZAV	5.0×10^7	5.0×10^3	-4.0
D	CZAV	C/E	6.3×10^7	1.2×10^6	-1.7
		C/A	1.3×10^8	4.3×10^6	-1.5
		C/E-RAV-49	5.3×10^7	7.5×10^5	-1.9
		C/E-CZAV	5.0×10^7	4.5×10^3	-4.1
	SR-H	C/E	1.3×10^7	5.4×10^4	-2.4
		C/E-CZAV	3.5×10^6	3.5×10^2	-4.0
E	RAV-O	Quail	4.6×10^6	1.6×10^5	-1.5
		C/E	4.6×10^6	$<3 \times 10^0$	<-6.2

[a] From Boettiger et al. (1975).

[b] Leukosis virus indicates resistance of host cells as a result of viral interference.

[c] PFU per milliliter; t_0: unheated viral stock; t_{60}: pseudotype stock heated at 45°C for 60 minutes.

the resistance and interference of avian cells analyzed with VSV pseudotypes bearing envelope antigens of subgroups A–E avian tumor viruses.

In addition, the infectivity of VSV pseudotypes with surface antigens provided by representatives of the five ATV subgroups was examined for plating efficiency on mammalian cells (Boettiger et al., 1975). The pseudotypes with envelope antigens provided by members of the A, B, and E subgroups were infectious only for susceptible avian cells, but not for mammalian cell cultures. Various members of the C subgroup behaved differently; the VSV pseudotype with the envelope of the sarcoma virus, B77, could infect both chicken and rat cells, but a similar pseudotype prepared with a transformation-defective derivative of the same virus infected only chick cells. All the VSV pseudotypes with glycoproteins of members belonging to the D subgroup were infectious for both chicken and mammalian cells (Table III). Several species of mammalian cells were examined (rat, mouse, hamster, and human) and all showed very similar efficiencies of plating for C and D subgroup pseudotypes. The efficiency of penetration of VSV pseudotypes into mammalian cells was 100–1000 times higher than the efficiency of transformation of these cells by corresponding sarcoma viruses. Furthermore, avian sarcoma viruses do not replicate in mammalian cells. Thus there are also apparently postpenetration blocks for avian oncornavirus expression in mammalian cells (Boettiger, 1974).

2. Mammalian C-Type Viruses

Murine leukemia and sarcoma viruses are the best known mammalian C-type viruses. The defectiveness of murine sarcoma virus (MSV) and pseudotype formation in the presence of helper MuLV resembles to some extent that seen with avian RNA tumor viruses. Murine leukemia viruses are competent for replication, and in tissue cultures of mouse fibroblasts they grow without causing cytopathic effect or transformation. Most strains of MSV are defective and they produce pseudotype particles, capable of inducing foci of transformed cells. The pseudotype particles show specific surface properties corresponding to the helper (Hartley and Rowe, 1966; Huebner et al., 1966). Most MSV strains are defective in several other functions, such as reverse transcriptase and gs antigens, so that the MuLV helper provides more than the envelope antigens. Some defective clones of MSV produce noninfectious particles in the absence of helper virus (Bassin et al., 1971) while others are unable to assemble any particles (Aaronson and Rowe, 1970).

In addition to MuLV, C-type viruses of other mammalian species not closely related to MuLV can exert helper activity for defective MSV, producing pseudotypes with altered species specificity. This has been shown for feline (Fischinger and O'Connor, 1969), hamster (Kelloff et al., 1971; Bomford, 1971), rat (Aaronson, 1971), as well as primate C-type viruses (Henderson et al., 1974) and a C-type virus of putative human origin (Teich et al., 1975).

Various strains of mice have a specific host restriction for replication of some MuLVs. There are N-tropic mouse leukemia viruses, which replicate with high efficiency in NIH mouse cells (N-type cells), but only at low efficiency in BALB/c (B-type) mouse cells. The opposite holds true for B-tropic mouse leukemia viruses. Permissiveness is governed by the *Fv-1* locus, comprising two known alleles, *Fv-1*[b] and *Fv-1*[n] (Lilly, 1967; Hartley

TABLE III. ASSAY OF AVIAN LEUKOSIS AND SARCOMA VIRUS PSEUDOTYPES OF VSV ON AVIAN AND MAMMALIAN CELLS[a,b]

| Virus determining pseudotype | | Host cell | Chicken cells (PFU/ml) | | NRK cells (PFU/ml) | | Penetration ratio[c] |
Subgroup	Strain		t_0	t_{60}	t_0	t_{60}	NRK/chick (log$_{10}$)
A	RAV-1	GPh	7.3×10^6	5.6×10^4	2.5×10^6	2.0×10^0	-4.0
	RAV-5	GPh	4.6×10^7	5.3×10^5	NA	NA	
B	RAV-2	GPh	1.0×10^8	7.6×10^5	7.2×10^7	2.0×10^1	-4.4
	RAV-2	SP	2.1×10^7	7.0×10^4	2.9×10^6	3.0×10^1	-3.5
	MAV-2	GPh	7.2×10^7	7.7×10^5	3.1×10^7	3.0×10^0	-5.1
	MAV-2	SP	3.2×10^6	2.0×10^3	5.2×10^5	$<3 \times 10^0$	<-2.0
C	RAV-49	GPh	2.0×10^8	2.8×10^6	8.1×10^7	$<3 \times 10^0$	<-5.7
	RAV-49	SP	7.7×10^7	4.8×10^4	3.5×10^7	$<2 \times 10^2$	<-2.0
	NT-B77	GPh	6.2×10^7	2.6×10^6	2.8×10^7	3.0×10^1	-4.4
	NT-B77	SP	8.0×10^7	3.2×10^5	2.1×10^7	$<3 \times 10^0$	<-4.5
	RAV-7	SP	1.3×10^8	9.0×10^5	2.8×10^7	$<3 \times 10^0$	<-4.8
D	CZAV	GPh	1.2×10^8	4.2×10^6	6.4×10^6	3.6×10^3	-1.8
	CZAV	SP	9.0×10^7	3.4×10^5	1.1×10^7	2.9×10^3	-1.2
	RAV-50	GPh	4.9×10^7	7.4×10^3	6.5×10^6	6.0×10^1	-1.2
E	RAV-0	GPh	4.6×10^7–Q	1.6×10^6–Q	3.5×10^6	$<3 \times 10^0$	<-4.6
A-Sa	PR-A	SP	3.4×10^6	1.1×10^4	2.3×10^6	$<3 \times 10^0$	<-3.4
C-Sa	B77	RC	1.5×10^8	1.4×10^6	9.0×10^6	7.0×10^4	-0.1
	RBA	SP	1.0×10^8	6.8×10^4	9.0×10^6	7.0×10^4	-1.5
D-SA	SR-H	SP	6.7×10^6	1.1×10^4	4.8×10^6	2.0×10^3	-1.6
Control		GPh	8.0×10^7	2.3×10^3	8.3×10^6	$<3 \times 10^0$	<-6.9
		SP	2.6×10^7	9.0×10^2	4.2×10^6	$<3 \times 10^0$	<-5.1

[a] Sa, sarcoma viruses; GPh, golden pheasant; SP, SPAFAS chf chicken; RC, Reaseheath C-line chicken; Q, assay on quail cells; PR, Prague strain of RSV; B77, chicken passage of B77 virus derived from virus produced; NT-B77, nontransforming derivative of B77 virus; RBA, B77 virus rescued from rat cells; SR-H, Schmidt–Rupping strain virus from hamster tumor.

[b] From Boettiger et al. (1975).

[c] Penetration ratio is a measure of relative complementation on chicken cells, t_{60}/t_0, to the relative complementation on rat cells, t_{60}/t_0, for the particular pseudotype.

et al., 1970; Lilly and Pincus, 1973). In contrast to the specific resistance in chicken to ATV subgroups, the mouse resistance is dominant but not fully penetrant, and N- and B-tropic MuLV do not necessarily differ in neutralization, nor in viral interference (Sarma *et al.*, 1967). This indicates that the restriction may occur not at the step of viral penetration, but that it affects later events of MuLV replication. The plating of MSV(MuLV) pseudotypes on "resistant" cells strengthens this supposition (Yoshikura, 1973). The Moloney and Rauscher strains of MuLV and NB-tropic and grow equally well in N- and B-type mouse cells. Some endogenous murine C-type viruses are "xenotropic" (Levy, 1973), i.e., only cells of other species are permissive, including avian cells (Levy, 1975).

Defectiveness and pseudotype formation is not restricted solely to sarcoma viruses. It has also been shown that the Friend (Fieldsteel *et al.*, 1971) and Rauscher viruses (Bentvelzen *et al.*, 1972), which cause erythroleukemia, are defective and helper dependent. This is also true of avian erythro- and myeloleukemia viruses, which are similarly helper dependent (Moscovici and Zanetti, 1970).

The formation of VSV(MuLV) pseudotypes was first described by Závada (1972a). The pseudotype particles showed three specific properties: (a) they were resistant to anti-VSV, but were neutralized with anti-MuLV sera; (b) they initiated plaques in mouse cells, but not in chicken or human cells; and (c) preinfection of mouse fibroblasts with MuLV induced specific interference to VSV(MuLV), but not to pure-grown VSV.

Consistent with what was already known concerning the specific resistance of N-tropic MuLV, the corresponding VSV pseudotypes showed the same plating efficiency in both BALB/c and NIH mouse cells (Huang *et al.*, 1973; Krontiris *et al.*, 1973). This confirmed that the genetic resistance of mouse cells to MuLV (in contrast to that of chicken cells to ATV) was due to restriction at a later step of MuLV replication than virus penetration. Huang *et al.* (1973) employed wild-type VSV and their pseudotype was prepared using anti-VSV serum, whereas Krontiris *et al.* (1973) exploited the thermolability of the VSV *ts*(O)45 mutant.

In cultures of BALB/c mouse, two distinguishable endogenous murine C-type viruses can be induced with iododeoxyuridine: an ecotropic and a xenotropic virus (Stephenson *et al.*, 1974). In this instance, the host-range specificity of corresponding VSV pseudotypes was also eco- and xenotropic: the VSV (eco-MuLV) plaqued in mouse cells, while the VSV (xeno-MuLV) was infectious for rabbit, but not for mouse cells (Besmer and Baltimore, 1975). Interference patterns of these VSV pseudotypes with MuLVs showed that there is an interference among various ecotropic viruses, and among xenotropic ones, but apparently there is no cross-interference between viruses which belong to these two classes of tropism. Unfortunately, the titer of VSV(xeno-MuLV) produced so far is very low, so that the studies on host range and interference are not as definite as with VSV(ATV) pseudotypes.

3. *Herpes Viruses*

Vesicular stomatitis virus has been found to form pseudotypes most efficiently on mixed infection with herpes simplex virus, type 1 (Huang *et al.*, 1974). In these experiments the MP strain of HSV was used, which grows well in HEp-2 cells, but not in the Chinese hamster ovary cell line, CHO. The host specificity of the pseudotype VSV(HSV) corresponded to the HSV strain used, e.g., it was infectious for HEp-2, but not for CHO cells (see Fig. 2).

B. DETECTION OF SURFACE ANTIGENS OF DEFECTIVE OR INCOMPLETELY EXPRESSED VIRUSES

The detection of viral surface antigens engendered by latent virus infections appears to hold great promise as an application of the pseudotype technique. There are many diseases of various types with a suspected viral etiology in which complete virions cannot be regularly detected or in which their detection is very laborious and time-consuming. These include not only malignancies of humans and of domestic animals, but also various degenerative diseases and other chronic processes, such as defective hemopoiesis, chronic brain diseases, etc. It has often been suggested that in many of these diseases enveloped viruses could be involved. In such cases, the viral agent might be either genetically defective, or the synthesis of complete virions may be repressed, possibly by regulatory control of host cells. However, if viral surface glycoproteins were present, that might be sufficient to produce pseudotypes with VSV or other enveloped viruses. The presence of antibodies against such antigens present on pseudotypes could then be tested in patients and this could contribute, together with other methods, to the elucidation of the disease.

Probably all chick cells carry genetic information of an endogenous ATV, belonging to the E subgroup (T. Hanafusa *et al.*, 1972; Weiss 1975b). The virus is usually only partially expressed, and such expression includes the viral surface glycoproteins. These cells are called chick helper positive (chf +) because the endogenous C-type glycoproteins complement defective Bryan-RSV. It was shown that on infection with VSV, these cells yield VSV (chf) pseudotypes, as seen in Fig. 7 (Love and Weiss, 1974).

The requirement for surface glycoproteins but not for synthesis of complete ATV virions has also been demonstrated the other way round; no VSV pseudotypes were obtained in Japanese quail cells shedding defective

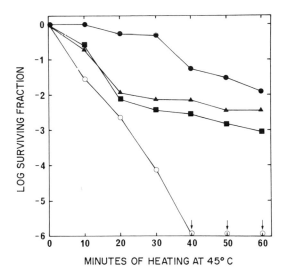

FIG. 7. Complementation of thermolabile VSV mutant *tl* 17 by RAV-O and by chick helper factor. VSV was inactivated at 45°C and samples taken at 10-minute intervals were titrated for plaque-forming units on golden pheasant cells. (O—O), VSV grown in chf-negative chick cells; (●—●), VSV grown in golden pheasant cells preinfected 11 days previously with RAV-O; (▲—▲), VSV grown in chf-positive Reaseheath C-line × I-line F₁ hybrid cells; (■—■), VSV grown in chf-positive Brown Leghorn cells. [From Love and Weiss (1974), with permission.]

Bryan-RSV, which lacks the surface glycoprotein (Ogura and Friis, 1975; Weiss, 1975a).

As mentioned above (Section II, C), in cultures that only partially express the herpes simplex virus (when inhibited with cytosine arabinoside) VSV pseudotypes were produced most efficiently (Huang *et al.*, 1974). Indeed, the yield of pseudotypes in this experiment was the highest so far observed; 30% of all VSV genomes were enclosed in HSV glycoprotein envelopes.

Viruses that are closely related if not identical with measles virus have been isolated from brain tissues of some patients with subacute sclerosing panencephalitis (SSPE). Some isolated agents appear to be defective since they were not transmissible as cell-free filtrates, although they spread in cell cultures by the formation of syncytia. In cultures containing such syncytia, antigens reacting with antimeasles sera have been detected by immunofluorescence. The VSV pseudotype particles that were neutralizable with antimeasles sera were obtained both in cultures infected with competent or with defective strains of measles virus isolated from SSPE patients (Wild *et al.*, 1976). It remains to be seen whether VSV will be useful in forming identifiable pseudotypes on direct infection of neural tissue of patients with SSPE or other neurological disorders for which a viral etiology is suspected.

Obviously, attempts have been made to detect antigens of hypothetical viruses present in human tumor cells by the formation of VSV pseudotypes. So far, there has been only limited success. A VSV pseudotype was produced in a permanent cell line, MaTu, reportedly derived from human breast cancer. No virions (including those of B- and C-types) could be found in these cells prior to infection with VSV. The VSV(MaTu) pseudotype showed a strict human specificity since it was infectious only for diploid cells derived from human embryos and for cell lines derived from human sarcomas. It was not infectious for mouse, hamster, or chicken cells, nor for human HeLa cells. Four hundred human sera from various donors (healthy, tumors other than breast carcinoma, and breast carcinoma) were tested for neutralization of this pseudotype in a double-blind experiment. Only two of them neutralized the VSV(MaTu) pseudotype, and both were identified as sera of patients after resection of breast carcinoma (Závada et al., 1972, 1974a,b). The neutralizing factor in the positive sera was IgG (G. Russ, personal communication). The origin of the envelope antigens detected in MaTu is not clear. Its primate origin is suggested both by the human host range of the pseudotype (which did not plate on rodent or avian cells) and, more importantly, the antibodies detected in two human sera. The MaTu cells have recently been found to contain alloantigens and glucose-6-phosphate dehydrogenase characteristic of HeLa cells. However, a similar VSV pseudotype could not be detected in any other stock of HeLa cells, whether shedding virions or not. One possibility is that, in the early history of deriving the line, slowly growing breast carcinoma cells were contaminated and overgrown by HeLa cells, which in turn picked up a coat donor from the breast carcinoma cells.

In most other cell lines originating from human tumors the results have been less encouraging, although in some of these there were indications that VSV could have produced a proportion of doubly neutralizable virus particles (Závada et al., 1974b). These, however, are less suitable for further studies than are pure pseudotypes.

C. Introducing Viral Genomes into Resistant Cells

Host restriction in some viruses in known to be due to the absence of specific receptors on the cell surface. In such instances the genome of a virus that is unable to enter a resistant cell as a normal virion with its own antigens could penetrate a resistant cell as a pseudotype, with surface antigens of another virus for which the cell possesses appropriate receptors. That was, in fact, the case in the very first instance of phenotypic mixing ever demonstrated (Novick and Szilard, 1951). Mutants B/2 and B/4 can be derived from *Escherichia coli* B, which are specifically resistant to T2 and T4 phages, respectively. These two phages are closely related, and in a mixed infection of their common host, *E. coli* B cells, they produce a high proportion of particles with T2 genomes but with tail fibers of T4 phage, and vice versa. The T2(T4) particles can infect B/2 cells and reproduce in these cells for one cycle of growth. Their progeny resembles the T2 parent, being infectious only for B/4, but not for B/2 cells. These particles therefore fail to produce plaques either in B/2 or in B/4 pure indicator cells, but they produce plaques in a mixed culture of both B/2 and B/4 cells. Therefore, a progeny of mixed infection of *E. coli* B culture with T2 + T4 phages gives significantly more plaques on a mixed indicator of B/2 + B/4 cells than is the sum of plaques counted on these two indicator strains separately (Novick and Szilard, 1951).

A similar situation was also found in phenotypic mixing between various members of the picornavirus group. Poliovirus RNA enclosed within the capsid of Coxsackie B1 virus was able to penetrate into mouse cells and multiply therein for one cycle, although mouse cells are resistant to normal poliovirus since they lack the appropriate receptors (Cords and Holland, 1964).

Phenotypically mixed particles with genomes of oncogenic RNA viruses and envelopes of other viruses might be of considerable interest. As mentioned above, the host range of oncogenic RNA viruses is often limited, due to absence of receptors in resistant cells (reviewed by Weiss, 1976). But phenotypically mixed particles containing surface antigens provided by other enveloped viruses with a broad host range might be infectious for many new hosts. Indeed, Weiss *et al.* (1974, 1976) were able to produce RSV(VSV) pseudotypes, showing an extended host range that was wholly neutralizable with anti-VSV sera (Table IV). VSV also acted as a helper virus for RSV variants defective for envelope antigen synthesis, but not for variants also defective for reverse transcriptase. Various VSV *ts* mutants were tested, of which the most efficient was *ts* T1026 with defective shutoff of cell synthesis. This was done in the following way: Chicken cells that had been transformed by and were producing Prague strain RSV of host-range subgroup A were superinfected with the VSV mutant *ts* T1026 at a multiplicity of 2 PFU/cell. The cells were incubated at 33°C and culture medium was harvested 16 hours later when a general cytopathic effect was evident. Cells were removed from the harvested medium by centrifugation and freeze-thaw treatment and serial dilutions of the medium were prepared following neutralization with VSV antiserum where indicated. The virus dilutions were plated on chicken cells that were genetically susceptible or resistant to penetration of RSV-A, as indicated in Table IV. Following adsorption for 1 hour at room temperature, the assay plates were overlaid with agar medium and were

TABLE IV

ASSAY OF PHENOTYPICALLY MIXED VIRUSES AFTER SUPERINFECTION OF RSV-PRODUCING
CELLS WITH VSV[a]

		Method of assay			
Plaques or foci	Pretreatment with VSV antiserum	Susceptibility of assay cells to RSV	Temperature of incubation (°C)	Virus assayed	Titer ($\log_{10}$ PFU or FFU/ml)
Plaques	−	Susceptible	33	VSV	8.1
Plaques	+	Susceptible	33	VSV(RSV)	6.2
Plaques	+	Resistant	33		<1.0
Foci	−	Susceptible	41	RSV	6.8
Foci	−	Resistant	41	RSV(VSV)	4.7
Foci	+	Resistant	41		<1.0

[a] From Weiss et al. (1974).

incubated at 33°C for the plaque assay of VSV and at 41°C for the transformed cell focus assay of RSV. Plaques were counted after 2 days and
foci 7 days after infection. It was necessary to use a ts mutant of VSV and to
incubate the focus assay above the permissive temperature for VSV replication in order to prevent the VSV from killing the focus assay cultures.

Vesicular stomatitis virus pseudotypes with envelope antigens of RSV
were revealed as plaque-forming units that resisted neutralization by VSV
antiserum. The envelope specificity of these pseudotypes was demonstrated
by their inability to plate on assay cells resistant to the RSV strain used,
although it was susceptible to VSV.

RSV pseudotypes with envelope antigens of VSV were revealed as focus-
forming units that infected cells resistant to the RSV strain used. The envelope specificity of these pseudotypes was demonstrated by neutralization
with antiserum specific to VSV (Weiss et al., 1976).

For such an expansion of the host range the formation of pure pseudotypes is not necessary. Particles bearing a mosaic of envelope antigens can
infect cells that possess receptors for only one of the two coat donors. This
was first claimed for the mixed pseudotype population RSV(RAV-1 +
RAV-2)(Vogt, 1967), and RSV(VSV) stocks also appear to be largely mosaic
(Weiss et al., 1976). RSV in mixed infection with parotitis virus produced a
proportion of particles doubly neutralizable with anti-RSV and antiparotitis
sera (Dyadkova and Kuznetsov, 1970). One would expect these RSV
particles to show a broader host range also. In this experiment, like RSV +
VSV, the phenotypic mixing was reciprocal; a high proportion of particles
with the parotitis virus genome were doubly neutralizable with antiparotitis
and anti-RSV sera. Formation of mosaic particles in mixed infections of two

competent viruses is apparently much more common than pure pseudo-types. Pseudotype formation between xenotropic murine C-type viruses and mouse or rat ecotropic viruses (Levy, 1976), and between eco- and xenotropic mouse viruses (Besmer and Baltimore, 1976) enabled the infection of nonpermissive host cells, provided the restriction occurs at the level of penetration.

IV. Discussion and Prospects

1. Advantages of Pseudotype Methods

Experiments with pseudotypes formed by interaction between enveloped viruses belonging to different families have several advantages over assays using some of the viruses alone. Pseudotype methods are very simple and inexpensive and could be used in any laboratory with elementary virological equipment, provided that the mixed infection and its products are not considered to constitute a potential or real biohazard, as discussed below. Other advantages are the accuracy, sensitivity, and rapidity inherent in plaque assays.

A major practical application of pseudotype techniques in the near future is that they may be adopted as routine for plaque assays for the detection of viruses causing no cytopathogenic alterations in tissue cultures, and for assaying antibodies directed against these viruses (e.g., for avian leukosis viruses). It is true that these can also be studied with comparable pseudotypes of defective strains of RSV, but while a focus assay with RSV pseudotypes lasts 7–10 days, 1–2 days is sufficient for plaque assay with VSV pseudotypes. The VSV plaque assay also requires less stringent standards of media and glassware than the RSV focus assay. Conceivably, similar VSV pseudotypes could be developed with surface antigens provided by some human noncytopathogenic enveloped viruses, and these could serve for the routine detection and assay of neutralizing antibodies in human sera. Ready-to-use defined preparations of VSV pseudotypes keep well when frozen at -70°C, and it is also possible to keep them lyophilized.

From Table I it can be seen that VSV successfully produced pseudotypes with representatives of most families of enveloped viruses. This shows that it is a rather common phenomenon. But it is evident that within one family, all viruses are not equally efficient producers of VSV pseudotypes. In general, there were no difficulties making pseudotypes with chick or mouse leukemia viruses, including endogenous, only partially expressed ones. With most ATV strains, about 1–10% of VSV genomes appeared as pseudotypes, but one strain, RAV-50, yielded less than 0.1% pseudotypes

(Boettiger *et al.*, 1975). MuLV was somewhat less efficient than ATV, with a pseudotype fraction of about 0.1% of the total VSV particles. There have been reports from several laboratories (see Závada, 1976) on the failure to produce VSV pseudotypes with envelope antigens of various oncorna-viruses. However, it now appears that in fact several more VSV pseudotypes can be obtained, but care must be taken to avoid loss of the pseudotypes due to their higher fragility or susceptibility to nonspecific heat labile serum inhibitors. These new VSV pseudotypes have neutralization antigens provided by Mason-Pfizer monkey virus (Altstein *et al.*, 1976), avian reticulo-endotheliosis viruses (Kang and Lambright, 1976), murine mammary tumor virus (C. Dickson and J. Závada, unpublished) and primate C-type viruses (Schnitzer, personal communication).

2. *Hazards and Unwelcome Pseudotypes*

Fear of new sorts of biohazards is nowadays a popular talking point, and besides genetic engineering, these biohazards might concern pseudotypes. The dangers should not be underestimated: the very fact that the RSV(VSV) pseudotype could be made without much difficulty is somewhat alarming. Such a wolf in sheep's clothing could obviously be a more harmful animal than the reciprocal sheep (wolf) pseudotype. Such a pseudotype will, how-ever, be dangerous mainly for people directly involved in the experiment and it will not spread further. Similar hazards apply even more to experi-ments with materials from various human fatal diseases.

One can even conceive a possible RSV × VSV recombinant with tumor determinants from RSV and with the host range, speed of growth, and contagion of VSV. We can only be encouraged that such a thing appears to be extremely improbable, but we shall never be absolutely sure that it will not happen. We have performed quite extensive experiments designed to obtain an RSV × VSV recombinant that would produce plaques, but carry surface antigens of RSV. There were never any positive results. The reci-procal recombinant of RSV carrying the envelope antigen determinant of VSV, which could more plausibly arise, since RSV is known to undergo recombination with C-type helper viruses, has also not been detected thus far (Weiss *et al.*, 1976).

On the other hand, pseudotypes are nothing novel, and for years thousands of virologists and cell biologists have probably been making them (and possibly more successfully than the present author) without ever suspect-ing it. There are hundreds of papers on viral interference; most different pairs of viruses have deliberately been put together and allowed to multiply in mixtures. No doubt, in very many of these dually infected cultures phenotypically mixed virions must also have arisen. Other people are every day growing various viruses in cultures of tumor origin, quite often in human tumor cell lines like HeLa. Any mixed propagation of cells, as in such widely used techniques as somatic cell hybridization or the implantation of human cells or tissues into immunodeficient mice, is likely to lead to the activation and amplification, if not the mixing, of latent viruses present in one or both species of host (Weiss, 1975a). Although nothing dreadful has happened yet, the lack of any reported disasters must not in any case be an excuse for careless work. Intentional experiments on pseudotypes should always be considered as potentially hazardous and everyone should take strict safety precautions.

3. Assembly of Cellular Components into Virions

It is known that enveloped viruses may assemble into their virions not only cellular lipids, but also various cellular proteins. Several cell enzymes associated with virions of enveloped viruses have been described and the following cellular antigens have also been detected: chick cell-specific antigens (Cartwright and Pearce, 1968), H-2 transplantation antigens from mouse cells (Hecht and Summers, 1972), and TSTA specific for SV40-transformed cells (Ansel, 1974). In these instances, the presence of cellular antigens was demonstrated by ways other than neutralization of the virus. But the neutralization of RSV with antisera against normal chick cell antigens has also been demonstrated (Aupoix and Vigier, 1975). This neutralization was complement dependent and its mechanism was immune virolysis.

One may therefore speculate whether there are also pseudotypes of VSV with normal cell glycoproteins on the surface. Although we have no definite answer, this speculation does not appear to be very likely. In experiments with model viruses, such as VSV controls produced in normal cells, no viral fraction, resistant to anti-VSV serum, or heat-stable in the case of tl 17 mutant, has been encountered except that ascribable to latent viral expression in the normal cells. Probably cell surface antigens, even when assembled in the virion envelope, cannot substitute the function of a viral glycoprotein, which no doubt has a complicated structure with several specific functions connected with adsorption, penetration, uncoating, etc. These functions can probably only be performed by viral antigens.

4. Prospects

The most attractive application is the search for antigens of unknown, obscure, defective, or partially expressed antigens involved in human malignancies and degenerative diseases. One hopes that the technology of making pseudotypes may be further improved and that such applications will become practicable.

One point which has largely been neglected so far is the analysis of the structure of pseudotypes. The most straightforward way for this analysis was pointed out several years ago by McSharry *et al.* (1971). It will be interesting to examine more pseudotypes formed between other enveloped viruses. Of course, molecular biology provides us with a range of other methods that might be useful in solving these problems, but at present the specificity of the biological assays remains more powerful than biochemical methods for detecting and analyzing pseudotypes that usually represent only a small proportion of the total virus population.

Phenotypic mixing and formation of pseudotypes might be useful not only for solving practical problems, such as diagnostics, but it may also contribute to a better understanding of viral structure, and the mechanisms of assembly and budding, of which we know so little at present.

ACKNOWLEDGMENT

The author is most grateful to Dr. Robin Weiss of the Imperial Cancer Research Fund Laboratories, London, not only for critically reviewing the manuscript, but also for reconstructing numerous paragraphs into appropriate English pseudotypes which, it is hoped, will give this contribution a broader host range of adsorption and penetration.

REFERENCES

Aaronson, S. A. (1971). *Virology* **44**, 29.
Aaronson, S. A., and Rowe, W. P. (1970). *Virology* **42**, 9.
Ansel, S. (1974). *Int. J. Cancer* **13**, 773.
Altstein, A. D., Zhdanov, V. M., Omelchenko, T. N., Dzagurov, S. G., Miller, G. G., and Závada, J. (1976). *Int. J. Cancer* **17**, 780.
Aupoix, M., and Vigier, P. (1975). *J. Gen. Virol.* **27**, 151.
Bassin, R. H., and Phillips, L. A., Kramer, M. J., Haapala, M. J., Peebles, P. T., Nomura, S., and Fischinger, P. J. (1971). *Proc. Natl. Acad. Sci. U.S.A.* **68**, 1520.
Bentvelzen, P., Aarsen, A. M., and Brinkhof, J. (1972). *Nature (London), New Biol.* **239**, 122.
Besmer, P., and Baltimore, D. (1975). *Proc. Int. Congr. Virol., 3rd, 1975* Abstract W 31.
Besmer, P., and Baltimore, D. (1976). *J. Virol.* (submitted for publication).
Bishop, D. L. H., Repik, P., Obijeski, J. F., Moore, N. F., and Wagner, R. R. (1975). *J. Virol.* **16**, 75.
Black, D. R., Counell, C. J., and Merigan, T. C. (1973). *J. Virol.* **12**, 1209.
Boettiger, D. (1974). *Cold Spring Harbor Symp. Quant. Biol.* **39**, 1169.
Boettiger, D., Love, D., and Weiss, R. A. (1975). *J. Virol.* **15**, 108.
Bomford, R. (1971). *Int. J. Cancer* **8**, 53.
Breitman, M., and Prevec, L. (1976). *Virology* (in press).
Burge, B. W., and Pfefferkorn, E. R. (1966). *Nature (London)* **210**, 1397.
Cartwright, B., and Pearce, C. A. (1968). *J. Gen. Virol.* **2**, 207.
Choppin, P. W., and Compans, R. W. (1970). *J. Virol.* **5**, 609.
Compans, R. W. (1974). *J. Virol.* **5**, 1307.
Cords, C. E., and Holland, J. J. (1964). *Virology* **24**, 492.
Crittenden, L. B. (1968). *J. Natl. Cancer Inst.* **41**, 145.
Crittenden, L. B., Stone, H. A., Reamer, R. H., and Okazaki, W. (1967). *J. Virol.* **1**, 898.

De Giuli, C., Kawai, S., Dales, S., and Hanafusa, H. (1975). *Virology* **66**, 293.
Deutsch, V. (1975). *J. Virol.* **15**, 798.
Dodds, J. A., and Hamilton, R. I. (1976). *Adv. Virus Res.* **20**, 33.
Duff, R. G., and Vogt, P. K. (1969). *Virology* **39**, 18.
Dyadkova, A. M., and Kuznetsov, O. K. (1970). *Neoplasma* **17**, 59.
Fieldsteel, A. H., Dawson, P. J., and Kurahara, C. (1971). *Int. J. Cancer* **8**, 304.
Fischinger, P. J., and O'Connor, T. E. (1969). *Science* **165**, 714.
Fraenkel-Conrat, H. (1956). *J. Am. Chem. Soc.* **78**, 882.
Fujita, D. J., Chen, Y. C., Friis, R. R., and Vogt, P. K. (1974). *Virology* **60**, 558.
Hanafusa, H. (1965). *Virology* **25**, 248.
Hanafusa, H., and Hanafusa, T. (1966). *Proc. Natl. Acad. Sci. U.S.A.* **55**, 532.
Hanafusa, H., and Hanafusa, T. (1971). *Virology* **43**, 313.
Hanafusa, H., Hanafusa, T., and Rubin, H. (1963). *Proc. Natl. Acad. Sci. U.S.A.* **49**, 572.
Hanafusa, H., Hanafusa, T., and Rubin, H. (1964). *Proc. Natl. Acad. Sci. U.S.A.* **51**, 41.
Hanafusa, H., Baltimore, D., Smoler, D., Watson, K. F., Yaniv, A., and Spiegelman, S. (1972). *Science* **177**, 1188.
Hanafusa, T., Hanafusa, H., Miyamoto, T. and Fleissner, E. (1972). *Virology* **47**, 475.
Hartley, J. W., and Rowe, W. P. (1966). *Proc. Natl. Acad. Sci. U.S.A.* **55**, 780.
Hartley, J. W., Rowe, W. P., and Huebner, R. J. (1970). *J. Virol.* **5**, 221.
Hecht, T. T., and Summers, D. F. (1972). *J. Virol.* **10**, 578.
Henderson, I. C., Lieber, M. M., and Todaro, G. J. (1974). *Virology* **60**, 282.
Hill, M., and Hillová, J. (1972). *Nature (London), New Biol.* **237**, 35.
Hložánek, I. (1975). *In* "Advances in Acute Leukaemia" (F. J. Cleton, D. Crowther, and J. S. Malpas, eds.), pp. 229–296. ASP-Biol. Med. Press BV, Amsterdam.
Huang, A. S., Besmer, P., Chu, L., and Baltimore, D. (1973). *J. Virol.* **12**, 659.
Huang, A. S., Palma, E. L., Hewlett, N., and Roizman, B. (1974). *Nature (London)* **252**, 743.
Huebner, R. J., Hartley, J. W., Rowe, W. P., Lane, W. T., and Capps, W. I. (1966). *Proc. Natl. Acad. Sci. U.S.A.* **56**, 1164.
Ishizaki, R., and Vogt, P. K. (1966). *Virology* **30**, 375.
Issinger, O. G., Beier, H., and Hausmann, R. (1973). *Mol. Gen. Genet.* **122**, 81.
Kang, C. Y., and Lambright, P. (1976). *J. Virol.* (submitted for publication).
Kelloff, G. J., Huebner, R. J., and Gilden, R. V. (1971). *J. Gen. Virol.* **13**, 289.
Kimura, Y. (1973). *Jpn. J. Microbiol.* **17**, 373.
Krontiris, T. G., Soeiro, R., and Fields, B. (1973). *Proc. Natl. Acad. Sci. U.S.A.* **70**, 2549.
Levy, J. A. (1973). *Science* **182**, 1151.
Levy, J. A. (1975). *Nature (London)* **253**, 140.
Levy, J. A. (1976). *Virology* (submitted for publication).
Levy, J. A., and Vogt, P. K. (1976). *Virology* (submitted for publication).
Lilly, F. (1967). *Science* **155**, 461.
Lilly, F., and Pincus, T. (1973). *Adv. Cancer Res.* **17**, 231.
Livingston, D. M., Howard, T., and Spence, C. (1976). *Virology* **70**, 432.
Love, D. N., and Weiss, R. A. (1974). *Virology* **57**, 271.
McSharry, J. J., Compans, R. W., and Choppin, P. W. (1971). *J. Virol.* **8**, 722.
Moscovici, C., and Zanetti, M. (1970). *Virology* **42**, 61.
Neff, J. M., and Enders, J. F. (1968). *Proc. Soc. Exp. Biol. Med.* **127**, 260.
Novick, A., and Szilard, L. (1951). *Science* **113**, 34.
O'Connor, T. E., and Rauscher, F. J. (1964). *Science* **146**, 787.
Ogura, H., and Bauer, H. (1976). *Arch. Virol.* **52**, 233.
Ogura, H., and Friis, R. (1975). *J. Virol.* **16**, 443.
Payne, L. N., and Biggs, P. M. (1966). *Virology* **29**, 190.

Piraino, F. (1967). *Virology* **32**, 700.

Pringle, C. R. (1975). *Curr. Top. Microbiol. Immunol.* **69**, 85–116.

Rettenmier, C. W., Dumont, R., and Baltimore, D. (1975). *J. Virol.* **15**, 41.

Rubin, H. (1961). *Virology* **13**, 200.

Rubin, H. (1965). *Virology* **26**, 270.

Sarma, P. S., and Gazdar, A. F. (1974). *Curr. Top. Microbiol. Immunol.* **68**, 1–28.

Sarma, P. S., Cheong, M. P., Hartley, J. W., and Huebner, R. J. (1967). *Virology* **33**, 180.

Scheele, C. M., and Hanafusa, H. (1971). *Virology* **45**, 401.

Stephenson, J. R., Crow, J. D., and Aaronson, S. A. (1974). *Virology* **61**, 411.

Svoboda, J. (1960). *Nature (London)* **186**, 980.

Svoboda, J., and Dourmashkin, R. (1969). *J. Gen. Virol.* **4**, 523.

Teich, N. M., Weiss, R. A., Salahuddin, S. Z., Gallagher, R. E., Gillespie, D. H., and Gallo, R. C. (1975). *Nature (London)* **256**, 551.

Temin, H. M. (1962). *Cold Spring Harbor Symp. Quant. Biol.* **27**, 407.

Temin, H. M., and Rubin, H. (1958). *Virology* **6**, 669.

Tooze, J. (1973). "Molecular Biology of Tumor Viruses." Cold Spring Harbor Lab., Cold Spring Harbor, New York.

Vogt, P. K. (1967). *Virology* **32**, 708.

Vogt, P. K., and Ishizaki, R. (1965). *Virology* **26**, 664.

Vogt, P. K., and Ishizaki, R. (1966). *Virology* **30**, 368.

Weiss, R. A. (1969a). *J. Gen. Virol.* **5**, 511.

Weiss, R. A. (1969b). *J. Gen. Virol.* **5**, 529.

Weiss, R. A. (1975a). *Nature (London)* **255**, 445.

Weiss, R. A. (1975b). *Perspect. Virol.* **9**, 165–205.

Weiss, R. A. (1976). *In* "Cell Membrane Receptors for Viruses, Antigens and Antibodies, Polypeptide Hormones, and Small Molecules" (R. F. Beer and E. G. Bassett, eds.), pp. 237–251. Raven, New York.

Weiss, R. A., and Wong, A. C. H. (1976). *Virology* (submitted for publication).

Weiss, R. A., Boettiger, D. E., and Love, D. N. (1974). *Cold Spring Harbor Symp. Quant. Biol.* **39**, 913.

Weiss, R. A., Boettiger, D. E., and Murphy, H. M. (1976). *Virology* (in press).

Wild, F., Cathala, F., and Huppert, J. (1976). *Intervirology* **6**, 185.

Yoshikura, H. (1973). *J. Gen. Virol.* **19**, 321.

Závada, J. (1972a). *J. Gen. Virol.* **15**, 183.

Závada, J. (1972b). *Nature (London) New Biol.* **240**, 122.

Závada, J. (1976). *Arch. Virol.* **50**, 1.

Závada, J., and Rosenbergová, M. (1968). *Acta Virol. (Engl. Ed.)* **12**, 282.

Závada, J., and Rosenbergová, M. (1972). *Acta Virol. (Engl. Ed.)* **16**, 103.

Závada, J., and Závodská, E. (1973–1974). *Intervirology* **2**, 25.

Závada, J., Závadová, Z., Malír, A., and Kočent, A. (1972). *Nature (London) New Biol.* **240**, 124.

Závada, J., Závadová, Z., Widmaier, R., Bubeník, J., Indrová, M., and Altaner, Č. (1974a). *J. Gen. Virol.* **24**, 327.

Závada, J. Bubeník, J. Widmaier, R., and Závadová, Z. (1974b). *Cold Spring Harbor Symp. Quant. Biol.* **39**, 907.

5 Methods for Assaying Reverse Transcriptase

Daniel L. Kacian

I. Introduction

The RNA tumor viruses contain a DNA polymerase in the virus particle (Temin and Baltimore, 1972). It is believed to catalyze the reaction

$$\text{viral RNA} \rightarrow \text{viral RNA:DNA} \rightarrow \text{DNA:DNA}$$

producing a double-stranded DNA molecule for integration into the host genome (Guntaka *et al.*, 1975; Gianni *et al.*, 1975). Considerable biochemical and genetic evidence shows that the polymerase performs a necessary function in the viral life cycle (Temin, 1971; Hanafusa *et al.*, 1972; Verma *et al.*, 1974).

In vitro the enzyme is able to use RNA, DNA, and RNA : DNA hybrids as templates, and it is especially active with certain homopolymer template-primer combinations (Duesberg *et al.*, 1971; Kacian *et al.*, 1971; Spiegelman *et al.*, 1970b,c). Like other DNA polymerases, it is unable to initiate a polynucleotide chain and must be provided with a 3′-OH-terminated primer molecule complementary to its template (Baltimore and Smoler, 1971). It requires all four deoxynucleoside triphosphates when copying heteropolymeric templates, a divalent cation (Mg^{2+} or Mn^{2+}), and it is stabilized by the presence of nonionic detergent and a sulfhydryl-reducing agent (Temin and Baltimore, 1972). The enzyme molecule also carries a nuclease activity that, *in vitro*, degrades the RNA portion of an RNA:DNA hybrid (Mölling *et al.*, 1971; Baltimore and Smoler, 1972; Keller and Crouch, 1972; Leis *et al.*, 1973). Genetic evidence has been presented that this activity is a part of the enzyme itself and is not a contaminating nuclease (Verma, 1975a; Verma *et al.*, 1974).

The assay of reverse transcriptase is frequently complicated by the presence of other DNA polymerase activities, some of which will also accept RNA templates under certain conditions (Weissbach, 1975); therefore, it is necessary to demonstrate both RNA-instructed DNA polymerase activity and to establish, where necessary, the viral origin of the enzyme. Rigorous proof that a reverse transcriptase is present in viruses or cells is not trivial, except where fairly large amounts of easily purified virus are available. It is possible, however, to accumulate with present technologies reasonably compelling, if not conclusive, evidence for the existence of the enzyme.

The assay of reverse transcriptase requires the demonstration of several known properties of the enzyme, its product, and the virus particle. Generally accepted as necessary and sufficient criteria for a positive result are the following. (1) The polymerase activity should be found at the density characteristic of RNA tumor virus particles when banded in isopycnic gradients. (2) Endogenous DNA polymerase activity should be demonstrable. It should require disruption of the virus, be sensitive to RNAse in the presence of high salt, and require all four deoxynucleoside triphosphates and a divalent cation. (3) Proof should be offered that the product is DNA, that it is associated with 60–70 S viral RNA at *early* stages of the reaction, and that it can be hybridized to the RNA species characteristic of the virus if known, or to that from other, presumptively related, viruses. The hybridization must be shown to be specific for heteropolymeric regions of the RNA and DNA. (4) The response of the enzyme to homopolymer template–primer combinations should follow patterns characteristic of the known reverse transcriptases. (5) The activity should be inhibited by antisera against the enzyme from related RNA tumor viruses.

Fulfilling the majority of these criteria requires isolation and purification of the virus particles. Purification using a combination of sedimentation velocity and equilibrium density gradient centrifugation is applicable to all types of starting material and affords, in most cases, extensive purification with high yield. The use of these techniques will, therefore, be considered in detail.

II. Isolation and Purification of Virus Particles from Tissue Culture Media, Tissue Homogenates, and Biological Fluids

The design of useful procedures for isolating RNA tumor viruses requires a knowledge of the properties of the viruses and of subcellular components.

A. PROPERTIES OF RNA TUMOR VIRUSES

The physical and chemical properties of the RNA tumor viruses have been discussed in numerous reviews (Beard, 1963; Vogt, 1965; Temin, 1971, 1974; Bader, 1969). They consist of about 60–70% protein, 20–30% lipid, 2% carbohydrate, and 1–2% RNA. The sedimentation coefficient of the particle is about 500–600 S. The buoyant density of C-type viruses is about 1.15–1.16 gm/ml in most gradient materials. B-Type viruses are slightly more dense and band at about 1.18–1.19 gm/ml. The infectivity of the viruses is stable between pH 5 and 9. The virus is enclosed by an outer lipid membrane, which is sensitive to ether, chloroform, deoxycholate, nonionic detergents, and lipases. The particle is about 100 nm in diameter. Within the viral envelope is a protein nucleocapsid containing the viral RNA and reverse transcriptase. The RNA sediments at 60–70 S but may be melted by heat or denaturing solvents to give major components sedimenting at 35 S. Several smaller RNA components, including tRNAs, are also found in the 70 S complex. One of the tRNAs has been shown to serve as the primer molecule for DNA synthesis in the virion (Dahlberg et al., 1974).

B. PROPERTIES OF SUBCELLULAR COMPONENTS

The sedimentation coefficient and buoyant density of various cell components are shown in Fig. 1. The sedimentation values are corrected to water at 20 °C. Most of the buoyant densities were determined in cesium salt gra-

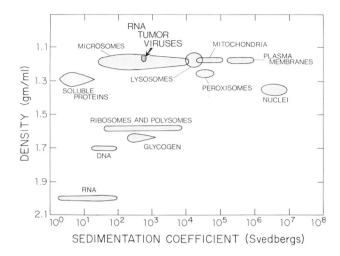

FIG. 1. Buoyant density and sedimentation coefficient of subcellular components and RNA tumor viruses. [After Anderson (1966).] Other references: Perry and Kelley (1966), Reid (1972).

dients. In other gradients different densities may be obtained; however, the relative densities of the various components generally will remain the same.

It is apparent from Fig. 1 that a combination of velocity and equilibrium density gradient centrifugation will separate the virus from most subcellular species. The major contaminant is the microsome population since a portion of this material possesses the same S value and density as the virus. An additional source of contamination results from aggregation of the virus particles with soluble proteins and other species.

C. DESIGN OF GRADIENT PROCEDURES FOR RNA TUMOR VIRUS PURIFICATION

There are several points to be considered in designing gradient procedures for purifying RNA tumor viruses.

1. Cellular contaminants are present having sedimentation coefficients both greater than and less than the RNA tumor viruses. It is therefore clear that sedimenting the material through a column of relatively low density sucrose or glycerol fails to remove a major class of contaminant.

2. Pelleting of viruses is to be avoided, if possible. Centrifuging viruses against the walls of centrifuge tubes produces aggregates of virus and cellular debris that confound further purification steps. In addition, particles may be damaged, resulting in reduced yields.

3. Cell homogenization procedures should be utilized that minimize the production of microsomes of the same size and density as the virus. Tissue culture techniques and the handling of biological fluids such as blood should be designed to minimize production of cell debris. Extremes of pH or osmolarity, freezing and thawing, excessive centrifugal force for pelleting cells, and transfer of fluids using syringes that may generate high shear forces should be avoided to minimize cell disruption. Cells should be removed without undue delay from culture fluids or blood and always before the material is frozen for storage.

4. Gradients should be designed to maximize differences between the virus particles and microsomal material. Extremes of pH, osmolarity, temperature and time of centrifugation should be avoided. Ionic conditions should be chosen to discourage aggregation.

D. CONCENTRATION OF VIRUS PARTICLES

For virus grown in tissue culture or found in biological fluids such as milk or blood, the initial steps in purification are low-speed centrifugations to remove cells and debris. It is usually advantageous to perform two spins—one to remove whole cells and a second to remove smaller debris. If a single, high-speed centrifugation is used, cells may be ruptured, releasing debris that could otherwise be removed. For cells and tissues, homogenization of the material is required. Tissues are chopped and dispersed into cells by methods appropriate to the material. The cells are broken, and the nuclei are removed by low-speed centrifugation. It is important to design procedures that allow the nuclei to be removed intact in order to minimize contamination with nuclear DNA polymerases and cellular DNA. A second low-speed centrifugation removes membrane fragments and other debris. The removal of mitochondria at this stage is important for the same reasons as removal of nuclei.

Most workers centrifuge at 1000 g for 10 minutes to remove cells and nuclei and at 20,000 g for 20 minutes to remove mitochondria and cell debris. Longer centrifugation times may be required with unusually large volume rotors. The supernatant may then be applied directly to sedimentation or equilibrium gradients if the volume is small; however, in most cases the titer is low and large amounts of fluid must be processed. The most common concentration procedure is ultracentrifugation.

Frequently used conditions for this operation are (1) Beckman type 19 rotor at 19,000 rpm for 90–120 minutes; (2) Beckman SW 27 rotor at 25,000 rpm for 60 minutes; (3) Beckman type JA10 rotor at 8000 rpm for 6 hours; and (4) Beckman type SW 41 rotor at 40,000 rpm for 30 minutes. All centrifugations are performed at 2 °–4 °C. If the volumes being processed are sufficiently small, it is preferable to use swinging bucket rotors and to deposit the virus onto a cushion of dense sucrose or glycerol. This results in less aggregation and makes unnecessary the rather difficult resuspension of virus pellets. For larger volumes, angle rotors are required, and the virus is deposited on the bottom of the tube. In order to obtain maximum purity and yields in subsequent steps, it is *essential* that the virus be thoroughly dispersed. A Dounce homogenizer will be found very useful for suspending the virus after centrifugation; however, brief sonication of the preparation is also necessary for best results. The sonication step is performed directly in the homogenizer tube, which is immersed in an ice bath. We sonicate Rous sarcoma virus (RSV) preparations for 30 seconds at 100 W, using a microtip on a Bronson sonifier. In our hands, when sonication is omitted, virus yields drop an average of 50%. Care must be taken to avoid damaging the virus particles by excessive sonication.

An alternative to high-speed centrifugation for virus concentration is precipitation with ammonium sulfate (Duesberg *et al.*, 1968; Fan and Baltimore, 1973; Teramoto *et al.*, 1974) or polyethylene glycol (Bronson *et al.*, 1975; Syrewicz *et al.*, 1972). Precipitation permits the use of large volume, low-speed centrifuge rotors and thus makes possible the processing of much greater amounts of material. Most procedures in the literature have applied these methods to tissue culture media; however, they have also found use with other sources of virus.

For precipitation with ammonium sulfate, an equal volume of cold, saturated salt solution (adjusted to neutrality with ammonium hydroxide) or the solid salt to 50% saturation are added slowly with stirring in an ice bath. After 30 minutes, the precipitate is collected by centrifugation at 8000 rpm for 15 minutes in the Sorvall GSA rotor. The pellets are resuspended in 10 mM Tris-HCl, pH 7.5, and further purified by gradient centrifugation. If solid ammonium sulfate is used for the precipitation, the pH should be kept near neutrality by the addition of 0.05 N NaOH together with the salt.

For precipitation of RNA tumor viruses from tissue culture fluids, solid polyethylene glycol is added after removal of cells by low-speed centrifugation, and the virus is allowed to precipitate in the cold for 16 hours. Both 6000 and 20,000 MW polymer have been used, and one is more suitable than the other in some situations. The concentration of the polyethylene glycol is generally between 2 and 8% (w/v). The precipitated virus is collected by centrifugation. The time of centrifugation is variable since the composition of the precipitate depends upon the starting material.

Ammonium sulfate precipitation has, in general, not given satisfactory results with murine viruses. These agents are apparently damaged by the procedure. It should also be noted that polyethylene glycol is a relatively potent inhibitor of reverse transcriptase activity; therefore, attempts to monitor the virus content of the crude precipitates by reverse transcriptase assays may give false results.

E. SEDIMENTATION VELOCITY TECHNIQUES

A simple linear sucrose gradient can be used to remove contaminants that sediment differently than the virus. If volumes are relatively small, a useful procedure is to run a 5–20% (w/v) sucrose gradient in the Beckman SW 41 rotor (Smith and Bernstein, 1973). About 1–1.5 ml of virus suspension are layered onto each tube, and the gradients are spun at 30,000 rpm at 4 °C for 20 minutes. Maximum acceleration and braking are used in the L3-50 centrifuge; slight alterations in the time of the run may be necessary in the L2-65B and L2-75B machines because of the longer acceleration time. The virus is found as a diffuse band approximately half-way down the tube.

For larger volumes, Bolognesi and Bauer (1970) have devised an extremely useful procedure, which combines in a single step both concentration of the virus particles and sedimentation velocity purification. In addition, the procedure helps to break up aggregates of viral and nonviral material. The following procedure is slightly modified from the original.

The gradients consist of 7 ml 50% (w/v) sucrose, 11 ml of 35% (w/v) sucrose, and 10 ml 20% (w/v) sucrose in an appropriate buffer. The solutions are layered carefully over one another to give sharp interphases. The tube is held at an angle and the solutions pipetted down the side. The gradients should be made up from cold solutions within an hour before use for optimal results. SW 27 cellulose nitrate tubes are used. Ten milliliters of virus suspension are layered carefully atop each gradient so that a sharp interphase is formed. The gradients are spun at 27,000 rpm for 60 minutes at 4 °C. The virus will emerge as a band in the middle of the 35% sucrose layer. It may be located by reference to a marker virus run on a parallel gradient. The band patterns with crude tissue homogenates are quite reproducible using this method; therefore, once experience is gained, the viral band can be identified without the necessity of the marker virus.

The principles involved in the gradient are straightforward. The virus particles sediment rapidly through the initial loaded volume until they strike the 20% sucrose layer. As they enter the more dense and viscous sucrose solution, they sediment more slowly; hence, the virus tends to concentrate and passes through the 20% sucrose in a considerably narrower band than that originally applied to the tube. In addition, striking the 20% layer applies a mild shock to the particles. This shock is not sufficient to damage RNA tumor virus particles, but does help to break up some types of aggregated material. Further concentration and disaggregation occur when the virus enters the 35% sucrose layer. The end result is to produce a band of concentrated virus in the middle of the 35% layer that is relatively free of material with different sedimentation rates. The 50% layer catches cell debris and viral aggregates. In the original procedure, designed for use with relatively pure tissue culture viruses, the 50% layer served to catch virus aggregates in case the material had not been adequately dispersed after pelleting. It may be useful to remove the 50% layer when the gradients are used with cell homogenates so as to allow increased capacity and resolution.

F. Equilibrium Centrifugation Techniques

Following sedimentation velocity separation, the virus band is removed and layered onto a gradient for equilibrium separation. Gradients from 15–60% (w/v) sucrose or 20–70% (w/v) sucrose are ordinarily employed. Depending upon the type of sedimentation gradient that has been used, it may be necessary to dilute the virus prior to applying it to the equilibrium gradient. Since the gradients concentrate the material, larger volumes may generally be applied than is the case for the usual sedimentation run. Depending upon the type of separation required, it may be useful to use a more shallow gradient (e.g., 25–55% sucrose) than is normally used.

Equilibrium gradients may also be prepared in discontinuous fashion, and in some cases resolution is superior to continuous gradients. We routinely use gradients consisting of 1.75 ml each of 60%, 51%, 42%, 33%, 24%, and 15% sucrose. These are run in the SW41 rotor at 36,000 rpm for 3 hours at 4°C. The stepwise configuration may aid in dispersing aggregates by the same mechanism discussed above.

The quality of the separation achieved in an equilibrium gradient may vary with the time of centrifugation because the components of the mixture are not completely stable. Prolonged exposure to the gradient conditions frequently results in changes in both the virus and the contaminating species that shift the material from one density to another. In some cases, this shift may be beneficial and in others detrimental. Since the sedimentation coefficient of the virus is large, it will generally reach its equilibrium position after a few hours' centrifugation. When the purification procedure for a new type of material is being worked out, it is advisable to try several different run times for the gradient in order to determine which gives the optimal purification.

G. Choice of Gradient Conditions

Although sucrose is the most commonly used solute, other gradient materials may be more suitable for some virus purifications. When cells are disrupted, membranes (both those at the cell surface and those enclosing various cellular organelles) are usually fragmented. Membrane fragments may possess the same sedimentation value as RNA tumor virus particles. The density of membrane pieces that have not formed vesicles depends on their state of hydration, which is in turn a function of the type of gradient solution. Membranes that have formed vesicles will have a density that depends upon the composition of the material within the vesicle, the density of the membrane itself, and the proportions of each. The composition of the intravesicular material will vary with the gradient composition and the permeability of the membrane forming the vesicle. Most homogenization procedures used for cells and tissues (sonication, osmotic shock, detergent treatment, nitrogen decompression, Dounce homogenizer, Potter homogenizer) do result in the production of many membrane vesicles.

Sucrose gradients have high osmolarity. Since the sugar does not pass through the membrane, the vesicle or virus particle becomes progressively shrunken as it sediments through the gradient. The exact amount of water present in the interior of the particle at equilibrium will depend upon what molecules were trapped in the vesicle when it was formed; however, in general the collapse is essentially complete. Membrane vesicles, virus particles, and open membrane fragments will thus all be found at approximately the same density, 1.12–1.17 gm/ml.

Gradients of Ficoll or other polymers have low osmolarity, and these materials also do not enter the interior of the vesicle. If the gradient solution is of low osmotic activity, the closed vesicles will tend to swell. They thus band at lower densities (1.04–1.08 gm/ml) than in sucrose. Open membrane fragments will still be found at heavier positions.

In gradients of small, uncharged molecules, such as glycerol, that freely pass the membrane, both open membranes and vesicles tend to behave the same and band at 1.15–1.19 gm/ml. The situation is more complex with salt gradients since the presence of charges on the membrane and within the vesicle will affect the density of the particles. In general, the RNA tumor viruses band in CsCl and potassium tartrate gradients at the same or slightly higher densities than in sucrose.

These considerations suggest that gradients of Ficoll or other inert polymers would be preferable to those of sucrose, glycerol, and probably cesium salts and potassium tartrate since they should separate viruses from open membrane fragments. A number of workers (Lyons and Moore, 1965; Oroszlan et al., 1965; Stromberg et al., 1972) have used such gradients and report that they obtained better results than with other gradient materials. This type of gradient should be particularly useful in purifying crude cell and tissue homogenates prior to attempting reverse transcriptase assays since these will contain a large proportion of membrane fragments.

In many cases, however, the production of open membrane fragments appears to be minimal, accounting for the continued popularity of the sucrose gradient for RNA tumor virus purification. When performing the initial experiments with new material, the virus band from sucrose should be taken and rebanded on a Ficoll gradient to see if additional purification can be obtained. If the Ficoll gradient appears to be superior, the gradients should then be performed in reverse order to make certain that both are not required to achieve the same degree of purification.

Many viruses, particularly those of murine origin, appear to be sensitive to prolonged exposure to high ionic strength; therefore, gradients of cesium chloride or other salts must be used with caution to avoid damage to the particles and loss of material.

Sedimentation rates of different particulates will also vary in gradients of different composition, but the effects are somewhat more difficult to predict. A major factor is the difference in viscosity among various gradient materials. Ficoll solutions have a much higher viscosity than sucrose solutions of the same density; sucrose in turn produces more viscous gradients than do cesium salts. Depending upon their size, shape, and surface properties, molecules that sediment at identical rates in one gradient medium may not do so in another. Since in some cases the density of the particle changes as it sediments through the gradient, its rate of sedimentation will not be constant; therefore, the quality of the separation obtained may depend upon the time of the run and on how far through the gradient the virus particles have moved. Effects of this type have been observed in studies on the purification of Rauscher leukemia virus (Oroszlan *et al.*, 1965).

Generally, workers have used Ficoll gradients of 5–25% (w/v) for sedimentation velocity purification and 5–40% (w/v) for equilibrium density centrifugation. Because of the high viscosity of Ficoll, some workers prepare the gradients in D_2O or D_2O with some sucrose to achieve higher densities at lower concentrations of polymer (Calafat and Hageman, 1968; Duesberg and Blair, 1966).

Gradient buffers should avoid extremes of pH and osmolarity. Generally, gradients have been run at pHs between 7.0 and 8.8. It has been found in some cases that the use of isotonic saline solutions increases the extent of aggregation of viruses and cellular components. It does not appear that salt is necessary for the integrity of the virus particles; therefore, higher salt concentrations are probably best avoided.

H. OTHER PURIFICATION PROCEDURES

In 1957 Bryan and Moloney reviewed attempts to purify RSV from solid tumors and concluded that the final preparations contained at best less than 1% virus. Since that time density gradient centrifugation has become the most widely used procedure for RNA tumor virus purification. Little has been done to apply newer concepts of column chromatography, aqueous polymer phase separation, and other advances in purification techniques to the problem of isolating the virus from cells and tissues; thus, for the present, gradient centrifugation remains the most reliable method for purifying virus for reverse transcriptase assays.

III. Procedures for the Assay of Reverse Transcriptase Associated with Virus Particles

A. ASSAY OF ENDOGENOUS DNA POLYMERASE ACTIVITY

Principle

The assay measures the conversion of radioactive nucleotide to an acid-insoluble form. The reaction mixture (total volume 0.1 ml) contains:

Tris-HCl, 50 mM, pH 8.3
$MgCl_2$, 8 mM, or $MnCl_2$, 0.5–1.0 mM
Dithiothrietol, 10 mM
dATP, dTTP, dGTP, 0.8 mM each
[^{3}H]dCTP, 0.8 mM, 500–1000 cpm/mole for purified viruses, higher specific activities for systems where virus concentrations may be low
Nonionic detergent, 0.002–0.2% (v/v), amount titered individually for each virus preparation
Virus, 10–50 μg of purified virus protein or larger amounts of material from crude systems

Procedure

Samples are taken at intervals and mixed with an equal volume of TCA reagent (see Appendix). After 10 minutes at 0°C, the acid-precipitable radioactivity is collected on membrane filters. The filters are washed 10 times with cold 5% (v/v) trichloroacetic acid, dried under an infrared lamp, and counted in a toluene-based scintillation fluid.

Kinetics of the reaction vary widely among different virus preparations. Some syntheses are linear for less than 1 hour, whereas others continue for up to 20 hours. The rate of reaction is usually proportional over a reasonable range to the amount of viral protein added irrespective of the type of kinetics exhibited.

Reaction conditions have been similar for all viruses examined (Temin and Baltimore, 1972). Since the polymerase is located in the core, disruption of the virus envelope is necessary to elicit synthesis. The amount and type of detergent required for optimal activity varies with the type of virus and among different preparations of the same virus. Virus particles that have been stored for long periods or that have been repeatedly frozen and thawed are usually disrupted and require less or no detergent to show activity. Murine viruses show maximum incorporation at about 0.01% Nonidet P-40 or Triton X-100, whereas the avian viruses require 0.1–0.2%. The former are markedly inhibited by excess detergent, whereas the latter agents are resistant to even a 10-fold excess. The enzyme itself is not damaged by the higher detergent concentrations since dilution and addition of an exogenous template restore activity. It appears that high detergent concentrations result in separation of the enzyme from its endogenous template. Some workers preincubate the virus with detergent before beginning the reaction; however, the time of treatment does not seem to be important. The nonionic detergents Nonidet P-40, Triton X-100, and Sterox SL are the most widely used. Stromberg (1972) has done a comparative study of detergent effects with avian myeloblastosis virus.

A divalent cation, Mg^{2+} or Mn^{2+}, is essential for synthesis by all reverse transcriptases examined. There is a broad optimum (5–10 mM) with Mg^{2+} and a much sharper (0.5–2.0 mM) optimum with Mn^{2+}. Which of the two cations affords maximum synthesis at its optimum concentration varies with the template and virus. Low concentrations (generally about 50 mM) of Na^+ or K^+ stimulate activity; however, monovalent cations do not appear to be required. Again, the optimum levels vary with the template and type of virus.

For most viruses tested, the enzyme displays maximum activity between pH 7.8 and 8.5. Activity falls off rapidly below pH 7.5 and much less rapidly above pH 8.5. The optimum reaction temperature is about 37°–40°C for mammalian viruses and about 40°–45°C for avian viruses, which may reflect the higher body temperature of the latter's hosts.

The requirement for all four triphosphates is frequently incomplete since reverse transcriptases are capable of homopolymer synthesis if the appropriate template–primer combination is available. If labeled dTTP is used, incorporation due to copying the polyadenylic acid portion of the viral RNA may be observed in the absence of the other triphosphates. In addition, the presence of nucleotide kinases and phosphotransferases may also result in inability to demonstrate the need for all four substrates.

The reaction is usually performed on all of the fractions of the equilibrium density gradient to show that the activity is found exclusively or predominantly where the virus bands. Since cell membrane vesicles and possibly cell membranes also band in this region, activity so located fulfills a necessary but not sufficient criterion for the presence of RNA tumor viruses. Care must be utilized in interpreting both positive and negative results. The greater concentrations of gradient medium (especially ionic species) in the higher density regions may be inhibitory to some DNA polymerase activities or to contaminating nucleases, phosphatases, and the like and thus shift the polymerase activity peak from its true position. The presence of inhibitors banding at specific regions of the gradient may also affect the results. The amount of protein added to the test reactions may also affect the position of the peak. We have frequently noted such shifts when gradients are performed with crude tissue extracts.

Where assays are repeatedly done on the same type of material, it is advisable to determine the locations of interfering activities. At the same time, determine the range of protein concentrations within which the assay is proportional. The range within which the position of the peak is constant should also be found. The effect of varying the time of incubation must be observed, and optimum concentrations of detergent, divalent cation, salt, and temperature should be determined. Since these parameters will also affect competing activities, determine the positions of these following optimization of other reaction conditions. The sensitivity and reliability of subsequent assays will be greatly improved if the system is well characterized prior to beginning work with large numbers of samples.

B. RIBONUCLEASE SENSITIVITY OF THE ENDOGENOUS REACTION

The endogenous reverse transcriptase reaction of oncornaviruses has been shown to be sensitive to ribonuclease. An aliquot of virus particles (10–50 μg of viral protein) is treated with RNase A and RNase T1 (10 μg/ml each) in the presence of 50 mM Tris-HCl, pH 8.3, 10 mM dithiothrietol, 200 mM NaCl, and the optimal amount of nonionic detergent for 30 minutes at 37°C. A control aliquot is run in parallel, omitting the nucleases. The ribonuclease should be heated at 100°C for 10 minutes at slightly acid pH to inactivate contaminating DNase. RNase A that has been lyophilized contains aggregates that sometimes precipitate during heating. The use of RNase A sold in solution is preferable.

The exact concentration of detergent must be chosen with care. Too little will open the virions sufficiently to allow the polymerase reaction to proceed; however, the viral RNA may still be relatively well protected from the nuclease. Too much detergent will inhibit the reaction, even after dilution, probably due to dissociation of the enzyme and its endogenous template. The treated virions are diluted into the standard assay mixture and processed as described above. The concentration of NaCl in the polymerase reaction should be less than 35 mM for maximum activity.

The purpose of including salt in the ribonuclease digestion reaction mixture is to eliminate selectively single-stranded regions while preserving RNA that is hybridized to DNA. This helps to rule out the possibility that RNase sensitivity of the reaction is due to the destruction of an RNA primer on a DNA template.

C. CHARACTERIZATION OF THE REACTION PRODUCTS

For characterization of the reaction products, the DNA is first freed of protein by extracting the reaction with sodium dodecyl sulfate (SDS) and phenol-cresol-chloroform (PCC—see Appendix). The chloroform is included as suggested by Perry *et al.* (1972) to ensure recovery of poly (A)-containing RNA molecules. The DNA is separated from unincorporated substrate by gel filtration and concentrated by lyophilization from volatile buffer salts. It is then subjected to a variety of chemical and physical tests to prove that the product is DNA and that it has been copied from an RNA template.

Procedure

The reaction is terminated by the addition of SDS (or sodium lauroyl sarcosinate if potassium ion is present or if application of large amounts of material to cesium gradients is contemplated—the potassium and cesium salts of dodecyl sulfate are appreciably less soluble than those of sarcosinate) to 0.1% (w/v), NaCl to 0.4 M, and an equal volume of PCC. The mixture is shaken at room temperature for 5 minutes, and the phases are separated by centrifugation at 3000 g for 5 minutes. The aqueous phase is re-extracted with an equal volume of PCC and then applied to a 1 × 10 cm column of Sephadex G-50 (coarse grade) equilibrated with 0.05 M triethylamine bicarbonate, pH 8.0 (see Appendix). Fractions are collected, and aliquots are counted in water-miscible scintillation fluid to locate the excluded peak, which contains the DNA product free of unincorporated triphosphate. The peak fractions are pooled, frozen in a polypropylene tube, and lyophilized to dryness. Triethylamine bicarbonate is a volatile salt; it is removed during lyophilization as triethylamine and carbon dioxide, leaving the triethylammonium salt of the nucleic acid.

Carrier DNA may be added if desired prior to PCC extraction. DNA adsorbs tenaciously to many types of surfaces, especially glass, and losses may be considerable. Glassware used with DNA products should be siliconized with a solution of dichlorodimethylsilane in benzene (Bio-Rad) to prevent adsorption. DNA used as carrier should be from a distantly related organism; *E. coli* DNA is inexpensive and will not compete in subsequent hybridization analysis. Excessive amounts of carrier DNA may interfere with assays using structure-specific nucleases; therefore, the amount added should be adjusted relative to the amount of product obtained. An amount of

10–20 μg is usually more than adequate to prevent losses during normal handling.

The DNA is taken up in an appropriate volume of 1 mM EDTA, pH 7.0, and divided into aliquots that are analyzed for RNase and DNase sensitivity, resistance to alkaline hydrolysis, buoyant density in Cs_2SO_4, sedimentation coefficient of the RNA:DNA complex, and hybridizability of the DNA to viral RNA.

1. Sensitivity to RNase and DNase

An aliquot of the purified product is treated for 15 minutes at 37°C with RNase A and RNase Tl (10 μg/ml each) that have been heated at 100°C for 10 minutes to destroy contaminating DNase. The reaction is carried out in 10 mM Tris-HCl, pH 7.5, and 1 mM EDTA, at a salt concentration of less than 50 mM. The reactions are terminated by the addition of an equal volume of TCA reagent, kept at 0°C for 10 minutes, and the acid-precipitable radioactivity is collected on membrane filters. A control sample omitting the nucleases is run in parallel.

Treatment with DNase I is performed under the same conditions except that 10 mM $MgCl_2$ is present. DNase I is treated with iodoacetate (Laskowski, 1966) to destroy contaminating RNase and used at a concentration of 10 μg/ml. Commercially available preparations of DNase I labeled "RNase free" often contain sufficient ribonuclease to affect the results and must be assayed for contamination.

2. Alkaline Hydrolysis of the Reaction Products

The DNA product is adjusted to 0.3 N NaOH and incubated at 37°C for 16–24 hours to hydrolyze RNA completely. The mixture is neutralized with 0.3 N HCl. Tris is added to 50 mM before adjusting the pH, and the neutral point is determined by spotting 0.5 μl aliquots onto pH paper. Acid-precipitable radioactivity is determined and compared with a control sample incubated at neutral pH. If the entire sample is to be used for this analysis, it is, of course, not necessary to neutralize the solution before TCA precipitation.

3. Sedimentation Coefficient of the RNA:DNA Complex

RNA tumor viruses contain RNA with a characteristic sedimentation coefficient of 60–70 S. The small (less than 10 S) DNA product that is synthesized in the endogenous reaction remains hydrogen-bonded to the viral RNA during the *early* stages of the reaction and does not significantly alter its sedimentation properties. It was pointed out (Spiegelman et al., 1970a) that the presence of a high molecular weight RNA:DNA complex among early reaction products would be diagnostic of reverse transcriptase activity.

Aliquots of the reaction mixture are extracted with PCC and layered onto 10–30% (v/v) linear glycerol gradients containing 10 mM Tris-HCl, pH 7.5, 1 mM EDTA, and 150 mM NaCl. Samples should include untreated reaction product, product that has been melted by heating at 100 °C for 2 minutes, and product that has been treated with RNase as described above. The materials are sedimented at 40,000 rpm in the Spinco SW 41 rotor at 4 °C for 3.5 hours. Fractions are collected dropwise and acid-precipitable radioactivity is determined. The native product should contain a substantial proportion of the radioactivity cosedimenting with the viral RNA. After melting and RNase treatment, the radioactivity should shift to less than 10 S.

The demonstration of a DNA reaction product sedimenting at 60–70 S is a sensitive indicator of the possible presence of RNA tumor viruses in crude material. Because two features of the reverse transcriptase reaction are assayed together, a positive result is more credible than demonstration of DNA synthesis alone (Schlom and Spiegelman, 1971). In addition, the putative viral DNA product is separated from the bulk of the cellular DNA product, resulting in a much greater percentage potentially hybridizable to the viral RNA. Thus preliminary assay and production of DNA product for further analysis are combined in a single procedure.

It is important to emphasize that the presence of a complex involving 70 S RNA and newly synthesized DNA is not sufficient to ensure that an RNA-instructed DNA synthesis has been identified. One must eliminate the possibilities of end addition reactions or nonspecific aggregation of the DNA with high molecular weight cellular RNA. These difficulties are particularly prone to occur in crude extracts from tumors or normal tissues. Thus in addition to demonstrating the RNase sensitivity of the complex, it is useful to show that its appearance requires the presence of all four deoxyriboside triphosphates. Finally, most convincing is a demonstration that the DNA product purified from the complex can form a specific hybrid structure with heteropolymeric regions of the RNA derived from a known related oncogenic virus.

In applying the procedure, the possibility that large differences among tissues with respect to nuclease and phosphatase levels may profoundly influence the outcomes should be considered. Negative results may be due to difficulties in maintaining the integrity of large viral RNA molecules or to destruction of triphosphates by phosphatases present in crude tissue extracts. Comparative studies between normal and neoplastic tissues to determine levels of reverse transcriptase activity must include controls to show that any differences observed are not due to interfering activities.

4. *Buoyant Density of the Reaction Products*

Aliquots of *early* reaction product—native, melted, and RNase-treated—are dissolved in 40 mM Tris-HCl, pH 7.5, 100 mM NaCl, and 6 mM EDTA. An equal volume of saturated Cs_2SO_4 is added (final density 1.55 gm/ml), and the solution is centrifuged, as described in Table I. Fractions are collected from the bottom of the tube, and the density of every fifth fraction is determined by measuring the refractive index. The samples are then precipitated by adding 0.5 ml yeast RNA (100 μg/ml) and 0.5 ml of 30% (w/v) TCA. After 10 minutes at 0 °C, the acid-precipitable radioactivity is collected on membrane filters.

Recovery of the cDNA is frequently poor due to adsorption to the walls of the tube. Cellulose nitrate tubes are especially to be avoided. Polyallomer tubes give considerably less trouble, especially if they are precoated with some heterologous DNA. The tubes are filled with a solution of DNA (e.g., *E. coli* DNA in 0.01 M NaOH) and allowed to stand for several hours. The DNA is poured off for reuse, and the tubes are rinsed thoroughly with distilled water. They are allowed to dry at room temperature.

A significant part (at least 25%) of the early reaction product should band in cesium sulfate gradients at a density of about 1.67 gm/ml, the density of the RNA template. As the size of the DNA product is increased, the density will shift toward 1.55 gm/ml, approximate density of a 1:1 RNA:DNA hybrid. Free DNA will band at about 1.43–1.47 gm/ml depending upon base composition.

D. Use of Exogenous Templates to Assay Virion DNA Polymerase Activity

In some instances, the purity of the viral RNA is subject to question. Preparations of viral RNA often contain smaller components that can sediment at 70 S with the viral genome. Some of these components may be

TABLE I

Conditions for Cesium Sulfate Equilibrium Density Gradients

Rotor type	Volume per tube (ml)	Speed (rpm)	Temperature (°C)	Run time (hours)
SW 56	3.6	31,000	25	60
SW 50.1	5.0	31,000	25	60
50Ti[a]	10.0	44,000	15	60
60Ti[a]	16.0	40,000	25	60

[a] These tubes are filled to the top with light paraffin oil to prevent contact between the cesium solution and the aluminum tube cap; otherwise, the cap will corrode.

cellular contaminants aggregating with the viral RNA. Back hybridizations to such RNA give results that for some studies are unreliable. The difficulty can be eliminated by offering the enzyme an exogenous natural RNA template from an unrelated source and showing that the DNA product is complementary to it. In addition, the ability of the reverse transcriptase to accept at high efficiency various synthetic homopolymer duplexes (Spiegelman *et al.*, 1970c; Mizutani *et al.*, 1970) has been used as a convenient, if not conclusive, method of assay. Some workers have used these templates with solubilized virions and looked for an increase in incorporation. Since many of these templates give a 100- to 1000-fold stimulation of the reaction, the results are significant. Others have first destroyed the endogenous activity with nucleases that will not attack the homopolymer duplex (e.g., RNase A for assays with dT:rA) or with nucleases that can be inactivated (e.g., micrococcal nuclease) once the endogenous template is destroyed. These methods can also be applied to assay exogenous DNA templates.

1. *Procedure for Exogenous RNA and DNA Templates*

Virus particles are treated with Nonidet P-40 for 10 minutes at $0\,°C$ in the presence of 10 mM dithiothrietol. They are then diluted 10-fold for digestion with micrococcal nuclease (240 μg/ml) in the presence of 50 mM Tris-HCl, pH 8.3, and 2 mM CaCl$_2$. After 30 minutes at room temperature, the nuclease is inactivated by chelating the Ca^{2+} with ethyleneglycol-bis(aminoethyl ether) tetraacetic acid (EGTA) at a final concentration of 4 mM. Nucleoside triphosphates, MgCl$_2$, and exogenous template are added, and the reactions are incubated at $37\,°C$. Aliquots are taken for analysis of acid-precipitable radioactivity, as described previously.

2. *Procedure for Exogenous Homopolymer Template–Primer Combinations*

The reaction mixture is identical to that for the endogenous reaction except that only one triphosphate, that utilized in copying the template strand, is included. The template–primer combination is included at approximately 1 μg per reaction.

Homopolymer template:oligonucleotide primer combinations have been widely used to assay reverse transcriptases because they give about 100- to 1000-fold higher incorporation of the labeled substrates than do natural RNA or DNA templates. The relative preferences of reverse transcriptases for certain templates have been used as a means for distinguishing them from contaminating cellular DNA polymerases, and their sensitivity makes them useful for an initial screening. Table II shows the approximate activities of enzymes of various types with different synthetic templates useful for this

TABLE II
RESPONSES OF DNA POLYMERASES TO VARIOUS TEMPLATES[a,b]

Template–Primer	Substrate	Relative incorporation			
		Reverse transcriptase	Mammalian DNA polymerases		
			α	β	γ
Activated DNA	[³H]dTTP, 3dXTP	100	100	100	100
Oligo(dT):poly(rA)	dTTP	200–2000	<5	5–10	100–1000
Oligo(dT):poly(dA)	dTTP	<5	<5	200	<5
Poly(rA)	dTTP	<1	<5	<1	<1
Oligo(dT)	dTTP	<1	<5	<1	<1
Oligo(dG):poly(rC)	dGTP	500	<5	<1	17–60
Oligo(dG):poly(rC$_m$)	dGTP	<1	<5	<1	<1
Oligo(dG)	dGTP	<1	<5	<1	<1
Poly(rC)	dGTP	<	<5	<1	<1

[a] For each enzyme, the incorporation is expressed relative to that obtained with activated DNA (=100) under commonly used assay conditions. The actual incorporation obtained with this template varies with each enzyme. Values are approximate and will vary if reaction conditions are changed. "Activated DNA" is prepared by limited digestion of double-stranded DNA by nucleases that produce 3′-OH groups. Poly(rC$_m$) is 2′-O-methyl-polycytidylic acid.

[b] References: Abrell and Gallo (1973), Baltimore and Smoler (1971), Bandyopadhyay (1975), Chang and Bollum (1972), DeRecondo et al. (1973), Fridlender et al. (1972), Gallagher et al. (1974), Gerard and Grandgenett (1975), Gerard et al. (1974, 1975), Gerard (1975), Goodman and Spiegelman (1971), Holmes et al. (1974), McCaffrey et al. (1973), Moelling (1974), Mondal et al. (1975), Scolnick et al. (1971), Smith and Gallo (1972), Spadari and Weissbach (1974), Verma (1975a), Wang et al. (1975), Waters and Yang (1974), Wu et al. (1974).

purpose. To date, every primer–template combination except oligo(dG): poly(C$_m$) claimed to be specific for reverse transcriptases has proved to be acceptable to some DNA polymerase of nonviral origin. In addition, the need to add oligonucleotide primers provides the necessary starting point for synthesis by terminal transferase-type activities. Thus, template specificities cannot be used to unambiguously prove the presence of reverse transcriptase activity.

In using these templates, it is important to remember that differences in salt conditions or the presence of various nuclease activities may affect the results. An RNase A-type nuclease activity, specific for pyrimidine stretches will affect reactions using oligo(dG):poly(rC) templates but not those using oligo(dT):poly(rA). The salt concentrations optimal for different enzymes and for the same enzyme with different templates are not identical and are an additional source of error.

E. SEROLOGICAL ASSAYS

Immunological assays provide a means of distinguishing viral reverse transcriptase from other DNA polymerases. The extremely sensitive, but not completely specific, homopolymer-templated reaction can be made considerably more useful if immunological criteria are applied to identify the activity (Parks *et al.*, 1972; Aaronson *et al.*, 1971). Viruses can be easily divided into family groups (avian, mammalian, and reptilian) since there is no cross reaction between them. The two major classes of mammalian viruses, the B-type mammary tumor virus and the C-type viruses, also are easily differentiated. More sensitive, quantitative assays are capable of distinguishing even finer relationships, and it is possible to classify the mammalian C-type viruses according to antigenic relatedness (Sherr *et al.*, 1975).

Procedure

A reaction (90 μl) is prepared containing the detergent, virus, DTT, divalent cation, Tris buffer, and salt at 10/9th the usual concentration for a 100 μl assay, and appropriate amounts of IgG or serum. After incubation for 5–10 minutes at 25 °C, 10 μl containing the appropriate amounts of template and labeled substrate are added. The reaction is incubated at 37 °C for 30 minutes, and the acid-precipitable radioactivity is determined as described above.

The success of the assay depends primarily on the quality of the antiserum used; therefore, it is essential that the protein used for inoculation be of high purity. A number of different animals—rabbits, rats, and guinea pigs—have been used to prepare antisera, and injections into the footpads or in multiple subcutaneous sites have been satisfactory. The primary injection is given in complete Freund's adjuvant with booster injections at 2- to 4-week intervals. Sera with sufficiently high titers are usually obtained after two or three injections. The enzymes are good antigens, and only a few micrograms of protein are needed for each immunization.

The specificity of the antibodies should be checked carefully, and the presence of interfering activities such as nucleases determined. If the serum is of high titer, it may be possible to dilute out nuclease activities. With rabbit serum, a 1000- to 100-fold dilution of crude serum is generally satisfactory. In most cases, and for the most careful work, it will be necessary to purify the IgG fraction. A frequently used procedure is to obtain a crude IgG fraction by ammonium sulfate precipitation (40% fraction) and further purify it by DEAE-cellulose column chromatography. The DEAE column step removes most of the RNase activity present in the serum.

Procedure

Rabbit serum (40 ml) is diluted with an equal volume of 10 mM sodium phosphate, pH 7.5, and 15 mM NaCl, and saturated ammonium sulfate (neutralized with ammonium hydroxide) is added to 40% saturation. After stirring for 10 minutes at room temperature, the precipitate is collected by centrifugation at 10,000 rpm in the Sorvall SS-34 rotor for 20 minutes. The precipitate is dissolved in 100 ml of the sodium phosphate buffer and again brought to 40% saturation. After centrifugation as before, the precipitate is dissolved in 20 ml phosphate buffer and dialyzed against the same buffer to remove the ammonium sulfate. The precipitate that forms during dialysis is removed by centrifugation, and the material is applied to a 1.5 × 5 cm column of DEAE cellulose equilibrated with the same buffer. The column is eluted with the same buffer, and the IgG peak eluting from the column is located by absorbance at 280 nm. The peak fractions are pooled and dialyzed against 0.1 M Tris-HCl, pH 8.0. The phosphate buffer must be removed from the antiserum since it inhibits the reverse transcriptase assay.

F. HYBRIDIZATION OF DNA PRODUCTS

DNA synthesized in an endogenous reverse transcriptase reaction should be hybridizable to the corresponding viral RNA. Product copied from exogenous nucleic acids should similarly reanneal to the proffered template.

Analysis of the hybrid structure must verify that the hybrid is formed between heteropolymeric regions of the RNA and DNA since the synthesis of DNA homopolymers is sometimes observed. These homopolymers may by synthesized by the reverse transcriptase itself, by terminal addition enzymes, by synthetic RNA-dependent DNA polymerases, which copy homopolymer stretches of RNA into DNA, but which are not true reverse transcriptases, or by cellular DNA polymerases.

Poly(dT) is the most likely source of difficulty since poly(A) stretches are present in the viral RNA. One method of controlling this problem is to measure the melting temperature of the complex on hydroxylapatite and compare it to that for a poly(dT):poly(rA) duplex. A true heteropolymer hybrid will have a mean melting point substantially higher than the homopolymer duplex. Another procedure is to label the product in a base other than thymidine and use structure specific nuclease to analyze the complex. Poly(dT):poly(rA) hybrids, being unlabeled, will not be scored. A third method is to use excess unlabeled poly(T) in the hybridization reaction to compete out labeled poly(T) in the DNA product.

Hybridization analyses should always include unrelated heterologous

RNAs to eliminate errors due to trapping of the DNA by protein contamination of the product or by precipitation of the RNA in Cs_2SO_4 or other sources of nonspecificity. In addition, challenging the relevant viral RNA with a heterologous DNA product rules out the possibility that some contaminant of the RNA preparation is responsible for nonspecific RNA:DNA complex formation.

Double-stranded DNA synthesis may also produce difficulties in interpreting hybridization assays. In the presence of excess homologous RNA, a significant fraction of the DNA radioactivity may not reanneal. With heterologous control RNAs, DNA duplexes may form and be scored incorrectly as RNA:DNA hybrids. To circumvent this problem, actinomycin D (100 μg/ml) may be added to the reaction. The antibiotic inhibits almost all DNA-directed DNA synthesis, while reducing RNA-directed DNA synthesis by only a small amount (Ruprecht et al., 1973). Since double-stranded DNA synthesis occurs frequently, many workers include actinomycin D routinely in all reaction mixtures.

Procedure

Because of the scarcity of most challenge RNAs, hybridizations are done in volumes of 10–20 μl in glass capillaries. DNA adsorbs tenaciously to many glass and plastic surfaces; therefore, the capillaries must be treated to prevent loss of the material. The best procedure is to siliconize the capillaries with a 2% solution of dichlorodimethylsilane in benzene. The solution is heated to 60°C, and the glassware is immersed in the hot liquid for several minutes. The surface should be clean and absolutely dry before being treated. The solution is poured off, and the glass is dried in an oven at 100°C. The procedure is then repeated a second time. The solution may be kept for reuse. The coating solution is toxic and flammable, and appropriate precautions should be taken during its use. In our hands, water-soluble siliconizing solutions have produced erratic results with respect to DNA adsorption protection.

The capillaries are drawn out in a flame to a point and cut with a carbide-tipped pencil. This serves to increase the capillarity and facilitates transfer of solutions from micropipettes. Assembly of the reactions frequently requires the pipetting of very small volumes. If this is done in a nonwetting tube, transfer from the pipette is frequently not quantitative, and losses due to evaporation can be quite considerable even in a few moments' time. If done in a wettable tube, recoveries of small volumes are poor. By transferring directly from the pipette to the glass capillary, it is easy to construct accurate reaction mixtures even under 5 μl total volume.

Hybridization reactions to be analyzed on hydroxylapatite columns are performed in 0.12 M sodium phosphate, pH 6.8, 1 mM EDTA, and 0.1%

(w/v) SDS. For analysis with structure-specific nucleases or by cesium sulfate density centrifugation, concentration of phosphate ion must be reduced because it is inhibitory to the enzymes and precipitates with cesium ion at high concentrations. For these procedures, the reaction contains 0.01 M sodium phosphate, pH 6.8, 0.1% (w/v) SDS, 3 mM EDTA, and 0.3 M NaCl. The detergent is included in the reaction mixture to inhibit nuclease activities. EDTA is present to chelate divalent cations that may catalyze the breakdown of the nucleic acids.

The cDNA is treated with NaOH and freed of salt and unincorporated substrate, as described above. The desired amount of cDNA is used in the reaction together with an excess of challenge RNA. When the RNA component of the reaction is in excess, the kinetics of annealing are dependent only on the RNA concentration, i.e., the reaction is pseudo-first order with respect to RNA concentration. If two reactions are run with different concentrations of RNA, C_1 and C_2, they will reach equivalent points on their kinetic curves when $C_1 t_1 = C_2 t_2$; i.e., when the product of RNA concentration and the time of the reaction is the same. Hybridization reactions with RNA tumor virus RNAs generally require that a $C_{RNA} \cdot t$ value of 0.3 mole nucleotide $\cdot$ liter^{-1} $\cdot$ second be attained before the reaction is complete. When the reaction is 50% complete, $C_{RNA} \cdot t = 0.015$ mole $\cdot$ liter^{-1} $\cdot$ second. The amount of RNA used in the reaction should be at least 100-fold greater than the cDNA because of frequent sequence heterogeneity in the latter. Reaction times should be kept conveniently short to minimize breakdown of reaction components; if possible, reaction times of less than 16 hours are recommended.

If the kinetics of the reaction with respect to time are to be measured, either the amount of RNA in the reaction or the time of incubation or both may be varied; however, if the RNA is varied, there is the danger that minor components of the RNA may exert a proportionately greater effect at higher $C_{RNA} \cdot t$ values. It is generally advisable to prepare a single reaction mixture, distribute aliquots to capillaries, and remove points at various times. These are frozen immediately in a Dry Ice–ethanol bath and kept for analysis when convenient. In choosing the time intervals, it is important to remember that the curves are normally plotted on a logarithmic scale; therefore, if the time points are evenly spaced, they will cluster at one end of the curve or the other. Ten to twelve points are normally adequate for a hybridization curve; if spaced so that the $C_{RNA} \cdot t$ values equal 3.2 or 10 or multiples thereof, a series of evenly spaced points on the graph will be obtained.

The capillaries are filled with the reaction components by direct transfer from a microliter pipette to the drawn out point of the reaction capillary. The pipette should be held perpendicular to the capillary; otherwise, if the contents of the pipette are expelled too rapidly, there is danger of introducing a bubble or even expelling the contents out the other end. After the reaction is assembled, a short length of narrow bore, thin wall rubber tubing is attached to the blunt end of the capillary to allow manipulation of the contents. By holding the tubing between the thumb and forefinger, it is possible to draw up or expel smoothly the liquid by moving the finger across the thumb. The reaction mixture is sucked back into the middle of the capillary, and the pointed end is then sealed in a flame. The tubing is then removed and the other end sealed. The capillary is allowed to balloon out *very slightly* under pressure of the entrapped air; this provides proof that a complete seal has been obtained.

The capillaries are then transferred to a boiling water bath and heated at 100 °C for 2 minutes to melt the RNA and DNA. They are then immediately transferred to a 68 °C bath. They are removed at intervals and frozen awaiting analysis.

The capillaries are removed from the freezer and laid across a piece of Dry Ice. While the reaction mixture is still solidly frozen, the capillary is quickly scored at both ends with a carbide-tipped pencil, and the ends are snapped off. This operation should not be performed with the reaction mixture thawed since the rapid motion needed to break the glass may result in loss of the sample. The contents are thawed by holding the capillary between the thumb and finger for a few seconds. The tubing is then attached to one end and used to expel the contents. For hydroxylapatite analysis, the entire contents are expelled into 1 ml of 0.12 M sodium phosphate, pH 6.8, 0.4% (w/v) SDS and for cesium analysis into 0.04 M Tris-HCl, pH 7.4, 6 mM EDTA. For assays with structure-specific nuclease, the reaction mixture is divided into two aliquots, one to be treated with the enzyme and the other to serve as an incubated control without nuclease. Again, it is necessary to make a quantitative transfer of a small volume. Two micropipettes are fitted into their holders and placed on the bench. We use Drummond Microcaps supported on a large rubber stopper into which several deep grooves have been cut. The tip of the reaction capillary is then brought into contact with the tip of the pipette and by manipulating the rubber tubing, as described above, the contents are transferred to the pipettes. Using this technique, it is easy to take two aliquots of 4 μl from a 10 μl hybridization reaction without loss of material or inaccuracies due to evaporation. The material in the pipettes is then transferred to the nuclease reaction mixtures.

1. Hybridization Analysis Procedures

a. *Cesium Sulfate Density Gradient Centrifugation.* The hybrid reaction mixture is diluted with 40 mM Tris-HCl, pH 7.4, 6 mM EDTA to one-half the volume of the tube being used. Gradients are run and processed as described above (Section III,C,4).

Depending upon base composition, the cDNA will band in Cs_2SO_4 at a density of 1.41–1.49 gm/ml. RNA will generally precipitate and form a very sharp band at about 1.60–1.69 gm/ml. RNA:cDNA hybrids will be found at the same density if the cDNA is small relative to the RNA or at intermediate densities between the two components, depending upon the lengths of the complexing species.

Because there may not be a complete separation between hybridized and unhybridized cDNA, the gradients may be difficult to analyze quantitatively. Also, radioactivity found in the RNA or hybrid regions may represent only a small duplex region attached to a longer unhybridized portion. The true amount of hybrid will therefore be overestimated. Further disadvantages of the procedure are its time, expense, and the necessity of fractionating an entire gradient to obtain a single value. Its major advantage is that it generally provides the lowest background value of all available methods, especially if the RNA is considerably larger than the cDNA being used. Thus, if extremely small amounts of hybridization are expected, the cesium gradient may provide the best method of assay.

It is important to be aware of the unusual behavior of poly(dT):poly(rA) duplexes in Cs_2SO_4. Unlike other pairs of deoxy- and ribohomopolymers, these band only slightly heavier than the DNA component (Szybalski and Szybalski, 1971). Thus, Cs_2SO_4 gradient centrifugation may not distinguish between poly(dT) and poly(dT):poly(rA), and hybridization of cDNA to poly(rA) does not afford a suitable assay for poly(T) in a cDNA preparation when analyzed in that way. If poly(dT) is a possible source of difficulty, excess unlabeled polymer should be included in the hybridization reaction mixture to eliminate it by competition.

b. Hydroxylapatite Column Chromatography. Hydroxylapatite column chromatography provides a considerably more convenient and less expensive method of monitoring hybridization than does cesium sulfate density gradient centrifugation. In addition, it permits the determination of the melting point of the hybrid, which provides a measure of the fidelity of base pairing. A completely complementary RNA:DNA duplex will generally have a melting temperature above 80°C when analysis is performed as described below. If hybrid formation is mainly the result of poly(dT):poly(rA) base pairing, the hybrid will elute from the column at temperatures below 75°C. Intermediate values suggest partial sequence complementarity.

Background values generally are low on hydroxylapatite, provided the cDNA is single-stranded. The starting buffer concentration usually ranges from 0.12 to 0.18 M sodium phosphate. The most suitable value depends upon the amount of secondary structure in the DNA. The lowest starting concentration within the given range should be used that provides an acceptable background level for the assay. The recovery of DNA from hydroxylapatite is dependent upon the buffer concentration used to load the DNA onto the column. In no case should starting buffer concentrations be less than 0.1 M phosphate; the use of lower concentrations leads to erratic behavior of the column and frequently to incomplete recovery of the applied radioactivity. Martinson has recently published an extensive series of papers on the binding of nucleic acids to hydroxylapatite, and these contain much valuable information for those using the technique (Martinson, 1973a–d; Martinson and Wagenaar, 1974a,b,c).

Procedure

A 1 cm inside diameter jacketed column is connected to a constant temperature circulator filled with a liquid permitting temperatures above 100 °C (e.g., 33% ethylene glycol in water). The column is brought to 60 °C, and 5 ml of 0.12 M sodium phosphate, pH 6.8, 0.4% (w/v) SDS are added. The required amount of dry, powdered hydroxylapatite (Bio-Rad HTP, *not* DNA grade) is mixed with 5 ml of buffer to give a uniform slurry and poured into the column. Any particles adhering to the sides are washed down with buffer, and air pressure (about 5–10 psi) is applied to pack the bed firmly. The column should not be permitted to run dry or channeling will occur. An additional 5 ml of buffer are applied and forced through the bed.

The reaction mixture is expelled from the capillary into 2 ml of buffer and applied to the column. After the solution has reached 60 °C, the sample is forced through the bed under pressure. The flow rate may be very fast. The column is then washed with additional aliquots (5 × 2 ml is generally adequate) until no more radioactivity is eluted. The temperature of the column is raised in 5 °C steps, and the column is washed with additional aliquots of buffer at each temperature. At 100 °C, the column is washed with the 0.12 M buffer and then with 0.4 M sodium phosphate, pH 6.8, 0.4% (w/v) SDS to remove very tightly bound material.

If the melting behavior of the hybrid is not monitored, simple fractionation into single- and double-stranded material, respectively, is performed by eluting first with the starting buffer and then with 0.4 M sodium phosphate, pH 6.8, 0.4% (w/v) SDS at 60 °C.

The amount of hydroxylapatite to be used depends upon the amount of RNA applied. The capacity of the material varies considerably from batch to batch; therefore, it is convenient to purchase a large bottle and measure the capacity for RNA and DNA. A good lot will bind approximately 100 μg RNA per gram of dry material. The capacity for DNA is generally about 500 μg of DNA per gram. A 0.5 gm column may be used for amounts of RNA less than a few micrograms since excess capacity does not result in poor recovery of the applied material. In testing the capacity of a particular lot, it is important to apply increasing amounts of RNA or DNA and monitor elution of the material as the column is washed. When overloaded, the hydroxylapatite will initially bind nucleic acid that elutes as the column is washed; therefore, the working capacity of the material is somewhat less than the initial binding capacity.

The hydroxylapatite should be weighed individually for each column, and suspended very gently in the starting buffer. Hydroxylapatite is crystalline calcium phosphate, and it is easily fractured into smaller crystals resulting in markedly reduced flow rates. If a stock slurry is prepared for use with many columns, the beds will not be uniform because of repeated mechanical damage to the crystals during resuspension.

c. Structure-Specific Nuclease Assays. The use of structure-specific nucleases provides the most reliable and convenient way of monitoring hybridization when the label is in the DNA strand. Unhybridized regions attached to duplex material are digested and not scored as complexed, whereas both cesium sulfate density gradient analysis and hydroxylapatite column chromatography are subject to this error. Since the viral RNA contains a stretch of poly(A) and the cDNA often contains poly(T), the opportunity for considerable uninteresting hybrid formation exists. By labeling a base other than thymidine, such hybrids are not counted using structure-specific nucleases; therefore, this source of difficulty is easily eliminated. In addition, the method provides a much faster way of handling large numbers of samples than do either gradient or column procedures, and melting point determinations may be done with nucleases as well as with hydroxylapatite.

A number of structure-specific nucleases are known, and nuclease S1, micrococcal nuclease, and *Neurospora crassa* endonuclease are available commercially (Miles Laboratories). The results obtained with each are essentially the same and do not vary by more than a few percent in parallel assays. The enzyme mechanisms are not completely known, and they may differ in the type of structure attacked and the mode of digestion.

The most frequent sources of difficulty in using structure-specific nucleases are (1) failure to use the correct amount of enzyme; (2) failure to control the amount of input nucleic acid, and (3) failure to take account of components in the assay mixture that may interfere with digestion. These problems are easily handled by the use of proper controls and need not cause error.

Like any enzyme reaction, the digestion of DNA by a DNase is described by the rate and extent of the reaction. Because hybridization assays measure the amount of DNA complexed, the rate component of the assay is often overlooked, and the choice of enzyme concentration is made on the basis of the plateau value alone. Thus, a cDNA or other labeled DNA is incubated with various levels of enzyme for what is chosen to be a convenient assay time, and the amount ultimately chosen for the assay is that which digests virtually all of the input DNA. The difficulty with this approach is that DNA molecules themselves often contain considerable secondary structure, sometimes as much as 30% (Shishido and Ando, 1972). Double-stranded DNA will be digested in the presence of excess nuclease; therefore, this procedure frequently results in the use of an incorrect level of enzyme. The method which follows avoids this difficulty by choosing enzyme concentrations that maximize the rate of digestion of the single-stranded component in the test DNA.

Procedure (Vogt, 1973; Kacian and Spiegelman, 1974b)

The assay contains, in a final volume of 0.2 ml:

S1 Nuclease
50 mM sodium acetate, pH 4.6
1 mM zinc sulfate
200 mM sodium chloride
2.0 μg denatured carrier DNA (e.g., *E. coli* DNA)
Labeled test DNA (e.g., the cDNA under study)
Hybridization reaction buffer (at the same concentration to be used when assaying the hybridization reactions)
S1 nuclease (DEAE fraction prepared according to Vogt, 1973)

Micrococcal Nuclease
50 mM Tris-HCl, pH 8.3
10 mM MgCl$_2$
200 mM NaCl
1 mM CaCl$_2$
2.0 μg denatured carrier DNA
Labeled test DNA
Hybridization reaction buffer
Bovine serum albumin (100 μg/ml)
Micrococcal nuclease

The hybridization reaction buffer is included in the same amount as will be present when sampling the actual hybridization reaction. The labeled test DNA is best included at the level actually used, and it should represent only a small fraction (less than 1%) of the denatured carrier DNA. Nucleases should be diluted into the buffer used for the assay containing 100 μg/ml BSA. If purer fractions of S1 nuclease are used, BSA should also be included in the reaction itself to prevent denaturation of the enzyme. The addition of carrier DNA is particularly important for S1 nuclease reactions; without it, digestion is frequently incomplete (Sutton, 1971). The reason for this unusual behavior is obscure.

The reactions are incubated at 45 °C, samples are taken at intervals, and the acid-precipitable radioactivity is determined. The optimal level of enzyme is the lowest amount that produces the maximum *rate* of digestion at short intervals (under 15 minutes). Under assay conditions, the enzyme remains active for several hours; therefore, reactions are run for 2 hours, affording the nuclease ample opportunity to degrade all of the single-stranded DNA present.

It is not clear why excess enzyme degrades double-stranded DNA. This phenomenon may be due to low levels of contaminating nucleases, to breathing of the paired regions, to the intrinsic rate difference between digestion of double- and single-stranded substrates, or to an effect of enzyme binding upon the stability of the duplex. If the level of enzyme chosen under the conditions described above does not completely degrade the cDNA (under 10% should generally remain acid-precipitable), there is probably structure in the DNA itself. This is not infrequently found in cDNA preparations, and occasionally background levels as high as 30% have been observed (Vogt, 1973). In such cases, the cDNA may be first fractionated into single- and double-stranded portions by hydroxylapatite column chromatography, as described above. Care should be taken that the procedure does not remove sequences of interest from the cDNA preparation.

Sodium dodecyl sulfate, phosphate ion, and EDTA are potent inhibitors of the nuclease reactions. Detergent concentrations should not exceed 0.01% (w/v). Phosphate ion concentration should be less than 1 mM. EDTA inhibits by chelating divalent cations required for enzyme activity. The problem may be partially overcome by adding excess cation; however, it is preferable to maintain EDTA concentrations at noninhibitory levels.

After the optimal concentration of enzyme has been determined, the enzyme is diluted so that the required amount may be dispensed in 5 μl. The remaining components of the assay are prepared in 191 μl; 4 μl of hybridization reaction mixture and the nuclease are added. Incubation is at 45 °C for 2 hours, and the acid-precipitable radioactivity is collected on membrane filters. The percentage of hybrid is taken as the fraction of counts resistant to nuclease divided by those in the control reaction without enzyme times 100. Recovery of radioactivity in the control hybridizations should be quantitative, and siliconized tubes should be used to prevent adsorption of the DNA.

2. *Preparation of High Molecular Weight RNA from Tumor Viruses*

The following procedure provides good quality RNA for hybridization with endogenous DNA reaction products or for use as exogenous template. In our laboratory the procedure, a modification of that described by Brahic *et al.* (1971), produces RNA of higher quality and with greater yields than does direct phenol extraction of the virus. The method has been successful with a large number of RNA tumor viruses from both avian and mammalian hosts.

Procedure

The virus is purified, as described above, including both sedimentation velocity and equilibrium density gradient centrifugation. The material removed from the equilibrium gradient is diluted with several volumes of 50 mM Tris-HCl, pH 7.4, 150 mM NaCl, 5 mM EDTA (TNE buffer), and pelleted. As taken from the equilibrium gradient, the virus is in a medium approximately equal to its density and will not pellet without dilution. Centrifugation is performed at 50,000 rpm at 4°C for 60 minutes in the Beckman 50Ti or 60Ti rotor.

The pellet is suspended thoroughly in TNE (0.9 ml per 10 mg virus). One-tenth ml 10% (w/v) SDS (British Drug House specially pure grade) is added per 0.9 ml solution, and the mixture is allowed to stand at room temperature for 5 minutes. The solubilized virus is layered onto a 5–20% (w/v) linear sucrose gradient in TNE containing 0.5% (w/v) SDS. Gradients are prepared at room temperature just before use. The gradients are centrifuged in the Beckman SW 41 rotor at 40,000 rpm at 20°C for 90 minutes. Fractions are collected from the bottom of the tube. Because of the detergent, the flow is rapid and some means of controlling the rate of drop formation is necessary. The Beckman universal fraction recovery system fitted with an adjustable clamp on the air inlet tubing is useful for this purpose. The outlet tubing is flared at the end so that excessively small drops are not formed. A small precipitate is usually present at the bottom of the tube and can be ignored.

The optical density at 260 nm is determined, and the high molecular weight RNA is pooled. The material is normally found about one-third of the way from the bottom of the tube. The peak will be broad because of the comparatively large volume loaded and because of diffusion. Protein and small nucleic acids will be found near the top of the gradient. Normally, 8–10 fractions out of 30 are pooled. Sodium chloride is added to 0.4 M followed by two volumes of absolute ethanol. The solution is mixed thoroughly, and the RNA is allowed to precipitate at −20°C overnight in a siliconized tube. The RNA is centrifuged at 12,000 rpm at 0°C for 30 minutes in the Sorvall HB-4 rotor, and the supernatant is removed by inverting the tube. The pellet is allowed to drain thoroughly, and the RNA is dissolved in 0.15 ml 1 mM EDTA, 0.5% (w/v) SDS for each 10 mg original virus. The solution is heated at 80°C for 3 minutes and allowed to cool to room temperature. In order to avoid precipitation of the detergent the solution is not put into ice. The material is layered onto SDS-sucrose gradients prepared as above and centrifuged as before for 3 hours. Fractions are collected, and the absorbance at 260 nm is determined. The high molecular weight RNA is pooled and precipitated with NaCl-ethanol, as described above. The pellet is suspended in cold 70% (v/v) ethanol and recentrifuged in order to remove residual salt and detergent. The pellet is drained thoroughly and dissolved in 0.1 mM EDTA before it dries. If the pellet is dried prior to dissolving the RNA, considerable aggregation occurs, and the material redissolves only with great difficulty. The RNA solution is frozen and lyophilized to eliminate residual ethanol. It is then taken up in a volume of water equal to the EDTA solution added previously. The concentration is determined from the optical density, and the RNA is divided into convenient aliquots. If the virus preparation is not very pure or if protein remains, the RNA should be extracted with PCC following the second gradient and prior to the final ethanol precipitation.

If the virus has not been harvested at short intervals from tissue culture media (generally less than 4 hours), a substantial portion of the RNA may be degraded (Bader and Steck, 1969; Erikson, 1969). On the first gradient it may still sediment at 70 S, but upon melting it runs at 10 S or less. In such cases, isolation of intact 35 S RNA is impractical, and the procedure is modified as follows: After isolation of the 70 S RNA component, as described above, the gradient fractions are pooled and extracted with an equal volume of PCC. If desired, the material may be treated with Pronase (Calbiochem, nuclease-free grade) at 100 μg/ml for 1 hour at 37 °C prior to PCC extraction. Sodium chloride is then added to 0.4 M, and the RNA is precipitated with ethanol and lyophilized, as described above. The heating step and the second gradient to isolate the 35 S component of the RNA are omitted.

IV. Assay of Solubilized Reverse Transcriptase

In many cases it is necessary to assay reverse transcriptase not in association with virus particles. The number of actual virions that can be isolated from a sample of tumor tissue is probably small. Once they have budded from the surface of the producing cell, virions probably enter the intercellular fluids and are carried throughout the body; therefore, they may not accumulate in great numbers in association with the involved tissue.

Considerable reverse transcriptase activity is found in cell- and virus-free extracts from tumor tissues taken from infected animals, such as chickens with avian myeloblastosis or mice infected with Rauscher leukemia virus. Thus, the soluble fraction of a cell or tissue extract may provide a better source of enzyme than the particulate fraction having the same density and sedimentation characteristics of virions.

There are essentially three procedures that are useful for assaying soluble reverse transcriptase: (1) assays with exogenous templates, particularly homopolymer template–primer combinations; (2) serological assays testing for inhibition of reverse transcriptase activity with homopolymer template–primer combinations; (3) radioimmunoassay. The exogenous template assays depend upon showing a response with certain template–primer combinations known to be copied efficiently by viral reverse transcriptases and poorly by cell enzymes. The assays are extremely useful because of their great sensitivity (up to 1000-fold higher than reactions with endogenous templates or natural RNA), but, as discussed above, they are not completely specific for reverse transcriptases. Thus, the assays should only be considered initial indicators of the presence of reverse transcriptase.

By combining the homopolymer template–primer assay with serological tests, the specificity of the assay for viral reverse transcriptase is considerably enhanced. The assay is performed as described above for detergent-treated virions.

Radioimmunoassay provides a third method of monitoring the presence of reverse transcriptase. Panet *et al.* (1975) have described an assay for the DNA polymerase from avian myeloblastosis virus. They estimate the sensitivity of the assay to be one-seventh to one-ninth that of the activity neutralization assay. Radioimmunoassay has the disadvantage of measuring both active and inactive protein; however, it is able to detect the presence of enzyme in association with inhibitors that make activity assays impossible. Since at the present time no other radioimmunoassays for oncornavirus reverse transcriptases have been published, the reader is referred to the cited paper for details.

V. Purification of Reverse Transcriptase

The presence of contaminating activities and nucleic acids frequently limits the amount of information obtainable with crude systems. In order to obtain an enzyme fraction able to provide a sufficient amount of cDNA or reverse transcriptase activity for adequate characterization, partial or complete purification of the enzyme may be necessary.

Reverse transcriptase has been extensively purified from avian myeloblastosis virus (AMV) (Kacian *et al.*, 1971; Kacian and Spiegelman, 1974a; Hurwitz and Leis, 1972), Rous sarcoma virus (Faras *et al.*, 1972), and murine leukemia viruses (Hurwitz and Leis, 1972; Moelling, 1974; Verma, 1975a,b) by essentially similar procedures. Avian myeloblastosis virus is the most economical source, and the purification of its enzyme will be detailed to illustrate the methodology.

1. Assay Procedure

The assay measures the incorporation of radioactively labeled deoxy-nucleoside triphosphate into acid-insoluble material. The reaction mixture (total volume 0.1 ml) contains:

Tris-HCl, pH 8.3, 50 mM
MgCl$_2$, 8 mM
dATP, 0.2 mM
[3H]Methyl-dTTP, 0.2 mM, 25 cpm/pmole
oligo(dT):poly(rA) (Miles Laboratories), 0.4 μg
Enzyme (0.1–1.5 Units)

Samples are incubated at 37°C for 10 minutes, and the reaction is terminated by the addition of 0.1 ml TCA reagent (see Appendix). After standing at 0°C for 10 minutes, the sample is filtered through a membrane filter, washed 10 times with 4 ml cold 5% (w/v) TCA, and dried under an infrared lamp. The acid-insoluble radioactivity remaining on the filter is determined by scintillation counting.

Unincubated controls are run in parallel with all assays.

A unit of enzyme activity converts 1 nmole of dTTP to an acid-insoluble form under the conditions described.

Assays using natural RNA and DNA templates are prepared identically except that they contain 0.2 mM each of three unlabeled nucleoside triphosphates and 0.004 mM of the fourth triphosphate labeled with tritium at 500 cpm/pmole. Templates are used at 1–5 μg per reaction.

The reaction with oligo(dT):poly(rA) is very rapid and usually exhibits linear kinetics until about 10 minutes. Assays are proportional to the amount of added enzyme using approximately 1–20 μl of the diluted crude extract, DEAE-cellulose column pool, and phosphocellulose column pool, and from 0.5 to 5 μl of the glycerol gradient pool. The kinetics obtained with natural RNA and DNA templates vary greatly depending upon the template itself; linear kinetics may be obtained for several hours in many cases.

For determining recovery, the amount of detergent and salt should be adjusted to equal that present when assaying the crude extract since these materials result in a significant stimulation of the reaction.

2. Protein Determination

Protein determination using the purification procedure described is problematic. Most of the buffer components interfere with the Lowry method (Lowry et al., 1951), whereas the very high absorption of Nonidet P-40 in the ultraviolet precludes use of the method of Warburg and Christian (1942).

Precipitation of the protein with TCA in order to remove interfering substances and to permit use of one of the standard methods is not satisfactory
since the detergent itself precipitates and interferes with quantitative
precipitation of the protein. Extraction with water-immiscible solvents to
remove the detergent also gave poor results.

The fluorescamine procedure developed by Bohlen *et al.* (1973) does not
suffer these drawbacks. There is little interference by the buffer components,
including the detergents, and by appropriate standardization the error is
easily controlled. The amount of protein present may be determined easily
in the later stages of the purification without sacrificing large amounts of
material. The procedure is rapid and gives results that are in excellent agreement with those obtained by other methods applied to this system. A disadvantage is the requirement for a spectrophotofluorimeter.

Procedure

The protein sample in buffer is diluted with water to 1 ml. While the tube
is agitated on a vortex mixer, 0.5 ml of fluorescamine (Hoffmann-La Roche)
solution (300 μg/ml spectrograde acetone) is added. The reaction is complete within seconds, as is the destruction of excess reagent. The fluorescence
is determined (390 nm excitation, 475 nm emission) and compared with
bovine serum albumin standards prepared in equivalent buffer concentrations.

3. Preparation of Ion Exchange Celluloses

Whatman DE52 is suspended in about 5 volumes of distilled water and
allowed to settle. Fine particles are removed by aspiration of the supernatant
liquid. The operation is repeated until the particles settle in 5 minutes,
leaving a clear liquid above. The cellulose is then poured into a Buchner
funnel lined with fine mesh polypropylene screen. A gentle vacuum is
applied to remove most of the water. The cellulose is then transferred to a
beaker containing 5 volumes of 1.0 M potassium phosphate, pH 7.2. It is kept
in suspension by gentle intermittent stirring with a glass rod. After 10
minutes, the material is returned to the funnel and washed with 15 volumes
of 10 mM potassium phosphate, pH 7.2, under suction. It is then resuspended
in the same buffer to give a 50% (v/v) slurry.

Whatman phosphocellulose P11 is treated as described above to remove
fine particles. The bulk of the water is removed, and the cellulose is transferred to a beaker containing 5 volumes 0.5 M NaOH. It is allowed to stand
for 30 minutes (no longer!), poured into the funnel, and washed with distilled
water until the pH of the filtrate is that of the water. The use of polypropylene mesh instead of filter paper is strongly recommended to permit rapid
washing. The cellulose is then transferred to a beaker containing 5 volumes

of 0.5 N HCl. After 30 minutes, it is poured into the funnel and washed with water as before. The cellulose is equilibrated with potassium phosphate, pH 7.2, as above, except that it is washed with 15 volumes of 100 mM potassium phosphate, pH 7.2, on the funnel after being first suspended in 1.0 M buffer. After washing, the material is suspended in buffer to make a 50% (v/v) slurry.

The celluloses may be stored in the cold after addition of sodium azide to 0.2% (w/v). Storage longer than 6 months, even in the presence of preservative, is not recommended.

The columns are poured in the following manner. The appropriate starting buffer is added to the column and allowed to run out to dislodge bubbles, in the support screen. With only a small amount of buffer remaining in the column itself, the cellulose slurry is added all at one time. The column outlet is immediately opened and the column permitted to pack undisturbed. The columns are then washed with the complete starting buffers containing dithiothrietol, glycerol, and Nonidet P-40 (a minimum of 5 column volumes) just before use.

4. Purification of AMV from Plasma of Infected Birds

The frozen plasma (obtained from Dr. J. Beard, Life Sciences, St. Petersburg, Florida) is thawed under cool, running water or in the cold room overnight. As the material thaws, the bottles are squeezed to break the mass into small chunks. The bottles are put into an ice slurry just before they completely thaw so that the temperature does not rise above 10°C.

One gram of kieselguhr is added to each 100 ml of plasma and suspended by stirring gently with a glass rod. The plasma is then centrifuged at 2000 rpm for 10 minutes at 4°C in the Sorvall GSA rotor. The rotor is allowed to stop without braking, and the supernatant is taken. Generally the pellet is firm, and the supernatant can be removed easily. Occasionally, it is gelatinous and adheres poorly to the bottle. In such cases, an additional gram of kieselguhr is added per 100 ml supernatant, and the centrifugation is repeated.

A Buchner funnel and vacuum flask are chilled. A 10 cm diameter funnel is adequate for about 500 ml plasma. For other quantities, the area of the filter should be scaled proportionately. A piece of fine-grade filter paper (Whatman No. 1 or S&S White Ribbon) is put into the funnel and moistened with 0.05 M Tris-HCl, pH 8.3, 0.1 M NaCl, and 1 mM EDTA (TSE buffer). A gentle vacuum is applied, and a dilute slurry of kieselguhr in TSE is poured into the funnel to deposit a uniform layer about 5 mm thick. The kieselguhr layer is washed carefully with several portions of TSE, and the washes are discarded.

The plasma is then filtered. Excessive vacuum should not be used in order to minimize foaming. After all the supernatant has been passed through the funnel, the filter is washed with several small aliquots of TSE. The filtration step removes fibrinogen strands and other debris from the plasma. The filtrate is transferred to centrifuge bottles, and the virus is pelleted in a Beckman Type 19 rotor at 18,000 rpm for 90 minutes at 4 °C. The supernatant is poured off, and the bottles are allowed to rest at an angle in ice to permit additional liquid to drain away from the pellet. After removal of the remaining supernatant with a pipette, the pellets are resuspended in TSE (10 ml per 100 ml original plasma). A Dounce homogenizer with a tight-fitting pestle is used; otherwise, it is very difficult to disperse the gelatinous pellets. It is extremely important is suspend the virus completely to obtain quantitative recovery at the sedimentation velocity gradient step.

The virus suspension is sonicated briefly to ensure complete dispersion. A period of 30–60 seconds, using a Branson sonifier with microtip at 100 W output, is adequate for volumes up to 60 ml. The tube is kept in an ice slurry during the sonication to avoid temperature rise.

The virus suspension is then layered over discontinuous sucrose gradients in SW 27 cellulose nitrate tubes. The gradients consist of 7 ml 50% (w/v) sucrose in TSE, 11 ml 35% (w/v) sucrose in TSE, and 10 ml 20% (w/v) sucrose in TSE. Each gradient is loaded with 10 ml virus suspension. The gradients are centrifuged at 27,000 rpm for 60 minutes at 4 °C. The virus will be found as a heavy band in the middle of the 35% sucrose layer. There should only be a small amount of material on top of the 50% layer, which consists of virus aggregates and debris. The presence of a heavy band indicates that the virus was not thoroughly suspended. This material should be removed, resonicated after dilution, and rerun on the gradients.

The virus bands at the middle of each of the 35% sucrose layers are removed and pooled. They are diluted to three times their volume with TSE and layered over discontinuous sucrose gradients consisting of 4.7 ml each 60%, 51%, 42%, 33%, 24%, and 15% (w/v) sucrose in TSE (or a continuous 15–60% gradient if preferred). The gradients are spun at 27,000 rpm in the Beckman SW 27 rotor at 4 °C for 5 hours. The virus band (density = 1.16 gm/ml) is removed, and the yield is determined by reading the optical density at 260 nm. One A_{260} unit corresponds to approximately 243 μg virus (Smith and Bernstein, 1973). If the virus bands from the sedimentation gradients are carefully removed so that the volume is kept to a minimum, the virus from two of the three step gradients may be layered onto one equilibrium gradient. This is best done by side puncture with a hypodermic syringe.

The virus is aliquoted into convenient portions and stored at −70 °C. It is more stable in the presence of high sucrose; therefore, if removal of the sucrose is necessary for subsequent use, it is accomplished after the virus is taken from the freezer. The material is generally suitable for enzyme isolation or RNA extraction for over 1.5 years.

5. *Enzyme Isolation*

The procedure is described for 100 milligrams of virus purified as described above. Other quantities are easily handled by scaling all steps proportionately.

Ten ml AMV (10 mg/ml) are mixed in order with 1.0 ml Nonidet P-40, 1.0 ml 10% (w/v) sodium deoxycholate (Merck), and 3 ml 4 M KCl until homogeneous. The suspension is kept at 0 °C for 15 minutes and then centrifuged at 16,000 g for 10 minutes. The pellet is discarded, and the supernatant is diluted to 10 times its volume with 10 mM potassium phosphate, pH 7.2, 2 mM dithiothrietol, 10% (v/v) glycerol, 0.2% (v/v) Nonidet P-40.

The solution is applied to a 1.5 × 11.0 cm column of DEAE-cellulose equilibrated with the same buffer. The column is washed with 100 ml of the same buffer containing 50 mM potassium phosphate, and eluted with the buffer containing 300 mM potassium phosphate. The flow rate is maintained at 40 ml per hour, and fractions of 2 ml are collected.

The active fractions from the DEAE-cellulose column are pooled and diluted to three times their volume with the same buffer containing 10 mM potassium phosphate, pH 7.2. The material is loaded onto a 1.5 × 7.0 cm column of phosphocellulose equilibrated with the buffer containing 100 mM potassium phosphate. The column is washed with 24 ml of 100 mM potassium phosphate buffer and eluted with an 80 ml linear gradient from 100 mM potassium phosphate buffer to 600 mM potassium phosphate buffer. The flow rate is kept at 40 ml per hour, and 1 ml fractions are collected.

The peak fractions from the phosphocellulose column are pooled into a dialysis bag (Spectrum Spectrapor No. 1) and dialyzed against 200 mM potassium phosphate, pH 7.2, 2 mM dithiothrietol, 0.2% Nonidet P-40. Aquacide II (Calbiochem) is then packed around the bag to concentrate the protein to 0.2 ml. The solution is layered over a 10–30% (v/v) linear glycerol gradient in the same buffer. The protein is sedimented at 50,000 rpm in the Beckman SW 50.1 rotor for 18 hours at 1 °C. Fractions are collected from the bottom of the tube. The peak activity is pooled, adjusted to contain 50% (v/v) glycerol, and stored at −20 °C.

Gel filtration may be used instead of glycerol gradient centrifugation. The phosphocellulose column fractions are pooled, concentrated if necessary, and applied to a 0.9 × 60 cm column of LKB Ultrogel AcA 44 equilibrated with 200 mM potassium phosphate, pH 7.2, 2 mM dithiothrietol, 0.2% NP-40, 10% glycerol. The column is eluted at a flow rate of 2.5 ml per hour, and fractions of 0.5 ml are collected. The peak fractions are pooled, brought to 50% (v/v) glycerol, and stored at −20 °C.

For maximum yields, the procedure should be completed within 2.5 days. The DEAE column is loaded early on the first day, and the enzyme is applied overnight to the phosphocellulose column. The column is eluted on the second day, the enzyme concentrated if necessary, and the gel filtration or glycerol gradient centrifugation step allowed to proceed overnight. The enzyme purification is completed on the morning of the third day.

The procedure is relatively short and contains no unnecessary steps. If removal of contaminating material does not occur at the proper stage, it frequently is not eliminated at a subsequent step. A common source of difficulty is failure to prepare the ion exchangers properly. Another problem arises when the same reservoir and tubing used to load the crude extract onto the DEAE column are used for the wash buffer without being cleaned in between. If the wash buffer is simply added to the reservoir after loading, it becomes contaminated and will not effectively remove unwanted activities from the column.

The use of nonionic detergent in all the buffers markedly stabilizes the activity, resulting in a several-fold improvement in yield (Faras *et al.*, 1972). Some difficulty may be experienced in preparing the buffers described since a fine precipitate forms when they are warm. The precipitate readily dissolves when the buffers are cold, and a clear solution is obtained.

The purified enzyme retains full activity for over 1 year when stored in 50% (v/v) glycerol at $-20\,^{\circ}\mathrm{C}$. The enzyme also retains activity when stored frozen at $-70\,^{\circ}\mathrm{C}$ or in liquid nitrogen; however, it is rapidly inactivated by repeated freezing and thawing.

Moelling (1974) has described purification of reverse transcriptase from Friend murine leukemia virus by essentially the same procedure. A number of authors have also used very similar procedures for isolating the enzyme from cells or tissues (Bandyopadhyay, 1975; Mondal *et al.*, 1975; Abrell and Gallo, 1973; Sarngadharan *et al.*, 1972). Affinity chromatography, using various nucleic acids bound to cellulose or agarose or immunoadsorbants, has also proved to be a valuable technique for isolating the enzyme from a variety of sources (Kacian *et al.*, 1971; Gerwin and Milstein, 1972; Gerwin and Bassin, 1973; Livingston *et al.*, 1972a,b; Marcus *et al.*, 1974; Grandgenett and Rho, 1975).

Appendix

The preparation of certain reagents mentioned in the text is detailed below.

1. *Phenol-Cresol-Chloroform (PCC)*

PCC is prepared by mixing 500 gm phenol, 70 ml *m*-cresol, 0.5 gm 8-hydroxyquinoline, and equilibrating with 100 m*M* Tris-HCl, pH 7.5. After equilibration, a volume of chloroform equal to the organic layer is added.

The solution is stored in the dark under a layer of buffer at 4 °C. The phenol and cresol are distilled and stored at 4 °C or colder until used. The phenol is weighed and liquified by adding water so that the color of the compound can be checked. Both the phenol and cresol should be completely colorless. Even a slight pink appearance indicates the presence of undesirable breakdown products, and the material must be redistilled.

2. *Triethylamine Bicarbonate Buffer* (1.0 *M*)

Freshly distilled triethylamine (139.5 ml) is placed into a 2-liter round bottom flask with side arm, and 800 ml distilled water are added. An egg-shaped magnetic stirring bar is placed in the flask, and a fritted glass gas dispersion tube is inserted into a silicone rubber stopper and positioned into the solution through the side arm. A reflux condenser is attached to the flask, and water at about 2 °C is circulated through it. While stirring vigorously, purified CO_2 is bubbled slowly through the mixture until all of the triethylamine has gone into solution. It is then transferred to a 1-liter volumetric flask, brought to volume, and returned to the round bottom flask. The pH is adjusted to the desired value by sparging further with CO_2. The reagent is stored in tightly closed, amber glass bottles at 4 °C. The pH should be checked frequently and adjusted if necessary.

3. *TCA Reagent*

TCA reagent is prepared by mixing equal volumes of 100% (w/v) trichloroacetic acid, saturated sodium phosphate (monobasic), and saturated sodium pyrophosphate. It is stored in the cold in a dark bottle. It is stable for 6 months to 1 year, but it should be discarded if a brown residue develops.

References

Aaronson, S. A., Parks, W. P., Scolnick, E. M., and Todaro, G. J. (1971). *Proc. Natl. Acad. Sci. U.S.A.* **68**, 920–924.

Abrell, J. W., and Gallo, R. C. (1973). *J. Virol.* **12**, 431–439.

Anderson, N. G. (1966). *Natl. Cancer Inst., Monogr.* **21**, 253–283.

Bader, J. (1969). *In* "The Biochemistry of Viruses" (H. B. Levy, ed.), pp. 293–328. Dekker, New York.

Bader, J. P., and Steck, T. L. (1969). *J. Virol.* **4**, 454–459.

Baltimore, D., and Smoler, D. (1971). *Proc. Natl. Acad. Sci. U.S.A.* **68**, 1507–1511.

Baltimore, D., and Smoler, D. (1972). *J. Biol. Chem.* **247**, 7282–7287.

Bandyopadhyay, A. K. (1975). *Arch. Biochem. Biophys.* **166**, 83–93.

Beard, J. W. (1963). *Adv. Cancer Res.* **7**, 1–127.

Bohlen, P., Stein, S., Dairman, W., and Udenfriend, S. (1973). *Arch. Biochem. Biophys.* **155**, 213–220.

Bolognesi, D. P., and Bauer, H. (1970). *Virology* **42**, 1097–1112.

Brahic, M., Tamalet, J., and Chippaux-Hyppolite, C. (1971). *C. R. Hebd. Seances Acad. Sci.* **272**, 2115–2118.

Bronson, D. L., Elliott, A. Y., and Ritzi, D. (1975). *Appl. Microbiol.* **30**, 464–471.

Bryan, W. R., and Moloney, J. B. (1957). *Ann. N.Y. Acad. Sci.* **68**, 441–453.

Calafat, J., and Hageman, P. (1968). *Virology* **36**, 308–312.

Chang, L. M. S., and Bollum, F. J. (1972). *Biochemistry* **11**, 1264–1272.

Dahlberg, J. F., Sawyer, R. C., Taylor, J. M., Faras, A. J., Levinson, W. E., Goodman, H. M., and Bishop, J. M. (1974). *J. Virol.* **13**, 1126–1133.

DeRecondo, A.-M., Lepesant, J.-A., Fichot, O., Grasset, L., Rossignol, J.-M., and Cazillis, M. (1973). *J. Biol. Chem.* **248**, 131–137.

Duesberg, P., Helm, K. V. D., and Canaani, E. (1971). *Proc. Natl. Acad. Sci. U.S.A.* **68**, 2505–2509.

Duesberg, P. H., and Blair, P. B. (1966). *Proc. Natl. Acad. Sci. U.S.A.* **55**, 1490–1497.

Duesberg, P. H., Robinson, H. L., Robinson, W. S., Huebner, R. J., and Turner, H. C. (1968). *Virology* **36**, 73–86.

Erikson, R. L. (1969). *Virology* **37**, 124–131.

Fan, H., and Baltimore, D. (1973). *J. Mol. Biol.* **80**, 93–117.

Faras, A. J., Taylor, J. M., McDonnell, J. P., Levinson, W. E., and Bishop, J. M. (1972). *Biochemistry* **11**, 2334–2342.

Fridlender, B., Fry, M., Bolden, A., and Weissbach, A. (1972). *Proc. Natl. Acad. Sci. U.S.A.* **69**, 452–455.

Gallagher, R. E., Todaro, G. J., Smith, R. G., Livingston, D. M., and Gallo, R. C. (1974). *Proc. Natl. Acad. Sci. U.S.A.* **71**, 1309–1313.

Gerard, G. F. (1975). *Biochem. Biophys. Res. Commun.* **63**, 706–711.

Gerard, G. F., and Grandgenett, D. P. (1975). *J. Virol.* **15**, 785–797.

Gerard, G. F., Rottman, F., and Green, M. (1974). *Biochemistry* **13**, 1632–1641.

Gerard, G. F., Loewenstein, P. M., Green, M., and Rottman, F. (1975). *Nature (London)* **256**, 140–143.

Gerwin, B. I., and Bassin, R. H. (1973). *Proc. Natl. Acad. Sci. U.S.A.* **70**, 2453–2456.

Gerwin, B. I., and Milstein, J. B. (1972). *Proc. Natl. Acad. Sci. U.S.A.* **69**, 2599–2603.

Gianni, A. M., Smotkin, D., and Weinberg, R. A. (1975). *Proc. Natl. Acad. Sci. U.S.A.* **72**, 447–451.

Goodman, N. C., and Spiegelman, S. (1971). *Proc. Natl. Acad. Sci. U.S.A.* **68**, 2203–2206.

Grandgenett, D. P., and Rho, H. M. (1975). *J. Virol.* **15**, 526–533.

Guntaka, R. V., Mahy, B. W. J., Bishop, J. M., and Varmas, H. E. (1975). *Nature (London)* **253**, 507–511.

Hanafusa, H., Baltimore, D., Smoler, D., Watson, K. F., Yaniv, A., and Spiegelman, S. (1972). *Science* **177**, 1188–1191.

Holmes, A. M., Hesslewood, I. P., and Johnston, I. R. (1974). *Eur. J. Biochem.* **43**, 487–499.

Hurwitz, J., and Leis, J. (1972). *J. Virol.* **9**, 116–129.

Kacian, D. L., and Spiegelman, S. (1974a). *In* "Methods in Enzymology" (L. Grossman and K. Moldave, eds.), Vol. 29, pp. 150–173. Academic Press, New York.

Kacian, D. L., and Spiegelman, S. (1974b). *Anal. Biochem.* **58**, 534–540.

Kacian, D. L., Watson, K. F., Burny, A., and Spiegelman, S. (1971). *Biochim. Biophys. Acta* **246**, 365–383.

Keller, W., and Crouch, R. (1972). *Proc. Natl. Acad. Sci. U.S.A.* **69**, 3360–3364.

Laskowski, M. (1966). *Proceed. Nucleic Acid Res.* **1**, 85–101.

Leis, J. P., Berkower, I., and Hurwitz, J. (1973). *Proc. Natl. Acad. Sci. U.S.A.* **70**, 466–470.

Livingston, D. M., Parks, W. P., Scolnick, E. M., and Ross, J. (1972a). *Virology* **50**, 388–395.

Livingston, D. M., Scolnick, E. M., Parks, W. P., and Todaro, G. J. (1972b). *Proc. Natl. Acad. Sci. U.S.A.* **69**, 393–397.

Lowry, O. H., Rosenbrough, N. J., Farr, A. L., and Randall, R. J. (1951). *J. Biol. Chem.* **193**, 265–275.

Lyons, M. J., and Moore, D. H. (1965). *J. Natl. Cancer Inst.* **35**, 549–557.

McCaffrey, R., Smoler, D. G., and Baltimore, D. (1973). *Proc. Natl. Acad. Sci. U.S.A.* **70**, 521–525.

Marcus, S. L., Modak, M. J., and Cavalieri, L. F. (1974). *J. Virol.* **14**, 853–859.

Martinson, H. G. (1973a). *Biochemistry* **12**, 139–145.

Martinson, H. G. (1973b). *Biochemistry* **12**, 145–150.

Martinson, H. G. (1973c). *Biochemistry* **12**, 2731–2736.

Martinson, H. G. (1973d). *Biochemistry* **12**, 2737–2746.

Martinson, H. G., and Wagenaar, E. B. (1974a). *Anal. Biochem.* **61**, 144–154.

Martinson, H. G., and Wagenaar, E. B. (1974b). *Biochemistry* **13**, 1641–1645.

Martinson, H. G., and Wagenaar, E. B. (1974c). *Can. J. Biochem.* **52**, 267–271.

Mizutani, S., Boettiger, D., and Temin, H. M. (1970). *Nature (London)* **228**, 424–427.

Moelling, K. (1974). *Virology* **62**, 46–59.

Mölling, K., Bolognesi, D. P., Bauer, H., Busen, N., Plassmann, H. W., and Hausen, P. (1971). *Nature (London) New Biol.* **234**, 240–243.

Mondal, H., Gallagher, R. E., and Gallo, R. C. (1975). *Proc. Natl. Acad. Sci. U.S.A.* **72**, 1194–1198.

Oroszlan, S., Johns, L. W., and Rich, M. A. (1965). *Virology* **26**, 638–645.

Panet, A., Baltimore, D., and Hanafusa, T. (1975). *J. Virol.* **16**, 141–145.

Parks, W. P., Scolnick, E. M., Ross, J., Todaro, G. J., and Aaronson, S. A. (1972). *J. Virol.* **9**, 110–115.

Perry, R. P., and Kelley, D. E. (1966). *J. Mol. Biol.* **16**, 255–268.

Perry, R. P., LaTorre, J., Kelley, D. E., and Greenberg, J. R. (1972). *Biochim. Biophys, Acta* **262**, 220–226.

Reid, E. (1972). *In* "Subcellular Components. Preparation and Fractionation" (G. D. Birnie, ed.), pp. 93–118. Univ. Park Press, Baltimore, Maryland.

Ruprecht, R. M., Goodman, N. C., and Spiegelman, S. (1973). *Biochim. Biophys. Acta* **294**, 192–203.

Sarngadharan, M. G., Sarin, P. S., Reitz, M. S., and Gallo, R. C. (1972). *Nature (London), New Biol.* **240**, 67–72.

Schlom, J., and Spiegelman, S. (1971). *Science* **174**, 840–843.

Scolnick, E. M., Aaronson, S. A., Todaro, G. J., and Parks, W. P. (1971). *Nature (London)* **229**, 318–321.

Sherr, C. J., Fedele, L. A., Benveniste, R. E., and Todaro, G. J. (1975). *J. Virol.* **15**, 1440–1448.

Shishido, K., and Ando, T. (1972). *Biochim. Biophys. Acta* **287**, 477–484.

Smith, R. E., and Bernstein, E. H. (1973). *Appl. Microbiol.* **25**, 346–353.

Smith, R. G., and Gallo, R. C. (1972). *Proc. Natl. Acad. Sci. U.S.A.* **69**, 2879–2884.

Spadari, S., and Weissbach, A. (1974). *J. Biol. Chem.* **249**, 5809–5815.

Spiegelman, S., Burny, A., Das, M. R., Keydar, J., Schlom, J., Travnicek, M., and Watson, K. (1970a). *Nature (London)* **227**, 563–567.

Spiegelman, S., Burny, A., Das, M. R., Keydar, J., Schlom, J., Travnicek, M., and Watson, K. (1970b). *Nature (London)* **227**, 1029–1031.

Spiegelman, S., Burny, A., Das, M. R., Keydar, J., Schlom, J., Travnicek, M., and Watson, K. (1970c). *Nature (London)* **228**, 430–432.

Stromberg, K. (1972). *J. Virol.* **9**, 684–697.

Stromberg, K., Litwack, M. D., and Desmukes, B. (1972). *Proc. Soc. Exp. Biol. Med.* **141**, 215–221.

Sutton, W. D. (1971). *Biochim. Biophys. Acta* **240**, 522–531.

Syrewicz, J. J., Naso, R. B., Wang, C. S., and Arlinghaus, R. B. (1972). *Appl. Microbiol.* **24**, 488–494.

Szybalski, W., and Szybalski, E. H. (1971). *Proced. Nucleic Acid Res.* **2**, 311–354.

Temin, H. M. (1971). *Annu. Rev. Microbiol.* **25**, 609–648.

Temin, H. M. (1974). *Adv. Cancer Res.* **19**, 47–104.

Temin, H. M., and Baltimore, D. (1972). *Adv. Virus Res.* **17**, 129–186.

Teramoto, Y. A., Puentes, M. J., Young, L. J. T., and Cardiff, R. D. (1974). *J. Virol.* **13**, 411–418.

Verma, I. M. (1975a). *J. Virol.* **15**, 121–126.

Verma, I. M. (1975b). *J. Virol.* **15**, 843–854.

Verma, I. M., Mason, W. S., Drost, S. D., and Baltimore, D. (1974). *Nature (London)* **251**, 27–31.

Vogt, P. K. (1965). *Adv. Virus Res.* **11**, 293–385.

Vogt, V. M. (1973). *Eur. J. Biochem.* **33**, 192–200.

Wang, T. S., Fisher, P. A., Sedwick, W. D., and Korn, D. (1975). *J. Biol. Chem.* **250**, 5270–5272.

Warburg, O., and Christian, W. (1942). *Biochem. Z.* **310**, 384–421.

Waters, L. C., and Yang, W.-K. (1974). *Cancer Res.* **34**, 2585–2593.

Weissbach, A. (1975). *Cell* **5**, 101–108.

Wu, A. M., Sarngadharan, M. G., and Gallo, R. C. (1974). *Proc. Natl. Acad. Sci. U.S.A.* **71**, 1871–1876.

6 *Methods for Studying Viroids*

T. O. Diener, A. Hadidi, and R. A. Owens

I. Introduction

The term "viroid" has been introduced to denote a recently recognized class of subviral pathogens (Diener, 1971b). Presently known viroids consist solely of a short strand of RNA with a molecular weight of 75,000–125,000.* Introduction of this low molecular weight RNA into susceptible hosts leads to apparent replication of the RNA and, in some hosts, to disease. Viroids are the smallest known agents of infectious disease (Diener, 1974).

The first viroid was discovered in efforts to purify and characterize the agent of the potato spindle tuber disease, a disease which, for many years, had been assumed to be of viral etiology (Diener and Raymer, 1971). Diener and Raymer (1967) reported that the infectious agent of this disease is a free RNA and that virus particles, apparently, are not present in infected tissue. Later, sedimentation and gel electrophoretic analyses conclusively demonstrated that the infectious RNA has a very low molecular weight (Diener, 1971b) and that the agent, therefore, basically differs from conventional viruses.

Three additional plant diseases, citrus exocortis (Semancik and Weathers, 1972), chrysanthemum stunt (Diener and Lawson, 1973), and chrysanthemum chlorotic mottle (Romaine and Horst, 1975) are now known also to be caused by low molecular weight RNAs, i.e., by viroids. With a fifth disease, cucumber pale fruit, Van Dorst and Peters (1974) reported the existence of unpublished results which indicate that this disease is caused by a viroid. Recent evidence suggests that a sixth plant disease, coconut cadang-cadang, also may be of viroid causation (Randles, 1975).

Study of viroids poses problems because no viruslike particles (virions) are associated with viroid infection. Conventional methods of virus purification are thus not applicable; instead, methods used for the isolation and purification of nucleic acids must be used. Viroids constitute a very small fraction of the total nucleic acid complement of cells. Consequently, to obtain viroids free of cellular nucleic acids, highly selective techniques of nucleic acid separation are required.

In this chapter, emphasis will be placed on methods developed in our laboratory for the study of the potato spindle tuber viroid but, in cases where methods significantly different from ours have been developed by other workers, these will also be described.

---

*Although Singh et al. (1974) reported that a smaller infectious form of potato spindle tuber viroid (in the range of tRNA) exists in extracts from infected *Scopolia sinensis.*

II. Viroid Culture and Bioassay Procedures

A. PROPAGATION HOSTS AND ENVIRONMENTAL FACTORS

1. Potato Spindle Tuber Viroid

Potato spindle tuber viroid (PSTV) is conveniently propagated in tomato (*Lycopersicon esculentum* Mill., cultivar "Rutgers") (Raymer and O'Brien 1962). To obtain early symptom expression, plants are inoculated at the cotyledonary stage and maintained in a greenhouse at 30°–35°C. During winter months and in locations with low light intensity, supplementary lighting is essential. Vigorously growing plants develop symptoms earlier than slow growing plants; thus, application of fertilizer may be advantageous. Under optimal conditions and with a high titer inoculum, first symptoms regularly appear 10–14 days after inoculation (Fig. 1). Leaves are harvested 4–8 weeks after inoculation and are stored frozen until used.

PSTV may also be propagated in Cheyenne tomato (Singh and Bagnall, 1968) or in *Scopolia sinensis* Hemsl. (Singh *et al.*, 1974) plants.

2. Citrus Exocortis Viroid

Citrus exocortis viroid (CEV) is propagated in Etrog citron (*Citrus medica* L.) (Semancik and Weathers, 1968) or *Gynura aurantiaca* DC (Semancik and Weathers, 1972).

FIG. 1. Rutgers tomato plants. Left: Plant with symptoms of PSTV, 20 days post inoculation. Right: Uninoculated control plant of same age.

3. *Chrysanthemum Stunt Viroid*

Chrysanthemum stunt viroid (CSV) may be propagated either in *Senecio cruentus* DC (florists' cineraria) or *Chrysanthemum morifolium* (Ramat.) Hemsl. cultivar "Mistletoe" (Brierley, 1953). Cineraria leaves are harvested just as they begin to show leaf curl and distortion; highest infectivity is recovered 6–8 weeks after inoculation. In *Chrysanthemum* leaves, high infectivity titers occur at any time during fall and winter months (Diener and Lawson, 1973).

4. *Chrysanthemum Chlorotic Mottle Viroid*

Chrysanthemum chlorotic mottle viroid (ChCMV) appears to have a narrow host range and is propagated in *Chrysanthemum morifolium* cultivar "Deep Ridge." Leaves are harvested 3–5 weeks after inoculation (Romaine and Horst, 1975).

B. Inoculation Procedures

All viroids described so far are mechanically transmissible, either readily or with some difficulty. Although other means of transmission (such as grafting or by dodder bridges) are possible, all recent work was based on mechanical transmission.

With PSTV, CSV, and ChCMV, standard inoculation procedures used with conventional viruses and viral RNAs are suitable.

In our work with PSTV, tomato plants are dusted with 600-mesh carborundum and are inoculated by lightly rubbing cotyledons and terminal portions of the plants with cotton-tipped applicators dipped into viroid solution in 0.02 M phosphate buffer, pH 7.

With CEV, slashing of stems or petioles with a razor blade or knife dipped into inoculum is more efficient than inoculation by rubbing of carborundum-dusted leaves (Garnsey and Whidden, 1970). Often, a combination of razor-slashing and rubbing has been used (Semancik and Weathers, 1968).

C. Bioassay of Viroids

Estimation of viroid concentration is essential in efforts to purify and characterize the infectious molecules. In crude extracts and partially purified preparations, such estimates can only be obtained by measuring biological activity of viroids, i.e., symptom formation in susceptible plants. As with conventional plant viruses, infectivity assays rely on the percentage of inoculated plants becoming systemically infected or, in cases where a hypersensitive host is available, on the number of local lesions produced as a consequence of infection. Although local lesion assays are inherently more accurate than systemic assays, the latter type of assay has been used more often than the former in viroid work because local lesion hosts either were not available or because assays based on local lesions posed difficulties.

1. Systemic Bioassay of Viroids

The use of systemic infections for the determination of the relative concentration of a biologically active entity is complicated by many factors (see Brakke, 1970), particularly if relatively small differences in titer are to be determined. Fortunately, much work with viroids does not require an accurate knowledge of relative viroid titer. Often, certain procedures result in large effects and approach all-or-nothing responses.

For example, when a PSTV preparation that will induce symptoms in at least some plants inoculated at a dilution of 1/10,000 is incubated with pancreatic ribonuclease at low concentration, and is then inoculated into tomato plants, none of the plants will develop symptoms, even if inoculated with the undiluted preparation. Clearly, no statistical analysis is required to verify the significance of such a result. Similarly, if one studies the elution properties of a viroid from a chromatographic column, for example, or its electrophoretic mobility in a polyacrylamide gel, a bioassay based on systemic infection is adequate to obtain the desired information.

With experiments of this nature, it is usually sufficient to make a series of 10 fold dilutions of each preparation to be assayed and to inoculate each dilution to 3–5 plants. To prevent cross-contamination of samples, clean pipettes should be used for each dilution step and with each preparation, and inoculation should proceed from the most dilute to the least dilute sample. Assay plants must be inoculated by as uniform a technique as possible.

Several criteria may be used to evaluate results. The least desirable is the dilution end point because a large sampling error is associated with the small number of plants infected at the dilution end point (Brakke, 1970). Other criteria are percentage of inoculated plants that become infected at each dilution and time of symptom appearance.

For the assay of PSTV in tomato, we developed an empirical "infectivity index" which takes into consideration the dilution end point, the percentage of plants infected at each dilution, and the time required for symptom expression (Raymer and Diener, 1969). These three factors are used in the index as a means of estimating relative viroid concentration.

As soon as the first plant in an experiment begins to develop symptoms, readings are initiated and are continued at 2-day intervals until no further increase in the number of plants with symptoms has occurred for two or three consecutive readings. The experiment is then terminated. The type

TABLE I

PSTV Symptom Expression in Rutgers Tomato Plants as a Function of Dilution and Time[a]—Calculation of an Infectivity Index

Dilution	Days after inoculation								Sum[c]	Multiplier[d]	Product[e]
	10	12	14	16	18	20	22	24			
10^{-1}	1/3[b]	3/3	3/3	3/3	3/3	3/3	3/3	3/3	22	1	22
10^{-2}				3/3	3/3	3/3	3/3	3/3	15	2	30
10^{-3}					3/3	3/3	3/3	3/3	12	3	36
10^{-4}						1/3	2/3	2/3	5	4	20
10^{-5}							2/3	2/3	4	5	20
10^{-6}							2/3	2/3	4	6	24
10^{-7}							1/3	1/3	2	7	14
10^{-8}								0/3	0	8	0
								Total = Infectivity index:			166

[a] Data were obtained with a virus concentrate obtained by extraction of infected leaf tissue with 0.5 M K_2HPO_4, chloroform, and n-butanol, followed by phenol treatment and ethanol precipitation.

[b] Infectivity = number of plants with symptoms/number of plants inoculated.

[c] Sum of all plants showing symptoms at all dates for each dilution.

[d] Negative log of the dilution.

[e] Sum × multiplier for each dilution.

of data obtained by this method is illustrated in Table I, together with the method used to compute an "infectivity index" based on such data.

Evidently, the more concentrated the viroid preparation, the earlier symptoms appear. The index is computed by adding together the number of plants infected at each dilution over the entire recording period, multiplying this figure by the negative log of the dilution, and adding together these products for all the dilutions tested. This index permits discrimination between treatments that might achieve the same dilution end points but differ in both the earliness with which symptoms are expressed and in the number of plants infected. Undoubtedly, with this method, the difference in viroid titer is underestimated, since a 10-fold difference in dilution is represented by a difference of only 1 in the multiplier. On the other hand, the infectivity index gives relatively little weight to individual plants that express symptoms late, after having been inoculated with a highly diluted viroid preparation. Thus, much of the variability inherent in an assay of this type is "dampened out."

With careful and consistent inoculation techniques, however, spurious results are rare and, in these cases, a more realistic index may be computed by using the actual dilution factor as multiplier (Diener, 1971b) instead of the negative logarithm.

Semancik and Weathers (1972) expressed relative infectivity titers of CEV preparations by determining the total number of "infected plant days" on 3–5 *Gynura* plants inoculated with one concentration of inoculum only. The authors considered this procedure satisfactory, since the dilution–response curve was found to be reasonably linear within a certain range of "relative infectivity units" (Semancik and Weathers, 1972).

2. *Local Lesion Assay of Viroids*

So far, local lesion hosts have been discovered only for CSV and PSTV. With CSV, local starch lesions are detected on inoculated leaves of *Senecio cruentus* (florists' cineraria) 12–18 days after inoculation, provided the plants are grown with light intensities not higher than 2000 fc and at 18°–26°C (Lawson, 1968). Increased starch lesion formation is favored by an 18-hour light period with 500 fc fluorescent illumination and a constant temperature of 21°C. Variation in lesion counts among cineraria plants and between leaves on a single plant, however, precludes their use for detecting small quantitative differences in CSV concentration (Lawson, 1968).

With PSTV, Singh (1971) reported that *Scopolia sinensis*, a solanaceous plant species, produces necrotic local lesions 7 to 10 days after inoculation with the severe strain and 10 to 15 days after inoculation with the mild strain. Singh later stressed that local lesion development is critically dependent on environmental conditions (Singh, 1973). Local lesions developed best on leaves of plants at 22°–23°C; plants maintained at 28°–31°C developed mostly systemic symptoms without conspicuous local lesions. A low light intensity of 400 fc favored local lesion development.

III. Tests to Determine Viroid Nature of Unknown Pathogen

To determine whether the agent of an infectious disease of unknown etiology has properties typical of viroids, several simple exploratory tests may be performed. By necessity, these tests are based on the properties of presently known viroids; and it must be stressed that newly discovered viroids may differ in some of their properties from the few viroids so far investigated. It is by no means certain, for example, whether all viroids are mechanically transmissible or whether all are composed of RNA. Also, depending on the host in which viroids replicate, they may exhibit different characteristics in crude extracts.

A. CRITERIA FOR SUSPECTING VIROID NATURE OF PATHOGEN

Viroid etiology of an infectious disease of unknown causation should be considered if all of the following observations have been made:

1. No microorganisms are consistently associated with the disease;

2. No viruslike particles are identifiable by electron microscopy of extracts or in sections from infected tissue;

3. Much of the pathogen in extracts cannot be pelleted by ultracentrifugation; and

4. The agent is inactivated by either ribonuclease or deoxyribonuclease. Evidently, the last two tests presuppose that the pathogen is mechanically transmissible and that a suitable assay host is available.

None of these observations constitute conclusive evidence for viroid etiology. The first two observations are negative evidence and improved techniques may lead to opposite conclusions; the third observation may indicate the presence of virus particles of low density, such as lipid-containing virions; and the fourth observation could indicate the presence of virus particles with a loose protein shell which permits access of nucleases [such as cherry necrotic ringspot (Diener and Weaver, 1959), cucumber mosaic (Francki, 1968), or apple chlorotic leaf spot (Lister and Hadidi, 1971) viruses].

On the other hand, detection of viruslike particles in sections from infected tissue does not necessarily rule out viroid etiology of the disease in question; plants may be infected with a latent virus unrelated to the disease syndrome.

Also, some viroids are not readily released from host constituents or occur in extracts as aggregates of varying size (Diener, 1971a). In either case, much of the infectious material sediments faster than expected.

B. SEDIMENTATION PROPERTIES

Assuming that infectious extracts can be prepared from infected tissue and that a suitable bioassay host is available, accurate determination of the sedimentation properties of the infectious agent in such extracts should have high priority. This may conveniently be accomplished by subjecting an infectious extract together with markers of known sedimentation coefficients to velocity density gradient centrifugation, followed by fractionation of the gradient and bioassay of all fractions. Conventional techniques of sucrose density gradient centrifugation may be used, but precautions must be taken to assure freedom from nuclease contamination. Tissue extracts may be prepared in the presence of bentonite (ribonuclease inhibitor) and gradients should be made with ribonuclease-free sucrose. To obtain a reasonable estimate of infectivity distribution in the gradient, each fraction should be assayed undiluted and diluted 1/10 and 1/100.

Tissue extracts should be made with buffers of low and high ionic strength, such as 0.005 M and 0.5 M phosphate buffers, because some viroids are released from subcellular components only in high ionic strength medium (Raymer and Diener, 1969). Viroids usually are more stable in slightly alkaline than in acid media; thus, buffers of pH 8–9 are recommended.

Extracts to be analyzed by density gradient centrifugation may be clarified by shaking with 1 volume of a 1:1 mixture (v/v) of chloroform and n-butanol, followed by centrifugation and withdrawal of the aqueous phase. This treatment does not adversely affect the infectivity of viroids, but results in much cleaner preparations.

A sedimentation coefficient of 7–15 S of the bulk of the infectious material suggests that the unknown pathogen may be a viroid. In crude preparations of this type, however, some infectivity often is found in lower portions of the gradient; and it is not uncommon to find low levels of infectivity in most fractions. For this reason, it is important to assay 10-fold dilutions of the inoculum and to determine where in the gradient the bulk of the infectious material is located.

C. NUCLEASE SENSITIVITY

All known viroids are composed of RNA and are inactivated by incubation with pancreatic ribonuclease. At high ionic strength, some viroids are partially ribonuclease-resistant; thus, incubation should be made in buffers of low and high ionic strength, such as 0.01 M and 0.5 M phosphate buffers, pH 7. Sensitivity should be investigated in the range of 0.1–1 μg/ml of ribonuclease with incubations for 1 hour at 25 °C.

D. INSENSITIVITY TO TREATMENT WITH PHENOL

Treatment of buffer extracts from viroid-infected tissue with phenol or with phenol and sodium dodecyl sulfate (SDS) has little, if any, effect on the infectivity level of such preparations, or on the sedimentation properties of the infectious material. This is in contrast to extracts containing conventional viruses which, after such treatments, usually are much less infectious and where the infectious material sediments at lower rates than before treatment with phenol.

To make these comparisons, 10 gm of infected tissue are triturated in 20 ml of 0.5 M K_2HPO_4, 10 ml chloroform, and 10 ml n-butanol. The resulting emulsion is broken by low-speed centrifugation, and the aqueous phase is carefully withdrawn. One-half of the preparation is dialyzed versus 0.02 M phosphate buffer, pH 7, and, to the other half, one volume of water-saturated phenol is added. This preparation is shaken at room temperature for several minutes and is then centrifuged (20 minutes at 10,000 g). The aqueous (lower) phase is withdrawn and two volumes of ethanol are added. After storage for at least 30 minutes at 0 °C, the sample is centrifuged (15 minutes at 10,000 g) and the supernatant is poured off. The pellet is resuspended in 0.02 M phosphate buffer, pH 7. After one or two more ethanol precipitations and resuspensions, the nucleic acid is finally suspended in the original volume of 0.02 M phosphate buffer, pH 7.

Both the phenol-treated and the nontreated samples are bioassayed at serial 10-fold dilutions, and the sedimentation properties of the infectious material are determined by density-gradient centrifugation in sister tubes and bioassay of all fractions.

E. Electrophoretic Mobility

If the above tests give results consistent with a viroid nature of the pathogen, infectious extracts should be subjected to polyacrylamide gel electrophoresis, as described below. This can be accomplished by analyzing partially purified preparations in gels of low and high concentration (such as 2.4% and 20% polyacrylamide gels), followed by cutting of the gels into 1- to 2-mm thick slices and bioassay of each slice after crushing in 0.02 M phosphate buffer and serial dilution.

In the 2.4% gel, peak infectivity should be in the region between tRNA and 16 S ribosomal RNA, indicating a low molecular weight of the infectious particles. The infectious particles should be able to enter 20% gels and to move through such gels as a sharp band.

Alternatively, preparations may be more thoroughly purified before electrophoresis and the gels stained to detect the suspected viroid (see below).

IV. Viroid Purification

Two schemes for the extraction of nucleic acids, the first step in PSTV purification, have been used in our laboratory with essentially identical yields of total low molecular weight nucleic acid. Both procedures involve (1) an initial homogenization in the presence of high ionic strength buffers (Diener and Raymer, 1969; Diener, 1971b, 1972) and SDS to disrupt nucleoprotein complexes; (2) phenol extractions to remove protein; (3) extraction with ethylene glycol monomethyl ether to remove polysaccharides; (4) DNA digestion; (5) 2 M LiCl extraction to remove high molecular weight RNA; (6) gel filtration of total cellular low molecular weight RNA and DNA fragments; and (7) final purification by polyacrylamide gel electrophoresis. These procedures are described in detail below for purification of PSTV from 500 gm of tomato leaf tissue. All operations are peformed at $0°–4°C$ unless specifically stated otherwise; in all centrifugations, either Sorvall* GSA rotors (large volumes) or SS-34 rotors (small volumes) were used.

A. EXTRACTION OF NUCLEIC ACIDS

1. Direct Phenol Extraction

a. Combine 500 gm frozen ($-20°C$) leaf tissue, 500 ml 1 MK_2HPO_4, 10 gm SDS, 5 gm bentonite, and 1000 ml water-saturated phenol in a one-gallon stainless steel Waring Blendor container; grind 5 minutes at low speed.

b. Centrifuge 10 minutes at 6500 rpm, remove aqueous (lower) phase by aspiration, and precipitate nucleic acids by adding 2 volumes cold 95% ethanol. Store overnight at $-30°C$.

c. Carefully decant most of the upper ethanol phase before centrifuging 10 minutes at 6500 rpm. The nucleic acids are recovered as a tan "skin" at the interface between the lower phosphate-rich aqueous phase and the upper ethanol phase.

d. Total nucleic acids are dissolved in 100 ml of water and reprecipitated with 2 volumes of cold 95% ethanol overnight at $-30°C$.

e. Total nucleic acid is again recovered as a skin at the interface between the two phases after a 10-minute centrifugation at 6500 rpm. The nucleic acid is dissolved in approximately 40 ml of water.

f. Polysaccharide extraction: (1) Add 1 volume 2.5 M K_2HPO_4, 0.02 volume 85% H_3PO_4, and 1 volume ethylene glycol monomethyl ether. (2) Shake 2 minutes and centrifuge 5 minutes at 6500 rpm. (3) The upper phase, which contains the nucleic acid, is carefully removed without disturbing the interface and is dialyzed overnight against 2 liters of water, with at least one change during dialysis. (4) Nucleic acid is recovered by addition of 0.1 volume of 20% potassium acetate and 2 volumes of 95% ethanol and over-night storage at $-30°C$.

g. DNA digestion: (1) Collect nucleic acid by 10-minute centrifugation at 6500 rpm. (2) Dissolve nucleic acid in approximately 30 ml water and add 3 μl 1 M $MgCl_2$ and 0.01 ml 1 mg/ml DNase I (Worthington, electrophoretically purified) per 1 ml of nucleic acid solution. (3) Incubate 60 minutes at $25°C$. (4) Add 0.025 volume 20% (w/v) SDS, 1 volume water-saturated phenol, and shake 5 minutes at room temperature. (5) Centrifuge 10 minutes at 6500 rpm, remove the upper aqueous phase, add 0.1 volume of 20% potas-

---

* Mention of a commercial company or specific equipment does not constitute its endorsement by the U.S. Department of Agriculture over similar equipment or companies not named.

sium acetate and 2 volumes of cold 95% ethanol, and precipitate overnight at −30 °C.

h. LiCl fractionation of RNA: (1) Collect RNA by 10-minute centrifugation at 6500 rpm and dissolve in approximately 30 ml of water. (2) Measure the volume of nucleic acid solution and add sufficient solid LiCl to yield a final concentration of 2 *M* (i.e., 84.8 mg/ml). Add the LiCl slowly while stirring the nucleic acid solution in an ice-water bath to prevent excessive heating. Let stand overnight at 4 °C. (3) Centrifuge 10 minutes at 6500 rpm, save supernatant containing low molecular weight RNA. Add 2 *M* LiCl solution to the pellet, mix thoroughly, centrifuge as above, and save supernatant. Combine supernatants, add 2 volumes of cold 95% ethanol, mix thoroughly, and precipitate RNA overnight at −30 °C.

i. Collect the RNA by 10-minute centrifugation at 6500 rpm and dissolve it in 3 ml GM buffer (20 m*M* glycine, 3 m*M* $MgCl_2$, pH 9.0, with NaOH). Add 6 ml 95% ethanol and precipitate overnight at −30 °C.

j. Collect the RNA by 10-minute centrifugation at 6500 rpm and dissolve in 3 ml GM buffer for gel filtration.

2. Pretreatment with Chloroform-Butanol

The following is a modification of the method of Morris and Wright (1975).

a. Combine 500 gm frozen (−20 °C) tomato leaf tissue, 250 ml 16% (w/v) SDS, 250 ml GPS buffer (0.4 *M* glycine, 0.2 *M* Na_2HPO_4, 1.2 *M* NaCl, pH 9.5, with NaOH), 250 ml $CHCl_3$, and 250 ml *n*-butanol in a one-gallon stainless steel Waring Blendor container. Grind 3 minutes at high speed and recover aqueous (upper) phase by a 20-minute centrifugation at 8000 rpm.

b. Extract the aqueous phase three times with phenol-$CHCl_3$. Add 500 ml water-saturated phenol and 500 ml $CHCl_3$ to the aqueous phase and stir 20 minutes at room temperature on a magnetic stirrer. Recover the aqueous phase by 10-minute centrifugation at 8000 rpm.

c. Add 2 volumes of cold 95% ethanol to the aqueous phase and precipitate overnight at −30 °C. Collect the nucleic acid precipitate by 20-minute centrifugation at 8000 rpm.

d. Dissolve the precipitate in 150 ml TKM buffer (10 m*M* Tris-HCl, pH 7.4; 10 m*M* KCl; 0.1 m*M* $MgCl_2$) and dialyze overnight against 2 liters of TKM buffer, with at least one change of buffer during dialysis.

e. LiCl fractionation of nucleic acids: (1) Measure the volume of dialyzed nucleic acid solution and add sufficient solid LiCl to yield a final concentration of 2 *M* (i.e., 84.8 mg/ml). Add the LiCl slowly while stirring the nucleic acid solution in an ice-water bath to prevent excessive heating. Let stand overnight at 4 °C. (2) Centrifuge 10 minutes at 6500 rpm and recover low molecular weight RNA and DNA from supernatant by adding 2.5 volumes of cold ethanol. Precipitate overnight at −30 °C.

f. Polysaccharide extraction: (1) Collect nucleic acid by 10-minute centrifugation at 8000 rpm and dissolve in 25 ml water. (2) Measure the volume and add 1 volume 2.5 M K_2HPO_4, 0.02 volume 85% H_3PO_4, 1 volume ethylene glycol monomethyl ether. (3) Shake 2 minutes and centrifuge 5 minutes at 6500 rpm. (4) The upper phase, which contains the nucleic acid, is carefully removed and dialyzed overnight against 2 liters of water, with at least one change during dialysis. (5) The nucleic acid is recovered by addition of 0.1 volume of 2.2 M potassium acetate—0.1 M acetic acid and 2.5 volumes of cold 95% ethanol, and precipitation overnight at −30°C.

g. DNA digestion: (1) Collect nucleic acid by 10-minute centrifugation at 8000 rpm. Drain, dry the pellet with a N_2 stream, and dissolve in 10 ml water. Add 0.01 ml 1 M Tris-HCl, pH 7.5, 0.002 ml 1 M $MgCl_2$, and 0.01 ml 1 mg/ml DNase I per milliliter of nucleic acid solution. Incubate 60 minutes at 30°C. (2) Add 0.02 ml 500 mM Na_2EDTA, pH 8.1, 0.025 ml 20% (w/v) SDS, 0.1 ml Tris-HCl, pH 7.5, 0.5 ml water-saturated phenol, and 0.5 ml $CHCl_3$ per milliliter of solution. Shake 5 minutes at room temperature and centrifuge 10 minutes at 8000 rpm to recover the aqueous phase. (3) Precipitate nucleic acid by adding 0.1 volume 2.2 M potassium acetate—0.1 M acetic acid and 2.5 volumes of cold 95% ethanol. Let stand overnight at −30°C.

h. Collect the nucleic acid precipitate by 10-minute centrifugation at 8000 rpm. Dissolve in 1.5 ml GM buffer (20 mM glycine, 3 mM $MgCl_2$, pH 9.0, with NaOH) for gel filtration.

B. GEL FILTRATION

The next step in PSTV purification is gel filtration on long columns of either Sephadex or Ultrogel. We have found that gel filtration produces about an 8-fold enrichment of PSTV and a similar reduction in the numbers of polyacrylamide gels required for final purification.

Columns (1.6 × 170 cm) of either Sephadex G-100 (medium) or Ultrogel AcA44 (LKB Instruments) are prepared and equilibrated with GM buffer. Low molecular weight RNA prepared from 500 gm of tomato leaf tissue (approximately 850 A_{260} units) is applied to the column in a volume of 1.5–3 ml. The column is eluted with GM buffer at a flow rate of 8–12 ml/hour. The effluent is continuously monitored at 280 nm (to keep the entire column profile on scale). Fractions of 2.5–4.0 ml are collected. A comparison of the fractionation obtained on Sephadex and Ultrogel is presented in Fig. 2.

Polyacrylamide gel electrophoresis of samples across the column profiles shows that PSTV is located in the first peak of each column profile—the peak marked 2 in the Sephadex profile and the peak marked 6 in the Ultrogel profile. Five S ribosomal RNA is found in peaks 3 and 7, respectively.

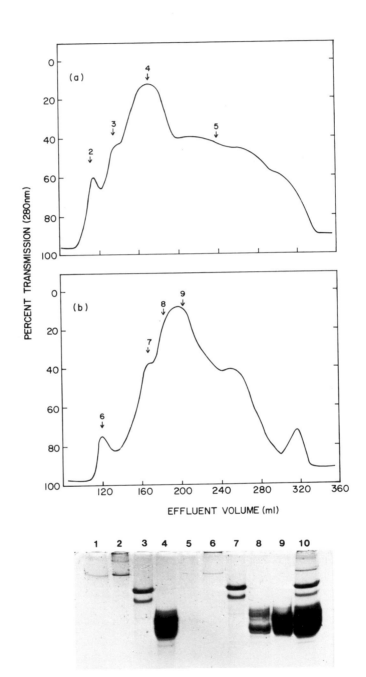

Chromatography on Ultrogel clearly yields superior resolution of PSTV from 5 S ribosomal RNA.

Fractions containing PSTV are pooled and 0.1 volume of 2.2 M potassium acetate—0.1 M acetic acid, and 2.5 volumes of 95% ethanol are added. After overnight precipitation at −30 °C, the RNA is collected by a 10-minute centrifugation at 8000 rpm, drained, dried with N_2, dissolved in 0.5–0.9 ml water, and stored at −70 °C before electrophoresis.

C. GEL ELECTROPHORESIS

Two electrophoresis procedures have proved satisfactory for the final purification of PSTV. Both involve extended electrophoresis on 20% acrylamide–0.5% Bis gels.

1. *Cylindrical Gels*

In the first procedure, the running buffer system of Loening (1967) is used (4.83 gm Tris, 1.64 gm sodium acetate, 0.37 gm Na_2EDTA per liter, adjusted to pH 7.2 with glacial acetic acid). RNA from 500 gm of tissue is dissolved in 0.45 ml H_2O, and 0.05 ml of 50% sucrose-5x concentrated running buffer is added. Fifty-microliter aliquots are layered on top of each of ten 0.6 × 9 cm cylindrical gels and electrophoresed for 16 hours at 4 mA/tube. After electrophoresis, each gel is scanned at 260 nm in a Gilford spectrophotometer equipped with a linear transport. Gel segments containing PSTV are excised with a razor blade and stored frozen at −70 °C until a sufficient number of slices have been accumulated for RNA elution.

2. *Slab Gels*

The second electrophoresis procedure uses the running buffer system of Peacock and Dingman (1967) (10.8 gm Tris, 5.5 gm H_3BO_3, and 0.93 gm

---

FIG. 2. Fractionation of low molecular weight nucleic acids isolated from PSTV-infected tomato tissue by gel filtration. Total cellular RNA was prepared from 1 kg of frozen tomato seedlings tissue by the method using pretreatment with chloroform-butanol. Extraction with 2 M LiCl yielded 1710 A_{260} units of low molecular weight nucleic acids. The nucleic acids were dissolved in 3 ml of GM buffer, and 1.5 ml aliquots were chromatographed on 1.6 × 170 cm (340 ml bed volume) columns of Sephadex G-100 (medium) and Ultrogel AcA44. Flow rates were 8–12 ml/hour. Two-hundred-microliter aliquots were removed from the indicated regions of the column profiles, 20 μl of 50% sucrose–100 mM Na_2EDTA, pH 8.1, were added, and the samples were electrophoresed on a 20% polyacrylamide slab gel in Tris–borate–EDTA running buffer for 16 hours at 4 °C. Slot 1 contained 3.5 μg of purified PSTV; slot 10 contained 11.4 A_{260} units of unfractionated low molecular weight nucleic acids. (a) Fractionation achieved with Sephadex G-100 (medium). (b) Fractionation achieved with Ultrogel AcA44.

Na$_2$EDTA per liter, pH 8.3). RNA from 500 gm of tissue is dissolved in 0.9 ml H$_2$O, and 0.1 ml 50% sucrose-5x concentrated running buffer is added. The sample is layered on a 10 (length) × 13 (width) × 0.24 (thickness) cm slab polymerized in the apparatus described by Studier (1973). After 60 minutes at 40 V, the voltage is increased to 100 V for the next 42 hours. The reservoir buffers are continuously recirculated. The PSTV zone is located by staining a thin strip cut from one side of the slab with 0.02% toluidine blue in 0.4 M sodium acetate–0.4 M acetic acid, excised, and stored at −70 °C until sufficient RNA has been accumulated for elution.

D. Recovery of Viroids from Gels

PSTV is eluted from polyacrylamide gels by a slight modification of the procedure of Diener (1973). From 20 to 40 cylindrical gel slices or 4 to 6 strips from a slab gel are homogenized for 2 minutes at 0 °C in 100 ml of GM buffer (20 mM glycine, 3 mM MgCl$_2$, pH 9.0, with NaOH) in a blender at full speed. The resulting slurry is centrifuged for 15 minutes at 500 g, and the sedimented gel particles are reextracted twice with 50 ml GM buffer. The combined supernatants are passed through a column of hydroxyapatite (Bio-gel HT, Bio-Rad Laboratories), previously equilibrated with GM buffer at room temperature. A column with 2-ml bed volume is sufficient to absorb 1 mg of nucleic acid. The flow rate should not exceed 2 ml per minute, and the effluent is monitored at 254 nm to detect any nucleic acid run-through.

After sample application, the column is washed with GM buffer until the effluent absorbance is essentially zero. PSTV is eluted as a sharp peak by washing the column with 20 mM glycine, 0.2 M Na$_2$HPO$_4$ (pH 9.0, with NaOH). The PSTV solution is frozen, thawed, and centrifuged for 15 minutes at 10,000 g to remove insoluble material. Two volumes of cold 95% ethanol are added to the supernatant, and the RNA is allowed to precipitate over-night at −20 °C. The PSTV is collected as a "skin" between the phosphate-rich lower phase and ethanol-rich upper phase following a 15-minute centrifugation at 5000 g. The drained PSTV is dissolved in a convenient volume of 1.25 M potassium phosphate, pH 8.0, and extracted for 2 minutes with 0.5 volume of ethylene glycol monomethyl ether. The upper phase is carefully removed after a 5-minute centrifugation at 5000 g and is reextracted with the lower phase from a parallel mock extraction to remove remaining gel impurity. PSTV is recovered from the upper phase after dialysis against water by addition of 0.1 volume of 2.2 M potassium acetate–0.1 M acetic acid and 2.5 volumes of cold 95% ethanol and overnight storage at −20 °C. PSTV is collected by a 15-minute centrifugation at 10,000 g, drained, dis-solved in water, and stored frozen at −70 °C. The UV spectrum of electro-phoretically purified PSTV is typical of RNA with an absorbance maximum of 258 nm, a minimum of 230 nm, and a maximum/minimum ratio of 2.1–2.4. The extraction with ethylene glycol monomethyl ether may be repeated if the maximum/minimum ratio is low. The recovery of PSTV from acrylamide gels ranges from 46 to 91%.

E. OTHER PURIFICATION PROCEDURES

Three other procedures for purification of viroid RNAs have been published. Purification of CEV (Semancik and Weathers, 1972; Semancik *et al.*, 1975) has several features in common with the procedure just described. Fresh tissue from CEV-infected *Gynura aurantiaca* (10 gm) is homogenized with 1 ml 5% SDS; 40 mg bentonite; 1 ml 0.1 *M* Tris-HCl, pH, pH 8.9; 1 ml 0.1 *M* EDTA, pH 7.0; and 20 ml water-saturated phenol. The aqueous phase is recovered by centrifugation and is reextracted with water-saturated phenol. Nucleic acids are recovered by ethanol precipitation at 4°C. Polysaccharides are removed by ethylene glycol monomethyl ether extraction, and the ethylene glycol monomethyl ether phase is dialyzed against TKM buffer.

High molecular weight RNA is removed by addition of an equal volume of 4 *M* LiCl and subsequent centrifugation. The LiCl supernatant is gently mixed with 3 volumes of absolute ethanol and the windable nucleic acids are removed on a glass rod. The remaining precipitate is collected by centrifugation, digested with DNase, and reprecipitated with ethanol.

CEV in the low molecular weight RNA fraction adsorbs to CF-11 cellulose (Franklin, 1966) in the presence of either 35% or 25% ethanol in 0.1 *M* NaCl, 1 m*M* EDTA, 50 m*M* Tris-HCl, pH 7.2, and is eluted with the above buffer in the absence of ethanol (Semancik and Weathers, 1972). The final steps in CEV purification are electrophoresis in 5% polyacrylamide gels, elution of the RNA from gel slices, and chromatography on columns of CF-11 cellulose-hydroxyapatite (1:1 w/w).

The major difference between this procedure for CEV purification and those previously described for PSTV purification are the use of CF-11 cellulose chromatography and electrophoresis in 5%, rather than 20%, polyacrylamide gels for final purification. Engelhardt (1972) has shown that chromatography on CF-11 cellulose can be used to discriminate between single-stranded RNA molecules in solution on the basis of their secondary structure. Elution in buffers which contain no ethanol is characteristic of double-stranded RNA and single-stranded RNA with a high degree of secondary structure. This behavior is consistent with the nuclease resistance and low-field nuclear magnetic resonance spectrum reported by Semancik *et al.* (1975). Although PSTV also shows a marked resistance to ribonuclease digestion (Diener and Raymer, 1969), its broad profile from CF-11 cellulose prevents the use of CF-11 cellulose chromatography in PSTV purification. Finally, we prefer electrophoresis on 20% rather than 5% polyacrylamide gels for final separation of PSTV because of better resolution of low molecular weight RNAs.

Morris and Wright (1975) have published a procedure for PSTV purification which has been modified in our laboratory and described above. Their homogenization buffer contains in addition 0.2 M Na$_2$SO$_3$ and 0.1% sodium diethyldithiocarbamate to minimize the formation of dark brown pigments by oxidation of plant phenolic compounds. We have found that these additions have no noticeable effect when PSTV is isolated from tomato seedlings. The procedure of Morris and Wright is intended to be used to diagnose potato spindle tuber disease, i.e., only small amounts of tissue are processed. For this reason, the polysaccharide extraction, DNA digestion, and gel filtration steps which we have used in our large-scale preparative procedures can be successfully omitted.

Cadang-cadang disease of coconuts is suspected to be caused by a viroid (Randles, 1975). The purification of a low molecular weight RNA species present only in nucleic acid extracts from diseased palm involves precipitation of macromolecules from the initial homogenate by addition of polyethylene glycol (PEG 6000) to a final concentration of 5%. Although polyethylene glycol precipitation is a commonly used step in virus and nucleic acid purifications, PSTV is not completely precipitated by addition of PEG 6000.

V. Radioactive Labeling of Potato Spindle Tuber Viroid

The availability of radioactively labeled viroid is essential for biochemical studies on viroid structure and replication. Viroids are present in extremely low concentration in infected cells (Diener, 1971c). Therefore, it is difficult to label viroids in plant tissues with radioactive isotopes. We have studied the conditions under which labeled PSTV is obtained in infected tissue or in isolated nuclei. The following methods have been used in our laboratory for labeling PSTV.

A. ^{32}P-Labeling (Hadidi, 1976)

1. Introduction of Label into Plants

Rutgers tomato seedlings to be used in labeling experiments are grown in a greenhouse with high light intensity and are kept at 30°–35°C. At the two- to four-leaf stage, both cotyledons of tomato seedlings and the terminal growth are inoculated mechanically with partially purified preparations

of PSTV. When the PSTV syndrome starts to appear on the young systemically infected leaves, plants are gently removed from soil without injuring
the root system and are then rinsed with tap water to remove the adhering
soil. Plants are immersed into glass vials (6.5 × 1.5 cm) which have been
coated with thin layers of paraffin and contain ^{32}P-orthophosphate (carrier-
free) solution and Hoagland nutrient solution (Hoagland and Arnon, 1950)
lacking phosphorus. Two tomato plants are placed into each vial containing 1 mCi ^{32}P solution and 7 ml of the nutrient solution; and the vials are
secured in racks. Vials are placed in a thick Plexiglas box (1/2 inch thick)
so that full protection from heavy radiation emitted from plants and vials is
afforded to workers. Plants are kept in the greenhouse under the same conditions of light and temperature favorable for PSTV synthesis. Nutrient
solution lacking phosphorus is added after all the isotope has been taken
up. After 5 days of exposure to ^{32}PO$_4$, plants are removed from vials, roots
are excised, and leaves are frozen until needed.

2. Nucleic Acid Extraction

Nucleic acid extraction from the ^{32}P-labeled leaves is done using a modification of the procedure of Morris and Wright (1975). All steps are performed at 4 °C. Fifteen milliliters of GPSS buffer (0.2 M glycine, 0.1 M Na$_2$
HPO$_4$, 0.6 M NaCl, 0.2 M Na$_2$SO$_3$, pH 9.5), and 2 ml of 10% SDS are added
to leaves obtained from 10 plants. Leaves are homogenized in a mortar with
a pestle. The homogenate is transferred to an Omni-Mixer, 30 ml of water-
saturated phenol is added, and the mixture is blended at maximum speed
for 1–2 minutes. In our hands, maceration of the tissue in a mortar with a
pestle is essential to facilitate the blending process in the Omni-Mixer. The
homogenate is centrifuged at 7000 rpm for 15 minutes. The aqueous (upper)
phase is removed and extracted by emulsification with an equal volume of
water-saturated phenol containing 0.1% 8-hydroxyquinoline. The solution is
held on ice for 15 minutes and is then centrifuged as above. The aqueous
(upper) phase is removed and mixed with 2 volumes of cold 95% ethanol.
The mixture is held on ice for 30 minutes to precipitate the nucleic acids
and is then centrifuged as above. The ethanol supernatant is discarded and
the nucleic acid pellet suspended in 10 ml TKM buffer (0.1 M Tris, 0.01 M
KCl, 4 mM MgCl$_2$, pH 7.4).

3. Detection of ^{32}P-Labeled PSTV

The suspension is dialyzed against 1 liter of TKM buffer for 48 hours
with frequent changes of the dialyzing buffer to completely remove the unincorporated ^{32}P. After dialysis the nucleic acid solution is treated with
DNase (100 μg/ml + 10 mM MgCl$_2$) at 37 °C for 60 minutes to degrade
DNA. The nucleic acid solution is mixed with an equal volume of 4 M LiCl

and the mixture is kept at 4 °C overnight. The LiCl soluble RNA fraction is obtained by centrifugation at 8000 rpm for 15 minutes. Two volumes of 95% ethanol are added to the LiCl-soluble fraction and the mixture is kept at −20 °C overnight to precipitate RNA. After centrifugation at 8000 rpm for 15 minutes, the supernatant is removed and the pellet is dissolved in about 0.2 ml of H_2O.

Electrophoresis in cyclindrical or slab gels is carried out as described under viroid purification. After electrophoresis, cylindrical gels are scanned at 260 nm. After scanning, the gels are cut into 2-mm slices with a razor-blade assembly; each slice is placed on Whatman No. 3 MM filter paper (1.0–1.5 cm²), placed into a scintillation vial, and the slices are dried at 100 °C for 2 hours. After cooling, 5 ml of a toluene-based scintillation fluid is added to each vial; and samples are counted in a liquid scintillation spectrometer. Alternatively, the location of ³²P-labeled PSTV in gel slices is determined without adding scintillation fluid by Cerenkov counting (Havi-

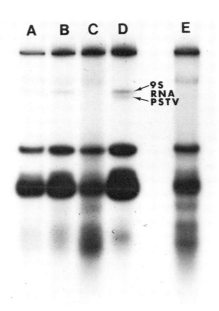

Fig. 3. Autoradiograph of ³²P-labeled low molecular weight RNA species from uninfected and PSTV-infected tomato leaves on a 10% polyacrylamide slab gel electrophoresed for 5 hours at 100 V. About 250–300 × 10³ cpm of each ³²P-labeled RNA sample was analyzed. (A and B) Uninfected tomato; (C) ³²P was introduced into infected plants 1 week after inoculation with PSTV; (D) ³²P was introduced into plants 2 weeks after inoculation when systemic symptoms of PSTV had just appeared; (E) ³²P was introduced into plants 3 weeks after inoculation when leaves had shown PSTV symptoms for 1 week.

land and Bieber, 1970). Slab gels are autoradiographed to determine the position of RNA species.

No differences in the number of labeled low molecular weight RNA species were detected between uninfected and PSTV-infected tomato when ^{32}P was introduced into infected plants 1 or 3 weeks after inoculation. However, when the systemic symptoms began to appear, it was incorporated into RNA with the electrophoretic properties of PSTV (Figs. 3 and 4).

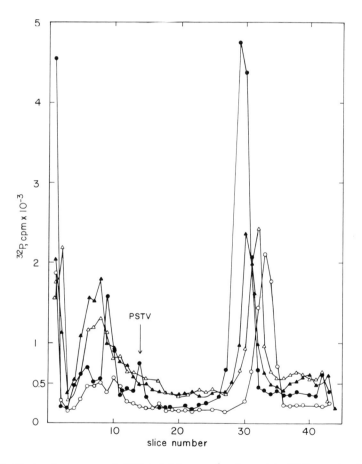

FIG. 4. Electropherograms of ^{32}P-labeled low molecular weight RNA species from uninfected and PSTV-infected tomato leaves in 20% polyacrylamide gels electrophoresed for 7 hours at 5 mA/gel. About 250–300 × 10^3 cpm of each ^{32}P-labeled RNA sample was used per gel. O–O–O, Uninfected tomato; ▲–▲–▲, PSTV-infected tomato, ^{32}P was introduced 1 week after inoculation; ●–●–●, PSTV-infected tomato, ^{32}P was introduced 2 weeks after inoculation; △–△–△, PSTV-infected tomato, ^{32}P was introduced 3 weeks after inoculation.

4. *Purification of ³²P-Labeled PSTV from Gels*

The location of different RNA species after electrophoresis in slab gels can be determined precisely by ³²P autoradiography. For autoradiography, wet slab gels are wrapped with Saran Wrap, covered with Kodak RP54 X-ray film, and placed between two pieces of flat wood in the dark. Depending on the level of radioactivity present and the intensity of blackening wanted, exposure time is usually between 24–72 hours. The film is developed in Kodak Rapid X-ray Developer, followed by fixing in Kodak X-ray Fixer. Specific ³²P-labeled bands may be cut out and other RNAs avoided. Figure 5 shows ³²P autoradiography of 8.8 × 10⁶ cpm of low molecular weight RNAs obtained from PSTV-infected leaves. Four RNA bands may be discerned; namely, 5 S RNA, PSTV, 9 S RNA, and high molecular weight RNA at the top of the gel. Each RNA band is excised and thoroughly homogenized in a glass homogenizer with 2 ml of 0.1 M glycine, 0.1 M NaCl, pH 9.5, buffer. The homogenate is centrifuged at 8000 rpm for 15 minutes and the supernatant is retained. This extraction procedure is repeated twice and the

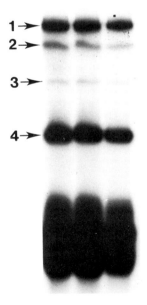

Fig. 5. Autoradiograph of ³²P-labeled low molecular weight RNA (total of 8.8 × 10⁶ cpm) from PSTV-infected tomato plants which incorporated ³²P 2 weeks after inoculation on 20% polyacrylamide slab gel electrophoresed for 24 hours at 100 V. (1) "High molecular weight RNA" which did not penetrate the gel; (2) 9 S RNA; (3) PSTV; (4) 5 S RNA.

TABLE II
Distribution of ^{32}P-Labeled RNAs[a]

RNA fraction	Cpm	Percentage of total low molecular weight RNAs
5 S	294,680	3.34
PSTV	35,000	0.39
9 S	64,720	0.70
Top of gel	102,240	1.16
Other RNAs	—	94.41

[a] After electrophoresis of 8.8 × 10⁶ cpm of a 2 M LiCl-soluble RNA preparation in a 20% slab gel for 24 hours. Each RNA was extracted from the gel and its total radioactivity determined (see text).

supernatants are combined. Two volumes of 95% ethanol and 100–200 μg of unlabeled RNA are added to the supernatant and the mixture is kept overnight at −80°C. After centrifugation at 12,000 rpm for 20 minutes, the pellet is dissolved in 0.5–1 ml of H_2O and total radioactivity is determined for each RNA species. Table II shows that ^{32}P-labeled PSTV is present at a very low concentration (0.4%) in the 2 M LiCl-soluble RNA of PSTV-infected tissue.

5. Determination of PSTV Homogeneity

The purity of each of the four RNA species was tested by electrophoretic analysis on cyclindrical gels. As shown in Fig. 6, each RNA band contains a single RNA species, except for the PSTV band which contains, in addition to PSTV, about 15% 5 S RNA (Fig. 6C). The 5 S RNA is removed from PSTV by electrophoresis in 20% cyclindrical gels for 14 hours, during which time 5 S RNA runs off the gel. The ^{32}P-labeled PSTV is located in gel slices by Cerenkov counting. Reisolation of ^{32}P-labeled PSTV from gels and analysis by electrophoresis (20% gels for 7 hours) reveals the presence of only one species of RNA (Fig. 7). Only at this step is ^{32}P-labeled PSTV considered electrophoretically homogeneous and suitable for biochemical studies on viroid structure.

B. ^{3}H-Labeling

1. In Vivo Labeling

Labeling with uracil-^{3}H is carried out using leaf strips from PSTV-infected tomato plants (Diener and Smith, 1975). One gram of infected leaves is cut with a razor blade into strips about 1 mm wide and 10 mm long. Leaf strips are submerged in 1.25 ml of 0.2 M ribonuclease-free

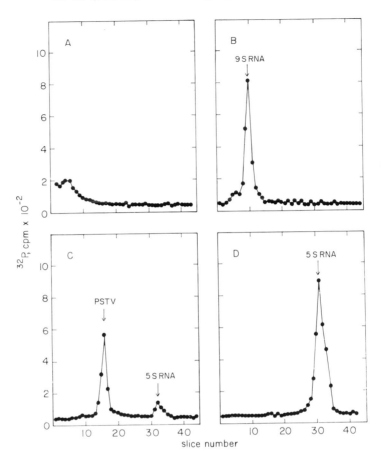

Fig. 6. Electrophoresis pattern of different species of RNA obtained from slab gels (Fig. 5) and extracted as described in the text. Electrophoresis was run in 20% polyacrylamide gels for 7 hours at 5 mA/gel. (A) High molecular weight RNA; (B) 9 S RNA; (C) PSTV + 5 S RNA contamination; (D) 5 S RNA.

sucrose, 0.05 M KH_2PO_4, pH 6.0, containing 125 μCi of uracil-^{3}H and 62.5 units each of penicillin and streptomycin. They are incubated for 8 hours at room temperature, after which time the leaf strips are thoroughly washed with tap water and quick-frozen in a Dry Ice–ethanol mixture.

Nucleic acids are extracted from leaf strips by the method of Diener and Smith (1975). Each sample is homogenized for 5 minutes at 0°C (Omni-Mixer, 50 ml stainless steel microcell, full speed) in 9 ml of glycine buffer (0.1 M glycine, 0.1 M NaCl, 0.01 M EDTA, 1% sodium dodecyl sulfate, pH 9.5), 1 ml bentonite (12 mg/ml), and 20 ml of water-saturated phenol. The

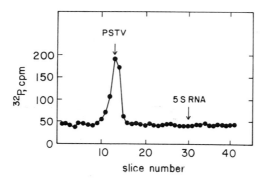

FIG. 7. Electrophoresis pattern of homogeneous preparation of ^{32}P-labeled PSTV electrophoresed in a 20% polyacrylamide gel for 7 hours at 5 mA/gel. PSTV-^{32}P was analyzed after removing the 5 S RNA contaminant (Fig. 6C) by electrophoresis in 20% cyclindrical gels for 14 hours.

emulsions are centrifuged for 15 minutes at 10,000 g; the aqueous phases are removed; and the nucleic acids are precipitated by addition of 2 volumes of ethanol followed by storage for 2–3 hours at 0°C. After centrifugation for 15 minutes at 10,000 g, each pellet is suspended in 2 ml of H$_2$O. The nucleic acids are again precipitated by addition of 2 volumes of ethanol and 3 drops of 3 M sodium acetate, pH 5.1. After centrifugation, each pellet is dissolved in 1 ml of H$_2$O. Nucleic acid concentration is determined by UV spectrophotometry. The nucleic acids are again precipitated by ethanol and are redissolved in the appropriate volume of electrophoresis buffer to give a final nucleic acid concentration of 2 mg/ml.

Each preparation is incubated for 1 hour at 25 °C with 0.1 mg/ml of DNase I and 0.01 M MgCl$_2$ to degrade DNA. To each sample, 1 volume of 4 M LiCl is then added. After storage for 16 hours a 4°C, the preparations are centrifuged for 15 minutes at 10,000 g; and the supernatant solution is retained. Each pellet is washed with 1 ml of electrophoresis buffer made 2 M with respect to LiCl, the suspensions are centrifuged, and the resulting supernatant solutions are pooled with the respective supernatants from the first centrifugation. Nucleic acids are precipitated by addition of 3 volumes of ethanol and 3 drops of 3 M sodium acetate, pH 5.1, followed by storage for 16 hours at −10°C. After centrifugation, each pellet is dissolved in 50 μl of electrophoresis buffer.

For the analysis of PSTV newly synthesized in the leaf strips, 20% gels are prepared and electrophoresis is run as described above. After electrophoresis, gels are scanned at 260 nm.

After scanning, gels are cut into 2-mm slices with a razor-blade assembly; each slice is placed into a scintillation vial, and the gel is dissolved by addition of 0.1 ml of 30% hydrogen peroxide to each vial (Tishler and Epstein, 1968). The vials are sealed with parafilm under the vial caps and are incubated for 16 hours at 50 °C. After cooling, 1 ml of NCS solubilizer and 10 ml of a toluene-based scintillation fluid are added to each vial, and the contents are mixed thoroughly. Samples are counted in a liquid scintillation spectrometer.

The radioactivity profile of low molecular weight RNAs from leaf strips (Fig. 8) shows the presence of PSTV-³H in the PSTV-infected leaf strips

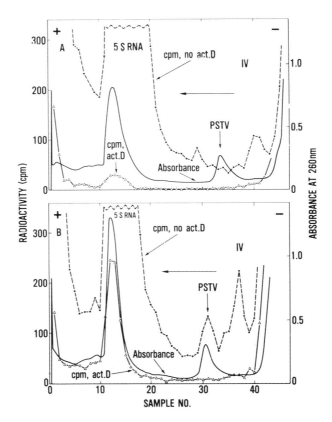

FIG. 8. Ultraviolet absorption (———) and radioactivity profiles of low molecular weight RNA preparations from healthy or PSTV-infected tomato leaves after electrophoresis in 20% polyacrylamide gels for 7.5 hours at 4 °C. Electrophoretic movement is from right to left. (A) Leaf strips from healthy plants infiltrated with water (● - ● - ●) or with 30 μg/ml of actinomycin D (△–△–△), followed by incubation for 8 hours at 25 °C with 125 μCi of uracil-³H. Unlabeled partially purified PSTV was added as internal marker. (B) Leaf strips from PSTV-infected leaves treated as described for (A). IV = unidentified minor component of cellular RNA (see Diener, 1972).

which coincides with the position of marker PSTV. The coincidence and absence of the component from identically treated preparations from uninoculated leaves constitues strong evidence that this component is the result of uracil-^{3}H incorporation into PSTV. Pretreatment of the infected leaf strips with actinomycin D prior to uracil-^{3}H incorporation results in an inhibition of the synthesis of PSTV-^{3}H.

2. In Vitro Labeling of PSTV in Cell Nuclei

The synthesis of PSTV-^{3}H in nuclei isolated from PSTV-infected tomato leaves is accomplished by the method of Takahashi and Diener (1975). The procedures used to purify cell nuclei and to assay for RNA- synthesizing activity are described below.

All operations for isolation of cell nuclei are carried out in a cold room at 4°C. One hundred grams of young, expanding leaves from PSTV-infected leaves are dipped into 300 ml of cold ether for 30 seconds. They are immediately rinsed twice with 200 ml of cold isolation medium (0.25 M sucrose, 0.01 M Tris, pH 7.2, 0.005 M MgCl$_2$, 0.003 M CaCl$_2$, and 0.005 M 2-mercaptoethanol). Leaves are homogenized in a Waring Blendor (1 quart stainless steel container) for 60 seconds at high speed. The resulting homogenates are filtered once through two layers of cheesecloth and then twice through two layers of Miracloth (Chicopee Mills, Inc., New York). The residue is reextracted by resuspension in 100 ml of isolation medium and by homogenization as described above. The combined filtrates are transferred to 90-ml centrifuge tubes and are centrifuged at 1000 g for 5 minutes in an International Centrifuge (Model K). The supernatants are discarded.

The green, nuclei-containing material is scraped off the starch pellet with a spatula and is resuspended in the isolation medium. To that, Triton X-100 (final concentration: 3%) is added. The suspension is allowed to stand for 30–40 minutes at 4°C. After centrifugation at 500 g for 5 minutes, the pellet is washed once with isolation medium. The crude nuclear suspension is layered onto 30 ml of 1.2 M sucrose containing 0.01 M Tris, pH 7.2, 0.003 M CaCl$_2$, and 0.005 M each of MgCl$_2$ and 2-mercaptoethanol. It is centrifuged at 1000 g for 5 minutes in a swinging bucket rotor. The resulting pellet contains semipurified nuclei. It is further purified by relayering of nuclei on the above medium containing 1.2 M sucrose, followed by centrifugation. The resulting pellet, which contains the nuclei, is taken up in a suitable volume of the assay medium. The above method for the isolation of nuclei is relatively rapid (usually 2 1/2 hours) and gentle. Storage of nuclear preparations for 10–40 days at −20°C results in a 60–70% reduction in RNA synthesizing activity.

To study RNA synthesis by purified nuclei, the following reaction mixture is assembled: 0.196 ml of assay medium [0.05 M TES (N-Tris (hydroxymethyl)methyl-2-aminoethanesulfonic acid), pH 7.8, 0.01 M MgCl$_2$, 0.004 M 2-mercaptoethanol, and 25 μg/ml of chloramphenicol] containing 50 nmoles each of ATP, GTP, and CTP; 1 mole of phosphoenol pyruvate; 2 μl of pyruvate kinase suspension; 2 μCi of UTP-^{3}H; and 0.3 ml of freshly prepared, purified nuclei suspended in assay medium. For each assay, nuclei equivalent to 5 μg of DNA are added. Samples are incubated for 60 minutes at 30 °C in a total volume of 0.5 ml.

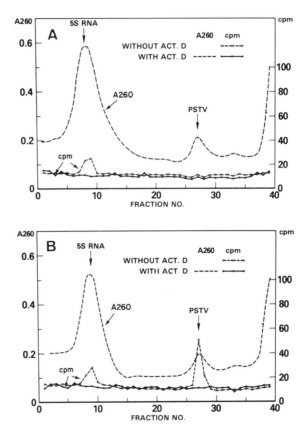

FIG. 9. Ultraviolet absorption and radioactivity distribution of low molecular weight RNAs isolated from incubation mixtures, each containing 8 μCi of UTP-^{3}H and nuclei equivalent to 20 μg of DNA, pretreated with either water or 50 μg/ml of actinomycin D, after electrophoresis in 20% polyacrylamide gels for 7.5 hours at 4 °C. Migration of RNAs is from right to left. DNA and high molecular weight RNAs were removed prior to electrophoresis. Conditions of electrophoresis were such that transfer RNA was run off the gels. Incorporation of UTP-^{3}H into 5 S RNA and PSTV amounted to 0.05–0.2% of total incorporation into RNA. (A) Nuclei isolated from uninfected leaves; (B) nuclei isolated from PSTV-infected leaves. Arrow indicates the position of added PSTV marker.

Incorporation reactions of UMP-^{3}H in a complete system are carried out on a scale four times larger than those described above. After incubation for 60 minutes at 30 °C, low molecular weight RNAs are prepared from the radioactive reaction product (see above) and are analyzed with added marker PSTV and 5 S RNA in 20% polyacrylamide gels. The radioactivity profile of low molecular weight RNAs from infected leaves shows two components, one of which coincides with the position of 5 S RNA, the other with that of PSTV. The radioactivity profile of low molecular weight RNAs from uninfected leaves shows one component that coincides with the position of marker 5 S RNA. Treatment with actinomycin D leads to inhibition of incorporation into both RNAs (Fig. 9).

VI. Physical and Chemical Methods for Determination of Viroid Structure, Replication, and Origin

A number of physical and chemical methods have been applied to purified viroids in efforts to elucidate their nature, structure, mode of replication, and possible origin. None of these methods, however, has been developed specifically for viroid research, and all have previously been used for the analysis of other nucleic acid molecules. These methods, therefore, fall outside of the scope of this chapter and are not described in detail.

In the following, some of these methods are enumerated, important results are summarized, and pertinent references are given.

A. Electron Microscopy

PSTV has been visualized by electron microscopy (Sogo et al., 1973). Several modifications of the Kleinschmidt protein monolayer spreading technique were used (Kleinschmidt and Zahn, 1959). Native PSTV, when spread in 8 M urea, was shown to consist of 500-Å long strands with the width of double-stranded DNA.

B. Thermal Denaturation Properties

Standard methods for the measurement of hyperchromicity of viroids upon heating have been applied to PSTV (Diener, 1972) and CEV (Semancik et al., 1975). In either case, thermal denaturation properties were shown to be unlike those expected with a perfectly matched, double-stranded RNA. The midpoint value of denaturation (t_m), however, was shown to be higher than that observed with single-stranded RNAs.

C. Base Composition

The base composition of purified PSTV (mole %) as determined by the Randerrath procedure is: G = 28.9, C = 28.3, A = 21.7, and U = 20.9 (Niblett et al., 1976). Thus, PSTV has a high G + C content (57%) and AMP/UMP and GMP/CMP ratios that approach unity. A minor component comprising 0.3% of the PSTV preparation was also detected and is being identified. Pure ^{32}P-labeled PSTV prepared as described above was shown to contain 55% G + C (Hadidi, 1976). The base composition of CEV as determined by polyacrylamide gel electrophoresis of nucleotides obtained from purified CEV by alkaline hydrolysis (Morris and Semancik, 1974) was shown to have a G + C content of 58% and AMP/UMP and GMP/CMP ratios that approach unity. ^{32}P-labeled CEV was shown to contain 56% G + C (Semancik et al., 1975).

D. Molecular Weight Determination

To obtain molecular weight estimates of PSTV independent of its conformation, purified PSTV was formylated by the method of Boedtker (1971) and analyzed, together with formylated marker RNAs, by polyacrylamide gel electrophoresis (Diener and Smith, 1973). A molecular weight estimate of 7.5–8.5 × 10^4 was obtained.

E. Messenger-RNA Properties

The ability of PSTV (Davies et al., 1974) and of CEV (Hall et al., 1974) to act as messenger RNAs for protein synthesis has been investigated in several cell-free protein-synthesizing systems. With either viroid, results were negative.

F. Aminoacylation Properties

The ability of CEV to serve as an amino acid acceptor was investigated by Hall et al. (1974). Results were negative.

G. Poly(A) or Poly(C) Sequences

Attempts have been made to detect poly(A) or poly(C) stretches in the PSTV molecule using the *Escherichia coli* DNA polymerase I system in the presence of oligo(dT)$_{10}$ or oligo(dG)$_{12-18}$ primer, as described by Modak et al. (1975). Additions of oligo dT$_{10}$ or oligo(dG)$_{12-18}$ to the assay mixture did not result in increased synthesis of DNA complementary to PSTV. The result indicates the absence of poly(A) or poly(C) in PSTV (Hadidi *et al.*, 1976a). The inability of CEV to hybridize with ^{3}H-labeled poly(U) indicates that CEV does not contain poly(A) sequences (Semancik, 1974).

H. *In Vitro* ^{125}I–Labeling

Purified PSTV and CEV have been labeled with ^{125}I by use of the Commerford reaction (Commerford, 1971; Prensky, 1975). Although this procedure leads to some random cleavage of RNA, both PSTV and CEV were shown to contain major components migrating at the position of the intact species in 10% polyacrylamide gels after iodination (Dickson *et al.*, 1975).

I. RNA Fingerprinting

Iodinated PSTV and CEV have been digested with either RNase T1 or with pancreatic RNase and fingerprinted by standard techniques (Brownlee and Sanger, 1969; Robertson *et al.*, 1973). Results indicated that each viroid has a complexity compatible with the size estimate of 250–350 nucleotides and that PSTV and CEV do not have the same primary sequence (Dickson *et al.*, 1975).

J. Molecular Hybridization

Iodinated PSTV has been annealed with fragmented cellular DNA from various plant species and with RNA from uninfected and PSTV-infected host plants as described by Gillespie *et al.* (1975). Results indicate that infrequent DNA sequences complementary to at least 60% of the PSTV molecule exist in both uninfected and infected cells of tomato plants (Hadidi *et al.*, 1976b) as well as in DNA of certain other uninfected solanaceous host plants. Phylogenetically diverse plants contain sequences related to less of the PSTV. No RNA complementary to PSTV could be detected in infected tomato cells. These results support the hypothesis that PSTV replication is DNA-directed and indicate that PSTV may have originated from genes in normal solanaceous plants.

References

Boedtker, H. (1971). *Biochim. Biophys. Acta* **240**, 448–453.
Brakke, M. K. (1970). *Annu. Rev. Phytopathol.* **8**, 61–84.
Brierley, P. (1953). *Plant Dis. Rep.* **37**, 343–345.
Brownlee, G. G., and Sanger, F. (1969). *Eur. J. Biochem.* **11**, 395–399.
Commerford, S. L. (1971). *Biochemistry* **10**, 1993–2000.
Davies, J. W., Kaesberg, P., and Diener, T. O. (1974). *Virology* **61**, 281–286.

Dickson, E., Prensky, W., and Robertson, H. D. (1975). *Virology* **68**, 309–316.

Diener, T. O. (1971a). *Virology* **43**, 75–89.

Diener, T. O. (1971b). *Virology* **45**, 411–428.

Diener, T. O. (1971c). *In* "Comparative Virology" (K. Maramorosch and E. Kurstak, eds.), pp. 433–478. Academic Press, New York.

Diener, T. O. (1972). *Virology* **50**, 606–609.

Diener, T. O. (1973). *Anal. Biochem.* **55**, 317–320.

Diener, T. O. (1974). *Annu. Rev. Microbiol.* **28**, 23–39.

Diener, T. O., and Lawson, R. H. (1973). *Virology* **51**, 94–101.

Diener, T. O., and Raymer, W. B. (1967). *Science* **158**, 378–381.

Diener, T. O., and Raymer, W. B. (1969). *Virology* **37**, 351–366.

Diener, T. O., and Raymer, W. B. (1971). "Descriptions of Plant Viruses," No. 66. Commonwealth Mycol. Inst. Assoc. Appl. Biol., Kew, Surrey, England.

Diener, T. O., and Smith, D. R. (1973). *Virology* **53**, 359–365.

Diener, T. O., and Smith, D. R. (1975). *Virology* **63**, 421–427.

Diener, T. O., and Weaver, M. L. (1959). *Virology* **7**, 419–427.

Engelhardt, D. L. (1972). *J. Virol.* **9**, 903–908.

Francki, R. I. B. (1968). *Virology* **34**, 694–700.

Franklin, R. M. (1966). *Proc. Natl. Acad. Sci. U.S.A.* **55**, 1504–1511.

Garnsey, S. M., and Whidden, R. (1970). *Phytopathology* **60**, 1292 (abstr.).

Gillespie, D. H., Gillespie, S., and Wong-Staal, F. (1975). *Methods Cancer Res.* **11**, 205–245.

Hadidi, A. (1976). *Beltsville Symp. Virol. Agric. 1976* Abstract, p. 24.

Hadidi, A., Modak, M. J., and Diener, T. O. (1976a). *Beltsville Symp. Virol. Agric. 1976* Abstract, p. 30.

Hadidi, A., Jones, D. M., Gillespie, D. H., Wong-Staal, F. and Diener, T. O. (1976b). *Proc. Natl. Acad. Sci. U.S.A.* **73**, 2453–2457.

Hall, T. C., Wepprich, R. K., Davies, J. W., Weathers, L. G., and Semancik, J. S. (1974). Virology **61**, 486–492.

Haviland, R. T., and Bieber, L. L. (1970). *Anal. Biochem.* **33**, 323–334.

Hoagland, D. R., and Arnon, D. I. (1950). *Calif., Agric. Exp. Stn., Circ.* **347**, 1–32.

Kleinschmidt, A. K., and Zahn, R. K. (1959). *Z. Naturforsch., Teil B* **14**, 770–779.

Lawson, R. H. (1968). *Phytopathology* **58**, 885 (abstr.).

Lister, R. M., and Hadidi, A. F. (1971). *Virology* **45**, 240–251.

Loening, U.E. (1967). *Biochem. J.* **102**, 251–257.

Modak, M. J., Marcus, S. L., and Cavalieri, L. F. (1975). *J. Biol. Chem.* **249**, 7373–7377.

Morris, T. J., and Semancik, J. S. (1974). *Anal. Biochem.* **61**, 48–53.

Morris, T. J., and Wright, N. S. (1975). *Am. Potato J.* **52**, 57–63.

Niblett, C. L., Hedgcoth, C., and Diener, T. O. (1976). *Beltsville Symp. Virol. Agric. 1976* Abstract, p. 27.

Peacock, A. C., and Dingman, C. W. (1967). *Biochemistry* **6**, 1818–1827.

Prensky, W. (1975). *Methods Cell Biol.* **13**, 121–152.

Randles, J. W. (1975). *Phytopathology* **65**, 163–167.

Raymer, W. B., and Diener, T. O. (1969). *Virology* **37**, 343–350.

Raymer, W. B., and O'Brien, M. J. (1962). *Am. Potato J.* **39**, 401–408.

Robertson, H. D., Dickson, E., Model, P., and Prensky, W. (1973). *Proc. Natl. Acad. Sci. U.S.A.* **70**, 3260–3264.

Romaine, C. P., and Horst, R. K. (1975). *Virology* **64**, 86–95.

Semancik, J. S. (1974). *Virology* **62**, 288–291.

Semancik, J. S., and Weathers, L. G. (1968). *Virology* **36**, 326–328.

Semancik, J. S., and Weathers, L. G. (1972). *Virology* **47**, 456–466.

Semancik, J. S., Morris, T. J., Weathers, L. G., Rodorf, B. F., and Kearns, D. R. (1975). *Virology* **63**, 160–167.

Singh, R. P. (1971). *Phytopathology* **61**, 1034–1035.

Singh, R. P. (1973). *Am. Potato J.* **50**, 111–123.

Singh, R. P., and Bagnall, R. H. (1968). *Phytopathology* **58**, 696–699.

Singh, R. P., Michniewicz, J. J., and Narang, S. A. (1974). *Can. J. Biochem.* **52**, 809–812.

Sogo, J. M., Koller, T., and Diener, T. O. (1973). *Virology* **55**, 70–80.

Studier, F. W. (1973). *J. Mol. Biol.* **79**, 237–248.

Takahashi, T., and Diener, T. O. (1975). *Virology* **64**, 106–114.

Tishler, P. V., and Epstein, C. J. (1968). *Anal. Biochem.* **22**, 89–98.

Van Dorst, H. J. M., and Peters, D. (1974). *Neth. J. Plant Pathol.* **80**, 85–96.

7 *Electron Microscopy of Viral Nucleic Acids**

Donald P. Evenson

*Some research and conclusions presented herein were based on work supported by grants from the National Cancer Institute Nos. NCI-CA 08748 and NIH-CA 16599.

I. Introduction

This chapter will outline methods to study isolated viral DNA and RNA under the electron microscope using the technique initially described by Kleinschmidt and Zahn (1959). The basis of their technique is that a solution of nucleic acid is mixed with a globular-basic protein, such as cytochrome c, that binds to the nucleic acid and also surface denatures to produce a network of unfolded polypeptide chains. Adsorption of the nucleic acid–protein complex to the surface-denatured protein film brings the nucleic acid molecules into a two-dimensional configuration required for electron microscopy.

The technique is simple and success can be obtained within a few hours; however, a number of its parameters are not yet well understood. Consequently, when artifacts do arise, the number of necessary controls for solving the problem can be overwhelmingly large. One leading expert in this field has said, "When the technique works, do the experiment and don't always ask why it may not have worked before." Since it is recognized that unknown variables and sometimes artistic aspects are involved in this technique, it is difficult to describe a guaranteed route to success in a written chapter; and, obviously, the best way to learn is in the laboratory under the guidance of an experienced investigator.

Hundreds of studies have been done by many investigators who have made numerous modifications of the original technique. For previous reviews, see Kleinschmidt (1968), Davis et al., (1971), Inman (1974), Younghusband and Inman (1974), and Heine et al. (1975a). Since it is the purpose of this volume to outline methods in detail without the need for additional references, only some of the modified techniques will be described here. However, where pertinent, references will be provided for readers who may want additional details.

Mention of specific equipment or a commercial company does not endorse it over similar equipment or companies not named, but is meant only to serve as a reference or example.

II. Materials and Equipment

A. Nucleic Acids

Viral DNA and RNA to be studied by electron microscopy can be isolated by standard procedures such as phenol extraction (Mandell and Hershey, 1960); however, numerous modifications and other methods exist for different viruses and for the prevention of nucleic acid degradation. For the Kleinschmidt technique, purification of the nucleic acid always requires particular attention. Although small amounts of contaminating viral or non-viral nucleic acids may not be detected by some biochemical techniques, virtually all species of nucleic acid molecules in a given sample can be visualized by the Kleinschmidt technique. A purified sample is sometimes difficult to obtain since even "highly purified" preparations of viral nucleic acid often contain a high percentage of cellular nucleic acid, e.g., samples of polyoma virus often contain a high percentage of pseudovirions (Pagano and Hutchison, 1971). Contaminating cellular nucleic acids often cannot be distinguished from viral nucleic acids, which may lead to misinterpretations. For example, molecules of nucleic acids isolated from RNA tumor virus samples, previously described as viral RNA (Granboulan et al., 1966), are now interpreted to correspond to contaminating DNA (Weber et al., 1975; Evenson et al., 1975). As will be described later, techniques using the electron microscope can now help distinguish whether a sample contains DNA and/or RNA. However, if procedures are applied that will fully extend single-stranded RNA molecules, DNA will be denatured to the degree that it may not be possible to distinguish between single-stranded DNA and RNA. Thus, whenever contamination can interfere with results, the desired nucleic acid should be purified from any contaminating species, e.g., by nuclease digestion and/or sedimentation in velocity or bouyant density gradients. The latter technique will also purify the nucleic acid from contaminating proteins that can interfere with the configuration of spread nucleic acids. Phenol-extracted DNA, after ether extraction, can be dialyzed extensively against 0.01 M phosphate buffer, 1 mM EDTA (to inhibit nucleases), or 0.1 SSC (SSC is 0.15 M NaCl, 0.015 M Na citrate, pH 7), or 0.02 M NaCl, 1 mM EDTA (pH 7) and stored for a number of weeks at 4 °C while samples are repeatedly used. With purified DNA, degradation due to exogenous nucleases is generally negligible. Alternatively, the solutions may be frozen at −20 °C or −70 °C for long-term storage. Ethanol-precipitated RNA, resuspended in NET (0.15 M NaCl, 0.01 M Tris, 1 mM EDTA), 0.1 SSC, or 1 mM dithiothreitol (DTT) (pH 7) should be frozen at −70 °C as soon as possible after collection to prevent possible ribonuclease degradation. Some RNAs, notably tumor virus RNAs, are especially labile and will even degrade at −70 °C after storing for 6 months (Heine et al., 1975b). If a given sample is to be used a number of times, it is preferable to divide it into small aliquots that can be stored in one-half dram vials. Very small sample volumes (5–10 μl) can be efficiently mixed stored in, and recovered from microconical vials (Wheaton, Millville, New Jersey). Single-stranded nucleic acids are less susceptible to freeze-induced breaks if the sample is frozen rapidly. Our procedure is to pipette the nucleic acid solution into a vial which is then immediately placed into a rack partially submerged in a Dry Ice–acetone bath; the vials are then stored in a −70 °C freezer.

B. GLASSWARE

Stock solutions and working samples can be stored in plastic containers or in Pyrex glassware. All glassware must be thoroughly washed with detergent and rinsed extensively either in a dishwasher or by hand, including a dozen rinses in single-distilled water followed by a dozen rinses with deionized or double-or triple-distilled water. All traces of detergents should be removed because they can significantly interfere with this technique, especially in the formamide procedure (Davis *et al.*, 1971). Up to a dozen small vials and test tubes can conveniently be rinsed with a hematocrit washer (Clay Adams, Parsippany, New Jersey), sequentially attached, by means of a Tygon tube to the tap of a single-distilled and finally double-distilled or deionized water supply for about 5 minutes each.

For the study of RNA, extra care must be taken to prevent RNase contamination, bearing in mind that RNase activity is present on human skin. Materials and equipment that come into contact with RNA solutions should be handled with sterile utensils while wearing plastic gloves. All glassware is pretreated with chromic–sulfuric acid solution for several minutes to inactive RNase activity and then rinsed extensively and dried. This glassware is then sterilized by autoclaving and/or placing in a dry-heat oven (180 °C) overnight. All other utensils that may come into contact with the RNA solutions, e.g., spatulas, weighing pans, stir bars, etc., must be treated for possible RNase contamination. In cases where the treatments listed above are not possible, materials can be treated with iodoacetate which attacks the active center of RNase (Egami and Nakamura, 1969). Submerge the item in 0.15 M sodium iodoacetate, 0.1 M sodium acetate in 2 X SSC, pH 5.4, incubate at 54 °C for 40 minutes, and then rinse well with water (Bøvre *et al.*, 1971). Dialysis tubing should be boiled for 30 minutes (no longer) in 1% sodium carbonate, 1 mM EDTA, and then washed several times with alkaline 1 mM EDTA, rinsed with distilled water until the washings are no longer alkaline, and subsequently stored in water in a glass bottle containing a few drops of chloroform at 4 °C (Brewer, 1974). Disposable plastic pipettes and vessels are considered clean and free from grease and RNase. However, they are sometimes contaminated with a small amount of lintlike debris and should be rinsed before use.

C. WATER

Excellent quality water is a must for success. Double- or triple-quartz distilled water is usually adequate; however, this may not always be the case. In our experience, double-distilled water was acceptable in two other laboratories, while in a third it was necessary to install a water deionization system (Millipore Corp., Bedford, Massachusetts) containing one carbon and two inorganic deionization columns following the house single-distilled water supply. After the water passed through the deionization system, which contained a terminal filter (0.22 μm), it was used directly for experiments. Without this treatment remaining ions apparently complexed with the nucleic acid in a way that prevented suitable staining with uranyl acetate and also produced oriented contours of the molecules.

D. Solutions

Solutions should be filtered through 0.10–0.45 μm pore filters to eliminate most particulate matter which would contribute to clumping of the nucleic acid molecules. Care should be taken, however, not to introduce detergents into the solutions from the filters. For example, if Millipore filters are used, use TF (triton-free) filters. Detergents can be removed from filters, however, by rinsing them several times with boiling water. Sweeney adapters fitted to a hypodermic syringe are suitable for filtering small volumes. For studying RNA, stock solutions should be autoclaved whenever possible. RNase-free sucrose is available commercially (Schwartz–Mann, Orangeburg, New York).

E. Grids

The choice of a grid type and mesh size depends on several factors, including: (1) the strength of the support film, (2) the manipulations required of the grid during the experiment, (3) the size of the molecules, and (4) the type of electron microscope used. A practical choice is copper- or nickel-mesh grids cleaned by rinsing in acetone before use. Films should always be placed on the same side of the grid, i.e., the shiny or dull side, so that no confusion will exist as to which side the film is on. For studying small molecules (1–2 μm in length), 300 mesh grids are suitable, but for larger molecules (30–80 μm in length) 50–100 mesh grids are recommended. With large mesh grids, breakage and/or stretching of the film can be a problem, especially for accurate determinations of molecular length. Very thin and fragile support films should be mounted on the smallest mesh possible, up to 1000 mesh (Fullam, Schenectady, New York). These very fine-meshed grids have, for most purposes, replaced the use of "holey films" (Ottensmeyer, 1969; Kleinschmidt, 1971) as secondary supports for very thin films. A "handle" of some type is needed on a grid when picking up a protein monolayer prepared sample (Section III,A,1), otherwise the forceps will touch and distort the protein film. Some investigators use circular grids and bend the edge of the grid to form a handle; this method works, but it risks damage to the supporting film on the grid. Copper mesh-handled grids (Pelco, Tustin, California) are preferred over circular grids for three reasons: (1) The handles can easily be bent at right angles to the plane of the grid without bending the grid itself. (2) Any handle formed must be flattened before the grid is metal-shadowed and/or placed in the electron microscope. Obviously, a preformed handle bent near its middle can be flattened more easily without risking damage to the sample than a circular grid bent on the edge. (3) The handle can be attached to a slide with double-stick tape (Scotch brand) which is a convenient method for holding grids in place during rotary shadowing procedures. The handle can be cut off with a fine scissors prior to insertion of the grid into an electron microscope.

Another type of grid that has a natural "handle" is the Siemens specimen mounts (Siemens America, Iselin, New Jersey), which are platinum cylinders measuring 2.3 mm in diameter and 0.9 mm in height and whose centers are drilled to form a "grid" on one end. These mounts are expensive; however, they can be reused after removing the old films by treatment with chromic–sulfuric acid followed by extensive rinsing in water. The 0.9 mm height of these mounts makes it easy to grasp them with a curved forceps, thus preventing distortion of the film (Kleinschmidt, 1968). These mounts are available containing from 1 to 21 holes 70 μm in diameter. The fewer the number of holes, the higher the ratio of flat surface to hole area that is favorable for maintaining intact fragile supporting films.

F. Supporting Films

The quality and type of supporting films are among the most important factors in the success of this technique. Parameters such as tensile strength, film composition, thickness, cleanliness, surface charges, age, etc., have paramount effects on the quality of the final preparation. Therefore, considerable emphasis will be placed here on the production and use of various supporting films.

1. Water-Cast Plastic Films

The most easily produced and widely used supporting film is Parlodion (collodion). A 2.9% (w/v) Parlodion solution is made by dissolving well-dried (90°C overnight) Parlodion strips in amyl acetate. Reagent quality and specially dessicated amyl acetate (Ladd, Burlington, Vermont) is recommended; furthermore, small bottles should be used to reduce hydration in frequently opened bottles. A number of grids can easily be coated with a Parlodion film by the method of Bradley (1965). First fill an acetone-cleaned dish measuring about 12 cm in diameter and 6 cm high about two-thirds full

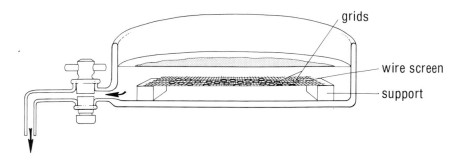

FIG. 1. Schematic of method for preparing plastic-coated grids.

with clean water. Place a 4 × 8 cm nickel-wire screen (15 gauge, 20 mesh) on two supports resting on the bottom of the dish as shown in Fig. 1. Place grids onto the screen with a clean pair of tweezers. Add one drop of the Parlodion solution to the water surface. As soon as the solvent evaporates, remove the film with a needle; this step cleans the surface. Immediately add another drop of Parlodion. Let the film stabilize for 2–3 minutes. Slowly withdraw the water from the dish until after the film has settled onto the grids. This is conveniently done if a stopcock drain is mounted on the bottom edge of the glass dish. The screen carrying the coated grids is then transferred to a 60 °C oven for 30 minutes. The grids are stored at room temperature, and preferably used the same day. Carbon can be evaporated onto the upper surface of the grid if desired. Filmed grids can be removed from the screen as needed by lifting them vertically off the screen with a fine-tipped tweezers.

2. Dipped Plastic Films

Plastic film can be formed on a glass slide and floated off in a trough of water. Advantages of this method are: (a) films are generally smoother; (b) carbon can be evaporated onto the plastic films before they are floated off; and (c) the films can be scored into slightly larger than grid-sized sheets and be picked up from the underside of the film. These films are prepared essentially according to Bradley (1965). First dip a standard microscope slide using a forceps into a solution consisting of three drops of a mild liquid detergent (Joy or liquid Ivory) in 250 mls of water. Gently wipe the slide dry with a Twill-Jean cloth (Ladd). Place the slide into a glass vessel slightly larger than the slide with a stopcock drain mounted on the bottom. Fill the vessel with 2% (w/v) Parlodion in amyl acetate and cover the top of the vessel with a petri dish. Immediately drain the Parlodion solution back into the stock bottle. Leave the slide in the amyl acetate vapors for about 5–60 minutes and then remove to dry completely. Carbon can now be evaporated onto the plastic film by a direct or indirect method (Section II, F, 3). The film can be floated off the glass slide and lowered onto grids, as previously described (Section II,F,1). Alternatively, the film can be scored into small squares which will float off as individual pieces. This method has the following advantages: (a) particular regions of the film can be selected, and (b) damage that may occur to the grid film during the "tearing" of the grid from a large film in the method above (Section II,F,1) can be avoided. To cut the film into small squares, first place the slide into a petri dish containing a filter paper onto which a pattern of appropriately sized grid squares has been drawn. While the slide is held in place over the grid pattern with a tool, such as blunt forceps, the grid pattern is scored on the slide with an acetone-cleaned single-edge razor blade. The film is then floated off by pushing the slide into a water-filled trough at an approximately 30° angle. The film should readily float off with many of the small film squares floating separately; these squares can best be seen for picking up if the water-filled trough is placed on a black surface and a strong light is obliquely shined on the water surface. The films are most easily picked up, and with the least distortion, using Siemens specimen mounts or bent-handled grids. A good alternative to Parlodion film is 1% (w/v) cellulose–acetate butyrate, 0.8% (w/v) Parlodion in ethyl acetate, which has a greater tensile strength than Parlodion alone (Mazaitis et al., 1976).

3. Carbon Films

For high resolution and/or high contrast work, including dark field microscopy, very thin carbon films (20–30 Å) are preferable for studying nucleic acids since, for a given tensile strength, they are more electron lucent than plastic films (Kleinschmidt, 1971). Thin carbon films are not easily produced and one has to accept a high percentage of unusable ones. For producing both moderately thick (80–100 Å) and thin (20–30 Å) carbon films, the best method is indirect carbon evaporation (Whiting and Ottensmeyer, 1972) on freshly cleaved mica, which is achieved as follows: 1 cm of the end of a 1/4 inch spectral-grade carbon rod is reduced to 1/8 inch in diameter with a special carbon rod sharpener (e.g., Fullam). This is placed in contact with the flat surface end of another 1/4 inch rod (Fig. 2a) by means of a standard spring-loaded holder. To produce an indirect deposition, a

FIG. 2. Schematic of method for preparing carbon-coated grids. (a) Carbon rods in stainless steel cylinder. (b) Position of carbon rods in relation to hole in cylinder. (c) Pattern of carbon deposition on mica sheet. (d) Mica–carbon sandwich resting on applicator sticks in hydrated petri dish. (e) Floating off carbon film onto surface of water in a Buchner funnel containing grids positioned on filter paper. (f) Side view position of carbon film, grids, filter papers, and funnel after removal of the water from the funnel. (g) Top view of carbon film on filter paper and grids.

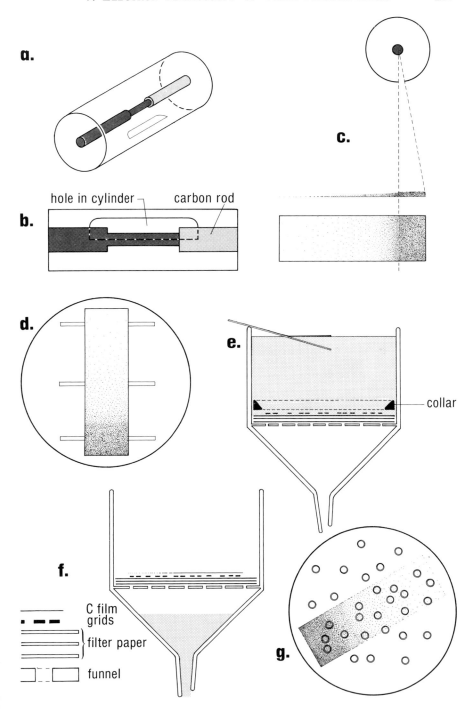

hole in cylinder   carbon rod

collar

C film
grids

filter paper

funnel

mechanical block must be situated between the carbon rods and the mica target (Whiting and Ottensmeyer 1972). We have found it practical to place a cylinder of stainless around the carbon rods which has a hole in the path between the carbon source and the site of deposition (Fig. 2a–c). The best substrate upon which to evaporate a layer of carbon, and from which the carbon can later be stripped, is mica, the smoothest natural surface available. A sheet of mica, about 1 × 3 inches, is cleaved by first starting a split in the sheet at one corner of the mica with a clean razor blade; from this point the sheet can be separated into halves using two fine-pointed tweezers. It is advantageous to use a relatively thick mica sheet (Pelco) so that the split halves are mechanically resistant to bending at the floating off stage. The mica is placed, cleaved side up, on a white piece of dehydrated filter paper below the carbon source, as shown in Fig. 2c. The bell jar is evacuated to a vacuum of about 10^{-6} torr, if possible; the better the vacuum during evaporation the better will be the quality of the carbon films. Furthermore, contamination from diffusion or forepump oil backstreaming into the vacuum chamber must be minimal, otherwise the carbon films will break up into many unusable pieces when one attempts to float them off the mica. The carbon rods are first outgassed by moderate heating. Carbon is evaporated, after the vacuum has reached about 10^{-6} torr, by resistance heating at about 25 A until the piece of filter paper under the mica is quite dark in the region exposed to direct evaporation (Fig. 2c). The mica–carbon sandwich is then removed from the evaporator. To facilitate separation of the carbon film from the mica, a few millimeters of the end portion of the mica receiving direct evaporation is first cut off with a scissors, then placed carbon side up in a petri dish containing hydrated filter paper and several wooden applicator sticks to elevate the mica off the filter paper (Fig. 2d). The petri dish is kept at 4 °C for an hour to several weeks before the mica–carbon sandwich is removed from the refrigerator and the carbon stripped off. This process is as follows: First place three pieces of Whatman # 1 filter paper on the bottom of a 10 cm Buchner funnel and then a stainless steel collar to help seal the edge, as diagrammed in Fig. 2e. Place the filter on a side-arm flask and fill the funnel about two-thirds full of water. Draw a small amount of water through the filter by applying a gentle vacuum to the side-arm flask. Deposit a number of grids, shiny side up, on the filter paper. Without disturbing the grids, fill the funnel to about 1 cm from the top with water. Remove the carbon–mica sandwich from the 4 °C humidified petri dish with forceps. Slowly insert the carbon–mica sandwich, thick carbon end first, at about a 30 ° angle into the water (Fig. 2e); the thicker, directly evaporated carbon facilitates stripping the thinner and more fragile region. After the carbon is stripped off, the mica strip is moved to the side of the funnel, farthest away from the floating film, and lifted out. Apply a very gentle vacuum until about 1–3 mm of water remains in the funnel with the carbon film floating over some of the grids; a hand vacuum pump (Nalgene, Rochester, New York) works well for controlling the vacuum. Then release the vacuum and allow the remaining water to drip through the filter, thereby allowing the carbon to settle *gently* onto the grids (Fig. 2f). From the thicker, easily visible region of the carbon film to the thinner region will be a gradient of film thickness from over 100 Å to about 20 Å. Even though the 20 Å film is invisible on the filter paper, the grids carrying these films can be located because of the visible orientation of the carbon film in the thicker regions (Fig. 2g). After 10–15 minutes, remove the three layers of filter paper with the attached grids and film from the funnel with a forceps and allow to dry slowly overnight, or more quickly by removing the upper filter paper and drying it in a 37 °C oven. The grids that were not overlayed with film may be cleaned and reused. The thicker films are used for routine sampling and the thinnest films, as fragile as they are, are used for high resolution. The following may be done to minimize thin film breakage: (a) use the smallest mesh grid possible, up to 1000 mesh; (b) after picking up the nucleic acid, immerse the grids into any additional solutions at an angle perpendicular to the liquid surface; and (c) if the carbon has a tendency to adhere poorly to the grids, the grids can be made sticky by dipping them into a solution of 0.5% polyisobutylene in toluene (Gordon, 1973) or 1% polybutene (Polysciences, Warrington, Pennsylvania), in xylene (Wellauer and Dawid, 1973), and then dried on filter paper before use.

Another way to prepare nucleic acid samples on moderately thin carbon films (> 70Å) is to first transfer a nucleic acid–cytochrome *c* film to a grid containing a carbon–Parlodion supporting film, followed by contrasting, if desired, of the nucleic acid. The grid is then placed, carbon film up, on a glass surface in a 180 °C oven for 10 minutes which will cause the Parlodion to retract to the periphery of the grid holes, leaving the carbon film intact (Kleinschmidt, 1971). Thus, the Parlodion film is used as a support for the thin carbon film during the initial manipulations and is removed prior to examination with the electron microscope.

The surface characteristics of a carbon film and its interaction with nucleic acids are poorly understood, and carbon films have been treated a number of ways in attempts to improve their adsorptive capacity for nucleic acids. For example, carbon films are sometimes treated with glow discharge (Section III,B,2,a) to carry positive charges which will enhance the adsorption of the negatively charged nucleic acids. Another alternative has been developed by Ruben and Siegel (1975) where aluminum is coevaporated with carbon to form the film; the aluminum provides a positive charge to the film. A significant advantage of these films is that they remain wettable for over a year after they are made. In our experience, these films are difficult to make and they adsorb salts, which is an undesirable feature in most experiments.

G. CYTOCHROME c

Most techniques for visualizing nucleic acid molecules under the electron microscope belong to a group of closely related procedures collectively called "protein monolayer" or "basic protein film" techniques (Kleinschmidt, 1968; Davis *et al.*, 1971). Although various proteins have been recommended to generate a film, cytochrome *c* is the most widely used. Horse heart cytochrome *c* (95–100%), type VI (Sigma, St. Louis, Missouri) yields good results. Stock solutions are made up in water to a 0.1% (w/v) final concentration. Since the quality of commercially prepared cytochrome *c* stocks vary, the actual protein content should be measured spectrophotometrically and adjusted to an actual 0.1% concentration using the following relationship:

$$E^{0.01\%\,\mathrm{cyto}\,c}_{\mathrm{O.D.408}} = 1.25$$

in 0.15 *M* ammonium acetate. The stock can be kept at 4°C for many months. After dilution to the desired concentration, the cytochrome *c* should be filtered through a 0.22 μm pore filter. The actual concentration of cytochrome *c* used is often critical, especially for the microdrop diffusion technique. For spreading on a hypophase, the general rule is one milligram protein per square meter of surface area (Kleinschmidt, 1968).

The configuration of the nucleic acid–cytochrome *c* complex is not well known. However, it should be pointed out that the nucleic acid remains in a mobile state in the protein film, as evidenced by the observation that partially denatured duplex DNA can renature after the molecules are spread on a hypophase within only a few minutes after spreading (Evenson *et al.*, 1972). Since the nucleic acid–protein film is in a fluid state, it is important when picking up this film with a grid to touch the film only with a straight-down motion; any lateral motion produces an oriented pattern of molecules.

III. Techniques for Transfer of Nucleic Acids from Solutions to Electron Microscope Grids

A. PROTEIN MONOLAYER TECHNIQUES

The two most commonly used techniques for the transfer of nucleic acids to electron microscope grids are the "spreading method" (Kleinschmidt and Zahn, 1959) and the "diffusion method" (Lang *et al.*, 1964), the latter usually by the microdrop version (Lang and Mitani, 1970).

These two techniques can be subdivided into the "aqueous" and "formamide" procedures. The aqueous procedure is used for visualization of molecules that are base paired over their entire length. The formamide procedure (Westmoreland *et al.*, 1969) is used for the visualization of single-stranded molecules or single-stranded portions within a duplex molecule, e.g., hetero-duplex molecules, or where single-stranded tails may be present, e.g., as when testing for terminally repetitive sequences by digestion of duplex DNA with exonuclease followed by annealing of the ends (Wadsworth *et al.*, 1976). These methods will be diagrammatically outlined here using duplex DNA as the experimental sample. Modifications of this technique applied to viral nucleic acids will be discussed later.

1. *Macrospreading Technique*

a. Aqueous. A 10 cm² by 2 cm deep plastic dish (Falcon, Oxnard, California) is slightly overfilled with a solution called the hypophase, usually consisting of ammonium acetate or Tris buffer at an ionic strength ranging from 0.01 M to 0.2 M (0.15 M is recommended as a general standard). The nature of the hypophase is related to the final contrast and resolution of the sample, e.g., spreading on water gives poor contrast compared to spreading on a salt and formamide containing hypophase. However, contrast is gained at the expense of a noisier background and lower resolution. After filling the dish, two Teflon-coated aluminum bars, or other clean hydrophobic bars (about 1 cm² by 15 cm long) are placed on the near edge of the dish and then both are pushed back to just past the center of the dish. The closer bar is then drawn forward to the edge of the dish; this sweeping action cleans the surface of the hypophase. A chromic–sulfuric acid-cleaned and well-rinsed microscope slide, either freshly dipped in 0.25 M ammonium acetate and air dried (Davis *et al.*, 1971) or wetted with the hypophase solution, is then placed with a forceps into the hypophase so that it rests on the back Teflon-coated bar at about a 30° angle. This method is diagrammed in Fig. 3.

The spreading solution (hyperphase) may consist of DNA at a concentration of 2–5 μg/ml in 0.4–4 M ammonium acetate, pH 8, and 0.01% cytochrome c. For a 0.5 M ammonium acetate spreading solution (recommended for standard procedures), a 0.15 M ammonium acetate hypophase is recommended. Fifty microliters of spreading solution are taken up with a micropipette and slowly deposited onto the slide, starting about 0.5 cm above the slide–hypophase interface and moving slowly up the slide in a serpentine motion. A variety of pipettes can be used; however, for small and constant volumes, automatic micropipettes work well. Alternatively, a Hamilton syringe with an attached piece of Teflon tubing is suitable (Davis *et al.*, 1971) since the flexibility of the tubing helps to prevent jarring of the slide which would disturb the spreading of the film. To minimize this disturbance, place a weight behind each end of the rear Teflon-coated bar to prevent its movement. The boundary of the protein film can be determined by dusting a few

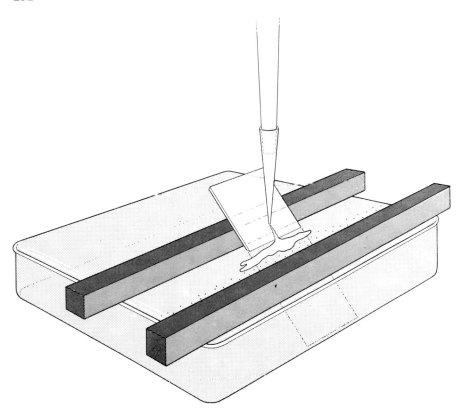

Fig. 3. Schematic of macrospreading technique. The spreading solution is being deposited on a glass slide resting at an angle on the rear Teflon bar. The nucleic acid–protein complex is forming a monolayer on the surface of the hypophase.

talc particles on the hypophase surface with the aid of a cotton-tipped applicator stick. The film is allowed to stabilize for 30–60 seconds and then three samples are picked up on filmed grids at about 0.5 cm in front of the slide–hypophase boundary in the middle and just inside the left and right edges of the glass slide. Anticapillary tweezers are helpful for manipulating the grids. Each sample is stained and/or dehydrated before the next sample is taken. In cases where repetitive film disturbances are disadvantageous or for any other reasons that all samples need to be taken at the same time, a multiple-grid holder can be used (Evenson *et al.*, 1972). The latter can easily be made by attaching two or more nibbed drawing pens to a holder so that the tips of the nibs are on a horizontal plane. The grids (handled or Siemens specimen mounts) are held in place by the nibs throughout the preparative procedure. This method is not preferred for routine sampling, however, because of the difficulty in touching the hypophase surface with even pressure on all grids.

b. Nonaqueous. The most useful nonaqueous solvent for spreading nucleic acids is formamide (Westmoreland *et al.*, 1969). The technique of formamide spreading is the same as that for the aqueous solution above except a standard spreading solution consists of 1 μg/ml DNA, 0.1 *M* Tris, pH 8.5, 0.01 *M* EDTA, 0.01% cytochrome *c*, and 40% formamide. The hypophase consists of 0.01 *M* Tris, pH 8.5, 1 m*M* EDTA, and 10% formamide (Davis *et al.*, 1971). Since the formamide will cause a rapid lowering of the pH, the hypophase should be prepared immediately before use and the spreading solution used within approximately 1 hour.

2. Microdrop Techniques

a. Single-Drop Spreading. Small quantities of nucleic acid in a small volume can be spread on a hypophase consisting of a single water drop (0.9 ml) setting within a hemispherical depression (1.9 × 0.1 cm) formed in a block of Teflon (Inman and Schnös, 1970). A clean, wet glass rod (0.3 cm diameter, drawn to a fine but rounded tip at one end) or cylinder with one end sealed, e.g., a sealed Pasteur pipette (Kung *et al.*, 1975), is held at an angle of 45°–60° in the hypophase. Five microliters of the spreading solution are applied to the surface of the glass rod by a 0.11 cm i.d. capillary pipette, as shown in Fig. 4. The rod is then carefully removed. The surface film can be slightly compressed by removing 0.1 ml of hypophase, using a syringe and fine needle. The sample is picked up by touching the surface of the drop with a filmed grid. This technique is rapid and reproducible and is especially useful for microquantities having unfavorable conditions (such as high salt concentrations) for microdiffusion techniques. For example, Inman and Schnös

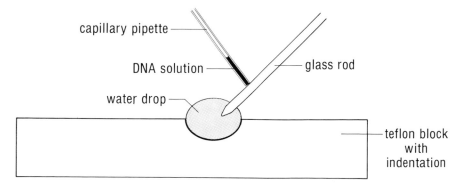

FIG. 4. Schematic of single-drop spreading technique. Reproduced from Inman and Schnös (1970) with permission.

(1970) have examined the DNA in a single drop isolated from a density gradient ($p = 1.71$) in which the concentration of DNA was low. Prior to spreading, the sample was admixed with cytochrome c (0.01%), HCHO (10%), and an equal volume of formamide.

A unique application of this technique for RNA was described by Kung *et al.* (1975) who spread urea formamide heat-treated RNA at an elevated temperature (up to 80 °C) in an oven in which the Teflon trough, hypophase, and all spreading materials had been preheated to the desired temperature.

b. Microdrop Diffusion. In the classic diffusion technique developed by Lang *et al.* (1964) a cytochrome c film is first *spread* over the surface of a nucleic acid solution followed by diffusion and irreversible attachment of the nucleic acid molecules to the cytochrome c film. This technique was miniaturized to a microdrop by Mayor and Jordan (1968) and Lang and Mitani (1970). In the microversion, drops are made from a solution *containing* cytochrome c, nucleic acid, and formaldehyde. Apparently, some cytochrome c binds to the nucleic acid and, in the presence of formaldehyde, other free cytochrome c molecules adsorb to the solution–air interface where unfolding of the polypeptide chains occurs, forming an insoluble film. Nucleic acid molecules diffuse by Brownian motion and are adsorbed to the surface film (Lang and Mitani, 1970) which is then transferred to a filmed grid by briefly touching the droplet surface. This technique is very rapid, simple, and allows the concentration of the nucleic acid per surface area of the film to be varied by increasing or decreasing the time for diffusion of the molecules to the surface. Nucleic acids adsorb to the protein film in proportion to (time)$^{0.75}$ (Lang and Mitani, 1970).

Table I shows the concentration of stock solutions and final dilutions needed for preparing 2 ml of working solution, or enough to make about 50 samples. When these stock solutions and other materials are available, a sample can be prepared for examination in the microscope in about 30 minutes.

Forty microliter drops are placed on a Teflon-coated surface which inhibits the spreading of the solutions so that a nearly spherical droplet is produced (Fig. 5a). Each drop serves as the source for one grid but many drops can be placed side by side on the Teflon surface. The drops are covered with an inverted petri dish to prevent air flow disturbances. An alternative to adding formaldehyde to the drop solution is to place a doughnut-shaped ring of filter paper under the petri dish and wet the ring with several drops of formaldehyde so that the vapors will aid the denaturation of the cytochrome c film (Brack *et al.*, 1972; Mazaitis *et al.*, 1976). After 10 minutes, which is the minimum time for the film to form and stabilize (Fig. 5b), and up to 16 hours for very low concentrations of nucleic acids, the handle of a filmed grid is bent (Fig. 5c) and the grid is touched to a drop (Fig. 5d). The sam-

TABLE I

Mixing Components and Final Dilutions for Microdrop Diffusion Technique

Stock solutions	Final concentrations
Double-stranded nucleic acid	
1.90 ml 0.2 M ammonium acetate, (pH 8.4)	0.19 M
0.02 ml EDTA (0.1 M)	1 mM
0.04 ml cytochrome c (1 $\times$ 10^{-2}%)	2 $\times$ 10^{-4}%
0.02 ml DNA (5–10 μg/ml)	0.05–0.1 μg/ml
0.05 ml 8% formaldehyde	0.2%
Single-stranded nucleic acid	
1.0 ml 0.02 M ammonium acetate (pH 8.4)	0.01 M
0.02 ml Tris (1 M)	0.01 M
0.02 ml EDTA (0.1 M)	1 mM
0.87 ml formamide (99%)	40%
0.04 ml cytochrome c (1 $\times$ 10^{-2}%)	2 $\times$ 10^{-4}%
0.02 ml DNA (5–10 μg/ml)	0.05–0.1 μg/ml
0.05 ml 8% formaldehyde	0.2%

ple is then rinsed, stained, and/or dehydrated, as described above (Fig. 5e–g). The grid is reflattened (Fig. 5h) and mounted on a glass slide with double-stick tape (Fig. 5i). After shadowing and/or storing on the slide, the handle of the grid is removed (Fig. 5j, k) prior to insertion of the grid into the microscope.

The advantages of this technique are that: (1) it is rapid, (2) as little as 2 $\times$ 10^{-5} μg nucleic acid in a single drop can be used to prepare a sample for electron microscopy, and (3) the concentration of nucleic acid on the filmed grid can be controlled by varying the diffusion time. The concentration of cytochrome c is critical to success, since too little cytochrome c prevents the formation of a film and too much will cause molecules, especially RNA, to artifactually shorten. The optimal concentration is the minimum amount to form a film.

B. Nonprotein Preparative Techniques

Since the approximate 100 Å diameter collar of cytochrome c around nucleic acids prepared by the protein monolayer technique often obscures small structural features, e.g., enzyme–nucleic acid complexes, attempts have been made to develop protein-free preparative techniques. Many of these techniques, some of which are described below, have demonstrated success with double-stranded nucleic acids; however, resolution of single-stranded nucleic acids is generally poorer than that obtained with protein monolayer methods.

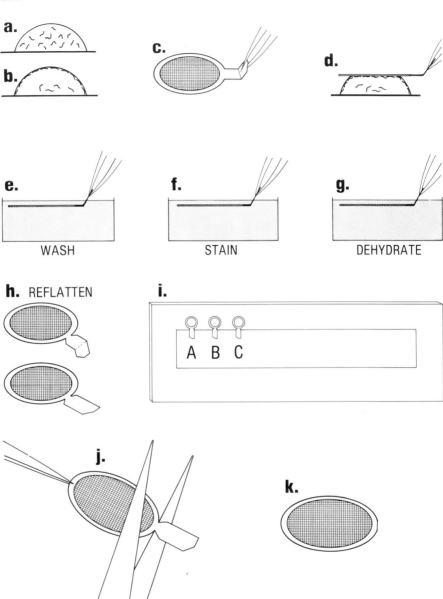

FIG. 5. Preparation of nucleic acid sample in a microdrop for electron microscopy. (a) Freshly prepared microdrop showing unadsorbed molecules. (b) Microdrop after adsorption of molecules to cytochrome *c* film. (c) Bending handled grid. (d) Touching drop surface with grid. (e) Washing of sample (f) Staining of sample. (g) Dehydration of sample. (h) Reflattening of grid handle. (i) Grids positioned on glass slide with double-stick tape. (j) Cutting off handle. (k) Grid ready for electron microscope examination.

1. BAC

Vollenweider *et al.* (1975) have demonstrated the use of benzyldimethylal-kylammonium chloride (BAC) for protein-free spreading of double- and single-stranded nucleic acid molecules. Our experience with this method confirms the report of Vollenweider *et al.*, however, we find that the method often produces tangled nucleic acid molecules. Also, the contrast of single-stranded molecules is rather poor, even after staining and shadowing. Nevertheless, the method has a significant potential for many types of experiments where chemicals as well as the procedures may interfere with a protein film. Furthermore, the resolution of fine structural features of the nucleic acid is significantly enhanced over cytochrome *c* preparations because the width of the double-stranded molecule after light shadowing with heavy metals is only about 50–60 Å, most likely as a result of the low molecular weight of BAC (340–368) compared to that of cytochrome *c* (about 13,000). This feature allows sufficient resolution to visualize nucleic acid–protein complexes, e.g., *E. coli* RNA polymerase–T7 DNA complex (Vollenweider *et al.*, 1975).

The method to prepare samples for electron microscopic observations is essentially the same as that used for protein monolayer techniques. A BAC mixture containing 60% $C_{12}H_{25}$ and 40% $C_{14}H_{25}$ in solid form (91.2%) is recommended (Bayer, Leverkusen, Germany; or Mobay, Industrial Chemicals Division, Pittsburgh, Pennsylvania). Alternatively, BAC is available as a 10% aqueous solution under the trade name Zephirol (Gallard–Schlesinger Carle Place, New York). A stock solution of 0.2% BAC in formamide is kept at room temperature. This solution is diluted 50-fold in the solvent to be used in the experiment. BAC can be used either by spreading on a hypophase or by a microdrop-diffusion technique. We find more tangling in the spreading method, perhaps due to surface action of the film on the hypophase. The method as described by Vollenweider *et al.* (1975) will be given in the following.

a. Spreading. Twenty microliters of a solution containing 0.1 μg/ml of DNA, 2.5×10^{-3}% of BAC, 13 m*M* ammonium acetate, 7 m*M* Tris-HCl, pH 8.3, and 77 m*M* KCl is spread [as described above (Section III,A,1,a)] on a hypophase of 0.3 *M* ammonium acetate (pH 9). The spreading of the film can be visualized by sprinkling a small amount of graphite or talc powder onto the hypophase. The film is picked up on fresh carbon-coated grids and washed by floating the grids on water for 5–10 minutes; this step is essential to minimize tangling. The sample can be contrasted by staining and/or metal shadowing as described for protein monolayer preparations. Adsorption of the DNA to the carbon films can usually be enhanced by floating the grids for 5–10 minutes on distilled water and blotting them dry on filter paper just before use (Vollenweider, personal communication).

For single-stranded DNA or RNA, a solution containing 1 μg/ml nucleic acid, 2.5 × 10^{-3}% BAC, 4 M urea, 8 mM Tris-HCl at pH 7.9, in formamide (99%) is spread on distilled water.

b. Microdrop. One hundred microliter droplets containing 0.1 μg of double- or single-stranded DNA per milliliter, 2.5 × 10^{-3}% BAC, 5.3 M urea, 1.3% formamide, and 3 mM triethanolamine, pH 7.9, are placed on a clean Teflon surface and covered with a petri dish. After 10 minutes the films are picked up by touching the surface of the drop with a carbon-coated grid, washed with H$_2$O, and contrasted as above. Figure 6 shows PM2 phage double-stranded DNA molecules prepared in the following ways: (a) with aqueous cytochrome c (Section III,A,1,a), uranyl acetate stain, platinum: paladium (Pt:Pd) shadow, molecule width is 180 Å; (b) with BAC microdrop (Section III,B,1,b), uranyl acetate stain, Pt:Pd shadow, molecule width is 80 Å; and (c) with BAC microdrop, uranyl acetate stain photographed in a dark field, molecule width is 40 Å. Figure 7 shows single-stranded fd phage DNA (a,b) and avian myeloblastosis virus RNA (c) prepared as follows: (a) urea-formamide (Section VII,B,2,a) with cytochrome c, uranyl acetate stain, Pt:Pd shadow; (b,c) BAC microdrop (Section III,B,1,b), uranyl acetate stain, Pt:Pd shadow.

BAC preparations for double-stranded nucleic acids have adequate contrast and enhanced resolution. Single-stranded nucleic acids prepared by this technique often lack good contrast. Contrast can probably be improved, however, by using high resolution shadowing procedures.

2. Direct Adsorption of Nucleic Acid Molecules to Supporting Films

a. Carbon Films. Carbon films placed in contact with nucleic acid solutions do not adsorb nucleic acids very well unless the films are positively charged; as a result the negatively charged nucleic acid molecules are selectively adsorbed. One method to charge carbon films is to subject them to glow discharge (Reissig and Orrell, 1970) as follows: Filmed grids are placed in a vacuum evaporator which is rough-pumped to a vacuum of 0.06–0.3 torr. Glow discharge can be done in air or in the presence of amylamine vapor (Dubochet et al., 1971) with seemingly similar results. Carbon-coated grids are placed within 10–15 cm of two separated plates across which a 10,000–20,000 V potential is placed for 5 minutes. The treated carbon films will bind DNA for about 20 minutes (Griffith, 1973). A drop of the sample containing nucleic acid is placed on the pretreated films for several minutes; the grid is then immersed into 100% ethanol for dehydration, touched to filter paper, air dried, and shadowed with metal. Limiting features of this technique are: (1) the molecules have a tendency to aggregate, and (2) the contrast is often poor in bright field electron microscopy, perhaps due in part from oil contamination during the glow discharge treatment. Better contrast has

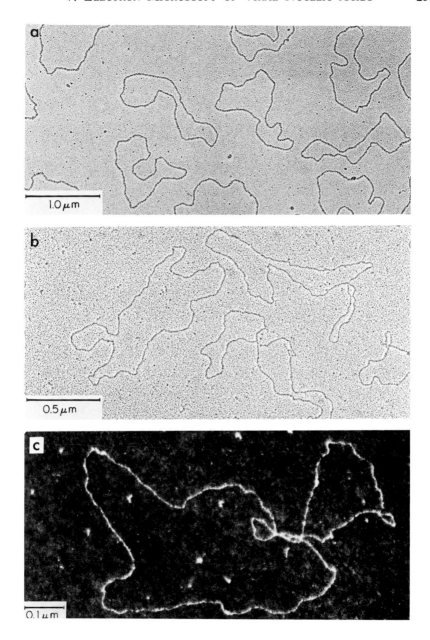

FIG. 6. Electron micrographs of PM2 phage double-stranded DNA prepared by: (a) aqueous cytochrome *c* technique; (b) BAC microdrop with uranyl acetate stain and Pt:Pd shadow; and (c) BAC microdrop, uranyl acetate stain, photographed in dark field.

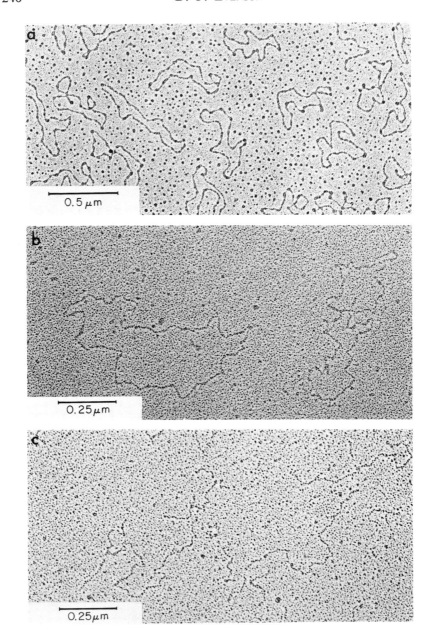

FIG. 7. Electron micrographs of single-stranded nucleic acids. (a) fd Phage DNA spread with cytochrome c by urea-formamide technique and contrasted with uranyl acetate and Pt: Pd shadow. (b) fd Phage DNA prepared by the BAC microdrop method and contrasted as in (a). (c) AMV RNA prepared by the BAC microdrop method and contrasted as in (a).

been obtained by mounting the nucleic acid sample on thin carbon films and using dark field electron microscopy (Dubochet et al., 1971).

b. *Protein–Carbon Films.* An alternative to glow discharge-treated films is the two-stage protein monolayer method of Harford and Beer (1972) and modifications of it by Abermann and Salpeter (1974). This method is not a classic protein–monolayer technique since the protein is first dried onto the grid, and therefore it was not discussed in Section III,A. According to Abermann and Salpeter (1974), droplets containing 20 μg/ml cytochrome c in 0.2 M sodium acetate, pH 5, are allowed to stand for 30 minutes on a clean Teflon surface. The cytochrome c films that form on the droplets are then picked up by briefly touching thin (30–50 Å) carbon-coated grids to the surface of the droplet. To remove the salt and excess cytochrome c, the grids are floated for 5 minutes on each of the following: distilled water (two changes), 50% ethanol and 100% ethanol (three changes). These grids can be used up to 4 days after preparation. The second step is to adsorb the DNA to the pretreated films by placing the grids on 50 μl drops of DNA (0.05–0.5 μg/ml) in sodium cacodylate–HCl buffer (pH 6) setting on a clean Teflon surface at room temperature. This pH is important to the positive charge of the cytochrome c films. After 30 minutes the grids are removed from the drops and most of the excess solution removed by touching them to filter paper. The grids are washed four times for 5 minutes each on water droplets (75 μl) and then for 5 minutes each in 50% ethanol and absolute ethanol. This washing procedure is crucial. The samples are then air dried and shadowed with metal (Section IV,B). Satisfactory resolution has not been obtained using single-stranded molecules with this technique; however, it has recently provided sufficient resolution to visualize and map the lactose repressor-operator complex in the hybrid phage φh80dlacUV[5] (Abermann et al., 1976).

IV. Contrasting Nucleic Acids

A. STAINING

1. General Considerations

Staining nucleic acids prepared by the protein monolayer technique often provides excellent contrast, even for single-stranded nucleic acids. Further contrasting by shadowing heavy metals is often not necessary and thus saves time in sample preparations. Uranyl acetate in organic solvents is superior to aqueous solutions for staining nucleic acids (Wetmur et al., 1966; Gordon and Kleinschmidt, 1968), giving both greater contrast and requiring lesser staining times and concentration of stain. Both the organic solvent of the stain and the nature of the supporting film influence the results of staining. For example, uranyl acetate in ethanol stains nucleic acids mounted on plastic films much more efficiently than nucleic acids mounted on carbon films. For nucleic acids mounted on carbon films, acetone is an effective staining solvent (Gordon and Kleinschmidt, 1968). The best contrast is obtained when the uranyl acetate stock solution and dilutions are relatively fresh. We routinely make up a fresh stock solution every 4 weeks and store it in darkness; however, a stock solution used the same day it is prepared provides the best results. The diluted solutions are used within 30 minutes of preparation and at room temperature. The following staining methods are recommended.

2. Nucleic Acids on Plastic Films

The stock solution is 0.05 M uranyl acetate in 90% ethanol and 0.05 M HCl which is diluted 1:100 to 1:1000 in 90% ethanol (Wetmur et al., 1966). The diluted solution should be filtered through a 0.45 μm filter before use. After picking up the sample, one should usually rinse the grid for 10–30 seconds in water or ethanol (up to 90%) to remove hypophase salts, dip it into the stain dilution for 30–60 seconds, rinse it for 10 seconds in 90% ethanol, touch it to filter paper to remove excess fluid, and then air dry it.

Under some conditions, for reasons which are not entirely clear, better contrast is obtained by skipping the rinsing step and placing the sample directly into the uranyl solution. Poor contrast may result from a loss of cytochrome c around the molecules during the water rinse; if so, a wash step in $>30\%$ ethanol may prevent this loss (if so desired) but still effectively wash out the hypophase salts. Staining directly without rinsing may result in "nonrelaxed" configurations of the molecules.

The final concentration of stain is determined by the amount required to gain sufficient contrast without causing an objectionable amount of uranyl contamination in the background. Sufficient contrast is usually attained by only staining samples that are prepared by aqueous or formamide procedures and picked up on plastic-coated grids.

3. Nucleic Acids on Carbon Films

The stock solution is $5 \times 10^{-3} M$ uranyl acetate in methanol which is diluted 50-fold with acetone and used within 30 minutes (Gordon and Kleinschmidt, 1968). The sample, after transference to the grid, is rinsed for 2–30 seconds in water, stained for 15 seconds or longer, rinsed briefly in absolute ethanol, and then immediately touched to filter paper to remove the excess ethanol prior to air drying.

B. METAL DEPOSITION

1. *Shadowing*

Although double- and single-stranded nucleic acids can usually be visualized by staining only, further contrast can be obtained by obliquely shadowing a heavy metal on the specimen immediately after staining and dehydration. The contrast results from a greater deposition of electron-opaque metal around the macromolecular complex than on the smooth substrate. Shadowing unstained specimens will also provide sufficient contrast in many types of preparations. For distinguishing between double- and single-stranded molecules, shadowing alone provides greater resolution than staining.

For most shadowing applications, the grids are attached by means of double-stick tape to a glass slide which in turn is attached by double-stick tape to a stage of an evaporator. It is important that the grids lie very flat on the slide. By this method, both the grids and the slide are secure on a stage that is positioned at any angle. A modified method is recommended when shadowing samples mounted on 20–30 Å carbon films; these films are so fragile that the movements required to stick and remove the grids to the tape will often produce enough stress to break the films. In such cases, the grids should be placed without tape directly on an upright stage. During evaporation, the stage is usually rotated at 30–60 rpm; however, shadowing without rotation can also be done from two or three directions with satisfactory results.

Although a variety of metal alloys has been used to increase the contrast of nucleic acids, one of the most commonly and successfully used is an 80:20 mixture of Pt:Pd. This alloy produces about as fine a grain as possible other than high melting point metals, e.g., tantalum tungsten, which require an electron beam for evaporation (Abermann *et al.*, 1972; Abermann and Salpeter, 1974).

The physical setup for shadowing is schematically diagrammed in Fig. 8. About 3 cm of 8 mil Pt:Pd wire are wrapped around the tip of a V-shaped 20 mil tungsten wire which is connected to the high current terminals of the evaporator unit. The source of metal evaporation should be at about a 6° angle to the specimen plane and 10–15 cm away from the center of rotation to avoid heat radiation damage.

The resolution of the shadowed material is dependent on the degree of vacuum during the evaporation (Holland, 1956). A liquid nitrogen trap between the diffusion pump and vacuum chamber should always be utilized to maintain the vacuum ($\gtrsim 10^{-5}$ torr) and minimize oil contamination during the evaporation. First, outgas the tungsten filament by heating it to a

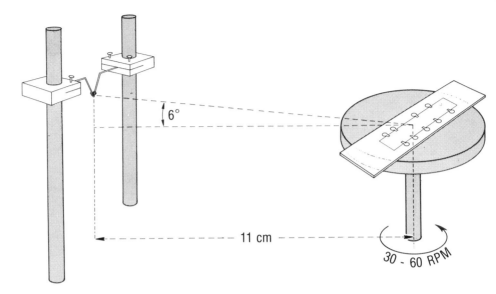

FIG. 8. Schematic diagram of metal shadowing procedure.

temperature below the melting point of the evaporation metal and then shut off the current to the filament. After optimal vacuum is reattained, slowly heat the tungsten filament until the Pt:Pd becomes molten (about 10–15 seconds). Then quickly raise the current so that within 5 seconds the metal evaporates and the tungsten filament breaks. Remove the sample from the evaporator. It is now ready for examination in the electron microscope.

For metal shadowing of nucleic acid samples and the production of carbon films, considerable emphasis needs to be placed on a clean evaporation unit. The most commonly used vacuum evaporators, and the most economical, have a mechanical forepump and an oil diffusion pump. Oil from both of these pumps can backstream into the vacuum chamber causing poor carbon film production and oil contamination of the specimen, leading to poor resolution. Some evaporators are fitted with dry-type pumps (e.g., cryo, ion, turbomolecular) and this obviously eliminates the oil contamination problem. However, if care is exercised with oil-pumped systems, potential problems can be minimized. Several points to be emphasized are: (1) to avoid backstreaming of oil from the forepump, the vacuum system should not be prepumped below 100 μm (0.1 torr); (2) the system should be as clean as practically possible so when the high vacuum valve is opened the pressure will quickly drop to the 10^{-5}–10^{-6} torr range; (3) the total system should be pumped to high vacuum prior to inserting a sample. During frequent usage, the evaporator should operate continuously; if shut off, the machine should preferably be run overnight before use; and (4) a cold trap placed between the oil diffusion pump and vacuum chamber should be utilized for each vacuum deposition.

2. Selective Nucleation

Another method of contrasting is by selective nucleation or "vapor phase positive staining," which is useful for defining nucleic acid–protein complexes adsorbed to glow discharge-activated carbon films, e.g., visualization of lambda repressor bound to the operator site of lambda DNA (Griffith *et al.*, 1971). The principle here is that a reactive metal such as evaporated tungsten binds selectively to DNA and protein rather than to the inert carbon substrate; the protein will bind more metal than the DNA. The method as described by Griffith (1973) is as follows: a straight piece of 20 mil tungsten wire is placed between two electrodes about 3 cm apart and 8 cm from the sample. Current is passed through the wire that will heat, evaporate, and break the wire in 5–7 minutes. The angle between the wire and the sample can vary between 8° and 45° with equal effectiveness because selective nucleation does not depend on low-angle deposition. The sample is rotated during evaporation.

V. Electron Microscopy

A. BRIGHT FIELD

For bright field electron microscopy, most samples can be initially scanned at 5000–10,000 × direct magnification at an accelerating voltage of 80–100 kV, using a 50–60 μm objective aperture. If the contrast is low, the accelerating voltage can be reduced to 60 kV and a 20–30 μm objective aperture may be used. However, since the astigmatism of the microscope is difficult to correct with very small apertures, the increase in contrast is gained at the expense of resolution. Furthermore, using a lower accelerating voltage often requires higher beam current and working close to crossover, which may contaminate the sample. In any case, a liquid nitrogen-cooled anticontamination device should always be used to reduce contamination of the specimen.

B. DARK FIELD

Dark field microscopy can be used to obtain additional contrast and resolution of nucleic acids and/or nucleic acid–protein complexes (de Harven *et al.*, 1971; Dubochet *et al.*, 1971; Kleinschmidt, 1971). The two main modes of dark field microscopy are (a) magnetic tilt of the electron beam (Henkelman and Ottensmeyer, 1971) and (b) cone mantle illumination using a ring aperture in the condenser lens (Dupouy *et al.*, 1966). The former can be obtained on some electron microscopes by use of a beam tilt device built into the microscope. This is most suitable for low magnification up to 20,000 × since the illumination by the condenser-ring aperture method is uneven at low mangification. The condenser-ring aperture method is best suited for high magnification observations since it provides more even and greater illumination than that obtained by the beam tilt method (Klein-schmidt *et al.*, 1972). Both modes of dark field can attain 3.4 Å linear resolution as demonstrated by Kleinschmidt *et al.* (1972) on graphite test specimens. Dark field microscopy requires that the mass of the nucleic acid complex to be visualized is high relative to the background film. Thus, if low contrast, high resolution potential samples are being studied, very thin carbon films must be used for support. Using such films, de Harven *et al.* (1971) demonstrated that unstained and unshadowed DNA molecules pre-pared by a cytochrome *c* monolayer technique and mounted on thin carbon films could be visualized with high contrast in dark field microscopy. Figure 6c illustrates a PM2 virus DNA molecule (about 40 Å diameter) that was prepared by the BAC method, lightly stained with uranyl acetate, and photo-graphed in dark field by the condenser-ring aperture method.

VI. Molecular Weight Determinations

A. Application

The electron microscope is a valuable tool for the evaluation of the molecular weight of viral nucleic acids. The sizes of DNA that have been measured range from 130 Å for restriction enzyme fragments of SV40 (Ruben *et al.*, 1975) to about 82 μm for herpes virus (Wagner *et al.*, 1974). For single-stranded RNA the range has been from 678 Å for tobacco ring-spot satellite virus RNA (Sogo *et al.*, 1974) to 5.2 μm for Newcastle disease virus (NDV) RNA (Kolakofsky *et al.*, 1974); for double-stranded RNA the range has been up to 8 μm for reovirus (Dunnebacke and Kleinschmidt, 1967; Vasquez and Kleinschmidt, 1968). Electron microscope methods are sometimes more reliable for the analysis of molecular weight than other methods. For example, the determination of molecular weights of RNA using polyacrylamide gel electrophoresis or sucrose gradient sedimenta-tion with marker RNAs of known molecular weight is reliable only when the marker and unknown polyribonucleotide chains are structural homo-logues, i.e., their Stokes' radius is a monotonic function of chain length

(Kolakofsky *et al.*, 1974). This is not always the case. For example, para-influenza RNAs and cellular rRNAs do not have the same hydrodynamic properties in 99% DMSO or after heating in 1.1 *M* formaldehyde, as shown by Kolakofsky *et al.* (1974). These workers demonstrated that sedimentation of Sendai and NDV RNA under denaturing conditions yielded molecular weight estimates of 2.3 to 2.6 × 10⁶, whereas length measurements in the EM yielded estimates of 5.2 to 5.6 × 10⁶, which must be closer to the real value.

B. INFLUENCE OF PREPARATIVE CONDITIONS

The contour lengths of DNA molecules vary greatly with experimental conditions, including: (a) the ionic strength of the hypophase (Lang *et al.*, 1967; Inman, 1967a; Bujard, 1970); (b) the type of protein or other substance used to generate the film (Hirt, 1969; A. Scotto and D. P. Evenson, unpublished observations); (c) the presence of denaturants, e.g., formaldehyde (Kleinschmidt *et al.*, 1965), formamide (Davis *et al.*, 1971), and urea (Sogo *et al.*, 1974) in the hypophase; and (d) pH and temperature (Lang *et al.*, 1967). The lengths of single-stranded nucleic acids are more sensitive to these conditions than double-stranded nucleic acids (Bujard, 1970).

C. REFERENCE NUCLEIC ACID MARKERS

The variables mentioned above (Section VI,B) plus other unknown factors may cause differences in the lengths of molecules from spread to spread, from grid to grid prepared from the same spread, and at different points on a single grid. Consequently, measured lengths of nucleic acid molecules will vary around a mean with a reproducible standard deviation. In addition, nucleic acid lengths are preferably measured as relative lengths to an internal standard which has been cospread with the unknown molecules so that both are subjected to the same calibration hazards. The marker should be of the same nucleic acid type with regard to RNA versus DNA, single-stranded versus double-stranded, and approximate size. If the molecules are of the same type and size, then a feature such as a circular molecule versus a linear molecule is desirable. If conditions are such that the experimental molecules cannot be distinguished from the marker molecules, the latter can be spread separately but immediately after the experimental molecules so that the spreading conditions are nearly identical.

For determining the molecular weight of single-stranded RNA, it can be cospread with single-stranded circular DNA, e.g., fd phage DNA. Although this may be a better alternative than separately spreading a known RNA marker, it is not clear whether RNA and DNA are extended to the same degree by the same conditions.

Single-stranded nucleic acids are not suitable markers for double-stranded nucleic acids. For example, single-stranded ϕX174 DNA is shorter than double-stranded ϕX174 DNA of the same original length in nucleotide pairs by a factor of about 0.88, as determined by comparative measuring of double- and single-stranded molecules present on the same photograph (Chamberlin *et al.*, 1975). The spreading conditions in this case were: hypophase, 57% formamide, 0.1 *M* Tris, 0.01 *M* EDTA spread on 17% formamide, 0.01 *M* Tris, and 1 m*M* EDTA, pH 8.5, Different spreading conditions will change the degree of this relationship, e.g., Bujard (1970) in a study on partial denaturation of duplex DNA, multiplied the lengths of the single-stranded regions by a factor of 1.4 to correct for the decrease in length on denaturation.

D. MICROSCOPY

1. *Random Sampling*

Due to the possible variations in linear density of nucleic acids in different areas of a grid, it is preferable to photograph a sufficient number of molecules in only one to a few grid squares without any realignment of the microscope. Thus, the concentration of nucleic acid should be adjusted, if possible, so that up to a hundred or more small molecules can be present and measured in a single grid square to minimize any variations. The fewer the number of molecules that are present, and the larger the molecule length, the greater the probability that the molecules will be only partially visualized on the photograph, which in effect selects for smaller molecules and produces a nonrandom sampling. One way to circumvent this problem is to take a series of photographs across a grid square and then develop the film. Any molecules that are only partially contained on the film can be relocated in the microscope, photographed in their entirety, and be included for measurement of all molecules present on the original series of photographs (Goddard and Cummings, 1975). For smaller molecules we draw a border around the photograph and measure all molecules that are partially or entirely within the border.

2. *Calibration*

The magnification of the microscope should be calibrated each time the microscope is used for a size-measurement study. A carbon replica of a cross-ruled diffraction grating (Fullam, 54,864 lines per inch) should be mounted in the same specimen cartridge and photographed immediately after photographing the molecules. The image of the grating replica should be brought into focus, if possible, by manipulating the grid into the same position as the sample rather than by changing the current of the objective lens.

E. Measuring Contour Lengths

Contour length measurements can be done by a variety of methods, all of which for reliability depend on measuring an enlargement of the original negative.

1. *Trace and Measure*

The most economical and widely used method is to project the negative onto a large piece of paper, trace the outline of the molecules, and measure the traced line. A Nikon profile projector, photographic enlarger, or slide projector can be used for projection. The projection image of a transparent plastic rule inserted along the emulsion side of the negative can be used to determine the magnification. For a more permanent record, we often print the negatives on 16 × 20 inch photographic paper, including the negative of the calibration grating replica. Bidirectional measurements of the printed grating replica indicate that Kodak paper does *not* shrink unidirectionally when dried on a heated roller drum. The lengths of the molecules can be determined by tracing the enlargement with a map ruler (e.g., Keuffel and Esser 620-300 map measurer).

2. *Graphics Calculator*

An easier method to measure molecule lengths is to trace the contours of the enlarged molecules with the arm of a graphics digitizer (e.g., Numonics Corporation, North Wales, Pennsylvania). One has the alternative of either tracing the contour from an enlarged print or tracing the outline of the molecule directly under a photographic enlarger without making a permanent copy. An instrument with a sensitivity of about 0.25 mm is recommended so that small loops in the molecule can be recorded. A more elaborate, but helpful modification would be to rear-project the image onto a horizontal screen and trace the molecules directly from the screen.

3. *Experimental Errors*

Depending on the homogeneity of the sample, approximately 100–1000 molecules should be measured to reach adequate statistical significance. The experimental errors in length measurement studies should not exceed 1–2% (Gordon, 1973). Perhaps the greatest potential for error in absolute length is an underestimation of the actual molecular length resulting from failure of the contrasting agents to resolve small kinks or bends in the molecule (Gordon, 1973).

4. *Presentation of Data*

Data are usually shown in a histogram as number of molecules versus length interval. However, another meaningful way of presenting the data, which would approximate the optical density profile of that nucleic acid sedimented in a velocity gradient, would be a weight average histogram. In this case, the ordinate represents the *total length* in each length interval and the abscissa represents the length interval.

Length measurements are often reported in daltons and/or units of 10^3 bases or base pairs (kilobases) (Sharp *et al.*, 1972). The symbol "kb" is used for the length of a duplex molecule in kilobase pairs or for the length of a single-stranded molecule in kilobases. Thus, the number of base pairs in a molecule is defined as daltons/662 (average residue weight for a base pair) (Sharp *et al.*, 1972).

The molecular weight of an unknown molecule is determined by a length ratio to a known molecule, using a well characterized standard. For example, the sodium salt of lambda DNA has been estimated to be 30.8 (± 1.0) $\times$ 10^6 daltons (Davidson and Szybalski, 1971), which corresponds to 46.5 $\pm$ 1.5 kb pairs. The ratio of lengths of ϕX174 RFII DNA to lambda DNA is 0.113 (± 0.003) (Sharp *et al.*, 1972). Therefore, ϕX-RF is calculated as 30.8 (± 1.0) $\times$ 10^6 daltons multiplied by the ratio 0.113 (± 0.003) to equal 3.48 (± 0.1) $\times$ 10^6 daltons, which corresponds to 5256 $\pm$ 170 base pairs (Sharp *et al.*, 1972).

The ratio of known to unknown must be determined for each experimental condition. However, as a general "rule of thumb," the mass per unit length for double-stranded molecules is *about 2* $\times$ 10^6 daltons/μm, and for single-stranded molecules 1.2 $\times$ 10^6 daltons/μm.

Table II lists the molecular weights of some molecules used as standards. Care should be taken, however, that a standard obtained does not represent a mutated form; thus, if in doubt, determine the ratio between two or more standards.

VII. Some Specific Techniques

A. VIRAL DNA

1. *Single-Stranded*

Single-stranded DNA tends to collapse unless a denaturant is used in the preparation of samples. Thus, single-stranded DNA can be prepared by any of the techniques outlined for single-stranded RNA (Section VII,B,2). However, for a start with an unknown nucleic acid, the formamide, micro-drop-diffusion method (Section III,A,2,b) is recommended. Several well-

TABLE II

MOLECULAR WEIGHTS OF SOME MOLECULES USED AS STANDARDS

Nucleic acid type	Molecular weight	References
Double-stranded DNA		
Lambda	$30.8\ (\pm 1.0) \times 10^6$	Davidson and Syzbalski (1971)
ϕX174 RF	$3.48\ (\pm 0.1) \times 10^6$	Sharp et al. (1972)
SV40	3.28×10^6	Wellauer et al. (1974)
PM2	6.4×10^6	Petterson et al. (1973)
Single-stranded DNA		
ϕX174[a]	$1.72\ (\pm .03) \times 10^6$	Wu et al. (1972)
fd	$2.1\ (\pm 0.1) \times 10^6$	Bujard (1970)
Single-stranded RNA		
VSV	$3.82\ (\pm 0.14) \times 10^6$	Repik and Bishop (1973)
MS2	$1.1–1.3 \times 10^6$	Fiers et al. (1965)
R17	$1.1\ (\pm 0.1) \times 10^6$	Mitra et al. (1963); Sinha et al. (1965)
Double stranded RNA		
ϕ6 Bacteriophage	3 segments: 4.78, 2.87, and 2.0×10^6	Semancik et al. (1973)
RNA–DNA duplex	No suitable standard	Wu et al. (1972)

[a]Sequencing data indicate that ϕX174 AM3 DNA contains 5375 nucleotides (Sanger et al., 1977).

characterized, single-stranded DNA molecules are available to serve as markers, e.g., fd and ϕX174 (cf. Table II).

2. Double Stranded

a. Partial Denaturation of DNA. Denaturation mapping has recently been reviewed by Inman (1974) and Inman and Schnös (1974) and therefore will only be summarized here. Under certain conditions, such as high formamide concentrations (Wolfson et al., 1972), high pH (Inman and Schnös, 1970), and high temperatures (Inman, 1967b), the DNA helix becomes stressed, and those regions richest in AT base pairs will denature. As the conditions favoring denaturation increase, regions progressively richer in GC base pairs will begin to denature. This method has shown that these regions of denaturation exist at the same site on DNA molecules isolated from a given type of virus, thus giving rise to a "denaturation map" for that set of molecules. The success of denaturation mapping for a given type of DNA depends on the magnitude of base content fluctuations along the molecule, the larger the base content fluctuation the more convincing will be the denaturation maps (Inman, 1974).

Partial denaturation permits the physical mapping of otherwise morphologically homogeneous molecules. Thus, maps of partially denatured molecules are useful in understanding base fluctuations along a molecule and in providing reference sites for the study of other features in a molecule. Examples are: (i) the treatment of DNA with restriction enzymes in combination with partial denaturation (Mulder and Delius, 1972), thus showing that restriction enzymes make specific cuts in the molecule; (ii) the determination of the origin and pattern of DNA replication (Bastia *et al.*, 1975); (iii) the determination of the position of deletion, insertion, and addition mutations (Chattoraj and Inman, 1972).

Although several methods exist for partial denaturation of duplex DNA, best results are usually obtained by alkaline denaturation (Inman and Schnös, 1970). To prevent the renaturation of the molecules before they are mounted on electron microscope grids, formaldehyde is added to the reaction mixture.

Although a number of variations exist, the standard conditions used by Inman (1974) are as follows:

Solution A: A solution of 67.8 mM Na$_2$CO$_3$, 10.7 mM EDTA, and 33.9% formaldehyde is adjusted with 5 M NaOH to make the solution pH of 9.9 rise to a pH level that achieves denaturation, as determined by optical density melting profiles and/or a trial series of increasing pH levels. As an example, lambda and P2 phage DNA will begin to denature at pH 11.0, but be almost completely denatured at pH 11.4 (Inman, 1974).

Solution B: Solution B consists of DNA in 20 mM NaCl and 5 mM EDTA.

Mix Solution A and B to give a final concentration of DNA of 0.01–0.03 OD$_{260}$ and 0.02 M Na$_2$CO$_3$, 3.2 mM EDTA, and 10% formaldehyde. Incubate the mixture for 10 minutes at 23 °C and then cool in an ice bath. Increasing pH and times of incubation are done until a satisfactory level of denaturation has been achieved. The solution is then diluted with an equal volume of formamide, adjusted to 0.01% cytochrome c, and prepared for electron microscopy by the single-drop spreading technique (Section III,A,2,a) or the microdrop diffusion technique (Section III,A,2,b) in which the droplet solution consists of 0.1 μg/ml DNA, 0.15 M NH$_4$ acetate, 0.07 M formaldehyde, 50% formamide, and 1.3 μg/ml cytochrome c, pH 7.0.

The mapping procedure is done essentially by the same method described for measuring molecule lengths (Section VI), i.e., the molecules are traced onto paper and measured with a map ruler or a graphics digitizer. In the regions of denaturation both strands of the duplex can be visualized and traced. Staining the sample with uranyl acetate significantly enhances the visualization of the single strands.

Since the lengths of single- and double-stranded DNA molecules are dependent on ionic strength (Bujard, 1970), a compensation factor is used to correct for shrinkage of the single-stranded regions. This can be calculated by determining the ratio of lengths of double-stranded DNA (e.g., SV40 DNA, form II) to denatured single-stranded SV40 DNA cospread with the experimental molecules.

Denaturation maps are often drawn with a horizontal line representing the full length of the native DNA and solid rectangles placed at the positions of the melted out loops. These maps can be arranged by trial and error or by computer programs; the latter is advantageous in the case of samples containing circular molecules with a high percentage of denaturation (Mayer *et al.*, 1975). The lack of exact alignment of denatured regions in different molecules may be due to differential stretching or other errors that are inherent in this technique. Thus, a large number of molecules may have to be mapped before a satisfactory pattern is obtained.

b. Heteroduplex Analysis. The principle of heteroduplex analysis is that the complementary nucleic acid strand of a wild-type virus and a mutated strand (e.g., by deletion, substitution, or addition) are hybridized so that the resulting heteroduplex will be a double-stranded molecule except at the sites of nonhomology where single strands will be visualized (Westmoreland *et al.*, 1969). By this method, precise physical maps of mutated, nonhomologous regions of DNA duplexes can be obtained; the standard deviation is frequently less than 2% (Westmoreland *et al.*, 1969). The heteroduplex method is also used for determining the level of genetic homology between related viruses, e.g., the sequence homology between lambdoid phages (Fiandt *et al.*, 1971; Simon *et al.*, 1971), coliphages (Godson, 1973), *Bacillus subtilis* phages (Chow *et al.*, 1972), T_3, T_7, and ϕII phages (Davis and Hyman, 1971; Hyman *et al.*, 1973), T even phages (Kim and Davidson, 1974), and substitution mutants of lambda (Mazaitis *et al.*, 1976; Henderson and Weil, 1975). The methodology has been described in a review by Davis *et al.* (1971); however, because of its great usefulness for studying viral DNA, some aspects of the method are summarized here.

Most experiments have been done with bacteriophage DNA. As described by Davis *et al.* (1971), the phage is purified by CsCl isopycnic centrifugation and stored in 4 M CsCl, 10 mM Mg^{2+}, 10 mM Tris, pH 8. Lysis of phage and denaturation of the DNA can be simultaneously accomplished by alkali treatment. If circular duplex DNA is being used, the circle can be broken in one of several possible ways: (i) a single strand break can be introduced into the duplex with DNAse (Hudson *et al.*, 1969) or (ii) with visible light in the presence of ethidium (Clayton *et al.*, 1970); however, these leave a mixture of linear and circular single strands. A better method would be to first cleave the duplex strands of isolated DNA with a restriction endonuclease that would make one cut per molecule and then denature the DNA.

For example, a restriction enzyme from *Haemophilus parainfluenzae* (HpaII) cleaves SV40 DNA at a single position (Sharp *et al.*, 1973).

Using lambda phage as an example, the following method (Davis *et al.*, 1971) for heteroduplex analysis is recommended. Lyse 5×10^{10} phage particles and denature the DNA (2.5 μg) simultaneously by incubation in 0.5 ml of 0.10 M NaOH, 0.02 M Na$_2$ EDTA for 10 minutes at room temperature. Then add 50 μl of 1.8 M Tris-HCl, 0.2 M Tris-OH, and 0.5 ml formamide (99%); this solution should have a pH of about 8.5. Renaturation should be at about the 50% completion level within 1–2 hours and stopped at this point by dialysis against 0.01 M Tris-OH, 1 mM Na$_2$ EDTA, pH 8.5 (HCl adjusted) at 4 °C. Alternatively a more direct method would be to chill the solution to 0 °C and prepare samples immediately for electron microscopy.

The sample can be prepared for electron microscopy by either a spreading or a microdiffusion technique. The aqueous technique can be used to easily locate the position of the single-strand nonpaired loop that will appear as a "bush"; however, a generally more useful technique includes the use of formamide (Westmoreland *et al.*, 1969) at a concentration that allows the single-strand portion of the DNA to stretch out. The formamide concentration in the spreading solution can be increased to 50–70% if duplex formation is to be minimized; the concentration of formamide in the hypophase is adjusted to 30% less than the spreading solution.

In the method described above, the samples need to be prepared for electron microscope analysis immediately after the renaturation step since the formamide-containing solution will become acidic. Thus, the method of Sharp *et al.* (1972) combined with the modification of Mazaitis *et al.* (1976) is recommended since the heteroduplex remains stable in solution for several weeks, at 4 °C, during which time a large number of samples can be prepared and studied. The method is as follows: 0.1 μg of each type of DNA is added to 70 μl of 0.3 M NaOH, which causes total denaturation of the DNA. Three to five minutes late, 30 μl of 1 M Tris-HCl and 100 μl of 0.02 M Na$_3$ EDTA, pH 8.5, are added and the solution is then dialyzed for 2 hours against 50 ml of 70% formamide, 0.25 M NaCl, 0.10 M Tris, and 0.01 M EDTA, pH 8.5, at room temperature; this is the renaturation step. After renaturation, the solution is dialyzed for 2 hours against 50% formamide, 0.1 M Tris, and 0.01 M EDTA, pH 8.5. Aqueous cytochrome c (50 μg/ml) can be added (immediately before spreading) and the solution spread onto a hypophase of 17% formamide, 0.01 M Tris, and 1 mM EDTA, pH 8.5.

So that this sample can be studied at a later time, the mixture, following dialysis against the renaturation solution (70% formamide, 0.25 M NaCl, 0.10 M Tris, 0.01 M EDTA, pH 8.5), is dialyzed for 4 hours against several changes of 0.01 M Tris, 1 mM EDTA, pH 8.5, at 4 °C; in this buffer the heteroduplexes are stable for several weeks at 4 °C (Mazaitis *et al.*, 1976). These heteroduplexes can be prepared for electron microscope analysis by the formamide spreading procedure (Section III,A,1,b) or by the microdrop-diffusion technique for single-stranded nucleic acid (Table I).

The procedure for measuring heteroduplex molecules is essentially the same as that for measuring molecule lengths (Section VI,E) and partially denatured duplex DNA (Section VII,A,2,a). A DNA length standard should be included at a concentration that permits both the heteroduplex and length standard to be on the same micrograph. The position of nonmatched base pairs is mapped in fractional units of the full length of the DNA.

B. VIRAL RNA

1. *Double-Stranded*

Double-stranded RNA can be prepared for electron microscope analysis by the same procedure used for duplex DNA except for the additional precautions needed to prevent RNase degradation. Examples of studies on duplex RNA include those isolated from *Reovirus* (Kleinschmidt *et al.*, 1964; Gomatos and Stoeckenius, 1964; Kavenoff *et al.*, 1975), bacteriophage $\phi6$ (Semancik *et al.*, 1973), and the replicative form of bacteriophage R17 (Granboulan and Franklin, 1966).

2. *Single-Stranded*

Single-stranded RNA spread by the aqueous technique is usually seen as a collapsed structure, and may even be "lost" in the cytochrome *c* background. In order to visualize partially or fully extended molecules, the preparation methods must include the use of nonaqueous solvents and/or denaturants. Some of the currently used methods are described below.

a. Urea-Formamide. The technique of Robberson *et al.* (1971) is useful for spreading nucleic acids so that some of the secondary structures may be retained. The method, slightly modified, is the following: 1 μl of RNA (10–15 μg/ml) in NET buffer or 1 mM DTT is diluted into 40 μl of a *freshly* prepared solution containing 1.2 gm urea (Schwartz–Mann ultrapure) in 4.2 ml formamide (Matheson, Coleman, Bell). The mixture, in a relatively thin-walled test tube, is heated in a water bath for 30 seconds at 53 °C. The mixture is cooled on ice for 2 minutes prior to adding 5 μl of cytochrome *c* (0.5 mg/ml) in 0.5 M Tris, pH 8.5, and 0.05 M EDTA. Then 40 μl of this mixture is prepared for electron microscopy by the macrospreading technique (Section III,A,1) on a hypophase of 0.15 M Tris, pH 8.5, at 4 °C. The cold subphase aids in the spreading of the protein monolayer. Samples are picked up after 1 minute on Parlodion-coated grids and stained with uranyl acetate (1:100 dilution) for 30–60 seconds, as previously described (Section IV,A,2), followed by shadowing with Pt:Pd (Section IV,B,1).

b. *Glyoxal-Formamide.* Glyoxal links to guanine residues in nucleic acids thereby inhibiting their hydrogen bonding (Hsu *et al.*, 1973) and reducing intramolecular interactions. Thus, glyoxal-treated RNA spread in the presence of weakly denaturing solvents, e.g., 40% formamide, 0.1 M Tris, is significantly extended. This method has been used successfully to study the structure of the endogenous C-type feline virus (RD-114) RNA (Kung *et al.*, 1974, 1975). The RNA (10 μg/ml) is first dialyzed against 0.5 M glyoxal in 0.01 M phosphate buffer, pH 7, for 1 hour at 35 °C and then diluted 20-fold into 40% formamide, 0.1 M Tris, 0.01 M EDTA, pH 8.2. Cytochrome c is added (30 μg/ml) and the sample is spread onto a hypophase of 10% formamide, 0.01 M Tris, 1 mM EDTA, pH 8.2, and immediately picked up on an appropriate grid. In addition, 5 μl can be spread by the single-drop method (Section III,A,2,a) or 50 μl can be spread by the macrotechnique (Section III,A,1,b). The sample can be contrasted as described above (Section VII,B,2,a).

c. *Formaldehyde–Formamide.* A more extensive denaturation of RNA can be obtained with the formaldehyde–formamide technique of Chi and Bassel (1974) than is generally obtained with the methods described above. A solution containing RNA (10 μg/ml), 4% formaldehyde, and PE buffer (0.01 M phosphate, pH 6.8, and 0.01 M EDTA) is heated at 65 °C for 15 minutes and then chilled on ice. Before use, the formaldehyde (reagent grade, 37%) is diluted to 33% with 10 X PE buffer and placed in a boiling water bath for 10 minutes.

Fifty microliters of spreading solution are prepared by sequentially admixing, at 0 °C, 15 μl PE buffer, 5 μl denatured RNA, 25 μl formamide, and 5 μl cytochrome c stock solution (1 mg/ml of 1 M ammonium acetate). Spread immediately onto a hypophase of 50% formamide, pH 5.4 (adjusted with glacial acetic acid), using the macrospreading technique (Section III,A,1). The sample can be picked up on grids of choice and stained and/or shadowed, as described above (Section VII,B,2,a).

d. *Bacteriophage T4 Gene-32 Protein.* Delius *et al.* (1973) have shown that all secondary structure of RNA can be removed by incubating the RNA in the presence of gene-32 protein (32 protein) prior to preparation for electron microscopy. The protein–RNA bond is rather weak; therefore glutaraldehyde is added to the reaction mixture to stabilize the binding before spreading the molecules. Examples of use have been the measuring of lengths of transcription products synthesized *in vitro* by *E. coli* RNA polymerase (Delius *et al.*, 1973) and the measuring of the size of RNA extracted from Rous sarcoma virus (Mangel *et al.*, 1974). Isolate and purify 32 protein according to Alberts and Frey (1970). The method of Delius *et al.* (1973) is as follows: Dilute the RNA (50 μg/ml) 10-fold into 0.01 M potassium phosphate, 2 mM EDTA, pH 7.0, containing 55 μg/ml 32 protein. Incubate for 3 minutes at 37°C, and then add glutaraldehyde to a final concentration of 0.01 M and incubate an additional 15 minutes at 37°C. Add this mixture to a solution consisting of 0.1 M Tris-HCl, pH 8.5, 30% (v/v) formamide, and 0.01% cytochrome c so that the final concentration of nucleic acid ranges from 0.1 to 1.0 μg/ml. Spread this mixture on a hypophase containing 10% (v/v) formamide and 0.01 M Tris-HCl pH 8.5, as described in Section III,A,1,b. The samples can be picked up on Parlodion- or carbon-coated grids and contrasted, as described above (Section VII,B,2,a).

e. Mapping of Secondary Structure. Just as partial denaturation studies have proved useful in mapping the physical structure of duplex DNA molecules, mapping of secondary structural features of single-stranded RNA molecules can serve to elucidate base sequence organization and can provide an orientation to other features. Wellauer and Dawid (1973), in an original and elegant study, elucidated the secondary structure of cellular ribosomal RNA using the technique of Robberson *et al.* (1971) (Section VII, B,2,a). Reproducible secondary structural features were first shown in tumor virus RNA in another elegant study by Kung *et al.* (1975), using essentially the same technique. When this same technique was applied to RNA isolated from avian myeloblastosis, Friend leukemia, Rous sarcoma and Rauscher leukemia virus, secondary structures were visualized, although unambiguous patterns have not yet been established (Evenson *et al.*, 1975, and unpublished).

In a recent study on MS2 phage, Jacobson (1976) showed that a meaningful pattern of secondary structure, including two types of loops, could be visualized to various degrees, depending on the $MgCl_2$ concentration in the spreading solution. At higher $MgCl_2$ concentrations, a striking symmetry between both ends of the molecule appeared. The pattern observed in Sendai virus RNA differed significantly from the MS2 pattern (Jacobson, 1976). The method is an adaption of the macrospreading formamide technique (Section III,A,1,b). Forty microliters of a solution containing 0.5–0.7 μg of RNA per milliliter, 50% (v/v) formamide, 0.01 M triethanolamine HCl, pH 8.5, 0.1 mg/ml of cytochrome c, and $MgCl_2$ at various concentrations (0.1–0.6 mM) are spread on a hypophase of distilled water as previously described. The samples can be stained with ethanolic uranyl acetate and shadowed with heavy metals, as described above (Section VII,B,2,a). The exact role Mg^{2+} plays in the formation and visualization of loops is unclear.

C. RNA–DNA Hybrids

RNA transcription products can be localized on their DNA template by using the aqueous, basic protein film technique (Section III,A,1,a). The RNA will appear as a bush under these conditions; the minimum size for an observable bush is about 400 nucleotides (Davis and Hyman, 1970). RNA–DNA hybrids, e.g., mRNA hybridized to one strand of denatured duplex DNA to localize the position of the gene for that mRNA, can also be localized by using the aqueous technique. In this case, the single-stranded DNA, not hybridized, will form a bush, but the hybridized portion will be delineated as a duplex region.

Another method is to spread the duplex by the formamide procedure (Section III,A,1,b); in this case, both the DNA–RNA portion and the single-stranded DNA will be extended. They can be distinguished, however, since the single-stranded DNA will appear kinky and somewhat thinner (Davis and Hyman, 1970).

D. Distinguishing between DNA and Single-Stranded RNA

In studies on viral RNA where high denaturation conditions are used to reduce the secondary structure, it is important to determine by another method whether any contaminating DNA exists in the sample, since denatured DNA could be mistaken for single-stranded RNA. The aqueous method (Section III,A,1,a) is not entirely satisfactory because the RNA may be so highly complexed that it may not be distinguished from the background. For example, Fig. 9a is an electron micrograph showing double-stranded DNA and 60–70 S Friend virus RNA prepared by the aqueous procedure. Better methods, allowing the visualization of both duplex DNA and single-stranded RNA, would be either a formamide procedure (40% spreading, 10% hypophase) (Section III,A,1,b), or a 4 M urea procedure (Weber et al., 1974). In the latter technique, the RNA sample (5–10 μg/ml) is initially diluted 1:5 with 4 M urea. To 25 μl of the diluted RNA, the following are added: 20 μl 1 M ammonium acetate and 5 μl cytochrome c (1 mg/ml); this mixture is spread by the macrospreading technique (Section III,A,1) on a hypophase of 0.015 M ammonium acetate. The sample is then stained with uranyl acetate and shadowed with Pt:Pd as described in Section IV. Figure 9b shows a preparation of 60–70 S Friend tumor virus RNA prepared by this technique, and demonstrates contaminating cellular DNA. The 60–70 S RNA complexes are partially opened and can be recognized as single-stranded nucleic acid as opposed to the smooth contoured duplex DNA. Figure 9c shows a sample of Friend tumor virus RNA prepared by the urea-formamide technique (Section VII,B,2,a). Any duplex DNA present would be denatured and would not readily be distinguished from viral RNA.

An easy method to determine whether an unknown virus contains single-stranded RNA or DNA would be to prepare the sample for electron microscopy before and after nuclease digestion. For RNase digestion, according

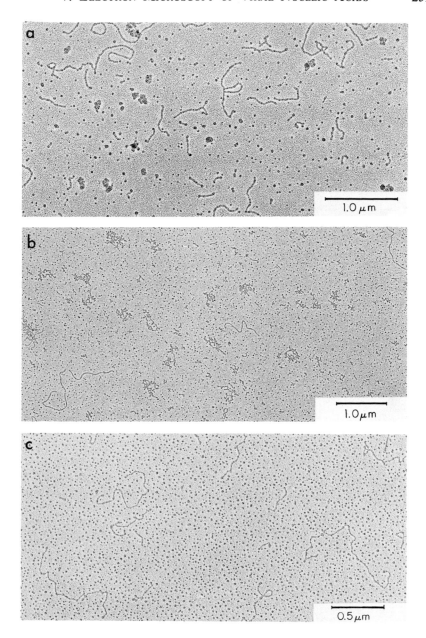

FIG. 9. Electron micrographs of nucleic acids isolated from a preparation of Friend leukemia virus. Nucleic acid was spread by: (a) aqueous procedure; (b) 4 *M* urea procedure; and (c) urea–formamide procedure.

to Sogo *et al.* (1974), first prepare a stock solution of RNase (1 mg/ml) in water, and heat to 90 °C for 10 minutes to inactivate any contaminating DNase activity. Dilute this stock to a working solution of 10 μg/ml containing 10 μg/ml of the nucleic acid in water; incubate at 25 °C for 1.5 hours. Then dilute the RNase incubation mixture with a solution containing 2.5 μg/ml of a known DNA sample (e.g., PM2 circular DNA) in 1 mM sodium acetate and 1 mM EDTA. Then dilute the sample with an equal volume of a freshly filtered solution of cytochrome c (0.02%) in 7.2 M urea and 0.1 M ammonium acetate, pH 8.0, and spread on a hypophase of 0.015 M ammonium acetate and 0.5% formaldehyde at pH 8.7. The visualization of the known DNA in the absence of single-stranded molecules provides evidence that the single-stranded molecules were RNA.

Alternatively, the sample can be incubated in the presence of 0.01% (w/v) chromatographically purified DNase I (Worthington, Freehold, N.J.07728), 0.1% w/v $MgSO_4 \cdot 7H_2O$, 0.01 M Na_2HPO_4, 0.01 MKH$_2$PO$_4$, pH 6.9–7.1 at 37 °C for 10–20 minutes. The DNase should be treated with sodium iodoacetate before use, according to Zimmerman and Sandeen (1966) to insure that any residual RNase activity is inactivated.

Finally, a sample suspected to contain both DNA and RNA can be alkaline digested which will visually remove the RNA strands. Add 10 μl of a sample to 40 μl 0.3 N KOH and incubate 16 hours at 37 °C. Then neutralize by adding 40 μl 0.3 N HCl. Prepare 5 μl of this sample for electron microscopy by the urea–formamide technique (Section VII,B,2,a).

E. DNA–PROTEIN INTERACTIONS

In the classic aqueous-spreading technique (Section III,A,1,a) the protein collar surrounding the DNA obscures any small protein attached to the DNA backbone. The most commonly used technique to mount nucleic acid–protein-specific complexes without a protein monolayer film is to adsorb the complex to glow discharge-treated carbon films (Dubochet *et al.*, 1971) prepared as described earlier (Section III,B,2,a). Among examples of this technique is that of Brack and Pirrotta (1975) who studied the interaction of the repressor of bacteriophage lambda and its operator DNA. Repressor bound to whole lambda DNA (2 μg/ml) was placed for 60 seconds on a glow discharge-treated thin carbon film. Excess fluid was removed by touching the grid to filter paper. For dark field studies, the sample was stained for 10–20 seconds with 2% aqueous uranyl acetate and then dried on filter paper. For bright field studies, the DNA–protein complex was stained with 0.5% uranyl formate, pH4, for 45–60 seconds, which makes it possible to visualize at the same time positively stained DNA and negatively stained protein molecules fixed on it (Brack, 1973; Brack and Delain, 1975).

Cremisi *et al.* (1976) applied viral DNA associated with cellular histones in a drop to a carbon-coated, glow discharge-treated grid. After 2 minutes, the liquid excess was removed and the complexes were stained for 1 minute with a 2% aqueous solution of uranyl acetate. The excess was removed with filter paper and the sample air dried and shadowed with Pt:Pd.

Another approach is to complex the protein of interest with an electron-opaque marker. For example, T-antigen binding sites on SV40 DNA can be localized by incubating SV40 (I) DNA successively with T antigen, anti-T γ-globulin, ferritin-conjugated antihamster γ-globulin, and ferritin-conjugated antihamster γ-globulin, as described by Reed *et al.* (1975). In such cases the ferritin core can be visualized under standard spreading conditions using 40% formamide in the spreading solution (Section III,A,1,b).

In those cases where the protein may be lost during preparation, the protein–nucleic acid complex can be fixed with glutaraldehyde, as described by Delius *et al.* (1972) for preparing the 32 protein nucleic acid complex.

VIII. Conclusion

The elegant technique introduced by Kleinschmidt and Zahn in 1959 is one of the most useful methods in molecular biology today to answer a variety of questions concerning nucleic acids. The introduction of formamide and other denaturing agents to allow the visualization of single-stranded nucleic acids was a significant modification of the original technique. The current methodology trends appear to be: (a) substitution of the cytochrome *c* film with potentially higher resolution films, e.g., BAC; (b) improvement of methods to adsorb naked nucleic acids to filmed grids so that base-specific proteins and chemicals can better serve as visual markers for genetic mapping; and (c) use of high resolution dark field electron microscopy for studying unstained and unshadowed nucleic acid molecules.

Acknowledgments

I would like to express my gratitude to Dr. A. K. Kleinschmidt who has always been enthusiastically willing to discuss various aspects of his original technique. I thank Dr. A. Mazaitis for kindly introducing some of the techniques to me and also for critically reviewing this chapter. I am grateful to Mr. A. Scotto for his skilled technical help and for making original improvements in some of the techniques. My thanks also go to Mr. J. Marchese for photography assistance and to Mrs. D. Saltzer for typing the manuscript. Finally, it is my pleasure to thank Dr. E. de Harven, whose dedication to high resolution electron microscopy is a continual inspiration to improve the resolution of this technique, for support during the writing of this chapter and for his critical review of it.

References

Abermann, R., and Salpeter, M. M. (1974). *J. Histochem. Cytochem.* **22**, 845.

Abermann, R., Salpeter, M. M., and Bachmann, L. (1972). *In* "Principles and Techniques of Electron Microscopy" (M. A. Hayat, ed.), Vol. 2, pp. 197–217. Van Nostrand-Reinhold, Princeton, New Jersey.

Abermann, R., Bahl, C. P., Marians, K. J., Salpeter, M. M., and Wu, R. (1976). *J. Mol. Biol.* **100**, 109.

Alberts, B., and Frey, L. (1970). *Nature (London)* **227**, 1313.

Bastia, D., Sueoka, N., and Cox, E. C. (1975). *J. Mol. Biol.* **98**, 305.

Berkowitz, S. A., and Day, L. A. (1974). *Biochemsitry* **13**, 4825.

Bøvre, K., Lozeron, H. A., and Szybalski, W. (1971). *In* "Methods in Virology" (K. Maramorosch and H. Koprowski, eds.) Vol. 5, p. 271. Academic Press, New York.

Brack, C. (1973). *Experientia* **29**, 768.

Brack, C., and Pirrotta, V. (1975). *J. Mol. Biol.* **96**, 139.

Brack, C., and Delain, E. (1975). *J. Cell Sci.* **17**, 287.

Brack, C., Delain, E., and Riou, G. (1972). *Proc. Natl. Acad. Sci. U.S.A.* **69**, 1642.

Bradley, D. E. (1965). *In* "Techniques for Electron Microscopy" (D. E. Bradley, ed.), p. 66. Davis, Philadelphia, Pennsylvania.

Brewer, J. M. (1974). *In* "Experimental Techniques in Biochemistry" (J. M. Brewer, A. J. Pesce, and R. B. Ashworth, eds.). p. 3. Prentice-Hall, Englewood Cliffs, New Jersey.

Bujard, H. (1970). *J. Mol. Biol.* **49**, 125.

Chamberlin, M. E., Britten, R. J., and Davidson, E. H. (1975). *J. Mol. Biol.* **96**, 317.

Chattoraj, D. K., and Inman, R. B. (1972). *J. Mol. Biol.* **66**, 423.

Chi, Y. Y., and Bassel, A. R. (1974). *J. Virol.* **13**, 1194.

Chow, L. T., Boice, L. B., and Davidson, N. (1972). *J. Mol. Biol.* **68**, 391.

Clayton, D. A., Davis, R. W., and Vinograd, J. (1970). *J. Mol. Biol.* **47**, 137.

Cremisi, C., Pignatti, P. F., Croissant, O., and Yaniv, M. (1976). *J. Virol.* **17**, 234.

Davidson, N., and Szybalski, W. (1971). *In* "The Bacteriophage Lambda" (A. D. Hershey, ed.), p. 45. Cold Spring Harbor Lab., Cold Spring Harbor, New York.

Davis, R. W., and Hyman, R. W. (1970). *Cold Spring Harbor Symp. Quant. Biol.* **35**, 269.

Davis, R. W., and Hyman, R. W. (1971). *J. Mol. Biol.* **62**, 287.

Davis, R. W., Simon, M. N., and Davidson, N. (1971). *In* "Methods in Enzymology" (K. Moldave and L. Grossman, eds.), Vol. 21, Part D, P. 413. Academic Press, New York.

de Harven, E., Leonard, K. R., and Kleinschmidt, A. K. (1971). *Proc. Electron Microsc. Soc. Am.* **29**, 426.

Delius, H., Mantell, N. J., and Alberts, B. (1972). *J. Mol. Biol.* **67**, 341.

Delius, H., Westphal, H., and Axelrod, N. (1973). *J. Mol. Biol.* **74**, 677.

Dubochet, J., Ducommun, M., Zollinger, M., and Kellenberger, E. (1971). *J. Ultrastruct. Res.* **35**, 147.

Dunnebacke, T. H., and Kleinschmidt, A. K. (1967). *Z. Naturforsch., Teil B* **22**, 159.

Dupouy, G., Perrier, F., and Verdier, P. (1966). *J. Microsc. (Paris)* **5**, 655.

Egami, F., and Nakamura, K. (1969). *In* "Microbial Ribonucleases," p. 28. Springer-Verlag, Berlin and New York.

Espejo, R. T., Conelo, E. S., and Sinsheimer, R. L. (1971). *J. Mol. Biol.* **56**, 597.

Evenson, D. P., Mego, W. A., and Taylor, J. H. (1972). *Chromosoma* **39**, 225.

Evenson, D. P., Scotto, A., Pavlovec, A., Hamilton, M., and de Harven, E. (1975). *Proc. Electron Microsc. Soc. Am.* **33**, 646.

Fiandt, M., Hradecna, Z., Lozeron, H. A., and Szybalski, W. (1971). *In* "The Bacteriophage Lambda" (A. D. Hershey, ed.), Cold Spring Harbor Lab., Cold Spring Harbor, New York.

Fiers, W., Lepoutre, L., and Vandendriessche, L. (1965). *J. Mol. Biol.* **13**, 432.
Goddard, J. M., and Cummings, D. J. (1975). *J. Mol. Biol.* **97**, 593.
Godson, G. N. (1973). *J. Mol. Biol.* **77**, 467.
Gomatos, P. J., and Stoeckenius, W. (1964). *Proc. Natl. Acad. Sci. U.S.A.* **52**, 1449.
Gordon, C. N. (1973). *J. Mol. Biol.* **78**, 601.
Gordon, C. N., and Kleinschmidt, A. K. (1968). *Biochim. Biophys. Acta* **155**, 305.
Granboulan, N., and Franklin, R. M. (1966). *J. Mol. Biol.* **22**, 173.
Granboulan, N., Huppert, J., and Lacour, F. (1966). *J. Mol. Biol.* **16**, 571.
Griffith, J., Huberman, J. A., and Kornberg, A. (1971). *J. Mol. Biol.* **55**, 209.
Griffith, J. D. (1973). *Methods Cell Biol.* **7**, 129.
Harford, A. G., and Beer, M. (1972). *J. Mol. Biol.* **69**, 179.
Heine, U. I., Cottler-Fox, M., and Weber, G. H. (1975a). *Methods Cancer Res.* **11**, 167.
Heine, U. I., Weber, G. H., Cottler-Fox, M., Layard, M. W., Stephenson, M. L., and Zamecnik, P. C. (1975b). *Proc. Natl. Acad. Sci. U.S.A.* **72**, 3716.
Henderson, D., and Weil, J. (1975). *Virology* **67**, 124.
Henkelman, R. M., and Ottensmeyer, F. P. (1971). *Proc. Natl. Acad. Sci. U.S.A.* **68**, 3000.
Hirt, B. (1969). *J. Microsc. (Paris)* **8**, 58a.
Holland, L. (1956). *In* "Vacuum Deposition of Thin Films." Chapman & Hall, London.
Hsu, M. T., Kung, H. J., and Davidson, N. (1973). *Cold Spring Harbor Symp. Quant. Biol.* **38**, 943.
Hudson, B., Upholt, W. B., Devinny, J., and Vinograd, J. (1969). *Proc. Natl. Acad. Sci. U.S.A.* **62**, 813.
Hyman, R. W., Brunovskis, I., and Summers, W. C. (1973). *J. Mol. Biol.* **77**, 189.
Inman, R. B. (1967a). *J. Mol. Biol.* **25**, 209.
Inman, R. B. (1967b). *J. Mol. Biol.* **28**, 103.
Inman, R. B. (1974). *In* "Methods in Enzymology" (L. Grossman and K. Moldave, eds.), Vol. 24, Part E, p. 451. Academic Press, New York.
Inman, R. B., and Schnös, M. (1970). *J. Mol. Biol.* **49**, 93.
Inman, R. B., and Schnös, M. (1974). *In* "Principles and Techniques of Electron Microscopy" (M. A. Hayat, ed.), Vol. 4, p. 64. Van Nostrand-Reinhold, Princeton, New Jersey.
Jacobson, A. (1976). *Proc. Natl. Acad. Sci. U.S.A.* **73**, 307.
Kavenoff, R., Talcove, D., and Mudd, J. A. (1975). *Proc. Natl. Acad. Sci. U.S.A.* **72**, 4317.
Kim, J. S., and Davidson, N. (1974). *Virology* **57**, 93.
Kleinschmidt, A. K. (1968). *In* "Methods in Enzymology" (L. Grossman and K. Moldave, eds.), Vol. 12, Part B, p. 361. Academic Press, New York.
Kleinschmidt, A. K. (1971). *Philos. Trans. R. Soc. London, Ser. B* **261**, 143.
Kleinschmidt, A. K., and Zahn, R. K. (1959). *Z. Naturforsch., Teil B* **14**, 770.
Kleinschmidt, A. K., Dunnebacke, T. H., Spendlove, R. S., Schaffer, P. L., and Whitcomb, R. F. (1964). *J. Mol. Biol.* **10**, 282.
Kleinschmidt, A. K.; Kass, S. J., Williams, R. C., and Knight, C. A. (1965). *J. Mol. Biol.* **13**, 749.
Kleinschmidt, A. K., Sato, T., and de Harven, E. (1972). *Inst. Phys. Cent. Ser.* **14**, 652.
Kolakofsky, D., Boy de la Tour, E., and Delius, H. (1974). *J. Virol.* **13**, 261.
Kung, H. J., Bailey, J. M., Davidson, N., Vogt, P. K., Nicolson, M. O., and McAllister, R. M. (1974). *Cold Spring Harbor Symp. Quant. Biol.* **39**, 827.
Kung, H. J., Bailey, J. M., Davidson, N., Nicolson, M. O., and McAllister, R. M. (1975). *J. Virol.* **16**, 397.
Lang, D., and Mitani, M. (1970). *Biopolymers* **9**, 373.
Lang, D., Kleinschmidt, A. K., and Zahn, R. K. (1964). *Biochim. Biophys. Acta* **88**, 142.
Lang, D., Bujard, H., Wolff, B., and Russell, D. (1967). *J. Mol. Biol.* **23**, 163.
Mandell, J. D., and Hershey, A. D. (1960). *Anal. Biochem.* **1**, 66.

Mangel, W. F., Delius, H., and Duesberg, P. H. (1974). *Proc. Natl. Acad. Sci. U.S.A.* **71**, 4541.

Mayer, F., Mazaitis, A. J., and Pühler, A. (1975). *J. Virol.* **15**, 585.

Mayor, H. D., and Jordan, L. E. (1968). *Science* **161**, 1246.

Mazaitis, A. J., Palchaudhuri, S., Glansdorff, N., and Maas, W. K. (1976). *Mol. Gen. Genet.* **143**, 185.

Mitra, S., Enger, M. D., and Kaesberg, P. (1963). *Proc. Natl. Acad. Sci. U.S.A.* **50**, 68.

Mulder, C., and Delius, H. (1972). *Proc. Natl. Acad. Sci. U.S.A.* **69**, 3215.

Ottensmeyer, F. P. (1969). *Biophys. J.* **9**, 1144.

Pagano, J. S., and Hutchison, C. A., III. (1971). *In* "Methods in Virology" K. Maramorosch and H. Koprowski, eds.), Vol. 5, p. 79. Academic Press, New York.

Petterson, E., Mulder, C., Delius, H., and Sharp, P. (1973). *Proc. Natl. Acad. Sci. U.S.A.* **70**, 200.

Reed, S. I., Ferguson, J., Davis, R. W., and Stark, G. R. (1975). *Proc. Natl. Acad. Sci. U.S.A.* **72**, 1605.

Reissig, M., and Orrell, S. A. (1970). *J. Ultrastruct. Res.* **32**, 107.

Repik, P., and Bishop, D. H. L. (1973). *J. Virol.* **12**, 969.

Robberson, D., Aloni, Y., Attardi, G., and Davidson, N. (1971). *J. Mol. Biol.* **60**, 473.

Ruben, G., and Siegel, B. M. (1975). *Proc. Electron Micros. Soc. Am.* **66**, 658.

Ruben, G., Jay, E., Wu, R., and Siegel, B. (1975). *Proc. Electron Microsc. Soc. Am.* **33**, 660.

Sanger, S., Air, G. M., Barrell, B. G., Brown, N. L., Coulson, A. R., Fiddes, J. C., Hutchison, C. A., Slocombe, P. N., and Smith, M. (1977). *Nature (London)* (in press).

Semancik, J. S., Vidauer, A. K., and Van Etten, J. L. (1973). *J. Mol. Biol.* **78**, 617.

Sharp, P. A., Hsu, M. T., Ohtsubo, E., and Davidson, N. (1972). *J. Mol. Biol.* **71**, 471.

Sharp, P. A., Sugdon, B., and Sambrook, J. (1973). *Biochemistry* **12**, 3055.

Simon, M., Davis, R. W., and Davidson, N. (1971). *In* "The Bacteriophage Lambda" (A. D. Hershey, ed.), p. 313. Cold Spring Harbor Lab., Cold Spring Harbor, New York.

Sinha, N. K., Fukimura, R. K., and Kaesberg, P. (1965). *J. Mol. Biol.* **11**, 84.

Sogo, J. M., Schneider, I. R., and Koller, T. (1974). *Virology* **57**, 459.

Vasquez, C., and Kleinschmidt, A. K. (1968). *J. Mol. Biol.* **34**, 137.

Vollenweider, H. J., Sogo, J. M., and Koller, T. (1975). *Proc. Natl. Acad. Sci. U.S.A.* **72**, 83.

Wadsworth, S., Hayward, G. S., and Roizman, B. (1976). *J. Virol.* **17**, 503.

Wagner, E. K., Tewari, K. K., Kolodner, R., and Warner, R. C. (1974). *Virology* **57**, 436.

Weber, G. H., Heine, U., Cottler-Fox, M., and Beaudreau, G. S. (1974). *Proc. Natl. Acad. Sci. U.S.A.* **71**, 1887.

Weber, G. H., Heine, U., Cottler-Fox, M., Garon, C. F., and Beaudreau, G. S. (1975). *Virology* **64**, 205.

Wellauer, P. K., and Dawid, I. B. (1973). *Proc. Natl. Acad. Sci. U.S.A.* **70**, 2827.

Wellauer, P. K., Reeder, R. H., Carroll, D., Brown, D. D., Deutch, A. Higashinakagawa, T., and Dawid, I. B. (1974). *Proc. Natl. Acad. Sci. U.S.A.* **71**, 2823.

Westmoreland, B. C., Szybalski, W., and Ris, H. (1969). *Science* **163**, 1343.

Wetmur, J. G., Davidson, N., and Scaletti, J. V. (1966). *Biochem. Biophys. Res. Commun.* **25**, 684.

Whiting, R. F., and Ottensmeyer, F. P. (1972). *J. Mol. Biol.* **67**, 173.

Wolfson, J., Dressler, D., and Magazin, M. (1972). *Proc. Natl. Acad. Sci. U.S.A.* **69**, 499.

Wu, M., Davidson, N., Attardi, G., and Aloni, Y. (1972). *J. Mol. Biol.* **71**, 81.

Younghusband, H. B., and Inman, R. B. (1974). *Annu. Rev. Biochem.* **43**, 605.

Zimmerman, S. B., and Sandeen, G. (1966). *Anal. Biochem.* **14**, 269.

8 Rapid Immune Electron Microscopy of Virus Preparations

Robert G. Milne and Enrico Luisoni

I. Introduction

Immune electron microscopy (IEM) is the term generally used for techniques that detect, by electron microscopy, the specific binding of antibody to antigen. These techniques can be broadly divided into two groups. First, there are those that detect immune reactions in fixed and sectioned material, and we are not concerned with them here. Second, there are the techniques that detect immune reactions in particulate material, especially virus preparations. We shall briefly review these methods and then describe some modifications that can save time and materials without sacrificing image quality.

II. Current Immune Electron Microscopic Methods

A. The Classical Technique

What we shall for convenience call the classical technique was developed by Anderson *et al.* (1961) and Lafferty and Oertelis (1961) following the pioneer work of Anderson and Stanley (1941), von Ardenne *et al.* (1941), and Williams and Wyckoff (1946). This technique has been reviewed by Almeida and Waterson (1969), Mandel (1971), and Doane (1974). A particularly clear recent example of the technique is the work of Bercks *et al.* (1974) using whole serum, IgG, and IgM.

The technique involves mixing the antigen (which we shall hereafter for simplicity refer to as the virus) with the antibody (either a purified antibody fraction or the crude antiserum) in suitable proportions and incubating the mixture for some hours, often overnight. The incubation time has sometimes been reduced to 1 hour, at 37 °C (Almeida and Waterson, 1969) or at room temperature (Feinstone *et al.*, 1974). A complex is formed due to bridging of the virus particles by the antibody molecules, and this complex is centrifuged to a pellet. The pellet is resuspended in a small amount of distilled water and mixed with neutralized phosphotungstic acid (PTA) before being applied to a support film and dried. The centrifugation step was omitted by Norrby *et al.* (1969) and by Chaudhary *et al.* (1971), but the first authors introduced a dialysis step while the second increased the incubation time at low temperature.

A positive result is seen in the electron microscope as a clumping of the virus particles and these clumps can often be detected when no virus was visible in the original suspension because of low concentration. If an excess of antibody is present, each virus particle is surrounded with a halo of antibody molecules and this halo effect, also called antibody coating, we shall refer to as decoration. Sometimes individual antibody molecules can be seen bridging the virus particles.

The advantages of the classical method are that (1) both the virus (or particular sites on the virus) and the antibody (usually IgG but also IgM or Fab fragments) can be visualized with good resolution (Fab fragments, being monopolar, can, of course, give decoration but not clumping); (2) the antibody can be used to trap a virus in recognizable clumps and specifically identify it (when antibody attachment is visible) in conditions where the virus concentration is too low to detect or recognize anything by straight electron microscopy; and (3) the pelleting step that follows clumping can be used to concentrate and partly purify the immune complex and so increase sensitivity even further.

The possible disadvantages of the method are that (1) incubation and centrifugation take some time, perhaps several hours; (2) the centrifugation step may cause nonspecific clumping; (3) the initial virus preparation must be suspended in a suitable medium, such as buffered saline, favorable to the reaction of antigen with antibody; and (4) the virus and antibody preparations must be rather clean, otherwise the centrifugation step will pellet not only the immune complex but also low-speed precipitating impurities that will degrade the image.

B. The Agar Technique

In an attempt to overcome the problem of impurities, and pursuing the ideas of Kellenberger and Arber (1957), Kelen *et al.* (1971) and Kelen and McLeod (1974) used agar surfaces to trap the virus–antibody complex and absorb some of the unwanted materials from the preparation. A drop of the unpurified immune complex was placed on an agar surface and a grid floated on top. Low molecular weight impurities as well as the water were absorbed into the agar but the clumped virus particles remained on the grid, to be negatively stained in the normal way. A rapid result was obtained with unpurified samples and the centrifugation step was avoided. Anderson and Doane (1973) and Doane (1974) also used agar, but primarily as a reservoir of antibody.

The agar method does indeed go some way to meet the impurity problem and saves on both time and manipulation with respect to the classical system; but the use of agar has not proved particularly popular, and, as we see below, washing can probably in most cases do the job more simply.

C. Leaf Dip Serology

Ball and Brakke (1968) and Ball (1971, 1974) introduced the leaf-dip method into plant virus serology. Essentially, they obtained a very rapid result (total elapsed time about 15 minutes) on material direct from leaves, but at the sacrifice of some resolution. The method is to place a drop of diluted antiserum on a filmed grid and hold the freshly cut edge of a virus-infected leaf in the drop for a second or two. The drop is then air dried or held at 40 °C for 5–10 minutes until dry. During this time the antiserum both clumps and decorates any specifically related virus, and the dried grid is then negatively stained.

The advantages, of course, are simplicity and rapidity and, as Brandes (1957) originally found, dipping a cut leaf into a drop of liquid on a grid leads to a surprisingly clean result. However, the final image does suffer in several respects: (1) the initial drying of the virus–antibody complex on the grid in the absence of negative stain leads to considerable distortion that would be avoided if drying occurred only once—in the presence of negative stain; (2) the image is inevitably degraded by salts, sugars, proteins, and other components from the leaf and the serum that are collected in the original drop; and (3) saline solutions normally used as the vehicle for immune reactions form disturbing crystals when dried on grids. To avoid this, Ball (1971) used 0.001 M ammonium acetate to dilute the serum, but she did point out that it was not an optimal medium and that in microprecipitin tests its use reduced serum dilution end points considerably. An only partly successful alternative was to use saline and then to float the grids on 20% glycerol to wash out the salt.

D. THE DERRICK TECHNIQUE

Derrick (1972, 1973a,b) made two important advances in technique. First, he realized that support films could be coated with antibody and that virus particles could then be specifically attached to and concentrated on such grids. Second, he found that the grids could with advantage be washed at two points—after antibody adsorption to the grid and after virus adsorption to the antibody. Washing removed salts and other components without disturbing the antibody or the virus, so that a medium could be used that was adapted to the immune reaction rather than to the needs of electron microscopy.

In practice, carbon-backed Parlodion-filmed grids were floated for 30 minutes at 24 °C on serum diluted 1:10 in 0.05 M Tris buffer, pH 7.2. The grids were then washed with the same buffer and used immediately. Crude virus extracts from plants were diluted in the Tris buffer containing 0.9% NaCl and the grids were floated on 0.1 ml drops of the virus for 1 hour at 24 °C. Grids were then washed with Tris-NaCl and with water and were finally air dried and shadowed.

Tests were made with tobacco mosaic virus (TMV) and potato virus Y (PVY) and their antisera. About 50 times more TMV and 20 times more PVY were trapped on specific grids than on control grids. Under given conditions, there was a linear relationship between the log of the number of virus particles trapped and the virus dilution. The number of particles trapped could be increased by longer incubation of the grid with the virus or by raising the incubation temperature to 37 °C.

More recently, Brlansky and Derrick (1974) have modified the procedure. Carbon-fronted Parlodion films were found superior to carbon-backed films, and the trapped virus particles were positively stained using 1% uranyl acetate in 95% ethanol for 1 minute, followed by washing in 95% ethanol. Brlansky and Derrick stated that Formvar films are unsuitable; we take this to mean carbon-backed Formvar. Carbon-fronted Formvar is routinely used by us and is quite satisfactory; we prefer Formvar to Parlodion (collodion) because it is tougher for a given film thickness (see Milne, 1972).

E. DECORATION

Yanagida and Ahmad-Zadeh (1970) introduced, for the purpose of locating bacteriophage structural proteins, what we shall call the decoration technique. With this method, virus is adsorbed onto grids and washed, then the antiserum is added; decoration, or the covering of antigenic sites by the antibody molecules, occurs but clumping does not since the virus particles are not free to move. Particular antigenic sites on a capsid can also be resolved if the appropriate serum is available (see, for example, Yanagida and Ahmad-Zadeh, 1970; Yanagida, 1972; Tosi and Anderson, 1973).

Yanagida and Ahmad-Zadeh added drops of purified phage suspension to carbon-coated collodion grids and after about 30 seconds, the unadsorbed phage was washed off with a drop of distilled water. The grids were floated specimen-side-down on diluted antiserum for periods of 30 minutes or 4 hours at 37°C and were then washed with 3 drops of distilled water and negatively stained in PTA.

III. Toward Simpler Techniques and Quicker Results

It is clear from all the work mentioned above that incubation times can be very short and that grids charged with antibodies and virus particles can without prejudice be washed to remove salts and other interfering substances. On this basis, two rapid IEM tests for plant virus material were developed based on clumping and decoration. These are applicable not only to purified preparations but also to crude or unpurified material, and are therefore useful where IEM is used in diagnosis (Milne and Luisoni, 1975). We have also slightly modified and shortened Derrick's method and have combined it with the decoration technique (Milne and Luisoni, unpublished).

Before we describe the techniques, it may be useful to discuss two related problems: which medium to use for the serological reaction and which negative stain to employ.

Most virologists still use PTA as their standard (indeed only) negative stain because it is simple to use and compatible with most preparations. However, PTA is destructive to a number of viruses and cell constituents (e.g., cucumber mosaic virus, all rhabdoviruses, and ribosomes) and for this reason we have preferred to use the less destructive 2% aqueous uranyl acetate (UA) routinely. UA also appears to give a clearer image than PTA and may offer higher resolution.

The serological reaction should take place near neutrality and phosphate-buffered saline has traditionally been used as the reaction medium. But UA precipitates in the presence of phosphate or at a pH much above 5, and to avoid such precipitation it is necessary to wash the reacted grid with water. In seeking a reaction medium more compatible with UA, we used a number of different buffers to dilute both viruses and antisera and reacted these in the decoration test (see below). The grids were then negatively stained in UA with or without an intervening water wash. The buffers used were 0.1 M phosphate, 0.1 M Tris-HCl, 0.1 M cacodylate, and 0.04 M veronal, all at pH 7 (0.1 M veronal cannot be prepared because of limited solubility). Saline was not incorporated because we have earlier found (Milne and Luisoni, 1975) that phosphate works, if anything, better when saline is omitted. Preliminary tests showed that when UA is mixed in equal volumes with these buffers, it immediately precipitates with phosphate and cacodylate but remains in solution with Tris and veronal.

The results established the following points: (1) the immune reaction proceeded efficiently in all buffers; (2) if the water wash was included, all results were good though perhaps phosphate was best; and (3) if the water wash was omitted, all results were poor, with unwanted deposits and bad distribution of stain.

In conclusion, it seems that washing with water is desirable in any case, and then buffers other than phosphate do not offer any advantage. We therefore use 0.1 M phosphate buffer, pH 7, to resuspend or dilute all viruses and sera used in IEM tests. The resulting grids are washed with water and negatively stained in UA. Haschemeyer and Myers (1972) recommended uranyl oxalate as a negative stain with properties similar to UA but tolerant of pH values between 5 and 7. This might be a useful negative stain for IEM, though it must be prepared in the laboratory and is rather unstable.

It is sometimes useful to fix virus preparations while retaining their serological activity. This avoids damage to the virus either in the presence of negative stains such as PTA or in the presence of serum, as reported for barley stripe mosaic virus by Ball (1971) and for maize rough dwarf virus by Milne and Luisoni (1975). Fixation of certain viruses in suspension is difficult because glutaraldehyde itself may damage them, but once such viruses are adsorbed to support films, fixation is easier (Luisoni *et al.*, 1975).

One further preliminary point is that the amount of virus and the distribution of stain on a grid are influenced by whether the support film is hydrophilic or hydrophobic. We routinely make the grids hydrophilic just before use by subjecting them to 10 seconds of high-voltage glow discharge at a pressure of 0.1 torr (see Williams and Fisher, 1974; Milne and Luisoni, 1975).

A. CLUMPING

The clumping method is essentially a version of the classical technique in which the incubation step is shortened, the centrifugation omitted, and the impurities removed by washing the complex once it is on the grid. The method has affinities with that of Ball and Brakke in that it can start with infected leaf material and give a result in 20 minutes, but because of the washing, there are no problems with salts and the image is of high quality.

The starting materials are an antiserum whose end point is, say, 1:1000, and a 2 mm square piece of infected leaf. The serum is diluted to a value 10 times less than the end point (i.e., in this case, 1:100) in 0.1 M phosphate buffer, pH 7, and a 10 μl drop is placed on a glass slide. The leaf sample is finely crushed in the drop of serum and the slide incubated in a humid box for 15 minutes at room temperature. A grid held in forceps is now touched to the drop and then washed with 20 drops of phosphate buffer, 30 drops of distilled water, and finally 5 drops of UA, before draining and drying. If a purified virus suspension is used, it is mixed with the diluted serum, incubated for 15 minutes, and then processed as before. Parallel samples are treated with normal or unrelated sera.

A positive result is indicated by three effects. First, the virus particles are clumped (Fig. 1); second, a halo of antibody molecules is visible around each particle (decoration); and third, there is an approximately 10-fold increase

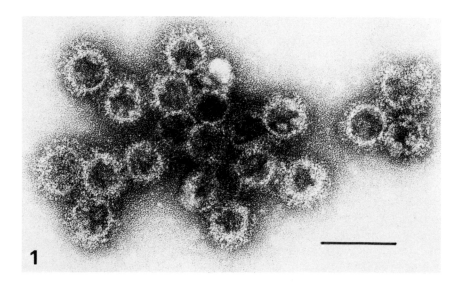

FIG. 1. Maize rough dwarf virus smooth cores clumped, using the method described in the text. The bar represents 100 nm.

in numbers of particles seen, compared with normal serum treatments. Clumping can, of course, occur at a serum dilution that is too great to give decoration and, conversely, if the serum concentration is too high, decoration may occur almost exclusively and clumping will be inhibited.

B. DECORATION

The main advantages of the decoration technique, apart from rapidity, are; (1) one can observe the site of antibody attachment without the complication introduced by clumping, and (2) one can start directly with virus-infected tissue or with fractions taken from sucrose or cesium chloride gradients, for example, as well as with purified virus.

If leaf material is to be tested, a 2 mm square sample is crushed in 10 μl of water on a glass slide and the mixture is touched to the grid. Samples from gradients or other kinds of virus suspensions, are likewise applied directly. After a few seconds, the grid, held in forceps, is rinsed with 20 drops of 0.1 M phosphate buffer, pH 7, and drained but not dried. A drop (2–4 μl) of diluted serum is added and the grid, still held in forceps, is incubated in a humid box for 15 minutes at room temperature. If the serum has an end point of about 1:1000, a convenient dilution is 1:100. The diluent is phosphate buffer, as before.

After incubation, the grid is rinsed with 20 drops of phosphate buffer, 30 drops of water, and 5 drops of UA, and is then drained and dried. If the reaction is positive, individual virus particles are seen to be decorated with shells of antibody (Figs. 2–4). Particular antigenic sites on a capsid can also be resolved if the appropriate serum is available (see Fig. 5, also Luisoni et al., 1975). It is interesting that double-stranded viral RNA can be decorated by anti-dsRNA serum, but normal serum also adsorbs to dsRNA in a similar way (Luisoni et al., 1975).

Using the decoration method, we compared the results obtained with unfractionated serum or purified gamma globulin obtained by ammonium sulfate precipitation (Stelos, 1967). Both serum and gamma globulin preparations had gel-diffusion end points of about 1:1000 and were diluted 1:100 in phosphate buffer. We used purified carnation mottle and tomato bushy stunt viruses together with their antisera or gamma globulin fractions of these sera. There was little difference in results between serum and gamma globulin. This is perhaps not surprising as dilution to 1:100 followed by washing on the grid is likely to have reduced to low levels very nearly all nonantibody materials in the crude serum. Complement was almost certainly no longer active in our sera and the thickening of the antibody halo in the presence of complement discussed by Almeida and Waterson (1969) was not observed.

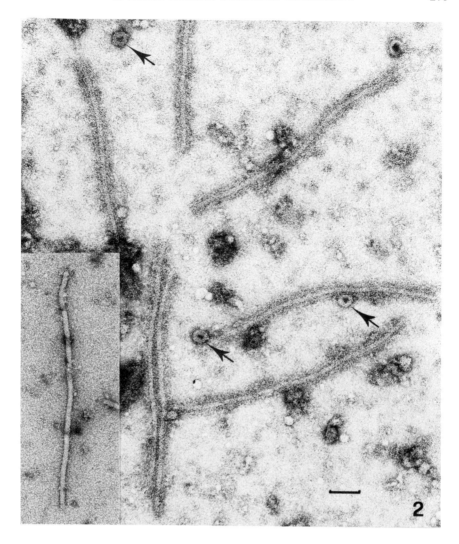

FIG. 2. A crude preparation from a carnation leaf, showing the rod-shaped carnation vein mottle virus decorated with antibody. The spherical carnation etched ring virus (arrows) is not decorated. Inset: A carnation vein mottle particle treated with normal serum. The bar represents 100 nm.

C. THE MODIFIED DERRICK SYSTEM

a. The General Method. We have verified with a number of viruses that the method of Derrick (1973b) works and that up to 50-fold increases in numbers of virus particles trapped are obtained. We have also found that

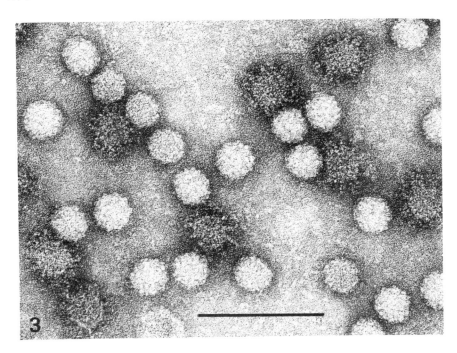

FIG. 3. A mixture of purified tomato bushy stunt and carnation mottle viruses treated according to the decoration technique (see text) using purified anti-carnation mottle virus gamma globulin. Only the carnation mottle particles are decorated with antibody. The bar represents 100 nm.

the method can be shortened and simplified, as described below (Milne and Luisoni, unpublished).

If the antiserum has a titer of about 1:1000, then a dilution of 1:10 or 1:100 in 0.1 M phosphate buffer, pH 7, is used. About 2 μl of the diluted serum is placed on a grid held in forceps and this is incubated in a humid box for 5 minutes at room temperature. The grid is then washed with 20 consecutive drops of buffer and drained but not dried, and a drop of the virus sample (2–4 μl) is added. The drop is incubated on the grid for 15 minutes as above and the grid is then washed consecutively with 20 drops of buffer, 30 drops of water, and 5 drops of UA. The preparation is then drained and dried (see Fig. 6).

Using a potexvirus, carnation mottle, cucumber mosaic, tomato bushy stunt, barley stripe mosaic, and tobacco mosaic viruses both in crude and purified preparations, together with their antisera, we have found that between 30 and 50 times more virus is trapped on the grid when it is coated with homologous rather than unrelated serum. The effect is the same when

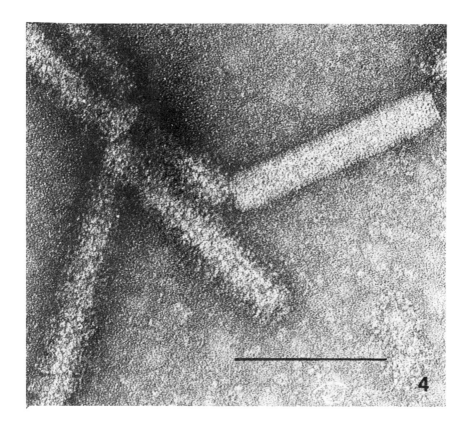

FIG. 4. A mixture of tobacco mosaic virus (TMV) and tobacco rattle virus treated as in Fig. 3 using anti-TMV serum. Only the TMV particles are decorated. The bar represents 100 nm.

gamma globulin fractions are used instead of serum. The final preparation is clean and the resolution visible on the grids is good. When grids are coated with unrelated serum, up to twice as many particles are seen as are visible on untreated grids.

In this modified procedure, Derrick's incubation times are reduced so that processing takes less than 25 instead of more than 90 minutes. We tested various periods of incubation of virus preparations on serum-coated grids held at room temperature and found that incubation for 1, 5, and 15 minutes trapped progressively more virus but that a 60 minute incubation gave little improvement over 15 minutes.

Derrick (1973b) contrasted his preparations by metal shadowing, and later, Brlansky and Derrick (1974) suggested that positive staining with ethanolic uranyl acetate was superior to both negative staining and shadow-

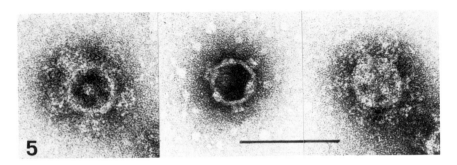

Fig. 5. Left and right, maize rough dwarf virus spiked cores decorated, using antispike serum. Center, an untreated spiked core. The bar represents 100 nm. Taken in part from Luisoni *et al.* (1975).

ing for the detection of icosahedral viruses and tomato spotted wilt virus. While we agree that positive staining may be superior to shadowing for some purposes, it is nevertheless well established that negative staining gives better preservation and better resolution of viral capsids, and both are indispensable for recognition of the virus in question. All the viruses we have tested with the Derrick method can be successfully negatively stained after the treatment, though it must be admitted that contrast is not always so uniform over the grid as that obtained with positive staining.

Derrick (1972) prepared grids by letting drops of antiserum dry down on them. The serum was first dialyzed against and then diluted with water. The grids could then be stored for later use. Derrick later found (1973a) that results were better if the antiserum was allowed to adsorb to the grids, the excess was washed off, and the virus was applied immediately. One further possible option we have tried is to adsorb the serum to the grids, wash off the excess with buffer, and wash the buffer off with water. The grids were then dried and stored. When a virus was to be tested, the grid was first wetted with phosphate buffer, the excess drained off, and the sample applied at once to the wet grid. The remaining steps followed normally. The amount of virus trapped specifically was found to be about the same as in the direct method, though in general, results were a little better with immediately prepared grids than with stored ones.

b. Particle Measurement. To establish the dimensions of a virus, relatively large numbers of particles should be photographed together with

Fig. 6. The same preparation of carnation mottle virus (A) adsorbed on an untreated grid, (B) adsorbed on a grid prepared according to the modified Derrick method, and (C) as (B), but followed by decoration. (A–C) are at the same magnification; the bar represents 1 μm. Inset in (C): Part of the field enlarged to show the decoration.

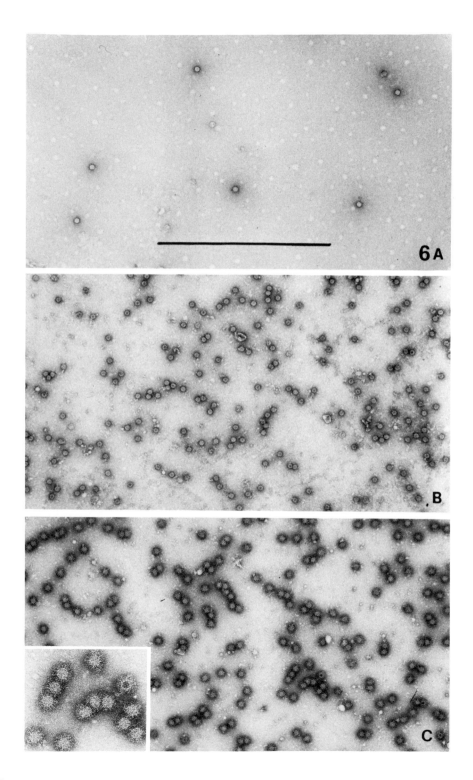

some internal standard, such as TMV particles or catalase crystals. With elongated viruses especially, it is preferable to use crude, unpurified material because purification and concentration can lead to particle aggregation or breakage. However, crude material often contains too few particles to photograph conveniently and the Derrick method can then increase the number, if an antiserum to the virus (or a related virus) is available (Milne and Luisoni, unpublished).

In practice, the grid is incubated with diluted antiserum for 5 minutes, washed with buffer, and drained. Crude material containing the virus and the standard is incubated on the grid for 15 minutes and is then washed off with buffer and water before negative staining. It is, of course, possible to trap not only more of the virus to be measured but also more of the standard by adding the appropriate serum in the original step.

An important question in this context is whether the dimensions of virus particles are changed by trapping them on serum-coated grids. As noted above, incubation with serum in suspension can sometimes lead to breakage of virus particles, but we have never observed increased breakage (or aggregation) after applying the modified Derrick method (in which the virus comes into contact only with a fixed carpet of antibody and not with antibody or serum in bulk). In theory, small dimensional changes seem possible but we have found that the normal lengths of TMV and a potexvirus are unchanged by the procedure. Further checks must establish whether the method can be used routinely.

D. THE DERRICK TECHNIQUE COMBINED WITH DECORATION

The coating or decoration of virus particles with a halo of antibody seems the best proof, obtainable in the electron microscope, of a specific immune reaction. Clumping or aggregation of virus particles may occur for a number of reasons and is therefore less sure; trapping virus particles with antibody-coated grids is subject to variations in the amount of virus trapped, the amount retained after washing, and the quality of the negative staining. Individual grids and grid squares may vary considerably in these respects due to factors not entirely under control.

As a useful diagnostic tool, we have therefore combined the Derrick and decoration techniques to trap more virus in the first step and to confirm which virus we have caught in the second (Milne and Luisoni, unpublished). The procedure is as follows.

A grid is incubated for 5 minutes with antiserum diluted to, say, 1:10 or 1:100, as described above. After a wash with 20 drops of phosphate buffer, the virus (either a purified preparation or material obtained directly by crushing a leaf) is added and left on the grid for 15 minutes. The grid is then rinsed once more with 20 drops of buffer and drained, and a drop of antiserum is added. The serum dilution used this time is 1:100. After a further incubation period of 15 minutes, the grid is washed with the routine 20 drops of buffer, 30 drops of water, and 5 drops of UA. It is then drained and dried.

The technique takes about 35 minutes and traps about 50 times more virus on the grid than is found with the normal serum control. In confirmation of the specific nature of the trapping reaction, each particle is decorated. Some results are shown in Fig. 6.

IV. Discussion

The principles of IEM are already well established, so here we have discussed the more rapid methods used for viruses in suspension and suggested some ways in which good results can be obtained by brief and simple routines. In the various methods described, there are many variables that can be tested, involving incubation times of the serum or the virus, their optimal concentrations, the temperature of incubation, and whether agitation of the incubated preparation increases sensitivity or reduces the time required for incubation. We have made some tests which, together with the results of others, suggest the following: (1) In the Derrick system, adsorption of antiserum to the grid is effective after 5 minutes and is not noticeably improved after 15 minutes. It may well approach completion in considerably less than 5 minutes. With serum whose end point is near 1:1000, a dilution of 1:10 or 1:100 is convenient. A more dilute serum may give a clearer result but may trap less virus. (2) In all systems, a 15-minute incubation of virus with antibody is almost certainly too short to give maximal combination but doubling this time traps less than twice as much virus. (3) All our experiments were done at room temperature (about 24°C); incubation at higher temperature (say, 40°C) would improve efficiency. (4) Using the Derrick system, we attempted to increase trapping efficiency by vibrating the drop of virus on the grid during a 15-minute incubation period. Up to twice as much virus was trapped but the results were not very consistent.

One further question is the sensitivity of the methods described, i.e., their ability to detect small amounts of virus. Using the modified Derrick system, we obtained about 10 particles per 400-mesh grid square using 2–4 μl of carnation mottle virus (a small isometric virus) at a concentration of 0.02 μg/ml. The sample contained about 5×10^6 particles. In our hands, the sensitivity of the clumping method was about the same. The classical technique again gave about the same virus dilution end point but we started with 1 ml of virus suspension rather than 2–4 μl.

These methods have only been tested with plant viruses and plant materials but they may prove useful in medicine or in other situations where there is a need for routine and rapid screening of many samples. One additional possibility would be the testing of the virus content of individual vector insects such as aphids, hoppers, or mosquitoes.

Glow discharge conditioning of the grids before use is not an absolute requirement and if it is omitted, all operations can be done without sophisticated equipment in "field" conditions. The completed grids can be stored at room temperature almost indefinitely so an electron microscope need not be immediately at hand. Grids precoated with antisera can, as we have noted, be prepared and stored or distributed before use.

In conclusion, we suggest that the decoration, clumping, and modified Derrick methods can, between them, rapidly and simply give answers to most IEM problems involving particulate (as opposed to embedded) virus material. However, they are not equivalent. The decoration method is best for high-resolution work where the site of antibody attachment has to be clearly seen. It is also the method of choice for IEM identification of virus bands taken directly from gradients, or of samples containing components that interfere with the immune reaction.

The clumping and modified Derrick methods are particularly useful in trapping viruses that are only available in low concentrations and amounts, and both methods can be used on crude material. The best method is probably the modified Derrick procedure followed by decoration, as it has the advantages of both systems.

REFERENCES

Almeida, J. D., and Waterson, A. P. (1969). *Adv. Virus Res.* **15**, 307.

Anderson, N., and Doane, F. W. (1973). *Can. J. Microbiol.* **19**, 585.

Anderson, T. F., and Stanley, W. M. (1941). *J. Biol. Chem.* **139**, 339.

Anderson, T. F., Yamamoto, N., and Hummeler, K. (1961). *J. Appl. Phys.* **32**, 1639.

Ball. E. M. (1971). *In* "Methods in Virology" (K. Maramorosch and H. Koprowski, eds.), Vol. 5, pp. 445–450. Academic Press, New York.

Ball, E. M. (1974). "Serological Tests for the Identification of Plant Viruses." Am. Phytopathol. Soc., St. Paul, Minnesota.

Ball, E. M., and Brakke, M. K. (1968). *Virology* **36**, 152.

Bercks, R., Querfurth, G., and Lesemann, D. (1974). *Phytopathol. Z.* **80**, 233.

Brandes, J. (1957). *Nachrichtenbl. Dsch. Pflanzenschutzdienstes (Braunschweig)* **9**, 151.

Brlansky, R. H., and Derrick, K. S. (1974). *66th Annu. Meet. Am. Phytopathol. Soc.* Abstract., p. 57, Vancouver.

Chaudhary, R. K., Kennedy, D. A., and Westwood, J. C. N. (1971). *Can. J. Microbiol.* **17**, 477.

Derrick, K. S. (1972). *Phytopathology* **62**, 753 (abstr.).

Derrick, K. S. (1973a). *Phytopathology* **63**, 441 (abstr.).

Derrick, K. S. (1973b). *Virology* **56**, 652.

Doane, F. W. (1974). *In* "Viral Immunodiagnosis" (E. Kurstak and R. Morisset, eds.), pp. 237–255. Academic Press, New York.

Feinstone, S. M., Kapikian, A. Z., Gerin, J. L., and Purcell, R. H. (1974). *J. Virol.* **13**, 1412.

Haschemeyer, R. H., and Myers, R. J. (1972). *In* "Principles and Techniques of Electron Microscopy" (M. A. Hayat, ed.), Vol. 2, pp. 99–147. Van Nostrand-Reinhold, Princeton, New Jersey.

Kelen, A. E., and McLeod, D. A. (1974). *In* "Viral Immunodiagnosis" (E. Kurstak and R. Morisset, eds.), pp. 257–275. Academic Press, New York.

Kelen, A. E., Hathaway, A. E., and McLeod, D. A. (1971). *Can. J. Microbiol.* **17**, 993.

Kellenberger, E., and Arber, W. (1957). *Virology* **3**, 245.

Lafferty, K. J., and Oertelis, S. J. (1961). *Nature (London)* **192**, 764.

Luisoni, E., Milne, R. G., and Boccardo, G. (1975). *Virology* **68**, 86.

Mandel, B. (1971). *In* "Methods in Virology" (K. Maramorosch and H. Koprowski, eds.), Vol. 5, pp. 375–397. Academic Press, New York.

Milne, R. G. (1972). *In* "Principles and Techniques in Plant Virology" (C. I. Kado and H. O. Agrawal, eds.), pp. 76–128. Van Nostrand-Reinhold, Princeton, New Jersey.

Milne, R. G., and Luisoni, E. (1975). *Virology* **68**, 270.

Norrby, E., Marusyk, H., and Hammarskjöld, M.-L. (1969). *Virology* **38**, 477.

Stelos, P. (1967). *In* "Handbook of Experimental Immunology" (D. M. Weir, ed.), 1st ed. pp. 3–9. Blackwell, Oxford.

Tosi, M., and Anderson, D. L. (1973). *J. Virol.* **12**, 1548.

von Ardenne, M., Friedrich-Freska, H., and Schram, G. (1941). *Arch. Gesamte Virusforsch.* **2**, 80.

Williams, R. C., and Fisher, H. W. (1974). "An Electron Micrographic Atlas of Viruses." Thomas, Springfield, Illinois.

Williams, R. C., and Wyckoff, R. W. G. (1946). *J. Appl. Phys.* **17**, 23.

Yanagida, M. (1972). *J. Mol. Biol.* **65**, 501.

Yanagida, M., and Ahmad-Zadeh, C. (1970). *J. Mol. Biol.* **51**, 411.

NOTE ADDED IN PROOF

Derrick and Brlansky [Derrick, K. S., and Brlansky, R. H. (1976). *Phytopathology* **66**, 815] have now slightly modified their technique. The method of Kellenberger and Arber (1957) is now improved [Kellenberger, E., and Bitterli, D. (1976). *Microscopica Acta* **78**, 131].

9 *Transfection Methods*

V. M. Zhdanov

I. Introduction

The term "transfection" was introduced as a description of the treatment (infection) of cells by nucleic acid isolated from a virus, resulting in the production of a complete virus. This term appeared as an analog to the term "transformation," i.e., introduction of a nucleic acid into a cell, resulting in acquisition by the latter of new hereditary properties, and to the term "transduction," i.e., seizure by a virus (the viral genome) of cellular genes and the transfer of these genes to cells.

After the phenomenon of integration of viral genomes by cellular genomes had been established for bacterial (Lwoff and Gutman, 1950) and animal viruses (Vogt and Dulbecco, 1960), the term transfection was often used in a narrower sense, i.e., as treatment of sensitive cells with cellular DNA containing integrated viral genomes (proviruses, DNA transcripts), resulting in the production of the corresponding virus.

References are given below for basic works on this subject, including those where this terminology is introduced or discussed. After it had been demonstrated with *Escherichia coli* bacteriophage T2 that only nucleic acid penetrates the cell and that viral proteins remain outside, it became evident that nucleic acid was the material substrate of heredity (Hershey and Chase, 1952). However, demonstration of infectivity of isolated viral nucleic acid occurred first in the RNA of tobacco mosaic virus (Gierer and Schramm, 1956; Fraenkel-Conrat, 1956), and later in the DNA of bacteriophages (Spizizen, 1957; Mahler and Fraser, 1959; Guthrie and Sinsheimer, 1960; Hofschneider, 1960; Meyer *et al.*, 1961) and the RNA of bacteriophages (Davis *et al.*, 1964; Strauss, 1964; Engelhardt and Zinder, 1964). The term transfection in its broad sense, i.e., infection of cells by isolated nucleic acid from a virus, thereby resulting in the production of a complete virus, had been introduced by Földes and Trautner (1964).

Transfection, in the narrow sense, i.e., treatment of sensitive cells with cellular DNA containing integrated viral genomes, resulting in the production of a virus, had first been carried out for bacterial DNA viruses by Rutberg and Rutberg (1971), for animal DNA viruses by Di Mayorca *et al.* (1959), for oncogenic RNA viruses by Hill and Hillova (1971), and for non-oncogenic RNA viruses by Zhdanov *et al.* (1974).

Transfection methods have been reviewed within the last 10 years by Spizizen *et al.* (1966), Tomasz (1969), Bharava and Shanmugam (1971), Trautner and Spatz (1973), Butel (1973), Notani and Setlow (1974), and in the second volume of this series, in a chapter concerned with the assay of infectivity of viral nucleic acids (Sarkar, 1967).

II. Transfection of Bacteria

Transfection of bacteria was accomplished with both DNA-containing (Spizizen, 1957) and RNA-containing bacteriophages (Davis *et al.*, 1964), and with DNA prophages (Rutberg and Rutberg, 1971). Before we describe these methods of transfection, we will discuss some of the basic mechanisms of transfection.

A. UPTAKE OF NUCLEIC ACIDS BY BACTERIAL CELLS

Intact bacterial cells, in spite of their rigid cell walls, are able to bind nucleic acids and take them up. The uptake of nucleic acids by bacterial cells sharply increases after destruction of the cell wall and the formation of spheroplasts (protoplasts).

During their passage through cell membrane and their further penetration into the cell, nucleic acids undergo degradation by nucleases. However, there sometimes exists a mechanism in cells that rescues the penetrating nucleic acids from this degradation. Such cells have been designated as competent cells (Tomasz, 1969).

Methods of measuring the fraction of competent cells in a culture have been reviewed by Notani and Setlow (1974), who mention four such methods: (1) comparison of the anticipated and the observed transformation frequencies using two unlinked markers (Goodgal and Herriot, 1961); (2) autoradiography (Javor and Tomasz, 1968); (3) survival after ultraviolet radiation, since competent cells in some cases are more sensitive than noncompetent ones (Beattie and Setlow, 1969); and (4) transfection by bacteriophage DNA (Riva and Polsinelli, 1968).

The efficiency of these methods is dependent on the growth stage of the bacterial population, the composition of the nutrient medium, the oxygen supply of the culture, and the presence in the medium of various compounds, including catabolites, and other factors that influence the physiological state of the culture. On the whole, compounds directly inhibiting protein synthesis or inhibiting protein synthesis through RNA synthesis usually increase the competence of the culture; competence also increases when aeration of the culture diminishes or when a rich nutrient medium is replaced by a poor one (Spencer and Herriott, 1965). Other factors increasing the competence of bacterial cells are linked with their influence on the state of the cell wall (Young et al., 1963).

A competence factor (CF) has been discovered in several systems that enables noncompetent cells to take up DNA (Pakula and Walczak, 1963). This factor has been isolated from *Diplococcus pneumoniae* and appears to be a protein with a molecular weight of about 10,000 (Tomasz and Mosser, 1966). A streptococcal CF also appears to be a protein with a molecular weight of about 5000 (Osowiecki et al., 1969). CF may be enzymic in character (Rudchenko et al., 1973).

Most studies of transfection in bacteria were carried out with DNA-containing bacteriophages, therefore uptake of DNA has been preferentially studied. Bacterial cells take up native double-stranded DNA most efficiently; the uptake is less than 1% of that of native double-stranded DNA (Miao and Guild, 1970). However, if conditions of the medium are changed (e.g., the addition of EDTA with low pH), there may be an opposite correlation toward uptake (Postel and Goodgal, 1966). Both linear and circular DNA molecules are taken up with the same degree of efficiency (Mandel and Higa, 1970). When large molecules are sheared, their uptake remains relatively stable, except when the molecular weight of these fragments reaches into the millions and tens of millions. Homologous and heterologous DNAs are taken up with similar efficiency (Lerman and Tolmach, 1957). DNA–RNA hybrids and RNA are taken up by bacterial cells to a lesser degree. However, this area has been insufficiently studied.

The number of sites on the bacterial cell through which DNA binds to the cell membrane and penetrates into the cell is limited (Fox and Hotchkiss, 1957; Stuy and Stern, 1964; Hotchkiss and Gabor, 1970). DNA molecules enter the cell lengthwise (MacAllister, 1970); they first pass through the cell wall, and then through the cell membrane. The latter process requires energy (Strauss, 1970).

In the uptake process of DNA, an important part of the DNA undergoes degradation. Thus, in *D. pneumoniae* cells, initial DNA, with a molecular weight over 3×10^7, is cut into fragments with molecular weights of 5×10^6 (Morrison and Guild, 1973).

B. MECHANISMS OF TRANSFECTION

1. *Transfection by Bacteriophage Nucleic Acids*

Transfection of competent cells by bacteriophage DNA has been performed with several systems: *E. coli* (Spizizen, 1957), *Bacillus subtilis* (Romig, 1962), *Hemophilus influenzae* (Harm and Rupert, 1963), *Mycobacterium smegmatis* (Tokunaga and Seller, 1964), *Staphylococcus aureus* (Sjöström *et al.*, 1972), and *Streptococcus* (Parsons and Cole, 1973). Transfection of *E. coli* was also performed with RNA of small bacteriophages—R17, MS2, and f2 (Paranchych, 1963; Davis *et al.*, 1964; Strauss, 1964; Engelhardt and Zinder, 1964).

The main criterion for the assay of efficiency of transfection is not the amount of bacteriophage particles produced by bacterial cells, but the number of cells yielding the bacteriophage (infective centers). The efficiency of transfection by DNA as compared with infection by bacteriophage particles (virions) is usually 10^{-5}–10^{-6} times lower and rarely increases above 10^{-3} (Horvath, 1969; Trautner and Spatz, 1973).

Generally, the efficiency of transfection increases with an increase in the concentration of DNA, which reflects the possible role of recombination processes in transfection (Spatz and Trautner, 1971), although such a relationship has not been observed in some cases (Notani and Setlow, 1974). A single molecule of DNA initiates transfection (Baltz, 1971). In experiments with ϕX174 bacteriophage DNA the number of infected protoplasts was dependent on the concentration of DNA molecules and protoplasts, according to the equation

$$\text{infected protoplasts} = k \, \text{DNA} \times \text{protoplasts}$$

where $k = 10^{-12}$, for a wide range of conditions (Guthrie and Sinsheimer, 1963).

Inasmuch as moderate bacteriophages of *H. influenzae* contain DNA with cohesive ends, multiple forms of bacteriophage genomes are obtained in these systems, and in this particular case the efficiency of transfection increases with an increase in concentration of DNA. However, such a relationship is not observed when *rec* 1 and *rec* 2 mutants of *H. influenzae* are used, which are defective in recombination systems (Boling *et al.*, 1972). This indicates that cellular recombination mechanisms are necessary for transfection of multiple forms of DNA. However, in experiments in the *B. subtilis* system, using a marker rescue, bacteriophage genes participate in recombination. Therefore *rec* 4 mutants of the host do not decrease the yield of the marker (Spatz and Trautner, 1971).

As has been mentioned, bacteriophage DNA is fragmented in the process of transfection and rapidly sedimenting replicative forms appear later on (Notani *et al.*, 1973). Since fragmentation and transfection rarely occur in *rec⁻* strains of bacteria, it is assumed that fragmentation reflects recombinative activity rather than nonspecific degradation.

Bacterial nucleases sharply limit the efficiency of transfection. The most destructive nuclease complex in *E. coli* is coded by the genes *rec B* and *rec C* (Oishi and Cosloy, 1972; Cosloy and Oishi, 1973; Pilarski and Egan, 1973; Wackernagel, 1973; Wackernagel and Radding, 1973). The nuclease complex *rec BC* has four characteristics, demonstrated *in vitro*: ATP-dependent double-stranded exonuclease, ATP-dependent single-stranded exonuclease, ATP-stimulated single-stranded endonuclease, and ATPase (Oishi, 1969; Wright *et al.*, 1971; Nobroga *et al.*, 1972; Goldmark and Linn, 1972). This enzyme complex destroys single- and double-stranded DNA in the 3′–5′ direction, including small oligonucleotides (Wright *et al.*, 1971; Goldmark and Linn, 1972), while circular DNA is resistant to the enzyme (Pilarski and Egan, 1973; Benzinger *et al.*, 1975). Therefore the efficiency of transfection in *rec⁻ BC*) strains is considerably higher than in *rec⁺* (*BC*) strains. Transfection of *E. coli* with DNA of amber mutations of bacteriophages T5 and BF23 decreases as compared with transfection with DNA of wild-type bacteriophages, since the amber mutations damage the genes, whose products defend bacteriophage DNA from the nuclease complex of the *rec BC* genes of *E. coli* (Benzinger and MacCorquodale, 1975).

It has been shown that transfection (transformation) of *E. coli* by moderate bacteriophages increases if cells are superinfected by helper bacteriophages (Kaiser and Hogness, 1960; Mandel, 1967). To be infectious, DNA of moderate phage must have free cohesive ends or at least one cohesive end (Strack and Kaiser, 1965; Kaiser and Inman, 1965), which should specifically react with the cohesive ends of the helper bacteriophage DNA (Mandel and Berg, 1968; Kaiser and Wu, 1968).

In transfection experiments with heteroduplex DNA obtained by hybridization of DNA from a wild-type and a mutant SPP1 bacteriophage of

B. subtilis, gene conversion was observed, i.e., both pure initial types and mixed types were revealed in progeny (Spatz and Trautner, 1970). The mechanism of gene conversion as applied to transfection may be explained by recombination phenomena that occur during transfection and which are provided (as has been mentioned) not by cellular, but by bacteriophage genes. Therefore gene conversion during transfection does not depend on the cellular *rec* system.

Transfected DNA is more sensitive to ultraviolet irradiation than virion DNA (Harm and Rupert, 1963). This may be caused by lower repairing activity of competent cells (Okubo and Romig, 1965), or by competition for repair of multiply damaged molecules (Setlow and Boling, 1972).

2. *Transfection by Prophage*

Transfection by prophage, i.e., by cellular DNA with integrated bacteriophage genomes, differs somewhat from transfection by bacteriophage DNA. For instance, transfection by prophage does not depend on concentration of DNA and it is low in both donor and recipient *rec⁻* mutants (Rutberg and Rutberg, 1971; Boling *et al.*, 1972). In *rec⁻* donors the process of excision of bacteriophage DNA is damaged; the *rec⁻* recipients recombination processes that occur during transfection are altered.

In comparing the data on transfection by bacteriophage and prophage DNA in *rec⁻* mutants, Notani and Setlow (1974) distinguish five types of *rec⁻* genes: (a) the *rec 1* type, which affects the recombination of phage and prophage DNA, the recombination of transforming DNA, market rescue, injected phage DNA, and induction; in other words, all recombination mechanisms; (b) the *rec 2* and *rec 4* types, which also affect both types of transfection, transformation, recombination of the injected phage DNA, marker rescue, but not induction; (c) the KB type, which affects both types of transfection, transformation, and marker rescue, but does not affect recombination of the injected phage DNA and induction; (d) the *rec B* type which affects only transformation and marker rescue; and (e) the *rec A* type, which affects prophage DNA transfection, transformation, and induction.

The presence of prophages in the tested cells creates cell immunity to homologous bacteriophage and therefore prevents transfection by bacteriophage DNA even if the prophage is defective (Okamoto *et al.*, 1968; Boling *et al.*, 1973).

C. METHODS OF TRANSFECTION

For the assay of transfection of bacteria by phage or prophage DNA, either intact bacteria or spheroplasts (protoplasts), i.e., bacterial cells deprived of a rigid cell wall, are used.

This technique with both types of DNA, has been described in detail in this series (Sarkar, 1967). Therefore modifications of this methods which have been published after 1967 will be given here.

1. Transfection of Intact Bacteria

Two main methods of transfection are by bacteriophage DNA with a helper phage and without a helper phage.

a. Transfection with a Helper Phage. This method (Kaiser and Hogness, 1960) is described in the second volume of this series (Sarkar, 1967). No essential modifications have appeared since then.

b. Transfection without Helper Phage. This method (Romig, 1962) is also described in the second volume of this series and also has not been essentially modified since then.

c. The Use of Calcium Ions. It has been found in studies of the influence of monovalent and divalent ions on the permeability of the cell wall of *E.coli* and the uptake of DNA that, in the presence of calcium ions, DNA of P2 and lambda bacteriophages may, without helper phages, infect cells (Mandel, 1967). Further study showed that the efficiency of transfection sharply increases at concentrations of Ca^{2+} from 0.01 M to 0.25 M. It reaches maximum at 0.025–0.125 M and slowly decreases in the area of 0.1–0.5 M. The optimal time of interaction of DNA with Ca^{2+} is 5–40 minutes at 37°C or at room temperature. Under these conditions both circular and linear bacteriophage DNA are infectious (Mandel and Higa, 1970).

The transfection assay was as follows. Bacteriophages lambda and P2 were purified by the method of differential centrifugation and DNA was extracted by phenol. The cells *E. coli* K12 strain C600 (K38) were grown in the P medium (Radding and Kaiser, 1963), containing 0.02 M potassium phosphate buffer, pH 7.0, 0.015 M $(NH_4)_2SO_4$, 0.001 M $MgSO_4$, and 1.8 × 10^6 M $FeSO_4$, at 37°C, with aeration until they reached the optimal density of 0.6 (10^9 cells per milliliter). The culture was then rapidly cooled. Cells were sedimented in the centrifuge, resuspended in 0.5 volume of cold $CaCl_2$ solution (0.1 M), kept in the cold for 20 minutes and were again centrifuged and resuspended in 0.1 volume of cold $CaCl_2$ solution.

DNA was dissolved in 0.1 ml of cold solution standard saline citrate (SSC) × 1 (0.15 M NaCl, 0.015 M trisodium citrate, pH 7.0), and added to 0.2 ml of the cell suspension. The mixture was incubated at 32°C for 20 minutes. Thereafter the mixture was treated with DNase at 37°C for 5 minutes, then dilutions were made and plated on appropriate indicator media. Under these conditions the authors (Mandel and Higa, 1970) obtained 10^5–10^6 plaques per 1 μg of DNA.

The same method with some modification was used by Wackernagel (1973). *Escherichia coli* cells were grown in TB4 broth to the titer of 2 × 10^8

cells/ml, were washed, treated with 0.03 M CaCl$_2$, and were concentrated as described above (Mandel and Higa, 1970). The assay mixture contained 0.1 ml of precooled DNA solution in Tris buffer and 0.2 ml of CaCl$_2$-precipitated cells. The mixture was kept on ice for 15 minutes, then at 37 °C for 20 minutes. Thereafter 0.1 ml was plated on two petri dishes with a selective medium for the determination of the resistance to streptomycin.

d. *The Use of Freeze-Thawed Cells.* It has been observed that bacteriophages can multiply in freeze-thawed cell paste (Mackal *et al.*, 1964; Weinstein *et al.*, 1969, 1971). However, Dityatkin *et al.* (1972) consider that under these conditions the virus-forming complex is formed in an intact cell.

The method of assay was as follows. DNA of 1 ϕ7 phage (related to ϕX174) was extracted with phenol from purified and concentrated phage suspensions containing $5 \times 10^{12} - 1 \times 10^{13}$ virions per milliliter. The cells of *Proteus vulgaris* were washed twice with Tris buffer (0.01 M, pH 7.4–7.6) and suspended in medium containing 0.5% bactopeptone, 0.002 M MgCl$_2$, and 0.01 M Tris (pH 7.4–7.6). To 0.9 ml of the bacterial culture (2×10^9–5×10^9 cells/ml) 0.1 ml of the DNA solution in Tris buffer was added.

The samples were frozen at −70 °C in a mixture of solid carboxidethanol and after 2–5 minutes were thawed in a water bath at 37 °C. Thereafter the samples were incubated at 37 °C for 5–10 minutes, and dilutions were prepared and plated by the agar layer method, using *E. coli C* as indicator.

2. *Transfection of Spheroplasts*

a. *The Use of Spheroplasts.* The technique of transfection of spheroplasts is considerably more efficient than transfection of intact bacteria (Meyer *et al.*, 1961; Guthrie and Sinsheimer, 1963; Brody *et al.*, 1964; Young and Sinsheimer, 1967). In experiments with intact bacteria the maximal efficiency was not more than 4×10^{-5} per 1 input DNA molecule, whereas the efficiency of transfection of spheroplasts in the same experiments was 1×10^{-2}, i.e., two and one-half orders higher (Wackernagel, 1972). At present, therefore, transfection of spheroplasts has almost fully replaced transfection of intact bacteria. The methods of preparation of spheroplasts with the use of lysozyme (Zinder and Arndt, 1956; Spizizen, 1957; Guthrie and Sinsheimer, 1960, 1963) are described in the second volume of this series (Sarkar, 1967).

In experiments by Young and Sinsheimer (1967) the following techniques were employed. *Escherichia coli* K12 cells were used as a source of spheroplasts and bacteriophage as a source of DNA. The bacteriophage was purified by differential centrifugation and CsCl density-equilibrium centrifugation. DNA was extracted by phenol. Precautions were taken to prevent shearing DNA molecules (see below).

Bacterial cells were grown to a concentration of 3×10^8 cells/ml. A total of 20 ml were sedimented by low-speed centrifugation (5000 g, 5 minutes, at room temperature). The pellet was resuspended in 0.7 mole of 1.5 M sucrose, and the following solutions at 25 °C were added, in order of gently mixing: 0.34 ml of 30% bovine albumin, 0.10 ml of lysozyme (2 mg/ml), 0.08 ml of 4% EDTA, and 10 ml of PA medium. The suspension was incubated at 25 °C for 15 minutes and 0.40 ml of 0.1 M MgSO$_4$ was added. This was the spheroplast stock.

For assay of infectivity 0.30 ml of spheroplast stock was added to 1.20 ml of a DNA solution in 0.05 M Tris (pH 8.1) and incubated at 30 °C for 15 minutes. Thereafter 1.50 ml of PA in medium was added and the suspension incubated at 30 °C for 6 hours. Infective centers were then measured by the method of agar plating.

PA medium contained, in 1 liter: 10 mg casamine acid, 10 mg metricut broth, 1 gm glucose, and 100 mg sucrose. PAM was PA medium plus 0.2% MgSO$_4$ added after autoclaving.

b. *The Use of Protamine Sulfate.* It has been shown that basic proteins stimulate the infectivity of poliovirus RNA (Smull and Ludwig, 1962), and among these protamine (protamine sulfate) has been widely used as a stimulator of transfection in bacteriophages (Hotz and Mauser, 1969; Benzinger et al., 1971). Benzinger and co-workers state this action is caused by binding of host DNA, which inhibits transfection. However, Sabelnikov et al. (1973) have shown that the main stimulatory action of protamine is related to a sharp increase of adsorption of phage DNA by spheroplasts, which increases the penetration of DNA into spheroplasts. This increase of absorption is due to structural and charge changes on the surface of the spheroplasts (Sabelnikov et al., 1974).

Benzinger et al. (1971) have shown that protamine sulfate at a concentration of 25 μg/ml stimulates transfection of E. coli spheroplasts by DNA of several bacteriophages. As compared with a control untreated with protamine sulfate spheroplasts, the efficiency of transfection of lambda bacteriophage DNA increased 10^4 times, having reached 10^{-3} infectious centers per 1 DNA molecule; for the replicative form of ϕX174 bacteriophage DNA efficiency increased 300 times for 10^{-6} infections centers per 1 DNA molecule; and for T7 bacteriophage DNA efficiency increased 300 times for 3×10^{-7} infectious centers. Native DNA was noninfectious in the absence of protamine sulfate. However, it was infectious in its presence for DNA of bacteriophages T4 (10^{-5}), T5 (3×10^{-6}), and P22 (3×10^{-9}).

From other basic polymers only spermine at a concentration of 40 μg/ml considerably facilitated transfection of E. coli spheroplasts. In some bacteriophages a cumulative effect was observed when both compounds were used. Histones, DEAE-dextran, polyarginine and polylysine, and many other polyamines were only slightly effective (Henner et al., 1973).

c. *Technique of Spheroplast Transfection.* According to Benzinger et *al.* (1971) various strains of *E. coli* were grown in 200 ml of a medium at 25 °C for 36 hours until the concentration of cells reached 5×10^8 cells/ml. The cells were sedimented in a centrifuge and the pellet was resuspended in 3.5 mole of 1.5 M sucrose, to which was added 1 ml of 30% bovine albumin and 0.2 ml of lysozyme (2 mg/ml in 0.25 M Tris buffer, pH 8.1). One minute later 0.4 ml of 4% EDTA was added for 40–120 seconds, and thereafter 95 ml of PA medium was added. The mixture was allowed to stand for 10 minutes at room temperature, then 2 ml of 10% $MgSO_4$ and 0.25 ml of 1% protamine sulfate was added. The suspension was placed on ice and 30 minutes later bacteriophage DNA dissolved in 0.05 M Tris buffer (pH 8.1) was introduced. The bacteriophage was titrated by the plaque method after 4 hours of incubation. This technique was used for transfection by the virion or by the replicative DNA of bacteriophage ϕX174.

For bacteriophages with high molecular weight DNA (λ, T7, T5, P22, and T4) the method was modified in the following way. Bacteriophage DNA was dissolved in 0.01 M Tris buffer (pH 8.1), since its ionic strength over 0.02 M inhibits its transfection (Meyer *et al.*, 1961). Equal volumes of spheroplast suspension and DNA solution were mixed and kept at 30 °C for 8–10 minutes. The infected cells were then mixed with 3 ml of soft agar (which contained 1% bovine albumin and phage indicator bacteria) and plated at 37 °C.

All experiments were conducted with controls that indicated the absence of infectious bacteriophages from the spheroplasts, the media, and the phage DNA species. The presence of mature bacteriophage 8–10 minutes after addition of spheroplasts was ruled out in similar samples treated with chloroform.

A description of a recent modification of this technique (Wackernagel, 1972) is given below. Two bacteriophages λ and T4, were assayed for infectious DNA by transfection of spheroplasts of the corresponding strains of *E. coli*.

The bacteriophage preparations were concentrated by the two-phase technique using polyethylenglycol and dextran sulfate (Albertson, 1967) and were purified by differential centrifugation followed by CsCl density-gradient centrifugation. DNA was extracted using the slightly modified method of Kaiser and Hogness (1960). Equal volumes of bacteriophage suspension (2×10^{12} particles/ml) and phenol equilibrated with Tris buffer (Tris HCl, 0.05 M, pH 7.9) were gently mixed at 4 °C for 1 minute. The aqueous phase was collected and phenol deproteinization was repeated. The DNA preparation (about 250 μg/ml) was dialyzed against Tris buffer with 0.001 EDTA, then 10^{-5} M spermine was added to stabilize the DNA molecules (Kaiser *et al.*, 1963). The preparations were stored at 4 °C for over 1 year without loss of biological activity. All further manipulations of the DNA were performed only with graduated 1 ml pipettes to reduce shearing.

For preparation of spheroplasts cells grown in TBY (trypten broth with yeast extract; $3-4 \times 10^8$ cells/ml) were sedimented in the centrifuge at room temperature and resuspended in 0.05 ml of 0.5 M Tris buffer plus 0.175 ml of 0.5 M sucrose solution. To this, 0.018 ml of a freshly prepared solution of lysozyme (2 mg/ml) was added and the suspension was incubated at 25 °C for 1 minute and gently swirled. Then 0.01 ml of 4% EDTA was added and the mixture was incubated for 8.5 minutes. Thereafter 1 ml of TBY containing 2% bovine serum albumin, 10% sucrose, 0.12% glucose, and 0.1 $MgSO_4 \cdot 7H_2O$ (TBSS) was mixed with the suspension within 5 minutes, and 3.75 ml of the same TBY medium, without bovine serum albumin (TBS), was added. The preparation was checked by phase contrast microscopy (99% of the cells were converted to spheroplasts). Thereafter 0.25 ml of aqueous solution of protamine sulfate (1 mg/ml) was added to 4.75 ml of spheroplast suspension, and the spheroplasts were used within 30 minutes to 1 hour of their storage on ice.

Bacteriophage DNA dissolved in Tris buffer was mixed with the buffer and spheroplast suspension (1:3:2.5). The mixture was incubated at 30 °C for 30 minutes to allow DNA adsorption. Then the mixture was diluted with TBS and, after incubation for 5 hours, the bacteriophage was titrated.

A similar technique was used for transfection of *E. coli* by half and inverted molecules of lambda bacteriophage DNA (Wackernagel and Radding, 1973).

III. Transfection of Plant Cells

Since for plant viruses the integration process is not known or, at least, not proved, the term transfection can be applied in its initial sense, i.e., infection with viral nucleic acid, resulting in the production of a virus. It is worthwhile to remember, that the vast majority of plant viruses are RNA-containing viruses and that infectivity of nucleic acids had first been established for the RNA of tobacco mosaic virus (Gierer and Schramm, 1956; Fraenkel-Conrat, 1956). The assay methods for infectivity of RNA in tobacco mosaic virus and other plant viruses are described in the second volume of this series (Sarkar, 1967). Therefore we will describe here several new techniques.

In a review of the structure of viruses Colter and Ellem (1961) mention 7 plant viruses whose RNA appears to be infectious. Sarkar (1967) has summarized the data on infectivity of 15 plant viruses, and this can be considered complete. Recently attention has been drawn to plant viruses that have multicomponent genomes (Jaspars, 1974; Lane, 1974). The majority of viruses of this group have genomes composed of 2 to 4 fragments of RNA. Each of these components is noninfectious, however, a mixture of 3 components is infectious in brome mosaic virus (Lane and Kaesberg, 1971), cowpea chlorotic mottle virus (Bancroft and Flack, 1972), and cucumber mosaic virus (Lot *et al.*, 1974). In alfalfa mosaic virus, citrus leaf rugose virus, and tobacco streak virus, infectivity is connected with, all 4 components (Bol *et al.*, 1971; Gonsalves and Garnsey, 1975; Van Vloten-Doting, 1975).

The technique of assay of infectivity is as follows (Gonsalves and Garnsey, 1975). The leaves of red kidney beans (*Phaseolus vulgaris*) grown in shadow are treated with carborundum. Viral RNA dissolved in phosphate buffer (0.06 M potassium phosphate, pH 8.0) with bentonite (400 μg/ml) is then deposited with a swab. After incubation the leaves are washed with distilled water. Further procedures for observation and registration of results are the same as those described in the second volume of this series (Sarkar, 1967).

Two new techniques have been devised for transfection by viral nucleic acids using tissue cultures and protoplasts.

A review of early publications is given in the first volume of this series (Kassanis, 1967), and recent data are reviewed by Zaitlin and Beachy (1974). Virus multiplication in callus tissues grown *in vitro* does not essentially differ from that in plant leaves. Advantages and disadvantages of these systems are analyzed by Zaitlin and Beachy. Callus tissues were used for transfection by viral RNA.

The use of protoplasts is a considerably more efficient method for the study of plant viruses (for a detailed description see the chapter by Sarkar in this volume). This technique may play the same revolutionary role for the study of plant viruses as the use of tissue cultures has for animal viruses. It is remarkable that the reproduction cycle of plant viruses in this system is measured not in days, as during infection of leaves, but in hours. Synchronization of infection can be easily attained in this system (Zaitlin and Beachy, 1974).

The use of protoplasts was applied for transfection with plant viral RNA (Aoki and Takebe, 1969; Motoyashi *et al.*, 1973). Since this technique is described in the present volume in more detail by Sarkar, we shall give only a short description of a method for preparation of protoplasts from the cells of leaves (Föglein *et al.*, 1975).

Peeled leaf pieces are floated on the surface of a mixture of 0.5% potassium dextran sulfate, 0.7 M mannitol (pH 5.8), and enzymes macerozyme R-10, 0.5%, and cellulase, 2%, which destroy the cell wall at 35°–36°C within 2–3 hours. The protoplasts thus formed are washed with the medium of Aoki and Takebe (1969). The washed protoplasts are infected with viral RNA and incubated in a medium of tobacco mosaic virus (Aoki and Takebe, 1969) at 25°C for 18 hours or more. Thereafter the virus is titrated in leaves by the standard methods.

IV. Transfection of Animal Cells

Transfection of animal cells was achieved by viral RNA (Colter *et al.*, 1957a,b; Wecker and Schäfer, 1957), DNA (Di Mayorca *et al.*, 1959), and with DNA proviruses of oncogenic DNA viruses (Boyd and Butel, 1972), RNA viruses (Hill and Hillová, 1971), and nononcogenic RNA viruses (Zhdanov *et al.*, 1974). Before we give a description of the methods of transfection of animal cells, we shall first discuss some basic questions related to them.

A. UPTAKE OF NUCLEIC ACIDS BY ANIMAL CELLS

Uptake of nucleic acids by animal and plant cells has been thoroughly analyzed in reviews by Ledoux (1965) and Bhargava and Shanmugam (1971). It is necessary to note, however, that some data by Ledoux on the uptake and integration of bacterial DNA by plants have been criticized (Kleinhofs *et al.*, 1975). The more recent experimental data of Farber *et al.* (1975) on uptake of nonviral (animal) DNA into animal cells are in accordance with previous data. Recipient culture has been shown to be more competent in the exponential growth phase than stationary culture. Polyornithine has enhanced the uptake of exogenous DNA more than DEAE-dextran, $CaCl_2$, latex spheres, spermine, polylysine, and polyarginine. About 25–30% of DNA inoculum became DNase-resistant and nucleus-associated 15 minutes after inoculation.

Uptake of viral nucleic acids by animal cells was studied for several viruses, particularly adenoviruses and papovaviruses, and the interaction of poliovirus RNA with animal cells has also been reviewed by Koch (1973).

Uptake and the further fate of adenovirus DNA by sensitive (KB) cells has been thoroughly studied by Groneberg *et al.* (1975). In these experiments purified and highly tritium-labeled DNA of adenovirus type 2 was introduced into an adsorption medium for long periods of time, from 2 to 24 hours. Within 2 hours the uptake of viral DNA was linear and 3–9% became associated with the cells. Between 2 and 6 hours part of the DNA was eluted and partly readsorbed. Within 0.5 to 2 hours DNA reached the nucleus and by 24 hours 70% of the DNA taken up was in the nucleus. The major part of it was cleaved into fragments sedimenting at 19–23 S both in the cytoplasm and in the nucleus. However, about 20% of the DNA became cell-associated or perhaps integrated into the cell genome.

In experiments by Kelly and Butel (1975) primary African green monkey kidney cells were exposed to DNA from SV40-transformed mouse cells (SV 3T3) in the presence of DEAE-dextran. When large amounts (10–50

μg) of high molecular weight (10^7) DNA were introduced (per 10^6 cells), less than 1 μg of it became cell-associated. When it was sheared to pieces of $1–3 \times 10^6$ daltons, larger amounts entered the cells, but not more than $5–10\%$ ($1–4$ μg) of the total input DNA. After 30 minutes of incubation about 60% of the cell-associated DNA became DNase-resistant. However, the production of the virus was higher when input DNA had a high molecular weight.

B. VIRAL INFECTIOUS NUCLEIC ACIDS

Based on the structure and function of viral genomes and taking into account previous considerations about RNA-containing viruses (Baltimore, 1971; Shatkin, 1974), Bukrinskaya and Zhdanov (1975) suggested the following classification of animal viruses.

1. RNA Viruses

a. Viruses whose genome functions as mRNA (plus-strand viruses): picornaviruses and togaviruses (arboviruses)

b. Viruses whose genome is complementary to mRNA (minus-strand viruses):

    i. the genome is continuous: rhabdoviruses and paramyxoviruses

    ii. the genome is fragmented: orthomyxoviruses and arenaviruses

c. Viruses with double-stranded (fragmented) RNA: reoviruses

d. Viruses with reverse transcription: oncornaviruses and other retraviruses.

Only the viruses of the first group (picornaviruses and arboviruses) possess infectious RNA, while all other viruses can start infection with transcription of the genome by virion-associated RNA or DNA polymerases. However, in these viruses not only virions but also nucleocapsids (Bukrinskaya, 1973) and nucleic acids from virus-infected cells (Orlova and Orlova, 1975) possess an infectious property.

2. DNA Viruses

a. Viruses with single-stranded linear DNA: parvoviruses

b. Viruses with double-stranded circular DNA: papovaviruses

c. Viruses with double-stranded linear DNA without virion transcriptases: adenoviruses and herpesviruses

d. Viruses with double-stranded linear DNA with virion transcriptases: poxviruses

Infectious DNA of virions was revealed in viruses of the second and third groups, though transfection was more difficult in viruses with a large genome (herpesviruses). Infectivity of parvovirus DNA has not yet been studied; However, it is anticipated that transfection will be successful. DNA of poxviruses does not possess infectivity and infection is possible if part of the genome is transcribed by the virion-associated transcriptase. A report by Abel and Trautner (1964) on the isolation of infectious DNA from vaccinia virus has not been confirmed.

Infectivity of virion RNA was established in many picornaviruses: Mengo virus (Colter et al., 1957a), poliovirus (Colter et al., 1957b), encephalomyocarditis virus (Huppert and Sanders, 1958), foot-and-mouth disease virus (Brown et al., 1958), Theilers GD VII virus (Ada and Anderson, 1959b), echovirus (Sprunt et al., 1959), Coxsackievirus (Holland et al., 1959), and rhinovirus (Sethi and Schwerdt, 1972). Extraction of nucleic acids was generally performed using the phenol method. The viral RNA preparations were introduced either intracerebrally into mice or in tissue cultures. Later the detergent–phenol method (Arlinghaus et al., 1966) and detergent method (Bachrach, 1966) were used.

Infectious titers of viral RNA's were several orders lower than those of virions. Thus, in experiments by Breindl and Koch (1972) the specific infectivity (PFU per microgram of RNA) of poliovirus was $1–2 \times 10^9$, whereas that of RNA was $1–5 \times 10^6$, i.e., three orders lower. In rhinovirus the infectious titer of RNA was 10^4 times lower than that of virions (Sethi and Schwerdt, 1972).

The replicative form and replicative intermediate RNA of poliovirus both possess infectious properties. These properties are preserved after treatment of RNA with RNase T1 and are decreased after treatment with RNase A in the presence of $0.3\ M$ NaCl. Exonucleases (RNase B, polynucleotide phosphorylase, and spleen phosphodiesterase) that rapidly destroy the virion RNA are not active toward the replicative form (Mittelstaedt et al., 1975).

Infectivity of RNA was determined for several arboviruses: West Nile encephalitis virus (Colter et al., 1957b), Eastern equine encephalomyelitis virus (Wecker and Schäfer, 1957), Semliki Forest virus (Cheng, 1958), Western equine encephalomyelitis virus (Wecker, 1959), Murray Valley encephalitis virus (Ada and Anderson, 1959a), dengue virus (Ada and Anderson, 1959b), tick-borne encephalitis virus (Sokol et al., 1959), Japanese B encephalitis virus (Nakamura, 1961), yellow fever virus (Nielsen and Marquardt, 1962), Venezuelan equine encephalomyelitis virus (Mika, 1963), Louping ill virus (Golubev and Polak, 1965), Chikungunya virus (Igarashi et al., 1967), and Sindbis virus (Dobos and Faulkner, 1969). In most cases RNA was isolated by the phenol or phenol–detergent methods and the assay of infectivity was performed in mice, chick embryos, and tissue cultures.

As stated, other RNA-containing viruses have no infectious RNA in their virions. Nevertheless several reports appeared on the isolation of infectious RNA from influenza virus preparations (Maassab, 1959; Portocala et al.,

1959; Sokol *et al.*, 1959; Smorodintsev *et al.*, 1961). These results have not been confirmed in follow-up experiments (Colter and Ellem, 1961).

It has been shown, however, that DNA–RNA complex isolated from cells infected with myxovirus possesses infectious properties (Orlova and Orlova, 1975). Further study (Menshikh *et al.*, 1975) showed that infectious nucleic acid complex consists of double-stranded RNA (the transcriptive and replicative complex) insensitive to pancreatic DNase and RNase. This complex is formed within 3 hours after infection. Its infectivity increases when the multiplicity of infection and incubation time increased (1–20 PFU/cell and 6–24 hours.)

Infectious DNA has been found in virions of papova viruses: polyoma virus (Di Mayorca *et al.*, 1959), Shope papilloma virus (Ito, 1960), SV40 (Gerber, 1962), and bovine papilloma virus (Boiron *et al.*, 1965). Infectious DNA of SV40 was isolated by the phenol method from the virions grown in African green monkey kidney cells. The activity of the preparations was preserved after treatment with RNase and was fully destroyed after DNase treatment (Gerber, 1962).

Infectious DNA of adenoviruses was obtained by Burnett and Harrington (1968) for C1 cells from the simian adenovirus SA7, and by Nicolson and MacAllister (1972) for the primary human embryo cells from the human adenovirus A1. Finally, infectious DNA was isolated and assayed from virons of herpesvirus (Graham *et al.*, 1973; Sheldrick *et al.*, 1973).

C. Transfection by Integrated Viral Genomes

Though four groups of DNA-containing viruses apparently possess infectious virion DNA, transfection of sensitive cells by a cellular DNA, containing an integrated provirus, was accomplished only for adenoviruses (Butel *et al.*, 1975). A series of papers appeared on the successful transfection of oncornaviral proviruses: Rous sarcoma virus (Hill and Hillová, 1971), avian myeloblastosis virus (Lacour *et al.*, 1972), and rodent sarcoma virus (Scolnick and Bumgarner, 1975). Finally, infectious provirus was found in nononcogenic RNA viruses: tick-borne encephalitis virus (Zhdanov *et al.*, 1974, Sindbis virus (Zhdanov and Azadova, 1976), SV5 (Zhdanov *et al.*, 1977), vesicular stomatitis virus (Zhdanov, 1975b), respiratory syncytial virus (Simpson and Iinuma, 1975), and influenza virus (Forman and Simpson, 1975). Though usually only a certain part of the viral genome is integrated in cells transformed by adenoviruses (Sharp *et al.*, 1974; Gallimore *et al.*, 1974), a system has been described for transfection of adenoviral provirus (Butel *et al.*, 1975). For this transfection tumors in hamsters injected with simian adenovirus SA7 were used. DNA–protein complex was obtained by a modified Hirt method, and BSC-1 cells were transfected by this complex after the treatment of the cells with DEAE-dextran. Infectivity can be abolished by exposure to DNase. A correlation between the persistence of complete virus and the presence of infectious DNA–protein complex was noted.

Hill and Hillová (1971, 1972a) transfected chick embryo fibroblasts by rat KC cell DNA transformed by Rous sarcoma virus. The virus that was recovered in chick embryo fibroblasts retained the properties of the initial virus. Similar results were obtained by Montagnier and Vigier (1972), and Hlozanek and Svoboda (1972). In another experiment hamster cells were transformed by a temperature-sensitive mutant of Rous sarcoma virus and chicken cells were transfected by DNA from the transformed hamster cells. The virus that was recovered in this system was the same temperature-sensitive mutant as the virus that had been used for transformation of hamster cells (Hill and Hillová, 1972b; Hillová et al., 1975). In one more series of experiments chick embryo fibroblasts were transformed by a helper-independent avian sarcoma virus and transfection was carried out by DNA from this cell system. Two viruses were isolated: the initial helper-independent virus and an endogenous transformation-defective virus of chick embryo fibroblasts (Hillová et al., 1974a). In a similar experiment Lacour et al. (1972) isolated avian myeloblastosis virus from a transfected tissue culture. In other experiments nuclear localization and covalent linkage of infective virus DNA to chromosomal DNA of Rous sarcoma virus-transformed cells was shown (Hillová et al., 1974b).

In experiments by Scolnick and Bumgarner (1975) BALB/c cells transformed by Kirsten sarcoma virus were taken as a source of infectious DNA. Mink fibroblasts were transfected and an endogenous xenotropic mouse oncornavirus was isolated.

All experiments by Hill and Hillová, until 1974, were done with DEAE-dextran techniques; the latest experiments (Hillová et al., 1975; Scolnick and Bumgarner, 1975) were carried out using the $CaCl_2$ technique.

Zhdanov and co-workers have found that tissue cultures can be chronically infected with measles virus (Zhdanov and Parfanovich, 1974), tick-borne encephalitis virus (Zhdanov et al., 1974), Sindbis virus (Zhdanov and Azadova, 1976), SV5 (Zhdanov et al., 1977), and vesicular stomatitis virus (Sovetova et al., 1976). Transfection was carried out by proviruses of tick-borne encephalitis virus, Sindbis virus, SV5, and vesicular stomatitis virus using both the DEAE-dextran and $CaCl_2$ techniques.

Several tick-borne encephalitis virus strains recovered from transfected culture were thoroughly studied (Gavrilov et al., 1974, 1975). Most of them preserved the characteristics of the initial virus that was used to obtain the chronically infected tissue culture, but one isolate appeared to be a mutant, which caused cytopathic changes of a sensitive culture (continuous pig kidney cell line) only in the presence of DEAE-dextran.

Double integration of two infectious viruses was found in chronically infected continuous mouse L cell lines: Sindbis virus plus SV5 (Zhdanov *et al.*, 1977) and vesicular stomatitis virus plus SV5 (Sovetova *et al.*, 1976). In the first case transfection was successful only for one partner, Sindbis virus. In the second case both viruses were isolated from transfected chick embyro fibroblasts and BHK-21 cells.

Similar results were obtained by Simpson and Iinuma (1975) who recovered infectious proviral DNA of respiratory syncytial virus by tranfection of sensitive cells with DNA from chronically infected tissue culture. Both initial and transfected viruses had similar temperature-sensitivity characteristics. Another wild-type strain was recovered in a similar way. The authors also stated (Forman and Simpson, 1975) that they had succeeded in isolating infectious influenza provirus. Possible applications of these data to the study of pathologic processes are discussed elsewhere (Zhdanov, 1975a,b).

D. Transfection of Mitochondria

Since mitochondria possess an autonomous system for the synthesis of proteins and nuclei acids (for references, see Boardman *et al.*, 1971; Linnane *et al.*, Borst, 1972; Schatz and Mason, 1974), attempts have been made to transfect them with a viral nucleic acid.

A series of papers have been published by Yershov and associates on replication of Venezuelan equine encephalitis (VEE) virus in isolated mitochondria. Mitochondria were isolated from rat liver and partly purified. RNA was isolated from purified VEE virus and added into the incubation mixture for the coupled synthesis of nucleic acids and proteins. About 50% of RNA that penetrated into mitochondria was found to be bound to mitochondrial ribosomes (Gaitskhoki *et al.*, 1971a,b). Replicative intermediate and newly synthesized virion RNA appeared within 40–120 minutes of incubation and increase of infectivity was observed by the end of this period. No mature virions appeared in this system and infectivity was connected with ribonucleoprotein structures (Yershov *et al.*, 1971a,b). The specificity of the virus-induced synthesis was tested in titration infectivity and serologic reactions (Zaitseva *et al.*, 1973; Yershov *et al.*, 1974a).

Experiments have been conducted to elucidate the possibility of using highly heterogeneous combinations of viral RNA and mitochondria. The treatment of rat liver mitochondria with tobacco mosaic virus RNA results in the synthesis of mature virions (Zhdanov *et al.*, 1971; Naroditsky *et al.*, 1973), and yeast mitochondria transfected with VEE virus RNA produce infectious ribonucleoproteins (Yershov *et al.*, 1974a).

Attempts have also been made to transfect hypothetical animal mitochondria by nucleic acid from patients with serum hepatitis. Though some evidence was obtained of possible persistence of replicating genomes in isolated mitochondria (Zhdanov, 1972a,b) the final results have been negative (Zhdanov *et al.*, 1973).

E. METHODS OF TRANSFECTION

Methods of transfection are generally the same for the assay of free viral nucleic acids or of DNA proviruses integrated by the cell genome. Some differences may relate to methods of extraction of nucleic acids or to the selection of sensitive cells or animals. Optimization of initial stages of transfection may differ, including changes of conformation of nucleic acid which facilitates its penetration into the cell and defends it from attacks by nucleases. Inhibition of the activity of nucleases, as well as inhibition of cellular protein synthesis and liberation of ribosomal systems for interaction with viral templates may vary. Also the creation of optimal conditions for transcription of viral nucleic acids will differ, along with other factors.

1. *Isolation of Nucleic Acids*

For isolation of RNA in picornaviruses and arboviruses, the phenol or detergent–phenol methods are widely used. An example of the phenol method is the following (Gierer and Schramm 1956): A suspension of tobacco mosaic virus in 0.01 M phosphate buffer or in other buffer, e.g., TNE (Tris HCl, 0.01 M, pH 7.4, NaCl, 0.1 M, EDTA 10^{-3} M), was mixed at $4°-5°C$ with an equal volume of buffer-saturated phenol, shaken vigorously, and the aqueous phase was separated from the phenol phase by low-speed centrifugation. The aqueous phase was retained using a Pasteur pipette and treated again with phenol (twice) until an interphase was free of coagulated proteins. The remnants of phenol were removed by ether extraction.

For storage RNA preparations are precipitated with two volumes of ethanol and kept at $-20°C$. If there are small amounts of nucleic acid in the preparation, monovalent ions (usually 2% sodium acetate) or a carrier RNA (usually yeast RNA), 25–50 μg per tube, are added.

Phenol extraction may be carried out either at $4°-5°C$, at room temperature, or at $50°-65°C$. The higher the temperature of extraction, the faster the RNA extraction; however, at high temperatures nucleic acid may decay. The main cause of degradation of DNA is admixtures of nucleases that remain in the aqueous phase after phenol extraction. Therefore modifications of the phenol method have been proposed.

For the stabilization of RNA and for more prolonged preservation of infectious properties in aqueous solutions it has been suggested that both isolation and storage of RNA be carried out in the presence of bentonite, which adsorbs proteins (Fraenkel-Conrat *et al.*, 1961).

Bentonite, which adsorbs nucleases and thus prevents nucleic acids from degradation (Bellett *et al.*, 1962), is also an adsorbent and carrier of viral nucleic acids and therefore increases transfection (Dubos, 1972).

Bentonite suspension is prepared in the following way: 2 gm of bentonite are suspended in 40 ml of water, centrifuged at 2500 *g* for 15 minutes, and then the pellet is discarded. The supernatant is centrifuged at 8500 *g* for 20 minutes and the pellet is collected, resuspended in 0.1 *M* EDTA (pH 7.0), and allowed to stay at room temperature for 2 days. Centrifugation at 2500 *g* and 8500 *g* is then repeated, and the last sediment is suspended (5%) in 0.01 *M* acetate and stored in the cold. In order to stabilize RNA 25 mg of bentonite per 1 ml of RNA solution must be added. To remove bentonite from RNA solution the latter is centrifuged at 30,000–40,000 *g* for 1 hour.

The detergent–phenol method is based on the use of sodium dodecyl sulfate (SDS), which separates proteins from nucleic acids and inhibits nucleases (Scherrer and Darnell, 1962). A virus preparation suspended in a buffer is treated at room temperature with 0.5–2% solution of SDS; then phenol is added and, after shaking, both phases are separated by low-speed centrifugation. The following extractions are performed without the addition of SDS.

A comparison of the efficiency of various methods of extraction of infectious RNA in arboviruses is given in Table I. More detailed comparative data are presented in Table II.

The same methods are used for isolation of nucleic acids from cells. RNA from virus-infected cells is usually extracted by hot phenol (50°–65°C). For the separation of RNA from DNA using the detergent–phenol method

TABLE I

INFECTIVITY OF RNA PREPARATION IN ARBOVIRUSES EXTRACTED BY VARIOUS METHODS

Virus	Method of extraction	Infectivity (log PFU/ml)	Reference
Sindbis	Detergent–phenol, 4 °C	4.5	Dobos and Faulkner (1969)
Eastern equine encephalomyelitis	Detergent–phenol, 60 °C	6.0	Zebowitz and Brown (1970)
Western equine encephalomyelitis	Detergent, 50 °C	3.0–4.0	Wecker and Richter (1962)
Semliki Forest	Phenol, 4 °C	3.0	Cheng (1958)
Chikungunya	Detergent–phenol, 50 °C	2.0–3.0	Igarashi *et al.* (1967)
Tick-borne encephalitis	Phenol, 4 °C	2.0–3.0	Bragina *et al.* (1964)
Japanese encephalitis	Phenol, 50 °C	3.0–5.0	Nakamura (1961)

TABLE II

INFLUENCE OF TEMPERATURE OF EXTRACTION BY PHENOL UPON
CONCENTRATION AND INFECTIVITY OF EXTRACTED RNA OF VENEZUELAN
EQUINE ENCEPHALOMYELITIS VIRUS[a]

Initial material	Temperature of extraction (°C)	Amount of RNA tracted (μg)	Infectivity (log PFU/ml)
Virus suspensions	4	80–100	1.9
with titer of	22	160–180	2.3
9.3 (log PFU/ml)	65	310–340	3.7

[a] From Zhdanov and Yershov (1973).

extraction is usually carried out at 50 °–65 °C, and under this condition RNA passes to the aqueous solution whereas DNA remains in the bound state. If extraction is carried out at room temperature, both RNA and DNA are revealed in the aqueous phase (Hiatt, 1962).

A modification of the detergent–phenol method is the addition of DEAE-dextran (0.5 mg/ml) to the extracted material. According to the data of Yershov (Zhdanov and Yershov, 1973), this method increases the yield of infectious RNA during its extraction from virus-infected cells (Table III). This is in good agreement with the data of Pagano and Vaheri (1965) and Maes *et al.* (1967) on the defence of viral nucleic acids from degradation by cellular nucleases.

The results allow one to recommend hot deproteinization of viral particles and infected cells by phenol with the additions of 0.5% SDS and 0.5 mg/ml DEAE-dextran.

To isolate nucleic acids from cells, digestion of protein by Pronase, a protease of *Streptomyces griseus*, has recently been performed. Pronase is prepared in phosphate buffer (pH 8.2) to a concentration of 2–5 mg/ml and incubated at 37 °C for 3 hours in order to destroy admixtures of nucleases that may occur in enzyme preparation. The prepared solution may be stored at −20 °C for several weeks or months. The cells or viral suspensions are treated with 0.2–0.5% SDS (final concentration), then Pronase is added to a concentration of 50–500 μg/ml, depending on the amount of protein in the preparation, and incubated at 37 °C for 1–5 hours. Thereafter phenol extraction is performed.

Further separation of DNA and RNA is possible with the use of various methods. Single-stranded RNA is precipitated with strong solutions of neutral salts. Usually 2.0 M NaCl (final concentration) at 0 °C for one day or overnight is used. Thereafter an additional centrifugation (15,000–20,000 g for 15–20 minutes) is needed for more reliable removal of the sediment. Admixtures of polysaccharides are also precipitated by the Tris method.

TABLE III

EFFICIENCY OF ISOLATION OF INFECTIOUS NUCLEIC ACIDS OF VENEZUELAN
ENCEPHALOMYELITIS VIRUS BY EXTRACTION WITH PHENOL, PHENOL-SDS
(0.5%), AND PHENOL-SDS-DEAE-DEXTRAN (0.5 mg/ml)[a,b]

No. of experiments	Material	65°C log			4°C		
		Phenol	Phenol-SDS	Phenol-SDS-DEAE-D	Phenol	Phenol-SDS	Phenol-SDS-DEAE-D
1	Purified virus 10 log PFU/ml	2.7	4.4	3.0	1.0	2.5	2.5
2	Purified virus 10 log PFU/ml	2.6	4.1	3.2	1.3	1.9	2.1
3	Purified virus 10 log PFU/ml	2.2	3.7	2.8	1.0	1.3	1.4
1	Culture fluid from virus-infected tissue culture 8 log PFU/ml	2.0	3.5	3.1	1.1	2.2	2.1
2	Culture fluid from virus-infected tissue culture 8 log PFU/ml	2.3	3.5	2.8	0.8	2.0	2.0
3	Culture fluid from virus-infected tissue culture 8 log PFU/ml	2.5	3.4	3.0	1.3	2.3	2.0
1	Infected cells, 5 hours post-infection	1.9	3.2	3.6	0	3.2	3.7
2	Infected cells, 5 hours post-infection	1.4	2.9	3.1	0.7	3.1	3.3
3	Infected cells, 5 hours post-infection	1.0	3.0	2.8	0	2.9	3.1

[a] RNA titers in log PFU/ml.
[b] From Zhdanov and Yershov (1973).

For the liberation of DNA from RNA or of RNA from DNA the treatment of preparations with pancreatic ribonuclease or deoxyribonuclease (50–300 μg/ml) is used. The concentration of the nucleases depends upon the concentration of the nucleic acids in the preparations. The nucleases for this purpose must have the highest degree of purification. It is recommended that ribonuclease preparations be heated at 95°–100°C for 1–2 minutes to destroy admixtures of deoxyribonucleases.

In order to isolate DNA, several specific extraction methods besides the methods described are applicable, particularly if a viral nucleic acid is isolated from nuclear structures (e.g., SV40) or if a high molecular weight viral nucleic acid is isolated (e.g., adenoviruses, herpesviruses, or poxviruses).

The method of Hirt (1967) is based on observations that if cellular DNA is treated with SDS in the presence of high concentrations of NaCl, the main part of the chromatin DNA will be sedimented whereas smaller molecules of papovaviruses will remain in the supernatant. The following is a description of a recent modification of this method (Chou and Martin, 1975).

Monolayers of CV-1 cells in 75 ml flasks are infected with SV40 at the m.o.i. of 5 PFU per cell and incubated at 33°C for 72 hours. Then the medium is removed, the monolayer is washed with phosphate saline buffer (PBS), and lysed by introduction into flasks containing 1 ml of Tris HCl, 0.01 M, pH 7.6, EDTA, 0.01 M, and SDS, 0.6%. To this is added 1 M of NaCl and the mixture is then incubated overnight at 4°C. Thereafter the material is clarified by centrifugation (20,000 g for 30 minutes) and the supernatant is dialysed overnight against the buffer Tris HCl, 0.01 M, pH 7.5, NaCl, 1 M, and EDTA, 1 mM. For further purification of DNA preparations viral DNA may be extracted by phenol from the supernatant after pelleting of cellular DNA by hypertonic NaCl solution. DNA may also be further purified in CsCl density gradients (Chen et al., 1975).

For isolation of adenovirus DNA the following procedure has been applied (Van der Vliet and Sussenbach, 1975). The culture of KB cells infected with adenovirus type 5 was washed with Tris buffer (Tris HCl, 0.01 M, pH 8.1, EDTA, 0.01 M) and the cells were suspended in the same buffer to a concentration of 3 × 10^6 cells/ml. Then SDS, 0.1%, and Pronase 0.5 mg/ml, were added and the mixture was incubated at 36°C for 15 minutes. Digestion was arrested by addition of SDS, 2%, for 2 minutes. NaCl, 1 M, was then added and the material was left at 4°C for 16 hours (overnight). The next day the material was centrifuged at 20,000 g for 30 minutes, and the supernatant containing adenovirus DNA was collected. Thus, this procedure is also a modification of the Hirt method.

High molecular weight (about 100 × 10^6) DNA of herpesviruses needs special precautions during isolation, since a simple vigorous shaking, usually done during extraction of nucleic acids, causes ruptures of herpesvirus DNA molecules. One of the recently recommended procedures in dealing with these molecules is as follows (Pritchett et al., 1975). Virus harvested in tissue culture (HR-1 or B95-8 cells) is freed from cell debris (2500 g, 15 minutes), sedimented (60,000 g, 2 hours), resuspended in Tris buffer (0.025 M, pH 7.4), and homogenized by 50 strokes of a tight-fitted Dounce homogenizer. The debris is removed (4000 g, 10 minutes) and the virus is purified by equilibrium centrifugation in a sucrose density gradient (10–50% w/w).

The fractions that correspond to the density of the virions (1.22–1.25 gm/ml) are collected, diluted to the density lower than 1.2 gm/ml, and the virus is pelleted by centrifugation (100,000 g, 1 hour).

For extraction of DNA the pellet is resuspended in TNE buffer (Tris HCl, 0.05 M, pH 7.4, NaCl, 0.1 M, EDTA, 1 mM) at 4 °C, sodium lauryl sulfate (SLS) is added to 1%, and sarcosyl NL97 to 2%; these are active and soft detergents. The mixture is heated to 37 °C for 2 minutes, phenol and chloroform with isoamyl alcohol (2%) are added and DNA is separated from proteins by a gentle rotation of the tube. Further purification is carried out by centrifugation of carefully collected supernatant layered on a sucrose density gradient, 10–30% w/w, in a neutral buffer, taking into account that the sedimentation coefficient of herpesvirus DNA is about 56–58 S. Using an SW 41 (Spinco) rotor, centrifugation is carried out at 20 °C, 40,000 rpm for 4 hours.

2. Transfection of Experimental Animals

Tissue cultures, because of their sensitivity, are the most frequently used systems for transfection, though chick embryos and laboratory animals may be used for this purpose, if they display sensitivity to the given virus.

Thus, for revealing the infectiousness of nucleic acids of picornaviruses and arboviruses the method of intracerebral inoculation of mice has been used. This method appeared to be suitable for transfection by several viruses: Mengo (Colter et al., 1957a,b), mouse encephalomyelitis (Franklin et al., 1959), Theilers (Ada and Anderson, 1959b), encephalomyocarditis (Liebenow and Schmidt, 1959), foot-and-mouth disease (Strohmaier and Mussgay, 1959), West Nile encephalitis (Colter et al., 1957b), Semliki Forest (Cheng, 1958), dengue (Ada and Anderson, 1959b), and tick-borne encephalitis (Sokol et al., 1959). In the case of poliomyelitis the reproduction of the virus was successful after its intracerebral introduction (Colter et al., 1957b). Therefore, a system in sensitive to a virus appears to be sensitive to its RNA.

The technique of intracerebral infection of mice by nucleic acid preparations does not differ from the usual virologic technique described in the first volume of this series.

This method was used in particular by Zhdanov and Azadova (1975) for transfection of Sindbis virus, and by Zhdanov et al. (1977) for the separation of two proviruses—Sindbis and SV5, which were integrated into the same genome of L cells.

Chick embryos have also been used for the assay of infectivity of viral nucleic acids of Eastern equine encephalomyelitis virus (Wecker and Schäfer, 1957), Western equine encephalomyelitis virus (Wecker, 1959), and Murray Valley encephalitis virus (Anderson and Ada, 1959).

The technique of infection of chick embryos is commonly known, but in light of recent data special measures should be taken to stabilize nucleic acids. This subject will be discussed later.

3. Transfection of Tissue Cultures

In spite of successful experiments with assay of infectivity of viral nucleic acids by introduction into experimental animals (mice and chick embryos) the preferable systems are tissue cultures, as their use allows not only standardization of all procedures, but also permits quantitation of the characteristics of the transfection process (Cooper, 1961; Sarkar, 1967).

Standard procedures for the assay of infectivity of ribonucleic acids of poliovirus and Mengo virus are described in the second volume of this series (Sarkar, 1967). These methods may be reduced to the following: For Mengo virus mouse cells may be used, and for poliovirus human HeLa or monkey kidney cells may be used. Cells grown in monolayer cultures are washed with a buffer (0.02 M potassium phosphate buffer, pH 7.3, and 0.14 M NaCl), and RNA is dissolved in it. Then 0.1 ml aliquots of the RNA preparation are layered on petri dishes in monolayers. They are allowed to be adsorbed for 20 minutes at room temperature, and then are covered by an agar overlay.

Another variant of a standard procedure is the introduction of an RNA solution into a cell suspension and the subsequent titration for infectivity. For the assay of infectivity of encephalocarditis virus RNA, Krebs II ascites tumor cells are used. A suspension is prepared from these cells, and to this RNA is added. These are then dissolved in a phosphate buffer with sucrose. The mixture is allowed to stand for 10 minutes at room temperature, then the cells are placed in dishes with an indicator culture and covered with an agar overlay.

At present these methods are not used as originally conceived, since their sensitivity is not high. However, sensitivity may be considerably increased with the use of compounds having various mechanisms of action.

In order to increase the sensitivity (competence) of cells to transfection, osmotic shock (Koch et al., 1960) and heating of cells to 42 °C before layering of nucleic acids (Moscarello, 1965) were used as well as treatment of tissue cultures with various compounds: monovalent cations (Di Mayorca et al., 1959; Holland et al., 1959; Koch et al., 1960; Carp and Koprowski, 1962), divalent cations (Wecker et al., 1962; Graham and Van der Eb, 1973), polyanions and polycations (Vaheri and Pagano, 1965; Koch et al., 1966), dimethylsulfoxide (Amstey, 1966), etc. Farber et al. (1975) showed that the uptake of exogenous (nonviral) DNA by competent cells is increased in the presence of DEAE-dextran, $CaCl_2$, latex spheres, spermine, polylysine, and polyarginine. However, the greatest enhancement of uptake of DNA was observed in the presence of polyornithine, particularly when the latter and DNA were mixed prior to inoculation.

As has been stated, the efficiency of transfection by viral nucleic acids is sharply increased after treatment of tissue cultures with hypertonic concentrations of monovalent ions which change RNA conformation, protect RNA from degradation by nucleases, and improve the conditions for its penetration into the cells. According to Zhdanov and Yershov (1973) optimal results were obtained when cells were treated with 1 M NaCl for 10–15 minutes before the introduction of Venezuelan equine encephalitis (VEE) virus RNA (Table IV).

In contrast to RNA, treatment of tissue cultures with NaCl before infection with virus decreases the titers of the latter. This is shown in Table V (Zhdanov and Yershov, 1973).

The procedure used by Yershov and co-workers (Agabalyan et al., 1971) for the titration of infectivity of VEE RNA was as follows. Monolayers of chick embryo fibroblasts were treated with a 1 M solution of NaCl prepared in Tris buffer (0.01 M, pH 7.4) at room temperature for 10–15 minutes. Then a viral RNA solution in the same buffer or in nutrient medium with 1 M NaCl

TABLE IV

INFLUENCE OF VARIOUS CONCENTRATIONS OF NaCl AND OF TIME OF EXPOSURE ON INFECTIOUS TITERS OF VEE VIRUS RNA[a,b]

Concentration of NaCl	Time of exposure (minutes)				
	5	10	15	20	30
0.5 M	1.3 ± 0.2	1.7 ± 0.2	1.5 ± 0.2	1.6 ± 0.3	1.5 ± 0.3
1 M	2.5 ± 0.1	3.9 ± 0.3	3.5 ± 0.3	2.0 ± 0.1	1.2 ± 0.3
1.5 M	2.4 ± 0.1	2.1 ± 0.3	1.5 ± 0.1	0.8 ± 0.2	0[c]
2 M	1.2 ± 0.2	0.7 ± 0.3	0[c]	0[c]	0[c]

[a] RNA titers in log PFU/ml.
[b] From Zhdanov and Yershov (1973).
[c] Destruction of cells.

TABLE V

INFLUENCE OF 1 M NaCl ON INFECTION AND TRANSFECTION OF VEE VIRUS AND ITS RNA[a]

Material	No. of experiments	Infectious titers (PFU/ml)	
		with 1 M NaCl	without 1 M NaCl
VEE	1	6.2	9.9
virus	2	6.1	8.7
	3	6.5	9.0
VEE viral	1	4.1	0
RNA	2	3.8	0
	3	3.9	0

[a] From Zhdanov and Yershov (1973).

was layered for 5 minutes. Thereafter the fluid was removed and the cultures were covered with an agar overlay, which, besides the usual components, contained DEAE-dextran (2–4 μg/ml) or protamine sulfate (0.2–0.6 mg/ml). The use of DEAE-dextran for treatment of cells is described later.

Hypertonic solutions of NaCl were applied for the titration of infectious DNA of SV40 (Gerber, 1962). The method of titration was as follows. Culture of African green monkey kidney cells was washed with 0.6 M phosphate buffer (pH 7.4). DNA dissolved in a hypertonic buffer (0.9 M NaCl, 8% sucrose, 0.01 M Tris HCl, pH 8.2) was then layered, 0.2 ml per tube. After adsorption at room temperature for 25 minutes the inoculum was removed and replaced by minimal Eagle's medium with 1% calf serum. The titer of the virus was 10.3 log TCD_{50}/0.2 ml, and the titer of the DNA was 5.8 log TCD_{50}/0.2 ml.

According to Wecker et al. (1962), hypertonic solutions of $MgSO_4$ sharply increase the efficiency of the assay of infectious RNA. In experiments by these authors poliovirus RNA, after sedimentation from ethanol, was dissolved in 0.7–1 M $MgSO_4$ prepared in 0.01 M Tris buffer (ph 7.5). The culture of rhesus monkey kidney cells was grown in Eagle's medium with 10% calf serum. After removal of the medium 2 ml of saline solution in 0.01 M phosphate buffer (pH 7.2) and 0.02 M $MgSO_4$ layered and the culture was incubated at 37°C for 15 minutes in a CO_2 incubator. Then the cultures were again washed by the same buffer and 0.2 ml of viral RNA in 1 M solution of $MgSO_4$ was layered. After incubation for 20 minutes at room temperature the plates were covered with an agar overlay and placed in an incubator with 4% CO_2 for 3 days. Thereafter the cultures were stained with neutral red and plaques were counted.

In order to increase the efficiency of transfection nucleic acids and cells have recently been treated with solutions of $CaCl_2$. According to Graham and Van der Eb (1973), the treatment of nucleic acid preparations with $CaCl_2$ solutions increases their resistance to nucleases and facilitates their penetration into cells. Scolnick and Bumgarner (1975) have slightly modified this method. A culture sensitive to a given virus was grown in dishes with Delbecco–Vogt medium and 10% fetal calf serum. The grown cells were washed with the same medium without calf serum. Cellular DNA, which has been used as a source of mouse sarcoma provirus was sheared 5 times in a 1.0 ml plastic pipette. Then DNA (30 μg) was added to 0.5 ml of buffer, which contained 0.5 M HEPES (pH 7.05), 0.05 M KCl, 0.14 M NaCl, 0.25 mg/ml $Na_2HPO_4 \cdot 7H_2O$, and 0.1% w/v dextrose. To this $CaCl_2$ was added to a final concentration of 0.125 M and fine precipitate was allowed to form for a period of 20 minutes at room temperature. Thereafter 0.5 ml of the mixture was placed on the surface of the 60 mm plate and was incubated for 30 minutes at room temperature. Then 3 ml of Dulbecco medium with 10%

fetal calf serum was added and the plates were incubated at 37 °C for 18 hours. The solution was then removed and incubated for 7 more days, the subcultures being transplanted 3 times with 1 week intervals.

Even more frequently used than the previous method has been that of transfection with cells treated with DEAE-dextran (Vaheri and Pagano, 1965).

DEAE-dextran sensitizes cells for transfection (Al-Moslih and Dubos, 1973), but if complexes of DEAE-dextran and RNA are formed, then cells become less sensitive (Dubos, 1974). It has been shown that DEAE-dextran more effectively enhances transfection if it is mixed with a viral nucleic acid before inoculation of cells (mixed-inoculum method), than if cells are pretreated (pretreatment method) before inoculation with nucleic acids (Pagano and Vaheri 1965; Koch et al., 1966; Pagano et al., 1967; Tovell and Colter, 1967).

However, Al-Moslih and Dubos (1973) showed that these interrelations are more complex. In experiments with echovirus type-7 and poliovirus type-1 RNA it was shown that the pretreatment method under optimal conditions gives a yield of viruses 1 to 1.5 orders higher than the mixed-inoculum method. The kinetics of cell sensitization actually depend on the concentration of DEAE-dextran and temperature. At a temperature of 18 °C and a DEAE-dextran concentration of from 0.3 to 1 mg/ml the optimal time of pretreatment is 15–20 minutes. This optimal time can be shortened with an increase in the concentration of DEAE-dextran.

Two effective methods of transfection of primate cells, the bentonite and the dextran methods, have been compared (Table VI). Dubos (1975) showed that these two methods are reciprocal.

It was also shown (Saborio et al., 1975) that DEAE-dextran inhibits the synthesis of proteins that block several steps, in particular, the transport of amino acids, which apparently causes an increase of cellular competence to transfection by viral RNA and DNA proviruses.

TABLE VI

Efficiency of Transfection of Chimpanzee Kidney Cells by Poliovirus with Bentonine and DEAE-Dextran[a]

Transfection method	No. of plaques per plate (4 experiments)
None	0–0.5
Bentonine	58–174
DEAE-dextran	65–714
Combination	0.5–18

[a] From Dubos (1975).

Pagano (1969) describes the following technique of titration of infectious DNA of SV40. DNA dilutions, 0.3 ml, are poured into dry glass tubes. To each tube 0.3 ml of DEAE-dextran (double the concentration for a given system, varying from 0.1 to 1 mg/ml) is added. Serial dilutions of DNA in a buffer (e.g., 0.14 M PBS with 0.02% EDTA) are prepared immediately before the addition of DEAE-dextran. One should not mix DEAE-dextran with undiluted DNA, as the latter may form a precipitate. Pipetting also should be done gently to prevent shearing of DNA and precautions should be taken to prevent contamination of DNA with nucleases from the fingers and mouth.

Two plates with primary cultures of African green monkey kidney cells are then prepared for each dilution of DNA. The monolayers are washed with 5 ml of PBS without antibiotics. After 5–10 minutes of incubation at room temperature the washing fluid is carefully removed with a pipette and 0.2 ml of mixtures of DEAE-dextran and DNA dilutions are then introduced in each plate. These are incubated at room temperature for 15–20 minutes. The infected monolayers are then washed with 5 ml of balanced Hanks' solution, which contains antibiotics, and the cultures are covered with an agar overlay, as is done for plaque titration.

The agar monolayer medium, warmed to 43°C, contains 0.5% agarose, prepared in minimal Eagle's medium with 0.11% crystalline bovine albumin. After pouring the agar overlay on the cell monolayer the system is incubated at room temperature until the agar hardens. Initially 5 ml of the agar overlay is layered, and then 7 days later 3 ml more is layered. Finally at the 12th or 13th day 3 ml of agar with 0.01% neutral red is layered for counting plaques. The solution may be preserved up to the 20th day in incubation.

Excessive amounts of DEAE-dextran may be removed not only by washing the culture but also by neutralizing this polyanion. For this purpose heparin may be used. Heparin is not toxic in doses up to 1 mg/ml or more for tissue cultures. It may be added to a culture either as a pure solution directly after the end of the adsorption of DNA or may be introduced with a solution used for washing the culture. From a theoretical point of view this compound should be particularly valuable for titration of SV40 DNA as it is known that DEAE-dextran inhibits plaque formation by SV40.

It is essential that the inhibition of plaque formation be observed at a relatively low residual concentration of DEAE-dextran. However, in practice, in each focus of infection formed by viral DNA, such large amounts of the virus are synthesized that any increases in the amounts of plaques due to heparin plays no important role.

Of more importance is the fact that the use of heparin allows one to increase the dose of DEAE-dextran if necessary. It is also possible that polyanions promote healing of minor damages to the external cell membrane cause by polycations. This phenomenon may be of great importance for transfection of cells by viral DNA.

As has been mentioned, infectious DNAs isolated from adenoviruses of monkeys (Burnett and Harrington, 1968) and man (Nicolson and MacAllister, 1972). The method for the assay of infectivity of adenovirus DNA has been improved (Graham and Van der Eb, 1973). Some details of this technique are given below.

KB cells infected with human adenovirus type 5 (20 PFU/cell) were harvested by shaking and centrifuging. The pellet was resuspended in a small volume of 0.01 M Tris (pH 8.1) and sonicated. The cell debris was removed by centrifugation, and the supernatant was treated with pancreatic DNase and RNase (50 μg/ml) at 37°C for 30 minutes in the presence of 6 mM $MgCl_2$ then cooled on ice and extracted twice with genetron. The cell debris was removed by low-speed centrifugation, and the aqueous phase was then centrifuged over a CsCl cushion (with a density of 1.45 gm/ml) at 25,000 rpm (SW 25 or SW 27 rotors on a Spinco centrifuge) for 40 minutes. The virus band was collected and adjusted to a density of 1.34 gm/ml with solid CsCl and centrifuged at 35,000 rpm (SW 50.1 rotor) for 22–23 hours at 2°C or 17°C. The main band of the virus was dialyzed overnight against SSC × 1 buffer at 2°C.

Purified virions were treated with Pronase (50 μg/ml, 37°C, 30 minutes), then 2% SDS and 1 M sodium perchlorate were added and DNA was extracted with chloroform:isoamyl alcohol (24:1). After dialysis against SSC × 0.1, CsCl was added to a density of 1.715 gm/ml, and DNA was centrifuged at 72,000 g, 25°C, for 60 hours. The DNA band was collected and dialyzed against basal Eagle's medium with 0.01 M Tris (pH 7.4) (BME-Tris).

For transfection, a primary culture of human embryo cells or KB cells was used. Monolayers were washed with PBS, then a solution of DEAE-dextran (0.5 mg/ml in BMS-Tris) was layered and incubated for 30 minutes at room temperature. The monolayer was then washed with the medium and the DNA solution (1 to 20 μg) was layered. Adsorption was permitted at room temperature for 40 minutes and the plates or flasks were tipped to facilitate the distribution of DNA on the surface of the culture. Thereafter Eagle's medium with 5% calf serum was added and incubated at 37°C for 2 hours. The medium was then changed and the cultures were incubated until cytopathic changes appeared (up to 6–8 weeks, or 3–4 weeks with an agar overlay).

Cytochalasin B, which inhibits phagocytosis (Davis et al., 1971; Malawista et al., 1971) and active membrane transport (Estensen and Plageman, 1972; Kletzien et al., 1972), increased the formation of infectious centers in tissue cultures infected with poliovirus (Deitch et al., 1973). Therefore, the influence of cytochalasin upon the yield of virus has been studied in cells transfected by poliovirus RNA (Koch and Opperman, 1975).

RNA from virions was extracted by the phenol method, and HeLa cell tissue culture was used as the competent cells. The cells were treated with various concentrations of cytochalasin at 37 °C for 15 minutes, then RNA solution was layered and virus production was determined by the plaque method. The optimal concentration of the compound was 1–5 mg/ml (2 × 10^{-6} to $1 \times 10^{-5} M$), and the optimal time of treatment of the culture was 5–15 minutes before infection. Under these conditions the yield of the virus was 30- to 80-fold greater than that of a control.

Unlike DEAE-dextran, which in small doses increases adsorption and penetration of viral RNA, cytochalasin decreases both adsorption and penetration. An increase in the competence of cells treated with cytochalasin for viral RNA has been explained by the authors as a result of a temporary decrease of protein synthesis in the cells due to the decay of polyribosomes, Such an event initiation of virus-specific synthesis.

A short description follows of methods that were used for transfection of cells by cell DNA with integrated viral genomes. In experiments with DNA provirus of Rous sarcoma virus integrated in the genome of transformed cells the following technique has been used (Hillová et al., 1974a,b). After trypsinization of 2-day-old cultures of chick embryo fibroblasts, these cells were seeded in 250 ml flasks with 20 ml of a modified Eagle's medium containing 10% tryptose phosphate broth and 5% heated calf serum (Macpherson and Stoker, 1962). After 24 hours of incubation the medium was poured out and the cells were washed with DEAE-dextran (100 µg/ml) in 10 ml of the same buffer at 37 °C for 15 minutes. Then DEAE-dextran was removed and 2 ml of the DNA solution in the same buffer was introduced into the culture and incubated at 37 °C for 15 minutes. Thereafter the cells were covered with the previously mentioned modified Eagle's medium and the cultures were observed for 6 days. Reimplantation of one-third of the culture was made every 6 days. Usually foci of transformed cells appeared at 1 to 3 passages (6–18 days). If the foci did not appear after 5 passages (30 days) the results were considered negative.

In experiments by Zhdanov et al. (1974) on transfection of pig kidney cells by tick-borne encephalitis provirus DNA, a HEp2 cell line chronically infected with tick-borne encephalitis virus was used as the source of DNA provirus. DNA was extracted by the SDS-Pronase-phenol method described above, washed 3 to 5 times with ethanol, and redissolved in medium 199. For sterilization DNA was stored in 80% ethanol at −20 °C for 1 day or more. The pig kidney cell monolayers were washed with medium 199. A solution of DEAE-dextran in medium 199 was introduced into the culture at room temperature for 15 minutes and then DNA solution (5–80 μg in 0.2–0.5 ml) was introduced for another 15 minutes. Thereafter fresh medium 199 with 10% calf serum was added and the culture was incubated until cytopathic changes appeared (within 2–5 days). If the plaque method was applied, DEAE-dextran and DNA solution were removed after the end of adsorption and the culture was covered with an agar overlay and observed, as described earlier.

In transfection experiments with VSV and SV5 proviral DNA (Sovetova et al., 1975), two methods were compared: the DEAE-dextran and the calcium methods. The latter appeared to be more efficient.

In experiments on transfection of simian cells by DNA from SV40 transformed hamster, mouse, and monkey cells, none of which yielded the virus even after fusion, three methods of DNA extraction were used: the Hirt method (1967), the Marmur method (1961), and the method used by Graham et al. (1972). Transfection was carried out in the presence of DEAE-dextran, with cellular DNA in a concentration of 10 μg/culture of 10^6 cells. Special experiments were performed to prove that SV40 DNA was not freed, but was rather integrated in the cellular DNA. Successful transfection has been tested using the methods of plaque neutralization, immunofluorescence, and electron microscopy (Boyd and Butel 1972).

Transfection of BSC-1 cells by integrated DNA of simian adenovirus SA7 (Butel et al., 1975) has also been performed using the DEAE-dextran method. The DNA–protein complex was isolated from the transformed cells by a modified Hirt method (1967).

4. Transfection of Mitochondria

Transfection of mitochondria by viral RNA has been performed with the following method (Gaitskhoki et al., 1971b; Yershov et al., 1971a,b; Zhdanov et al., 1971). Rat liver mitochondria were isolated by a standard procedure of differential centrifugation. The animals were sacrificed, and the liver tissues were minced with scissors and homogenized in a Potter homogenizer in 5 volumes of 0.25 M sucrose prepared in TNE buffer (Tris HCl, 0.01 M, pH 7.4, NaCl, 0.01 M, EDTA, 10^{-3} M). The homogenate was centrifuged at 1000 g for 10 minutes to remove nuclei and cell debris. The supernatant was then collected and centrifuged at 10,000 g for 20 minutes. The pellet was then resuspended in the same buffer with 0.25 M sucrose. The pellet of mitochondria was immediately used for experiments. All procedures were performed at 2°–4°C.

RNA from purified virus preparations was extracted twice with phenol and a third time with chloroform:isoamyl alcohol (24:1), was precipitated by ethanol with 0.2% sodium acetate, and was stored at −20°C. Before the experiment RNA was centrifuged (15,000 g, 20 minutes), ethanol was removed, and RNA precipitate was dissolved in Tris buffer (0.01 M, pH 7.4).

Incubation medium for mitochondria contained sucrose, 0.25%, Tris HCl, 0.05 M, pH 7.8, KCl, 0.07 M, MgCl$_2$, 0.05 M, 2 mercaptoethanol, 0.006 M, phosphoenolpyruvate, 0.005 M, pyruvate kinase, 50 μg/ml, 20 amino acids, 10^{-5} M each, and 4 nucleoside triphosphates, 10^{-6} M each. Usually suspension of mitochondria (2 to 10 mg in 2 to 10 ml) was preincubated at 32 °C for 15–20 minutes to destroy endogenous messenger RNAs. Then viral RNA (1–10 μg) was introduced into the system and incubation was prolonged from 30 minutes to 2–4 hours. The system was either aerated or shaken during incubation.

Synthesis was stopped by rapid cooling, mitochondria was sedimented by centrifugation (10,000 g, 20 minutes), and both pellet and supernatant were used for titration of infectivity. The latter was performed by a routine plaque method with agar overlay, since the final products of transfection of mitochondria are ribonucleoprotein complexes.

V. Concluding Remarks

The use of transfection techniques has already played an important role in the development of our knowledge of viruses. First of all, the use of this method has clearly shown that the genetic material of viruses are nucleic acids and, in many viruses, an isolated viral nucleic acid can operate inside the cell, just as if it were the whole virus particle. Even for those viruses whose nucleic acid is noninfectious, the first step toward infectivity is also due to the activity of the viral genome transmitted by a viral genome-associated enzyme (DNA-dependent RNA polymerase, RNA-dependent RNA polymerase, or RNA-dependent DNA polymerase).

Second, the transfection technique has revealed the phenomenon of integration of the viral genome with the cellular genome, thus opening for study a new class of virus–cell interaction for bacterial viruses, for oncogenic animal viruses, and, finally, for infectious animal viruses.

Finally, transfection techniques are useful for understanding the steps of replication of viruses and of the interaction of viral genomes with cellular systems for the biosynthesis of nucleic acids and proteins. This latter field of research is far from fully explored, and one may foresee the use of transfection techniques as a tool for studies of the anatomy of cellular supramolecular organizations and as an instrument for intervention into these organizations.

REFERENCES

Abel, P., and Trautner, T. A. (1964). Z. Vererbungsl. **95**, 66–72.

Ada, G. L., and Anderson, S. G. (1959a). Nature (London) **183**, 799–800.

Ada, G. L., and Anderson, S. G. (1959b). Aust. J. Sci. **21**, 259–260.

Agabalyan, A. S., Menshikh, L. K., and Yershov, F. I. (1971). Vopr. Virusol. **5**, 527–531.

Albertson, P. A. (1967). In "Methods in Virology" (K. Maramorosch and H. Koprowski, eds.), Vol. 2, pp. 303–321. Academic Press, New York.

Al-Moslih, M. I., and Dubos, G. R. (1973). J. Gen. Virol. **18**, 189–193.

Amstey, M. S. (1966). Fed. Proc., Fed. Am. Soc. Exp. Biol. **25**, 492–493.

Anderson, S. G., and Ada, G. L. (1959). Virology **8**, 270–271.

Aoki, S., and Takebe, I. (1969). Virology **39**, 439–448.

Arlinghaus, R., Polatnick, J., and Vande Woude, G. (1966). Virology **30**, 541–548.

Bachrach, H. (1966). Proc. Soc. Exp. Biol. Med. **123**, 939–941.

Baltimore, D. (1971). Bacteriol. Rev. **35**, 235–241.

Baltz, R. H. (1971). J. Mol. Biol. **62**, 425–437.

Bancroft, J. B., and Flack, I. H. (1972). J. Gen. Virol. **15**, 245–251.

Beattie, K. L., and Setlow, J. K. (1969). J. Bacteriol. **100**, 1284–1288.

Bellett, A. J. D., Burness, A. T. P., and Sanders, F. K. (1962). Nature (London) **195**, 874–891.

Benzinger, R., Kleber, I., and Huskey, R. (1971). J. Virol. **7**, 646–650.

Benzinger, R., and MacCorquodale, D. J. (1975). J. Virol. **16**, 1–4.

Benzinger, R., Enquist, L. W., and Skalka, A. (1975). J. Virol. **15**, 861–871.

Bharava, P. M., and Shanmugam, G. (1971). Prog. Nucleic Acid Res. Mol. Biol. **11**, 103–192.

Boardman, N. K., Linnane, A. W., and Smillie, R. M. (1971). North-Holland Publ., Amsterdam.

Boiron, M., Thomas, M., and Chenaille, P. (1965). Virology **26**, 150 153.

Bol, J. F., Van Vloten-Doting, L., and Jaspars, E. M. J. (1971). Virology **46**, 73–85.

Boling, M. E., Setlow, J. K., and Allison, D. P. (1972). J. Mol. Biol. **63**, 335–348.

Boling, M. E., Allison, D. P., and Setlow, J. K. (1973). J. Virol. **11**, 585–591.

Borst, P. (1972). Annu. Rev. Biochem. **41**, 333–376.

Boyd, V. A. L., and Butel, J. S. (1972). J. Virol. **10**, 399–409.

Bragina, T. M., Zasukhina, G. D., and Poniklenko, A. A. (1964). In "Viral Neuroinfection," pp. 240–242. Moscow.

Breindl, M., and Koch, G. (1972). Virology **48**, 136–144.

Brody, E., Coleman, L., Mackal, R. P., Werninghaus, B., and Evans, E. A. (1964). J. Biol. Chem. **239**, 285–289.

Brown, F., Sellers, R. F., and Stewart, D. L. (1958). Nature (London) **182**, 535–536.

Bukrinskaya, A. G. (1973). Adv. Virus Res. **18**, 195–255.

Bukrinskaya, A. G., and Zhdanov, V. M. (1975). Springer Tracts in Mod. Phys. (in press).

Burnett, J. P., and Harrington, J. A. (1968). Nature (London) **229**, 1245–1246.

Butel, J. S. (1973). Methods Cancer Res. **8**, 287–335.

Butel, J. S., Talas, M., Cagur, J., and Melnick, J. L. (1975). Intervirology **5**, 43–56.

Carp, R. I., and Koprowski, H. (1962). Virology **16**, 371–383.

Chen, M. C. Y., Chang, K. S. S., and Salzman, N. P. (1975). J. Virol. **15**, 191–198.

Cheng, P. Y., (1958). Nature (London) **181**, 1800.

Chou, J. Y., and Martin, R. G. (1975). J. Virol. **15**, 145–150.

Colter, J. S., and Ellem, K. A. O. (1961). Annu. Rev. Microbiol. **15**, 219–244.

Colter, J. S., Bird, H. H., and Brown, R. A. (1957a). Nature (London) **179**, 359–360.

Colter, J. S., Bird, H. H., Mayer, A. W., and Brown, R. A. (1957b). *Virology* **4**, 522–532.

Cooper, P. D. (1961). *Adv. Virus Res.* **8**, 319–378.

Cosloy, S. D., and Oishi, M. (1973). *Mol. & Gen. Genet.* **124**, 1–10.

Davis, A. T., Estensen, R., and Quie, P. G. (1971). *Proc. Soc. Exp. Biol. Med.* **137**, 161–164.

Davis, J. E., Pfeifer, D., and Sinsheimer, R. L. (1964). *J. Mol. Biol.* **10**, 1–9.

Deitch, A. D., Sawacki, S. G., Goldman, G. C., and Tanenbaum, S. W. (1973). *Virology* **56**, 417–428.

Di Mayorca, G. A., Eddy, B. E., Stewart, S. E., Hunter, W. S., Friend, C., and Bendich, A. (1959). *Proc. Natl. Acad. Sci. U.S.A.* **45**, 1805–1808.

Dityatkin, S. Y., Lisovskaya, K. V., Panzhava, N. N., and Iliashenko, B. N. (1972). *Biochim. Biophys. Acta.* **281**, 319–323.

Dobos, P., and Faulkner, P. (1969). *Can. J. Microbiol.* **15**, 215–221.

Dubos, G. R. (1972). *Arch. Gesamte Virusforsch.* **39**, 13–25.

Dubos, G. R. (1974). *Acta Virol. (Engl. Ed.)* **18**, 457–466.

Dubos, G. R. (1975). *Arch. Virol.* **48**, 271–274.

Engelhardt, D. L., and Zinder, N. D. (1964). *Virology* **23**, 582–587.

Estensen, R. D., and Plageman, P. G. W. (1972). *Proc. Natl. Acad. Sci. U.S.A.* **69**, 1430–1434.

Farber, F. E., Melnick, J. L., and Butel, J. S. (1975). *Biochim. Biophys. Acta* **390**, 298–311.

Föglein, F. J., Kalpagam, C., Batos, D. C., Premecz, G., Nyitrai, A., and Farkas, G. L. (1975). *Virology* **67**, 74–79.

Földes, J., and Trautner, T. A. (1964). *Z. Vererbungsl.* **95**, 57–65.

Forman, M., and Simpson, R. W. (1975). In preparation.

Fox, M. S., and Hotchkiss, R. D. (1957). *Nature (London)* **179**, 1322–1323.

Fraenkel-Conrat, H. (1956). *J. Am. Chem. Soc.* **78**, 882–883.

Fraenkel-Conrat, H., Singer, B., and Tsugita, A. (1961). *Virology* **14**, 54–58.

Franklin, R. M., Wecker, E., and Henry, C. (1959). *Virology* **7**, 220–225.

Gaitskhoki, V. S., Yershov, F. I., Kiseliov, O. I., Menshikh, L. K., Zaitseva, O. V., Uryvayev, L. V., Zhdanov, V. M., and Neifakh, S. A. (1971a). *Dokl. Akad. Nauk SSSR* **201**, 220–223.

Gaitskhoki, V. I., Yershov, F. I., Uryvayev, L. V., Menshikh, L. K., Zhdanov, V. M., and Neifakh, S. A. (1971b). *Vopr. Virusol.* **3**, 269–273.

Gallimore, P. H., Sharp, P. A., and Sambrook, J. (1974). *J. Mol. Biol.* **89**, 49–72.

Gavrilov, V. I., Bogomolova, N. N., Deryabin, P. G., Astakhova, A. V., Andzhaparidze, O. G., and Zhdanov, V. M. (1974). *Arch. Gesamte Virusforsch.* **64**, 248–252.

Gavrilov, V. I., Zhdanov, V. M., Bogomolova, N. N., Deryabin, P. G., and Andzhaparidze, O. G. (1975). *Vopr. Virusol.* **4**, 406–412.

Gerber, P. (1962). *Virology* **16**, 96–98.

Gierer, A., and Schramm, G. (1956). *Nature (London)* **177**, 702–703.

Goldmark, P. J., and Linn, S. (1972). *J. Biol. Chem.* **247**, 1849–1960.

Golubev, D. B., and Polak, R. Y. (1965). *Vopr. Med. Khim.* **11**, 26.

Gonsalves, D., and Garnsey, S. M. (1975). *Virology* **64**, 23–31.

Goodgal, S. H., and Herriott, R. M. (1961). *J. Gen. Physiol.* **44**, 1201–1227.

Graham, B. J., Ludwig, H., Bronson, D. L., Benyesh-Melnick, M., and Biswal, N. (1972). *Biochim. Biophys. Acta* **259**, 13–23.

Graham, F. L., and Van der Eb, A. J. (1973). *Virology* **52**, 456–461.

Graham, F. L., Veldhuisen, G., and Wilkie, N. M. (1973). *Nature (London), New Biol.* **245**, 265–266.

Groneberg, J., Brown, D. T., and Doerfler, W. (1975). *Virology* **64**, 115–131.

Guthrie, G. D., and Sinsheimer, R. L. (1960). *J. Mol. Biol.* **2**, 297–307.

Guthrie, G. D., and Sinsheimer, R. L. (1963). *Biochim. Biophys. Acta* **72**, 290–297.

Harm, W., and Rupert, C. S. (1963). *Z. Vererbungsl.* **94**, 336–348.

Henner, W. D., Kleber, I., and Benzinger, R. (1973). *J. Virol.* **12**, 741–747.

Hershey, A. D., and Chase, M. (1952). *J. Gen. Physiol.* **36**, 39–56.

Hiatt, H. H. (1962). *J. Mol. Biol.* **5**, 217–229.

Hill, M., and Hillová, J. (1971). *C. R. Hebd. Seances Acad. Sci., Ser D* **272**, 3094–3097.

Hill, M., and Hillová, J. (1972a). *Nature (London), New Biol.* **237**, 35–39.

Hill, M., and Hillová, J. (1972b). *Virology* **49**, 309–313.

Hillová, J., Dantchev, D., Mariage, R., Plichon, M. -P., and Hill, M. (1974a). *Virology* **62**, 197–208.

Hillová, J., Goubin, G., Gouland, D., and Hill, M. (1974b). *J. Gen. Virol.* **23**, 237–245.

Hillová, J., Mariage, R., and Hill, M. (1975). *Virology* **67**, 292–296.

Hirt, B. (1967). *J. Mol. Biol.* **26**, 365–369.

Hlozanek, I., and Svoboda, J. (1972). *J. Gen. Virol.* **17**, 55–59.

Hofschneider, P. H. (1960). *Z. Naturforsch., Teil B* **15**, 441–444.

Holland, J. J., MacLaren, L. C., and Syverton, J. T. (1959). *J. Exp. Med.* **110**, 65–80.

Horvath, S. (1969). *Arch. Gesamte Virusforsch.* **28**, 325–336.

Hotchkiss, R. D., and Gabor, M. (1970). *Annu. Rev. Genet.* **4**, 193–224.

Hotz, G., and Mauser, R. (1969). *Mol. & Gen. Genet.* **104**, 178–194.

Huppert, J., and Sanders, F. K. (1958). *Nature (London)* **182**, 515–517.

Igarashi, A., Fukai, K., and Tuchinda, P. (1967). *Biken. J.* **10**, 195–201.

Ito, Y. (1960). *Virology* **12**, 596–601.

Jaspars, E. M. J. (1974). *Adv. Virus Res.* **19**, 37–149.

Javor, G. T., and Tomasz, A. (1968). *Proc. Nat. Acad. Sci. U.S.A.* **60**, 1216.

Kaiser, A. D., and Hogness, D. S. (1960). *J. Mol. Biol.* **2**, 392–415.

Kaiser, A. D., and Inman, R. B. (1965). *J. Mol. Biol.* **13**, 78–91.

Kaiser, A. D., and Wu, R. (1968). *Cold Spring Harbor Symp. Quant. Biol.* **33**, 729.

Kaiser, A. D., Tabor, H., and Tabor, C. W. (1963). *J. Mol. Biol.* **6**, 141–147.

Kassanis, B. (1967). *In* "Methods in Virology" (K. Maramorosch and H. Koprowski, eds.), Vol. 1, pp. 537–566. Academic Press, New York.

Kelly, R. K., and Butel, J. S. (1975). *Arch Virol.* **48**, 279–287.

Kleinhofs, A., Eden, F. C., Chilton, M. D., and Bendich, A. J. (1975). *Proc. Natl. Acad. Sci. U.S.A.* **72**, 2748–2752.

Kletzien, R. F., Perdue, J. F., and Springer, A. (1972). *J. Biol. Chem.* **247**, 2964–2966.

Koch, G. (1973). *Curr. Top. Microbiol. Immunol.* **62**, 69–138.

Koch, G., and Opperman, H. (1975). *Virology* **63**, 395–403.

Koch, G., Koening, S., and Alexander, H. E. (1960). *Virology* **10**, 329–343.

Koch, G., Quintrell, N., and Bishop, J. M. (1966). *Biochem. Biophys. Res. Commun.* **24**, 304–309.

Lacour, F., Fourcade, A., Merlin, E., and Huynh, T. (1972). *C. R. Hebd. Seances Acad. Sci.* **274**, 2253–2255.

Lane, L. C. (1974). *Adv. Virus Res.* **19**, 151–220.

Lane, L. C., and Kaesberg, P. (1971). *Nature (London) New Biol.* **232**, 40–43.

Ledoux, L. (1965). *Prog. Nucleic Acid Res. Mol. Biol.* **4**, 231–267.

Lerman, L. S., and Tolmach, L. J. (1957). *Biochim. Biophys. Acta* **26**, 68–82.

Liebenow, W., and Schmidt, D. (1959). *Acta Virol. (Engl. Ed.)* **3**, 168–171.

Linnane, A. W., Haslam, J. M., Lukins, H. B., and Nagley, P. (1972). *Annu. Rev. Microbiol.* **26**, 163–198.

Lot, H., Marchoux, G., Marrou, J., Kaper, J. M., West, C. K., Van Vloten-Doting, L., and Hull, R. (1974). *J. Gen. Virol.* **22**, 81–93.

Lwoff, A., and Gutman, C. (1950). *Ann. Inst. Pasteur, Paris* **78**, 711–720.

Maassab, H. F. (1959). *Proc. Natl. Acad. Sci U.S.A.* **45**, 877–881.

MacAllister, W. T. (1970). *J. Virol.* **5**, 194–198.

Mackal, R. P., Werninghaus, B., and Evans, E. A. (1964). *Proc. Natl. Acad. Sci. U.S.A.* **51**, 1172–1177.

Macpherson, I., and Stoker, M. (1962). *Virology* **16**, 147–151.

Maes, R., Sedwick, W., and Vaheri, A. (1967). *Biochim. Biophys. Acta* **134**, 269.

Mahler, H. R., and Fraser, D. (1959). *Virology* **8**, 401–424.

Malawista, S. E., Gee, J. B. L., and Bensch, K. G. (1971). *Yale J. Biol. Med.* **44**, 286–300.

Mandel, M. (1967). *Mol. & Gen. Genet.* **99**, 88–97.

Mandel, M., and Berg, A. (1968). *Proc. Natl. Acad. Sci. U.S.A.* **60**, 265.

Mandel, M., and Higa, A. (1970). *J. Mol. Biol.* **53**, 159–162.

Marmur, J. (1961). *J. Mol. Biol.* **3**, 208–218.

Menshikh, L. K., Rudneva, I. A., Kornilova, G. K., Sokolov, M. I., and Zhdanov, V. M. (1975). *Vopr. Virusol.* **6**, 722–725.

Meyer, F., Machal, R. P., Tao, M., and Evans, E. A. (1961). *J. Biol. Chem.* **236**, 1141–1143.

Miao, R., and Guild, W. R. (1970). *J. Bacteriol.* **101**, 361–364.

Mika, L. A. (1963). *J. Infect. Dis.* **113**, 135.

Mittelstaedt, R., Oppermann, H., and Koch, G. (1975). *Arch. Virol.* **47**, 381.

Montagnier, L., and Vigier, P. (1972). *C. R. Hebd Seances Acad. Sci.* **274**, 1977–1980.

Morrison, D. A., and Guild, W. R. (1973). *Biochim. Biophys. Acta* **299**, 545—556.

Moscarello, M. A. (1965). *Virology* **26**, 687–693.

Moskovets, S. N., Zhuk, I. P., Didenko, L. F., and Gorbarenko, N. I. (1971). *Vopr. Virusol.* **4**, 430–434.

Motoyashi, F., Bancroft, J. B., Watts, J. W., and Burgess, J. (1973). *J. Gen. Virol.* **20**, 177–193.

Nakamura, M. (1961). *Nature (London)* **191**, 624–625.

Naroditsky, B. S., Dreizin, R. S., Katsman, N. L., Tikhonenko, T. I., and Zhdanov, V. M. (1973). *In* "Molecular Biology of Viruses." pp. 98–107. Moscow.

Nicolson, M. O., and MacAllister, R. M. (1972). *Virology* **48**, 14–21.

Nielsen, G., and Marquardt, J. (1962). *Arch. Gesamte Virusforsch.* **12**, 335.

Nobroga, F. G., Pola, F. H., Paselto-Nobrega, M., and Oishi, M. (1972). *Proc. Natl. Acad. Sci. U.S.A.* **69**, 15–19.

Notani, N. K., and Setlow, J. K. (1974). *Prog. Nucleic Acid Res. Mol. Biol.* **14**, 39–100.

Notani, N. K., Setlow, J. K., and Allison, D. P. (1973). *J. Mol. Biol.* **75**, 581–599.

Oishi, M. (1969). *Proc. Natl. Acad. Sci. U.S.A.* **64**, 1292–1299.

Oishi, M., and Cosloy, S. D. (1972). *Biochem. Biophys. Res. Commun.* **49**, 1568–1572.

Okamoto, K., Mudd, J. A., Mangan, J., Huang, W. M., Subbaiah, T. V., and Marmur, J. (1968). *J. Mol. Biol.* **34**, 413–428.

Okubo, S., and Romig, W. R. (1965). *J. Mol. Biol.* **14**, 130–142.

Orlova, T. G., and Orlova, N. G. (1975). *Dokl. Akad. Nauk SSSR* **220**, 740–742.

Osowiecki, H., Nalecz, J., and Dobrazanski, W. T. (1969). *Mol. Gen. Genet.* **105**, 16.

Pagano, J. (1969). *In* "Fundamental Techniques in Virology" (K. Habel and N. P. Salzman, eds.), Vol. 1, Chapter 17. Academic Press, New York.

Pagano, J. S., and Vaheri, A. (1965). *Arch. Gesamte Virusforsch.* **17**, 456–464.

Pagano, J. S., MacCatchen, J. H., and Vaheri, A. (1967). *J. Virol.* **1**, 891–897.

Pakula, R., and Walczak, W. (1963). *J. Gen. Microbiol.* **31**, 125–133.

Paranchych, W. (1963). *Biochem. Biophys. Res. Commun.* **11**, 28–33.

Parsons, C. L., and Cole, R. M. (1973). *J. Bacterial.* **113**, 1505–1506.

Pilarski, L. M., and Egan, J. B. (1973). *J. Mol. Biol.* **76**, 257–266.

Portocala, R., Boeru, V., and Samuel, I. (1959). *Acta Virol. (Engl. Ed.)* **3**, 172–174.

Postel, E. H., and Goodgal, S. H. (1966). *J. Mol. Biol.* **16**, 317–327.

Pritchett, R. F., Hayward, S. D., and Kieff, E. D. (1975). *J. Virol.* **15**, 556–569.

Radding C. M., and Kaiser, A. D. (1963). *J. Mol. Biol.* **7**, 225–233.

Riva, S., and Polsinelli, M. (1968). *J. Virol.* **2**. 587–593.

Romig, W. R. (1962). *Virology* **16**, 452–459.

Rudchenko, O. N., Likhachova, N. A., Timakova, N. V., and Ilyashenko, B. N. (1973). *Dokl. Akad. Nauk SSSR* **211**, 224–225.

Rutberg, B., and Rutberg, L. (1971). *J. Virol.* **8**, 919–921.

Sabelnikov, A. G., Ditjatkin, S. J., and Ilyashenko, B. N. (1973). *Biochim. Biophys. Acta* **299**, 492–495.

Sabelnikov, A. G., Moiseeva, T. F., Avdeeva, A. V., and Ilyashenko, B. N. (1974). *Biochim. Biophys. Acta* **374**, 304–315.

Saborio, J. L., Wiegers, K. J., and Koch, G. (1975). *Arch. Virol.* **49**, 81–87.

Sarkar, S. (1967) *In* "Methods in Virology" Vol. 2, Chapter 18, pp. 607–644. Academic Press, New York.

Schatz, G., and Mason, T. L. (1974). *Annu. Rev. Biochem.* **43**, 51–87.

Scherrer, K., and Darnell, J. E. (1962). *Biochem. Biophys. Res. Commun.* **7**, 486.

Scolnick, E. M., and Bumgarner, S. J. (1975). *J. Virol.* **15**, 1293–1296.

Sethi, S. K., and Schwerdt, C. E. (1972). *Virology* **48**, 221–229.

Setlow, J. K., and Boling, M. E. (1972). *J. Mol. Biol.* **63**, 349–362.

Sharp, P. A., Pettersson, U., and Sambrook, J. (1974). *J. Mol. Biol.* **86**, 709–726.

Shatkin, A. J. (1974). *Annu. Rev. Biochem.* **43**, 643–665.

Sheldrick, P., Laitheir, M., Lando, D., and Ryhiner, M. L. (1973). *Proc. Natl. Acad. Sci. U.S.A.* **70**, 3621–3625.

Simpson, R. W., and Iinuma, M. (1975). *Proc. Natl. Acad. Sci. U.S.A.* **72**, 3230–3234.

Sjöström, J. E., Lindberg, M., and Philipson, L. (1972). *J. Bacteriol* **109**, 285–291.

Smorodintsev, A. A., Chalkina, O. M., Burov, S. A., and Ilyin, N. A. (1961). *J. Hyg. Epidemiol. Microbiol. Immunol.* **5**, 60–68.

Smull, C. E., and Ludwig, E. H. (1962). *J. Bacteriol* **84**, 1035–1040.

Sokol, F., Libikova, M., and Zemla, J. (1959). *Nature (London)* **184**, 1581.

Sovetova, G. M., Yakhno, M. A., and Zhdanov, V. M. (1976). In "Viruses of Cancer and Leukosis," pp. 155–157. Moscow.

Spatz. H. C., and Trautner, T. A. (1970). *Mol. & Gen. Genet.* **109**, 84–106.

Spatz, H. C., and Trautner, T. A. (1971). *Mol. & Gen. Genet.* **113**, 174–190.

Spencer, H. T., and Herriott, R. M. (1965). *J. Bacteriol.* **90**, 911–920.

Spizizen, J. (1957). *Proc. Natl. Acad. Sci. U.S.A.* **43**, 694–701.

Spizizen, J., Reilly, B. E., and Evans, A. H. (1966). *Annu. Rev. Microbiol.* **20**, 371–400.

Sprunt, K., Rodman, R. W., and Alexander, H. E. (1959). *Proc. Soc. Exp. Biol. Med.* **101**, 604–608.

Strack, H. B., and Kaiser, A. D. (1965). *J. Mol. Biol.* **12**, 36–49.

Strauss, J. H. (1964). *J. Mol. Biol.* **10**, 422.

Strauss, N. (1970). *J. Bacteriol.* **101**, 35–37.

Strohmaier, K., and Mussgay, M. (1959). *Z. Naturforsch., Teil B* **14**, 171–178.

Stuy, J. H., and Stern, D. (1964). *J. Gen. Microbiol.* **35**, 391–400.

Tokunaga, T., and Seller, M. I. (1964). *J. Exp. Med.* **119**, 139–149.

Tomasz, A. (1969). *Annu. Rev. Genet.* **3**, 217–232.

Tomasz, A., and Mosser, J. L. (1966). *Proc. Natl. Acad. Sci. U.S.A.* **55**, 58–62.

Tovell, D. R., and Colter, J. S. (1967). *Virology* **32**, 84–92.

Trautner, T. A., and Spatz, H. C. (1973). *Curr. Top. Microbiol. Immunol.* **62**, 61–88.

Vaheri, A., and Pagano, J. S. (1965). *Virology* **27**, 434–435.

Van der Vliet, P., and Sussenbach, J. S. (1975). *Virology* **67**, 415–426.

Van Vloten-Doting, L. (1975). *Virology* **65**, 215–225.

Vogt, M., and Dulbecco, R. (1960). *Proc. Natl. Acad. Sci. U.S.A.* **46**, 365.

Wackernagel, W. (1972). *Virology* **48**, 94–103.

Wackernagel, W. (1973). *Biochem. Biophys. Res. Commun.* **51**, 306–311.

Wackernagel, W., and Radding, C. M. (1973). *Virology* **52**, 425–432.

Wecker, E. (1959). *Virology* **7**, 241–243.

Wecker, E., and Richter A. (1962) Cold Spring Harbor Symp. **27**, 137.

Wecker, E., and Schäfer, W. (1957). *Z. Naturforsch. Teil. B.,* **12**, 415–417.

Wecker, E., Himmeler, K., and Goetz, O. (1962). *Virology* **17**, 110–117.

Weinstein, B. I., Mackal, R. P., Werninghaus, B., and Evans, E. A. (1969). *Proc. Natl. Acad. Sci. U.S.A.* **62**, 420–427.

Weinstein, B. I., Mackal, R. P., Werninghaus, B., and Evans, E. A. (1971). *Virology* **43**, 185–197.

Wright, M., Buttin, G., and Hurwitz, J. (1971). *J. Biol. Chem.* **246**, 6543–6555.

Yershov, F. I., Gaitskhoki, V. S., Kiseliov, O. I., Menshikh, L. K., Zaitseva, O. V., Uryvayev, L. V., Neifakh, S. A., and Zhdanov, V. M. (1971a). *Dokl. Akad. Nauk SSSR* **200**, 1452–1459.

Yershov, F. I., Gaitskhoki, V. S., Kiseliov, O. I., Zaitseva, O. V., Menshikh, L. K., Uryvayev, L. V., Neifakh, S. A., and Zhdanov, V. M. (1971b). *Vopr. Virusol.* **3**, 274–280.

Yershov, F. I., Gaitskhoki, V. S., Kiseliov, O. S., Golubkov, V. I., Zaitseva, O. V. Menshikh, L. K., Neifakh, S. A., and Zhdanov, V. M. (1974a). *Biokhimiya* **39**, 1009–1013.

Yershov, F. I., Gaitskhoki, V. S., Menshikh, L. K., Kiseliov, O. I., Zaitseva, O. V., Neifakh, S. A., and Zhdanov, V. M. (1974b). *Vestn. Akad. Med. Nauk SSSR* **3**, 21–27.

Young, E. T., and Sinsheimer, R. L. (1967). *J. Mol. Biol.* **30**, 147–164.

Young, F. E., Spizizen, J., and Crawford, I. P. (1963). *J. Biol. Chem.* **238**, 3119–3125.

Zaitlin, M., and Beachy, R. N. (1974). *Adv. Virus Res.* **19**, 1–35.

Zaitseva, O. V., Menshikh, L. K., Kiseliov, O. I., Gaitskhoki, V. S., Yershov, F. I., Uryvayev, L. V., Zhdanov, V. M., and Neifakh, S. A. (1973). *Dokl. Akad. Nauk SSSR* **208**, 985–987.

Zebovitz, E. and Brown, A. (1970). *J. Mol. Biol.* **50**, 185–196.

Zhdanov, V. M. (1972a). *Vestn. Akad. Med. Nauk SSSR* **2**, 86–91.

Zhdanov, V. M. (1972b). *Vopr. Virusol.* **3**, 277–283.

Zhdanov, V. M. (1975a). *Nature (London)* **256**, 471–473.

Zhdanov, V. M. (1975b). *Intervirology* **10**, 129–132.

Zhdanov, V. M., and Azadova, N. B. (1976). *Mol. Biol.* **10**, 1296–1302.

Zhdanov, V. M., and Parfanovich, M. I. (1974). *Arch. Gesamte Virusforsch.* **3**, 225–234.

Zhdanov, V. M., and Yershov, F. I. ed; (1973). "Meditsina." Moscow.

Zhdanov, V. M., Tikhonenko, T. I., Bocharov, A. F., and Naroditsky, B. S. (1971). *Dokl. Akad. Nauk SSAR* **199**, 944–947.

Zhdanov, V. M., Ketiladze, E. S., and Paktoris, E. A. (1973). *Vestn. Akad. Med. Nauk SSSR* **1**, 3–8.

Zhdanov, V. M., Bogomolova, N. N., Gavrilov, V. I., Andzhaparidze, O. G., Zmieva, R. G., and Astakhova, A. V. (1974). *Arch. Gesamte Virusforsch.* **45**, 215–224.

Zhdanov, V. M., Azadova, N. B., Yakhno, M. A., Zakstelskaya, L. Ya., Guschin, B. V., Gofman, Y. P., and Guschina, Y. A. (1977). *Mol. Biol.* **11**, 69–73.

Zinder, N. D., and Arndt, W. F. (1956). *Proc. Nat. Acad. Sci. U.S.A.* **42**, 586–590.

10 Invertebrate Cell Culture Methods for the Study of Invertebrate-Associated Animal Viruses

D. L. Knudson and S. M. Buckley

I. Introduction

This chapter builds upon the foundation that was provided by Vago (1967), who introduced the topic of invertebrate tissue culture to this series in Volume I. His contribution remains as an excellent source of information on invertebrate tissue culture methodology and on the use of invertebrate tissue culture in virology. The past 10 years have seen significant improvements in culture media formulations and methods, and they have also witnessed the development of additional cell lines which have proved useful to virologists. This chapter discusses many of the newer methodologies for invertebrate tissue culture, describes some of the techniques that are used in this laboratory, and outlines a few of the productive experimental approaches that have been employed to study virus replication in invertebrate tissue cultures. Several review articles and books are available which the reader is encouraged to consult. The articles of Vaughn (1968), Brooks and Kurtti (1971), and Hink (1972a), and the texts edited by Vago (1971, 1972), Weiss (1971), and Maramorosch (1976) are indispensable compendiums on invertebrate tissue culture methods and their application to virological problems. An earlier review by Day and Grace (1959) is also recommended because it recounts the early history of invertebrate tissue culture.

The tissue culture terminology that is used herein is consistent with the proposals of the Committee on Terminology of the Tissue Culture Association (Federoff, 1967). For example, invertebrate tissue culture pertains to the study of cells, tissues, and organs explanted from invertebrates and maintained or grown *in vitro* for more than 24 hours. Tissue or organ cultures are simply cultures containing explants of tissues and/or organs. A cell culture results when the cells are no longer organized into tissues or organs. These initial cell, tissue, or organ cultures that are derived from explanted materials are all referred to as primary cultures. When cells are removed from the primary culture vessel and transplanted into another, i.e., subcultured, the resultant proliferating cell population becomes a cell line. If it can be demonstrated that the cell line has the capacity for indefinite growth, it is denoted as an established cell line.

Over a hundred cell lines have been developed from approximately 57 invertebrate species. Table I represents a taxonomic listing of the different species from which the presently existing invertebrate cell lines have been derived (Hink, 1972b, 1976). Perhaps the most striking feature of this tabulation is the number of cell lines that have been derived from animal tissues of the class Insecta. The insect cell lines have also proved the most useful from a virological standpoint. For example, many of the arthropod-borne viruses (arboviruses) replicate in the dipteran cell lines; some of the viruses which infect plants and which are transmitted by insects replicate in the homopteran cell lines (See Chapter 11 this volume, by D. V. R. Reddy); and the lepidopteran cell lines are becoming increasingly important in the study of viruses which are pathogenic for insects.

Invertebrate-associated animal viruses include those viruses that infect and replicate in cells of invertebrates, and these viruses can be broadly differ-

TABLE I

TAXONOMIC LISTING OF SPECIES FROM WHICH EXISTING INVERTEBRATE
CELL LINES HAVE BEEN DERIVED[a]

Arachnoidea	*Agallia quadripunctata*
Acarina	*Agalliopsis novella*
Boophilus microplus	*Colladonus montanus*
Rhipicephalus appendiculatus	*Dalbulus maidis*
Rhipicephalus sanguineus	*Macrosteles fascifrons*
	Macrosteles sexnotatus
Insecta	*Nephotettix apicalis*
Diptera	*Nephotettix cincticeps*
Aedes aegypti	Lepidoptera
Aedes albopictus	*Antheraea eucalypti*
Aedes malayensis	*Antheraea pernyi*
Aedes novo-albopictus	*Bombyx mori*
Aedes pseudoscutellaris	*Choristoneura fumiferana*
Aedes taeniorhynchus	*Estigmene acrea*
Aedes vexans	*Heliothis zea*
Aedes vittatus[b]	*Laspeyresia pomonella*
Aedes w-albus	*Lymantria dispar*
Anopheles gambiae	*Malacosoma disstria*
Anopheles stephensi	*Mamestra brassicae*[c]
Armigeries subalbatus	*Manduca sexta*
Culex molestus	*Papilio xuthus*
Culex quinquefasciatus	*Samia cynthia*
Culex salinarius	*Spodoptera frugiperda*
Culex tarsalis	*Spodoptera littoralis*[d]
Culex tritaeniorhynchus summorosus	*Trichoplusia ni*
Culex tritaeniorhynchus	Orthoptera
Culiseta inornata	*Blabera fusca*
Drosophila immigrans	*Blattella germanica*
Drosophila melanogaster	*Leucophaea maderae*
Musca domestica	*Periplaneta americana*
Hemiptera	
Triatoma infestans	Mollusca
Homoptera	Gastropoda
Aceratagallia sanguinolenta	*Helix aspersa*
Agallia constricta	*Biomphalaria glabrata*

[a] Expanded from Hink, 1976.
[b] Bhat and Singh, 1970.
[c] H. G. Miltenburger, personal communication.
[d] D. L. Knudson and T. Lescott, personal communication.

entiated into two groups, using a criterion based on the pathologic effect
the virus has on invertebrate and vertebrate animals (Gibbs, 1973). One
group is characterized by invertebrate vectors for biological transmission,
by replication in invertebrates with little or no observable deleterious effect

to the whole organism, and by replication in vertebrates with varying degrees of pathologic effect. The arthropod-borne viruses (arboviruses), including certain Togaviridae, Bunyaviridae, Reoviridae, Rhabdoviridae, and other unclassified viruses, represent this group in invertebrate-associated animal viruses. The second group is characterized by replication in invertebrates with gross pathologic effect and by the absence of demonstrable replication in vertebrates. The pathogenic-invertebrate animal viruses, including Baculoviridae, cytoplasmic polyhedrosis viruses of Reoviridae, entomopoxviruses of Poxviridae, and iridoviruses or Iridoviridae, comprise this latter group of invertebrate-associated animal viruses. Nodamura virus (Picornaviridae) is an exception to these definitions. It has little or no effect in a dipteran vector, *Culex tritaeniorhynchus*, but it causes mortality in *Apis mellifera* (honey bees), *Galleria mellonella* (greater wax moth), and mice (Bailey *et al.*, 1975).

By way of introduction to invertebrate-associated animal viruses, Tables II–X represent a compilation of data on the host range of these animal viruses in invertebrate cell lines. It may be noted that some viruses are included in the tabulation even though an invertebrate association has not been necessarily demonstrated. The virus nomenclature is in accordance with the recent report of the International Committee on Taxonomy of Viruses (Fenner, 1976) and the serologic groupings follow those outlined in the "International Catalogue of Arboviruses Including Certain Other Viruses of Vertebrates" (Berge, 1975). The positive demonstrations of virus replication in invertebrate cell lines are noted in the table as well as the negative attempts to demonstrate virus replication. The negative results, of course, are problematic, but they are included for completeness. The references indicate either the original demonstration or a source from which the information can be obtained. These tables are included for the reader who is contemplating using an invertebrate cell line for the study of an invertebrate-associated animal virus, and they merely indicate "what has been tried and what has worked."

These features of the invertebrate-associated animal viruses are the fodder for many speculative discussions concerning the evolution of animal viruses and they may suggest an apparent pivotal role for invertebrates in virology. These topics are not addressed herein, but it is hoped that the description of methods for the use of invertebrate cell cultures for the study of animal viruses will provide a means by which a few of these virological questions may be elucidated. Invertebrate cell lines have different handling characteristics when compared to vertebrate cell lines; this contribution, however, also emphasizes areas of similarities where methodologies developed for the study of viruses in vertebrate cell lines should be adaptable to invertebrate cell systems. An acquaintance with vertebrate tissue culture methodologies, therefore, would be beneficial.

TABLE II
ARBOVIRUSES: HOST RANGE IN INVERTEBRATE CELL LINES

| | Cell line[a] | | | | | | | | | | | |
| | Diptera | | | | | | | | | | Lepidoptera | |
Togaviridae	A. aegypti	A. albopictus	A. vexans	A. vittatus	A. w-albus	A. stephensi	C. quinquefasciatus	C. tarsalis	C. inornata	D. melanogaster	A. eucalypti	Ref.[b]
Alphavirus												
Mosquito-borne												
Bebaru	−	0	0	0	0	0	0	0	0	0	0	(1)
Chikungunya	+	+	0	+	+	+	−	0	0	+	−	(2,3,4,5)
Eastern equine encephalitis	+	+	−	0	0	0	−	0	−	0	−	(5,6,7,8,9)
O'nyong-nyong	−	+	0	0	0	+	0	0	0	0	0	(10)
Ross River	0	+	0	0	0	0	0	0	0	0	0	(11)
Sagiyama	0	0	0	0	0	0	+	0	0	0	0	(8)
Semliki Forest	+	+	0	0	0	0	0	0	0	0	0	(12,7)
Sindbis	(+)	+	0	0	0	0	−	0	0	+	0	(2,5,8)
Venezuelan equine encephalitis	+	+	0	0	0	0	0	0	0	0	0	(7)
Western equine encephalitis	+	+	−	0	0	0	0	0	−	0	−	(5,9)
Whataroa	+	0	0	0	0	0	0	0	0	0	0	(3)

[a] + = Virus replication demonstrated; − = no virus replication demonstrated; 0 = not tested; and ( ) = conflicting or unconfirmed reports.
[b] (1) Rehaček (1968), (2) Singh and Paul (1968a,b), (3) Berge (1975), (4) Hannoun and Echalier (1971), (5) Yunker and Cory (1968), (6) Peleg (1968a,b), (7) Buckley (1969), (8) Hsu (1971), (9) Sweet and Unthank (1971), (10) Buckley (1971a,b), (11) Raghow *et al.* (1973b), (12) Peleg (1969).

TABLE III
ARBOVIRUSES: HOST RANGE IN INVERTEBRATE CELL LINES

| | Cell line[a] | | | | | | | | | | | | | | | |
| | Acarina | | Diptera | | | | | | | | | | | | Lepidoptera | |
Togaviridae	R. appendiculatus	A. aegypti	A. albopictus	A. malayensis	A. pseudoscutellaris	A. vexans	A. w-albus	A. stephensi	C. molestus	C. quinquefasciatus	C. tarsalis	C. inornata	C. tritaeniorhynchus	D. melanogaster	A. eucalypti	Ref.[b]
Flavivirus																
Mosquito-borne																
Dengue 1	0	(+)	+	0	0	0	0	0	0	0	0	0	0	−	0	(1,2)
Dengue 2	0	+	+	+	0	0	+	0	0	+	0	0	0	−	0	(1,2,3,4,5,6,7)
Dengue 3	0	+	+	0	0	0	0	0	0	0	0	0	0	−	0	(1,2)
Dengue 4	0	+	+	0	0	0	0	0	0	+	0	0	0	−	0	(1,2,6)
Edge Hill	0	−	0	0	0	0	0	0	0	0	0	0	0	0	0	(8)
Japanese encephalitis	0	(+)	+	+	+	+	+	+	+	+	0	+	+	−	+	(2,3,4,5,6,7,9,10,11,12)
Kokobera	0	−	0	0	0	0	0	0	0	0	0	0	0	0	0	(8)
Kunjin	0	+	+	0	0	0	0	0	0	0	0	0	0	0	0	(8,13)
Murray Valley encephalitis	0	+	0	0	0	0	0	0	0	+	0	0	0	0	0	(8,6)

Virus																Reference[b]
St. Louis encephalitis	0	+	+	0	0	0	+	0	0	0	0	+	0	−	+	(9,14,2,10)
West Nile	+	+	+	+	0	0	+	0	0	0	0	+	0	+	0	(3,4,5,6,2,15)
Yellow fever	+	+	+	0	0	0	+	0	0	0	0	0	0	−	+	(2,16,17,18)
Tick-borne																
Kadam	−	−	0	0	0	0	0	0	0	0	0	0	0	0	0	(19)
Kyasanur Forest Disease	−	−	0	0	0	0	0	0	0	0	0	+	0	0	0	(4)
Langat	−	+	0	0	0	0	0	0	0	0	0	0	0	0	−	(17,10,15)
Louping ill	0	+	0	0	0	0	0	0	0	0	0	0	0	0	0	(15)
Powassan	0	−	0	0	0	0	0	0	0	0	0	0	0	0	−	(10)
No arthropod vector demonstrated																
Cowbone Ridge	0	−	0	0	0	0	0	0	0	0	0	0	0	0	0	(17)
Modoc	0	−	0	0	0	0	0	0	0	0	0	0	0	0	−	(17,10)
Rio Bravo	0	0	0	0	0	0	0	0	0	0	0	0	0	0	−	(10)

[a] + = Virus replication demonstrated; − = no virus replication demonstrated; 0 = not tested; and ( ) = conflicting or unconfirmed reports.

[b] (1) S. M. Buckley (personal communication); (2) Hannoun and Echalier (1971); (3) Singh and Paul (1968a), (4) Singh and Paul (1968b), (5) Varma et al. (1974a), (6) Hsu (1971), (7) Singh (1971), (8) Reháček (1968), (9) Sweet and Unthank (1971), (10) Yunker and Cory (1968), (11) Kitamura (1974), (12) Hsu et al. (1975), (13) Davey et al. (1973), (14) Hink (1976), (15) Varma et al. (1974b), (16) Converse and Nagle (1967), (17) Buckley (1969), (18) Varma et al. (1976), (19) Mugo and Shope (1972).

TABLE IV
ARBOVIRUSES: HOST RANGE IN INVERTEBRATE CELL LINES

| | Cell line[a] | | | | | | | | | | Ref.[b] |
| | Diptera | | | | | | | | | Lepidoptera | |
Bunyaviridae	A. aegypti	A. albopictus	A. vexans	A. gambiae	A. stephensi	C. quinquefasciatus	C. tarsalis	C. inornata	D. melanogaster	A. eucalypti	
Bunyavirus											
Bunyamwera group											
Batai	−	+	0	0	0	0	0	0	0	0	(1)
Bunyamwera	0	0	0	0	0	+	+	0	•	+	(2, 3)
Cache Valley	+	+	−	−	+	0	+	−	0	+	(1, 3, 4)
Calovo	−	+	0	0	0	0	0	0	0	0	(5, 6)
Ilesha	0	0	0	0	0	0	0	0	−	0	(7)
Lokern	0	0	0	0	0	0	+	0	0	0	(3)
Main Drain	0	0	0	0	0	0	+	0	0	0	(3)

	1	2	3	4	5	6	7	
C group								
Apeu	0	–	0	0	0	0	0	(8)
Itaqui	0	0	0	0	0	0	–	(9)
Maritúba	0	+	0	0	0	0	–	(8, 9)
Murutucu	0	+	0	0	0	0	0	(8)
California group								
California encephalitis	+	+	–	0	+	–	+	(10,4,3,9)
Inkoo	+	+	0	0	0	0	0	(11)
Jamestown Canyon	0	0	0	0	+	0	0	(3)
Snowshoe Hare	0	0	0	0	0	0	+	(9)
Tahyna	+	+	+	+	+	0	+	(5,6,7,12)
Simbu group								
Nola	–	+	0	0	0	0	0	(11)
Sathuperi	–	+	0	0	0	0	0	(1)
Unassigned								
Kaeng Khoi	0	+	0	0	0	0	0	(11)

[a] + = Virus replication demonstrated; – = no virus replication demonstrated; 0 = not tested; and ( ) = conflicting or unconfirmed reports.

[b] (1) Singh (1972), (2) Hsu (1971), (3) Hink (1976), (4) Sweet and Unthank (1971), (5) Danielová (1973), (6) Marhoul (1973a,b), (7) Hannoun and Echalier (1971), (8) Yunker and Cory (1975), (9) Yunker and Cory (1968), (10) Whitney and Deibel (1971), (11) Berge (1975), (12) Malková and Marhoul (1973).

TABLE V

ARBOVIRUSES: HOST RANGE IN INVERTEBRATE CELL LINES

| | Cell line[a] | | | | | |
| | Diptera | | | | Lepidoptera | |
Bunyaviridae	A. aegypti	A.albopictus	C. tarsalis	D. melanogaster	A. eucalypti	Ref.[b]
Bunyavirus-like						
Minor antigenic groups						
Turlock	0	0	+	0	–	(1,2)
Phlebotomus fever group						
Pacui	–	–	0	0	0	(3)
Sandfly Fever (Naples)	–	–	0	0	–	(4)
Sandfly Fever (Sicilian)	–	–	0	–	–	(4,5,2)
Tick-borne						
Congo-Crimean hemorrhagic fever	–	–	0	0	0	(6)
Ganjam	–	+	0	0	0	(7)
Kaisodi	–	–	0	0	0	(6)
Silverwater	–	–	0	0	0	(6)

[a] + = Virus replication demonstrated; – = no virus replication demonstrated; 0 = not tested; and ( ) = conflicting or unconfirmed reports.
[b] (1) Hink (1976), (2) Yunker and Cory (1968), (3) Buckley (1976), (4) Buckley (1969), (5) Hannoun and Echalier (1971), (6) Buckley (1971c), (7) Singh (1972).

TABLE VI

ARBOVIRUSES: HOST RANGE IN INVERTEBRATE CELL LINES

| | Cell line[a] | | | |
| | Diptera | | Lepidoptera | |
Reoviridae	A. aegypti	A. albopictus	A. eucalypti	Ref.[b]
Orbivirus				
Minor antigenic groups				
African horse sickness	–	+	0	(1,2)
Irituia	+	+	0	(3)
Epizootic hemorrhagic disease	+	+	0	(4)
Kasba	0	+	0	(5)
Palyam	–	+	0	(3)
Vellore	0	+	0	(5)
Tick-borne				
Chenuda	+	+	–	(6,7)
Kemerovo	+	+	0	(6,8)
Lipovnik	+	+	0	(6)
Tribec	+	+	0	(6)
Ungrouped mosquito-associated				
Lebombo	–	+	0	(3)
Ungrouped tick-associated				
Colorado tick fever	–	+	–	(9,10)

[a] + = Virus replication demonstrated; – = no virus replication demonstrated; 0 = not tested; and ( ) = conflicting or unconfirmed reports.
[b] (1) Singh and Paul (1968b), (2) Mirchamsy et al. (1970), (3) Buckley (1972), (4) Willis and Campbell (1973), (5) Berge (1975), (6) Buckley (1971c), (7) Yunker and Cory (1968), (8) Libiková and Buckley (1971), (9) Yunker and Cory (1969), (10) Buckley (1969).

TABLE VII
ARBOVIRUSES: HOST RANGE IN INVERTEBRATE CELL LINES

	Cell line[a]				
	Acarina	Diptera		Lepidoptera	
Unclassified virus groups	A. appendiculatus	A. aegypti	A. albopictus	A. eucalypti	Ref.[b]
Tick-borne					
Hughes	0	−	−	0	(1)
Soldado	0	−	−	0	(1)
Bandia	0	−	−	0	(1)
Qalyub	0	−	−	0	(1)
Johnston Atoll	0	−	−	0	(1)
Quaranfil	+	−	−	−	(1,2,3)
Ungrouped mosquito-associated					
Arkonam	0	0	+	0	(4)
Minnal	0	0	+	0	(4)
Ungrouped tick-associated					
Dhori	0	−	−	0	(5)
Matucare	0	0	0	−	(3)
Nyamanini	0	0	0	−	(3)
Upolu	0	−	−	0	(6)
Wanowrie	0	0	−	0	(4)

[a] + = Virus replication demonstrated; − = no virus replication demonstrated; 0 = not tested; and, ( ) = conflicting or unconfirmed reports.
[b] (1) Buckley (1971c), (2) Varma et al. (1974b), (3) Yunker and Cory (1968), (4) Berge (1975), (5) Singh (1972), (6) Buckley (1969).

TABLE VIII

ARBOVIRUSES AND RELATED VIRUSES: HOST RANGE IN INVERTEBRATE CELL LINES

	Cell line[a]						
	Diptera					Lepidoptera	
Other families	A. aegypti	A. albopictus	A. w-albus	A. stephensi	D. melanogaster	A. eucalypti	Ref.[b]
Arenaviridae							
Junin	−	−	0	0	0	−	(1,2)
Lassa	−	−	0	0	0	0	(3)
Machupo	0	0	0	0	0	−	(4)
Tacaribe	0	0	0	0	0	−	(2)
Picornaviridae							
Nodamura	+	+	0	0	0	0	(5)
Rhabdoviridae							
Lyssavirus							
Mokola	−	+	0	0	0	0	(6)
Obodhiang	−	+	0	0	0	0	(6,7)
Kotonkan	−	+	0	0	0	0	(6,7)
Lagos bat	−	−	0	0	0	0	(6)
Rabies (CVS)	−	−	0	0	0	0	(6)
Vesiculovirus							
Chandipura	+	+	+	+	0	0	(8,9)
Cocal	+	+	0	0	0	0	(10)
Piry	+	+	0	0	0	0	(11)
Vesicular stomatitis (Indiana)	+	+	0	0	+	+	(12,13,14)
Vesicular stomatitis (New Jersey)	+	+	0	0	0	0	(12)
Sigmavirus group							
Sigma	0	0	0	0	+	0	(15)

[a] + = Virus replication demonstrated; − = no virus replication demonstrated; 0 = not tested; and ( ) = conflicting or unconfirmed reports.
[b] (1) Mettler and Buckley (1973), (2) Hann and Clarke (1971), (3) Buckley et al. (1970), (4) Yunker (1971), (5) Bailey et al. (1975), (6) Buckley and Tignor (1975), (7) Buckley (1973), (8) Singh and Paul (1968a), (9) Singh (1971), (10) Artsob and Spence (1974), (11) Singh (1972), (12) Buckley (1969), (13) Printz (1973), (14) Yang et al. (1969), (15) Ohanessian (1971).

TABLE IX

PATHOGENIC-INVERTEBRATE VIRUSES: HOST RANGE IN INVERTEBRATE CELL LINES

| | Cell line[a] | | | | | | | | | | | | | | |
| | Diptera | | | | Lepidoptera | | | | | | | | | | |
Baculovirus	A. albo-pictus	A. ste-phensi	A. tri-taeni-orhyn-chus	A. euca-lypti	B. mori	E. acrea	H. zea	L. pom-onella	M. dis-stria	M. sexta	L. dispar	S. frug-iperda	S. litto-ralis	T. ni	Ref.[b]
Baculovirus															
Granulosis virus (GV)															
Heliothis armigera GV	0	0	0	0	0	0	–	0	0	0	0	–	0	–	(1)
Spodoptera frugiperda GV	0	0	0	0	0	0	–	0	0	0	0	–	0	–	(1)
Trichoplusia ni GV	0	0	0	0	0	0	–	0	0	0	0	–	0	–	(1)
Nuclear polyhedrosis virus (NPV) multiple embedded type															
Antheraea pernyi NPV	0	0	0	0	0	0	0	0	0	0	(+)	0	0	0	(1)
Autographa californica NPV	–	–	0	0	0	+	(+)	–	0	+	(+)	+	0	+	(1,2,3,4)
Bombyx mori NPV	0	0	+	0	+	0	0	0	0	0	(+)	0	0	0	(1,5,6)
Choristoneura fumiferana NPV	0	0	0	0	0	0	0	0	+	0	0	0	0	0	(7)
Galleria mellonella NPV	0	0	0	0	0	0	0	0	0	0	(+)	0	0	+	(1,6)

Virus										Reference[a,b]
Heliothis armigera NPV	0	0	0	+	0	0	0	0	–	(1)
Lambdina fiscellaria somniaria NPV	0	0	0	0	+	0	0	0	0	(7)
Lymantria dispar NPV	0	0	0	–	0	+	–	0	–	(1)
Orgyia pseudotsugata NPV	0	0	0	–	0	+	–	0	–	(1)
Spodoptera exempta NPV	0	0	0	0	0	0	+	0	0	(4)
Spodoptera exigua NPV	0	0	0	0	0	0	+	0	0	(4)
Spodoptera frugiperda NPV	–	0	0	–	–	0	+	+	+	(1,8,9)
Spodoptera littoralis NPV	0	0	0	0	–	0	+	+	+	(8)
Trichoplusia ni NPV	–	0	0	–	+	(+)	+	0	+	(1,10)
Nuclear polyhedrosis virus (NPV) single embedded type										
Heliothis zea NPV	0	0	0	+	0	0	–	0	–	(11,12,1)
Trichoplusia ni NPV	0	0	0	–	0	0	–	0	+	(1)

[a] + = Virus replication demonstrated; – = no virus replication demonstrated; 0 = not tested; and ( ) = conflicting or unconfirmed reports.

[b] (1) R. H. Goodwin (personal communication), (2) Granados (1976), (3) Vail *et al.* (1973), (4) D. L. Knudson (personal communication), (5) Raghow and Grace (1974), (6) Hink (1976), (7) Sohi and Cunningham (1972), (8) Knudson (1975), (9) Goodwin *et al.* (1970), (10) Faulkner and Henderson (1972), (11) Ignoffo *et al.* (1971), (12) Goodwin *et al.* (1973).

337

TABLE X
PATHOGENIC-INVERTEBRATE VIRUSES: HOST RANGE IN INVERTEBRATE CELL LINES

| | Cell line[a] | | | | | | | | | | |
| | Diptera | | | | | Lepidoptera | | | | | |
Other families	A. aegypti	A. albo-pictus	A. ste-phensi	D. mela-nogaster	A. euca-lypti	B. mori	E. acrea	L. dispar	S. frugi-perda	T. ni	Ref.[b]
Reoviridae											
Cytoplasmic polyhedrosis viruses (CPV)											
Aedes sollicitans CPV	0	+	0	0	0	0	0	0	0	0	(1)
Malacosoma disstria CPV	+	0	0	0	0	0	0	0	0	0	(2)
Trichoplusia ni CPV	0	0	0	0	0	0	+	+	+	+	(1,3)
Poxviridae											
Entomopoxvirus											
Amsacta moorei	0	0	0	0	0	0	+	+	0	0	(1,4)
Iridoviridae											
Iridovirus											
Chilo iridescent (type 6)	+	0	+	0	+	0	0	+	0	+	(5,1,6)
Mosquito iridescent (type 3)	+	0	0	0	0	0	0	0	0	0	(7)
Sericesthis iridescent (type 2)	+	0	0	0	+	0	0	0	0	0	(5,4)
Tipula iridescent (type 1)	+	0	0	0	+	0	0	0	0	0	(4,8)
Picornaviridae											
Picorna-like											
Cricket paralysis virus	0	+	0	+	0	0	0	0	-	0	(9)

[a] + = Virus replication demonstrated; − = no virus replication demonstrated; 0 = not tested; and, ( ) = conflicting or unconfirmed reports.
[b] (1) Granados (1976), (2) Kawarabata and Hayashi (1972). (3) Granados et al. (1974), (4) Hink (1976), (5) Kelly and Tinsley (1974a,b), (6) McIntosh and Kimura (1974), (7) Webb et al. (1974), (8) Hukuhara and Hashimoto (1967), (9) P. D. Scotti and J. F. Longworth (personal communication).

II. Methods for Invertebrate Organ, Tissue, and Cell Cultures

A. ORGAN AND TISSUE CULTURES

Organ or tissue culture can be distinguished from cell culture by an organizational criterion. Organ or tissue culture refers to the *in vitro* maintenance or growth of an explant such that it retains its organ or tissue form and functions in a manner that permits its differentiation. Cell culture, however, denotes the *in vitro* maintenance or growth of cells that do not exhibit the organ or tissue organization (Federoff, 1967).

Although organ or tissue culture represents a valuable tool for the investigation of embryological, endocrinological, genetical, physiological, and virological questions, these topics are not discussed here. There exist several comprehensive reviews which may be consulted for a detailed discussion of invertebrate organ or tissue culture, organ or tissue culture methods (Vaughn, 1971), organ culture of insects (Demal and Leloup, 1972), and organ culture of invertebrates other than insects (Bayne, 1976; Gomot, 1972; Hansen, 1976).

B. PRIMARY EXPLANT CULTURES AND CELL LINES

Methods for primary explant cultures have been described in great detail elsewhere (Hink, 1972a; Marks, 1973; Schneider, 1973a; Weiss, 1971; Vago, 1967) and they are not exhaustively covered in this chapter. Although the many primary explant methods differ in specific details, they share similar procedural steps. These common features are outlined, with attention given to possible troublesome areas which may influence the success rate of primary culture establishment.

1. *Balanced Salt Solutions and Culture Media*

Saline-based, balanced salt solutions are used in primary explant culture protocols. A number of different formulations have been reported (Hink, 1972a), but the Rinaldini formulation (Table XI) (Rinaldini, 1954) is, frequently, the balanced salt solution of choice. Rinaldini's salt solution has been used for dipteran and lepidopteran primary explant cultures (Hink, 1972a; Schneider, 1973a).

Hink (1972a) has reviewed the salient features in the development of culture media for invertebrate cells and tissues. In general, the choice of media for the primary culture depends upon the animal used. For example, the Mitsuhashi and Maramorosch (1964) medium, the Schneider (1969) medium, the Varma and Pudney (1969) VP_{12} medium are media which have been used for dipteran primary cultures. Grace's basal medium (Grace, 1962)

TABLE XI
Rinaldini's Salt Solution[a]

Compound	Quantity
NaCl	800 mg
KCl	20 mg
$NaH_2PO_4 \cdot H_2O$	5 mg
$NaHCO_3$	100 mg
$Na_3C_6H_5O_7 \cdot 2H_2O$	67.6 mg
H_2O	100 ml

[a] Rinaldini (1954).

and its several modifications, including Yunker and co-workers' modification (Yunker *et al.*, 1967), Hink's (1970) TNM–FH medium, and Gardiner and Stockdale's (1975) BML/TC10 medium have been used for lepidopteran cultures. Orthopteran cultures have been maintained in Marks' (1973) modification (M20) of the Landureau (1966) medium, while the Chiu and Black (1967) medium has been successful for homopteran cultures.

The recently developed medium, BML/TC10 (Table XII) (Gardiner and Stockdale, 1975), which is a simplification of the Grace (1962) formulation, may represent an excellent culture medium for lepidopteran tissues. Primary cultures have been prepared and cell lines have been established from pupal ovaries of *Spodoptera littoralis* directly in BML/TC10 (D. L. Knudson and T. Lescott, unpublished data). Moreover, the majority of lepidopteran established cell lines has been adapted to the medium (D. L. Knudson, unpublished data; Gardiner and Stockdale, 1975).

2. Selection of Tissues

The choice of the invertebrate species from which primary cultures are attempted is invariably dictated by the individual research requirements of the investigator. Once the animal has been selected, a decision must be made as to which tissues will be the most amenable to cultivation. Since there are many factors that may influence the success rate of primary culture establishment, the astute investigator will consult the literature to determine which tissues have been successfully cultured. One comprehensive source of information on cultivatable invertebrate tissues is Hink's chronological list of cell lines (Hink, 1972b, 1976). Neonate mosquito larvae (K. R. P. Singh, and U. M. Bhat, personal communication) and *Drosophila* embryos (Schneider, 1973a), for example, represent excellent sources of tissue for dipteran primary cultures. Cockroach and grasshopper embryos

TABLE XII
BML/TC10 Medium[a]

Substance	Quantity (mg/liter)
Bactotryptose broth	2600
KCl	2870
$CaCl_2 \cdot 2H_2O$	1320
$MgCl_2 \cdot 6H_2O$	2280
$MgSO_4 \cdot 7H_2O$	2780
$NaH_2PO_4 \cdot 2H_2O$	1140
$NaHCO_3$	350
Glucose	1000
L–Cystine	22
L-Tryosine	50
L-Tryptophan	100
L-Alanine	225
L-Arginine	550
L-Aspartic acid	350
L-Asparaginine	350
L-Glutamic acid	600
L-Glutamine	600
Glycine	650
L-Histidine HCl	3380
L-Isoleucine	50
L-Leucine	75
L-Lysine HCl	625
L-Methionine	50
L-Proline	350
L-Phenylalanine	150
L-Serine	550
L-Threonine	175
L-Valine	100
Thiamine HCl	0.02
Calcium pantothenate	0.11
Pyridoxine HCl	0.02
Para-aminobenzoic acid	0.02
Folic acid	0.02
Niacin	0.02
Isoinosital	0.02
Cyanocobalamin	0.01
Riboflavin	0.02
Biotin	0.01
Fetal bovine serum	100 ml
H_2O	900 ml
pH	6.2
mOsm/kg	320

[a] From Gardiner and Stockdale (1975).

have also been employed for orthopteran primary cultures (Kurtti, 1976; Marks, 1973). Lepidopteran primary cultures have been initiated using a variety of tissues including embryos, pupal ovaries, adult ovaries, pupal hemocytes, and adult hemocytes (Hink, 1972a,b, 1976), but the pupal ovaries may represent the tissue of choice for primary culture establishment.

3. *Animal Preparation*

Concomitant with the culture of animal tissues and cells is the need to eliminate or significantly reduce the microbial flora of the animal. This may be accomplished in two ways: either animals that have been reared in aseptic conditions may be used (Meynadier, 1971), or animals must be sterilized via chemical treatment with disinfectants and/or via antibiotic inclusion in the salt solutions. Chemical and antibiotic treatments are more frequently employed to control the microbial population of the animal, since aseptic rearing of animals is not always feasible. Since it is difficult to eliminate contaminating bacteria and fungi completely, it is advisable to set up a number of primary cultures at a given time and thereby increase the statistical probability of isolating an uncontaminated primary culture.

The animal sterilization method is dependent upon the nature of the starting material. Mosquito eggs, for example, may be the simplest to treat because the egg chorion withstands harsh treatment with organic solvents such as acetone and/or 70% (v/v) ethanol, and disinfecting solutions such as White's solution (White, 1931; K. R. P. Singh, and U. M. Bhat, personal communication). Larval, pupal, and adult stages of lepidopterans have been surface-sterilized using a variety of different disinfectants, and these are reviewed by Hink (1972a).

4. *Tissue Dissection and Dissociation*

The tissue to be cultured from the prepared animal is dissected and/or teased from the animal under aseptic conditions. The excised tissue is transferred to a small volume of an appropriate balanced salt solution containing antibiotics. Any remaining extraneous tissue, such as fat body tissue, is removed, and the tissue is rinsed by transferring it to fresh salt solutions. It is finally placed into a salt solution containing an enzyme that facilitates tissue dissociation. Although trypsin (0.1–0.25% w/v) is the enzyme normally employed in dissociation (Hink, 1972a; Vago, 1967), other enzymes, such as collagenase (0.3% w/v), have been used (Goodwin, 1975). The tissue is minced with scissors and incubated at either ambient temperature, 27 °C or 37 °C (10–45 minutes), as the tissue dissociates. A hyaluronidase (0.02% w/v) treatment of 10 minutes has also been used after the first enzymic treatment (Goodwin, 1975).

The dissociation process is hastened by vigorous pipetting of the sample, using a Pasteur pipette. When the tissue has been sufficiently dissociated, the tissue suspension is centrifuged for 5 to 10 minutes at 500–1000 g. This step removes the enzyme from the suspension and it is advisable to repeat the centrifugation several times, washing the pellet with the balanced salt solution. The final tissue pellet is resuspended in the culture medium and an appropriate volume is dispensed into a culture vessel.

5. *Primary Culture Maintenance*

Cultures are usually incubated at 27°–28°C, the temperature optimum for many insect cultures (Hink, 1972b, 1976). They are microscopically examined, at least weekly, and the extent of the cell proliferation is noted. The cultures are "fed" at 10- to 20-day intervals by removing part of the medium and replacing it with fresh medium. The amount of replacement medium is determined empirically, but frequently one-half to four-fifths is replaced.

6. *Cell Lines*

In time, the primary explant culture may exhibit extensive cell proliferation until the entire surface of the culture vessel is covered by a confluent sheet of cells. A confluent culture will not survive for any extended period of time without making adjustments in either the temperature, or the culture medium, or both. One may attempt to "maintain" the culture, i.e., retain the confluent condition with little or no loss in cell viability, but this is not, in practice, always feasible. Cells from the confluent culture are usually subcultured; cells are transferred to another culture vessel with the addition of fresh medium. Once the primary explant culture has been subcultured, the resultant cell population is denoted as a cell line.

Many cell types may be frequently observed in these initial subculturings of the cell line, and some of the cell types may be lost with continued subculturings. A selection pressure is provided by the subculture process which invariably selects for the cell type(s) that has either a higher frequency of appearance in the cell population or that has successfully adapted to the growth conditions.

7. *Pitfalls*

There are several potentially troublesome steps that may greatly influence the primary culture establishment success rate. Perhaps one of the most important factors is the compatibility of the animal tissue with the balanced salt solution or culture medium. Since a number of media formulations are reported for invertebrate tissues (Hink, 1972b, 1976), a cautious approach, though tedious, is to attempt the primary cultures using a variety of different salt solutions and media.

Some invertebrate cell lines exhibit a marked sensitivity to amphotericin B, an antifungal agent, at concentrations routinely used for mammalian cell lines (Stanley and Vaughn, 1967). When the lepidopteran cell line derived from *Spodoptera frugiperda* pupal ovaries (J. L. Vaughn, unpublished data) was incubated in BML/TC10, containing 2.5 μg amphotericin B/ml, a 50% mortality was observed within 3 days (Knudson, 1975). It is not known whether amphotericin B sensitivity is a general phenomenon for all invertebrate cell lines. Therefore, either alternative antifungal agents should be considered when setting up primary cultures, or the time for the tissue's exposure to amphotericin B should be limited. The latter can be accomplished simply by feeding the culture with fresh medium which does not contain the antifungal agent.

Elimination of the resident microbial flora from the animal tissues can also be problematic in primary culture establishments. If the disinfectants, the antibiotics, and the increased number of primary culture setups still result in an unacceptably high contamination rate, it may be necessary to rear the animals under aseptic conditions (Meynadier, 1971).

The extent of tissue mincing and dissociation is not clearly noted (Section II,B,4). This is an area where empiricism becomes important. The simple answer is that variations in extent of mincing and in time of enzymic treatment must be tried. There are, however, guidelines in the literature, and these should be consulted.

Another important area that is frequently neglected in discussions of primary culture establishments is the amount of tissue to be added per culture vessel. The notion of a "critical mass" of tissue is an important one and many failures may be the result of starting with too little tissue. For example, several hundred mosquito eggs are surface-sterilized and the resultant hatch of several hundred larvae are used to seed one 3-oz. flask for mosquito primary cultures (K. R. P. Singh and U. M. Bhat, personal communication). Goodwin (1975) used 15–25 lepidopteran pupae to yield sufficient ovarian tissue to seed a 30 ml plastic tissue culture flask (Falcon Plastics).

The first subculture of the primary explant culture is also a critical step. The tendency is to attempt the subculture prematurely. Although the culture vessel may appear to contain a substantial number of cells, an early subculture can mean failure and the loss of the primary culture, as well as the potential cell line. Apparently only a percentage of the cells in the primary culture will establish themselves upon subculture. Therefore, the more cells in the primary culture, the greater the likelihood of obtaining a cell line.

C. ESTABLISHED CELL LINES AND CLONED CELL LINES

1. *Established Cell Lines*

The concept of an established cell line is artificial because the inherent criterion of indefinite growth is undefinable. Nevertheless, a cell line is often regarded as established when it has undergone numerous subcultivations. Table I represents a taxonomic listing of species from which existing invertebrate established cell lines have been derived. Although cell lines from tissues of 57 invertebrate species have been established, there have been more than 100 distinct establishments. There are, for example, more than 10 distinct derivations of a *Drosophila melanogaster* cell line.

In general, the invertebrate cell cultures have similar subcultivation characteristics and these features are described briefly. Many of the cultures are routinely subcultured at weekly intervals. If the cells attach to the surface of the culture vessel, the spent medium is decanted and a volume of fresh medium is added. A uniform cell suspension is obtained by removing the attached cells with a rubber scraper and vigorous pipetting. If floating cells predominate in the culture, it is subcultured by removing any attached cells and pelleting the spent medium-cell suspension at 500–1000 g for 10 minutes. The spent medium is decanted and a volume of fresh medium is added. A uniform cell suspension is obtained and a count of viable cells is made employing a dye exclusion test. The dye, either trypan blue or erythrosine B, is added to an aliquot of the cell suspension in a ratio of 1:2 (Phillips, 1973), and a viable cell count is made using a standard hemacytometer. The statistical accuracy of the cell count may be simply assessed using the graphic method of Cassell (1965). Culture vessels are seeded with cells to give 1–5×10^5 viable cells/ml depending upon the cell line being subcultured. Final cell densities, ranging from 10^6–10^7 cells/ml, are observed after a week's incubation of the cells at their optimal growth temperature (Hink, 1972a,b, 1976).

Invertebrate cell cultures are routinely grown as monolayers in glass or plastic flasks, but preliminary evidence suggests that the cells are amenable to spinning suspension culture methods. Spinner cultures offer distinct advantages over monolayer or stationary suspension cultures and they represent an economical means of growing large quantities of cells. The *Aedes albopictus* cell line was one of the first invertebrate cell lines to be grown in spinner culture using the Mitsuhashi and Maramorosch (1964) medium containing 10% fetal bovine serum (Stevens, 1970). Suspension or spinner flasks were seeded with 5×10^5 cells/ml and incubated at ambient temperature. After a week, yields of approximately 0.7 to 2.5×10^6 cells/ml were recorded and clumping of the cells was also noted in the older spinner cultures. *Spodoptera frugiperda* cell line was demonstrated to grow in spinner culture using BML/TC10 medium with no apparent adverse effects (Gardiner and Stockdale, 1975). *Trichoplusia ni* cell line also grew in suspension culture using Hink's (1970) TNM–FH medium, but the spinner culture did not reach similar cell densities compared with stationary cultures (Hink and Strauss, 1976a). This effect may be corrected by aeration of the medium and the authors felt that O_2 tension was an important factor in *T. ni* spinner cultures. The cells also grew in clumps in the spinner culture, but the addition of 0.1% methycellulose (50 cps) reduced the clumping (Hink and Strauss, 1976a). *Choristoneura fumiferana, Heliothis zea, Laspeyresia pomonella, Lymantria dispar, S. frugiperda,* and *S. littoralis* cell lines grow in spinner cultures (up to 1 liter in volume), with the magnetic stirring bar turning at 150 rpm, using BML/TC10 medium, with no apparent difficulties and with no additional supplements (D. L. Knudson, unpublished data). Initial cell seeding densities range from 2 to 5×10^5 viable cells/ml and reach cell densities comparable with those observed for stationary cultures after incubation at 28 °C for a week.

2. *Cloned Cell Lines*

When a single, isolated cell undergoes mitoses and gives rise to other cells, the resultant cell population is regarded as a clone. A cloned cell line is a line which has descended from a clone and it does not necessarily imply homogeneity in the cell population (Fedoroff, 1967). Invertebrate cell lines are usually comprised of heterogeneous cell types due to the manner in which the primary cultures are made. Cloning of a given cell type from either a cell line or an established cell line attempts to eliminate much of the observed heterogeneity. The cell line that has undergone the fewest subcultures from the primary culture represents the best source for cloning of cell types (Section II,B,6).

A number of methods for cell cloning of vertebrate cell lines have been described and these include the capillary method (Sanford, 1973; Sanford *et al.*, 1948), the microdrop method (Lwoff *et al.*, 1955; Macpherson, 1969, 1973a), the dilution method (Cooper, 1973; Ham, 1973; Paul, 1961), and the soft agar suspension culture method (Macpherson, 1973b). The majority of the invertebrate cell lines are not cloned, but recently, the successful cloning of several cell lines has been described. A variety of different methods were used, including the capillary method for an *Aedes aegypti* cell line (Suitor *et al.*, 1966) and the *T. ni* cell line (Volkman and Summers, 1975, 1976), the dilution method for the *Antheraea eucalypti* cell line (Grace, 1968) and the *T. ni* cell line (Brown and Faulkner, 1975; Hink and Strauss, 1976b), and an agar suspension culture method for the *T. ni* cell line (McIntosh and Rechtoris, 1974).

The agar suspension culture method as reported by McIntosh (1976) may be particularly promising because several invertebrate cell lines form colonies in the soft agar. The *T. ni*, *Agallia constricta*, and *A. albopictus* cell lines, for example, all produced colonies in the agar with a colony-forming efficiency of approximately 1% (McIntosh, 1976). Briefly, the method entails the use of two agar layers; a basal nutrient layer containing 1.0%(w/v) agar is made and overlayed with cell suspension–soft agar mixture (McIntosh and Rechtoris, 1974). The base layer is prepared by adding 7.0 ml of mixture containing equal volumes of 2.0% (w/v) Special agar-Noble in heart infusion broth (Difco Laboratories, Detroit, Michigan) and twice concentrated cell culture medium to a petri dish (60 × 15 mm). One ml of a cell suspension containing 10^4 to 10^5 viable vells is added to 2 ml of 0.5% (w/v) Special agar-Noble in cell culture media; the cell–agar suspension is mixed, and the base layer is quickly overlayed. The petri dish is sealed to prevent dessication and incubated at 30°C. The dish is microscopically examined for colonies after 10–14 days. The colonies are counted, picked, and grown up following conventional procedures.

The agar suspension culture method may prove useful in the cloning of cell types from other invertebrate cell lines, but there is one feature of the method that may be troublesome. Hink and Vail (1973), for example, reported that certain agar overlays were harmful for the *T. ni* cell line at concentrations normally used for vertebrate cell lines. It may be necessary to try several commercial products until one is found which is compatible with the cells. Alternatively, agarose or methyl cellulose could be used as a substitute for agar. Since the agar suspension culture method employs 0.33% (w/v) agar in the cell overlay, the effects of agar upon the cells observed by Hink and Vail (1973), however, may have been concentration dependent. The important advantage of using this method for cell cloning is, perhaps, that there is a greater probability of isolating the infrequent cell type. End point dilution methods, however, are inclined to yield cell types that appear in the cell population with the greatest frequency.

D. CHARACTERIZATION AND IDENTIFICATION OF CELL LINES

Characterization of cell lines implies a defining of their biological, physical, and biochemical properties. These parameters, in turn, allow assessments to be made of the homogeneity of the constituents of the cell culture. The resultant characterization data are important because they are amenable to comparative analyses, and it is by this process that means for identification and/or distinction between invertebrate cell lines are developed. Technology is the greatest limitation to the degree of characterization of a given cell line, and cell lines, therefore, are generally defined in terms that are now measurable.

The characteristics of a given cell line that are commonly recorded include culture and growth characteristics, morphology, karyology, and isoenzyme analyses. These parameters are not sufficient by themselves to identify a given cell line. Nevertheless, some distinctions can be made when these characteristics are used in concert. The usefulness of these parameters is particularly apparent when making comparisons of the characteristics of cell lines derived from invertebrates of different orders. Yet, a comparison of cell lines within an order frequently reveals that they possess similar characteristics, making distinctions between the cell lines difficult, if not impossible (Greene *et al.*, 1972). The use of immunological techniques has proved useful in distinguishing and identifying vertebrate cell lines (Stulberg, 1973), and the techniques appear applicable to invertebrate cell lines. The *Aedes albopictus* and *Antheraea eucalypti* cell lines, for example, are distinguishable by immunodiffusion tests (Greene and Charney, 1971; Greene *et al.*, 1972), but this is an interspecies comparison and there are insufficient data to determine whether immunological techniques will be a productive approach for intraspecies comparisons of invertebrate cell lines. At the present time, the use of morphology, karyology, isoenzyme analyses, and immunology appears the only avenue by which characterization of invertebrate cell lines can be pursued practically.

The morphological characteristics of invertebrate cell lines have been discussed extensively (Hink, 1972a; Vago, 1971) and they are not covered here. A few of the karyological characteristics of invertebrate cell lines have been described (Schneider, 1973b; Vago, 1971; Hink and Ellis, 1971; Greene *et al.*, 1972; Nichols *et al.*, 1971) and methods for the study of the karyology of insect cell lines, as well as the inherent problems, have been described and discussed (Schneider, 1973b; Nichols *et al.*, 1971). Greene and Charney (1971) and Greene and co-workers (1972) have demonstrated that analyses of glucose 6-phosphate dehydrogenase, lactic dehydrogenase, and malic dehydrogenase isoenzyme profiles by starch gel electrophoresis of dipteran cell line extracts, *Aedes aegypti* and *Aedes albopictus*, are distinct when compared with a lepidopteran cell line extract, *Antheraea eucalypti*. Further investigation of other enzymes and those of other invertebrate cell lines, therefore, may provide an additional system by which invertebrate cell lines may be characterized. Methods for enzyme assay have been described by Greene and Charney (1971), Gartler and Faber (1973), and Shannon and Macy (1973b). An immunological method of immunodiffusion for the identification of invertebrate cell lines has been described by Greene and Charney (1973), and this methodology distinguished the cell lines of *A. aegypti* and *A. albopictus* from the *A. eucalypti* cell line. Undoubtedly, other immunological techniques, as described by Stulberg (1973), which include cytotoxic antibody, hemagglutination, mixed agglutination, fluorescent antibody, and complement-fixation tests may be adjunct systems to aid in cell line characterization.

The characterization of cell lines by examining their culture and growth characteristics, morphology, enzyme profiles, and immunological properties, unfortunately, suffers from one major criticism. These techniques measure or monitor phenotypic properties that reflect an expression of the cell's genotype. Since it is the genotype of a cell in which one is ultimately interested, more direct techniques which elucidate the genetic constitution of a cell should be considered as criteria for a cell line's characterization. For the moment, the available techniques distinguish at the interspecies level and not at the intraspecies level.

E. CONTAMINATION OF CELL LINES

Hirumi (1976) has recently reviewed the literature on contaminants of invertebrate cell cultures, and his discussion represents an excellent source of information on the associated problems. Perhaps the simplest approach in dealing with cultures that are recognized as contaminated, regardless of whether the contaminant is an extrinsic or intrinsic one, is to destroy the culture. Before the culture is destroyed, it is advisable to identify the agent because its identification may provide an indication as to the best means of avoiding further instances. Once the contaminant has been identified and the source of the agent has been deduced, a "clean" culture from liquid nitrogen stocks should be revived (Section II,F).

1. *Extrinsic Contaminants*

Many extrinsic contaminants can be prevented by use of a stringent aseptic technique during the handling of the cell cultures. Laminar flow cabinets and hoods are particularly useful in this regard and they not only diminish the possibility of extrinsic contamination of the culture, but some hoods are also designed to protect the operator from exposure to possible contaminants in, or carried by, the culture. Designated cell culture rooms equipped with UV decontaminating lights and operated with a positive air pressure using submicron air filters often suffice and greatly reduce the risk of extrinsic contamination.

a. Bacteria and Fungi. Extrinsic bacterial of fungal contaminants that are isolated and identified from a culture are invariably indicative of the local environment, whether inanimate or animate (Armstrong, 1973). If antiseptic conditions are not available or contaminated inoculums are introduced into cultures, then the inclusion of antibiotics into the culture medium is necessary, keeping in mind the potential pitfalls concerning the use, or overuse, of antibiotics (Coriell, 1973), i.e., there is no adequate replacement for stringent aseptic technique. Invertebrate cell culture media have been supplemented with penicillin (100 Units/ml), streptomycin (100 μg/ml), kanamycin (100 μg/ml), and/or gentamicin (100 μg/ml), with no apparent deleterious effect upon the cells. As mentioned previously (Section II,B,7), lepidopteran cells exhibit a marked sensitivity to the antifungal agent, amphotericin B, and fungal contamination of the lepidopteran cell lines can be particularly distressing. The effect of amphotericin B upon the cells can be minimized by pretreating the suspect inoculum with the antifungal agent, adding the inoculum to the culture, incubating it for a short period, and diluting out the antifungal agent by washing the cells with fresh medium which does not contain amphotericin B.

b. Microsporidia. The experimental propagation of a microsporidian parasite in cell cultures derived from the tissues of forest tent caterpillars (*Malacosoma disstria*) has been reported (Kurtti and Brooks, 1971), but microsporidia are not, in general, conspicuous inhabitants of the inanimate environment of the laboratory. They do not, therefore, represent a serious threat as an extrinsic contaminant. If, on the other hand, considerable primary explant culture work is being done, then microsporidia do represent a potential contaminant because they are often observed as a part of the resident microbial flora of certain invertebrates, such as lepidopterans.

c. Mycoplasma. Methods for the detection, isolation, identification, elimination, and prevention of mycoplasma in cell cultures have been presented in detail by Brown and Officer (1968) in Volume IV, of this series and more recently by Barile (1973a,b) and Kenny (1973). These methods are for mycoplasma of vertebrate origin which are also the likely extrinsic contaminant for invertebrate cell cultures. Barile (1973a) records that 99% of mycoplasmas isolated from 1063 vertebrate cell cultures were of human, bovine, or swine origin. Hirumi (1976) has recently reported the isolation of mycoplasma from 33 of 74 cultures of 12 different invertebrate cell lines. Six of the mycoplasmal isolates were identified serologically as *Acholeplasma* species, implicating the bovine serum medium supplement as the source of the agent.

d. Viruses. If an extrinsic virus, one which is inadvertently introduced into the cell culture, causes a pronounced cytopathic effect (CPE), then the associated problems of its detection are trivial. Many of the arthropod-borne viruses, on the other hand, do not cause a marked CPE in dipteran cell lines (Buckley, 1976). Further, some arboviruses, such as Chikungunya, are capable of establishing persistently infected dipteran cell cultures (Section V) in which only a small percentage of cells are infected (Buckley *et al.*, 1975), increasing the difficulty of its detection.

The procedure of Cunningham and colleagues (1975) outlines a feasible approach for the detection of arboviruses as cell culture contaminants. These viruses are particularly amenable to this approach because suitable indicator systems, such as mice and vertebrate cell cultures which develop a CPE, are available. Briefly, the suspect cell line can be stressed by temperature, i.e., intermittent incubation at 4°C and 37°C for periods of 30 minutes to 1 hour, to see if a CPE becomes apparent. Examination of thin sections of cell pellets by electron microscopy may also reveal the viral contaminant. Clarified freeze-thawed cell extracts are inoculated into infant and adult mice, and the extract is plated on African green monkey kidney cell monolayers, such as VERO and LLC-MK$_2$, under agar overlays (Sections III,B,1 and 2). To determine the percentage of infected cells, the suspect cells are plated onto either VERO or LLC-MK$_2$ cells and overlayed with agar following the protocol for the infectious center assay (Section III,B,2). Alternatively, the cell extract material can be concentrated and assayed as described above. For example, the suspect cells are sonicated and clarified by centrifugation at 10,000 rpm for 30 minutes. The supernatant is 10 times concentrated by Aquacide II-A (Calbiochem) treatment for 14 hours at 4°C. The cell extract concentrate is layered on 15–35% (w/w) sucrose gradients and sedimented by rate zonal centrifugation for 3 hours at 27,000 rpm. Fractions are collected and assayed *in vitro* as described above. If an agent is isolated, it can be identified using immunological techniques (Section III,C). The need for adequate controls is inherent in this procedure, particularly the control of an uncontaminated culture of the cell line in question.

Cocultivation is another approach, that may aid in the detection of extrinsic virus contaminants. This technique also relies on the use of an indicator system. Briefly, cells from the suspect cell line are mixed with cells of the indicator system and the mixed culture is examined daily for CPE. This procedure is essentially a simplified version of the infectious center assay (Section III,B,2).

This general approach should also be applicable to animal viruses other than arboviruses. The keystone of the method is the indicator system, whether an *in vivo* or *in vitro* one. Otherwise, the detection of a viral contaminant can be extremely difficult.

e. Cell Contaminants in Cell Lines.   There are numerous accounts in the literature of mammalian cell lines that have become contaminated by extraneous cells (Fogh, 1973). Greene and Charney (1971) and Greene and co-workers (1972) have demonstrated by immunological, karyological, and enzyme techniques that presumed isolates of dipteran cell lines of *Aedes aegypti, Culiseta inornata,* and *Aedes vexans* appear to be of lepidopteran origin, specifically, *Antheraea eucalypti.* These reports emphasize

two important points: a stringent regime must be followed when subculturing cell lines to prevent mix-ups, and methods for identifying and distinguishing cell lines are necessary to aid in the recognition of such an event.
The former constraint is soluble, but the latter may be problematic for
reasons inherent in cell line characterization (Section II,D). It is likely that
interspecies contaminants can be detected, but intraspecies distinctions may
not be possible, given present technology.

2. *Intrinsic Contaminants*

These contaminants are not introduced into the culture, but they were
present in tissues originally cultured. For example, microsporidian contaminants have been encountered during *in vitro* cultivation of lepidopteran
hemocytes from *Malacosoma disstria* (Sohi, 1971) and during the cultivation
of larval tissues of *Choristoneura fumiferana* (Sohi, 1973).

Mycoplasmas indigenous to invertebrate tissues (Vago, 1970) have been
recorded and they could represent an intrinsic contaminant in a cell culture.
Since mycoplasmas which grow in invertebrate tissues may have different
media requirements, it is conceivable that the mycoplasma culture techniques designed for the isolation of vertebrate mycoplasmas may not be
applicable. Therefore, alternative broth formulations designed specifically
for invertebrate mycoplasmas (Giannotti and Vago, 1971) should perhaps
be included in any mycoplasma screening of invertebrate cell cultures.

Sanders (1973) has discussed the problems associated with sources and
methods of detection of intrinsic viruses, which he termed "passenger,
indigenous, and cryptic viruses." Again the need for an indicator system that
exhibits a CPE when inoculated with extracts from suspect cell lines is
emphasized. Therefore, the comments made previously concerning extrinsic viruses (Section II,E,1,d) are applicable here.

F. PRESERVATION, STORAGE, RECOVERY, AND TRANSPORT OF CELL LINES

Shannon and Macy (1973a) have outlined procedures for the freezing
of vertebrate cell lines, for their storage in liquid nitrogen, and for their
recovery from a frozen state with good cell viability. These methods are, in
general, directly applicable to invertebrate cell lines.

Dipteran and lepidopteran cell lines are generally amenable to storage
in liquid nitrogen using 10% (v/v) dimethyl sulfoxide as the cryoprotective
agent in appropriate culture medium (S. M. Buckley, and D. L. Knudson,
unpublished data). Cells in their logarithmic growth phase are harvested
(Section II,B), viable cell counts are made, and the cell suspension is adjusted to a concentration of 2–6 million viable cells/ml. Dimethyl sulfoxide

is added to the adjusted cell suspension, giving a concentration of 10% (v/v) of the cryoprotective agent. Once the dimethyl sulfoxide is added, the suspension is cooled in an ice-water bath and 1 ml aliquots are dispensed into either glass or plastic 2 ml ampoules. The ampoules can be cooled to the storage temperature using several different types of apparatus, but, regardless of the type of equipment used, it should produce a cooling rate of 1° to 3°C per minute over a temperature range of +10° to −30°C, after which a more rapid cooling rate can be used (Shannon and Macy, 1973a). There are several commercially available units that work very well with excellent cooling rate control. In the absence of such equipment, cells can be frozen by placing the ampoules in a cotton-lined box made of expanded polystyrene. The box is placed in a −65°C electric freezer for at least 16 hours or, conveniently, overnight. The ampoules can then be transferred to liquid nitrogen refrigerator (−196°C) or liquid nitrogen vapor storage container (−170° to −180°C) (Shannon and Macy, 1973a).

To revive the frozen cells, an ampoule is thawed quickly at room temperature and diluted 1 to 10 in fresh medium. The cells are removed from the cryoprotective agent by centrifugation at 1000 rpm for 10 minutes. The pellet is resuspended in fresh medium and the cell suspension is dispensed into a culture vessel.

Cell cultures can be sent through the postal service, but a recent postal guide should be consulted for the proper packaging and necessary permits. It is often advisable to send two culture vessels, one with the cells near confluency and the other semiconfluent. The culture vessels are also completely filled with growth medium and the cap is well secured. Hayflick (1973) has discussed the problems concerning the transport and shipping requirements of cell cultures.

III. Methods for the Study of Animal Viruses

Methods for the study of arboviruses in invertebrate cell cultures will be compared with those for the study of pathogenic-invertebrate viruses throughout this section. There are fewer data on the pathogenic-invertebrate viruses than on arboviruses, for which extensive data have been collated in the "International Catalogue of Arboviruses" (Berge, 1975). The early development of reliable *in vivo* and *in vitro* infectivity assay systems for arboviruses and an anthropocentric attitude toward the invertebrate-associated arboviruses may account for the wealth of information that has accumulated on these viruses. This has led to the development of fairly standardized protocols facilitating comparative analyses among members of this group. In contrast to the field of arbovirology, the study of pathogenic-invertebrate viruses is in an early stage of development. In recent years, conscious attempts have been made to standardize *in vivo* assays of virus infectivity. Likewise, considerable success has recently been achieved in *in vitro* infectivity assays of pathogenic-invertebrate viruses, much of which is undoubtedly due to the increased research activity in the development of invertebrate cell cultures and associated methodologies. Since the field of pathogenic-invertebrate viruses is comparatively young, many of the techniques and methods for the study of this group of viruses are currently being developed. A comparison of methodologies presently used in studies on these two groups of invertebrate-associated animal viruses, therefore, may provide some insight into new, experimental approaches.

A. VIRUS STOCKS

1. *Isolation*

Guidelines establishing an agent as an arbovirus are delineated in the "International Catalogue of Arboviruses" (Berge, 1975). Perhaps the most important criterion is that the virus must be "maintained in nature through biological arthropod transmission." Only 42% of the 359 registered viruses listed in the catalogue are considered as "arbovirus or probable arbovirus," 7% are classified as "probably not or not arbovirus," with the remaining majority of the viruses denoted as "possible arbovirus" (Berge, 1975). These viruses have been isolated *in vivo*, most commonly in the field laboratories, by intracerebral (i.c.) or intraperitoneal (i.p.) inoculation of white, Swiss mice, either newborn or young adult. There are instances where vertebrate cell cultures have been used in primary isolation of arboviruses. *Aedes albopictus* cell cultures have recently been employed for primary isolation of dengue viruses (Casals and Buckley, 1973; Singh and Paul, 1969) and also the *Aedes pseudoscutellaris* cell line (Varma *et al.*, 1974a). Primary cultures of a tick, *Hyalomma dromedarii*, have been used in isolation of tick-borne encephalitis virus from a natural focus of the virus infection (Rehaček and Kozuch, 1969). Arbovirus isolates in general consist of infected mouse brain suspensions or fluids from infected vertebrate or invertebrate cell cultures.

Generally, pathogenic-invertebrate viruses have been isolated from invertebrate populations displaying a disease state or condition. The prognosis of the afflicted invertebrate is invariably poor, and as a result invertebrate populations in nature often exhibit high mortality rates. Since viruses of this group produce a dramatic effect upon the infected host, i.e., death, the accordant problems associated with virus detection in invertebrate populations are somewhat simplified. The isolation procedures for pathogenic-invertebrate viruses are conceptually analogous to those used for arbovirus isolations in that a susceptible host is required. The host of choice is that in which the virus disease was initially observed. Unfortunately, the use of the primary host is not always feasible because difficulties may be encountered in maintaining certain invertebrate colonies under laboratory conditions (Smith, 1966) and certain invertebrates are not continuously available due to their seasonal cycle in nature. It is advantageous to use a susceptible host that is readily colonized under laboratory conditions and, preferably, one that has been under observation in the laboratory over a number of generations, thus, insuring that the hosts are "healthy" (free of virus). The introduction of field-collected invertebrates into the laboratory is not a practice to be recommended because it potentiates the probability of introducing other undetected viral agents into the healthy invertebrate colony and/or eventually into the virus stock.

Since ingestion is a common pathway of pathogenic-invertebrate virus infections in invertebrates, virus stocks are produced by mixing or applying the inoculum, which consists of a triturated field sample in a small volume of H_2O or balanced salt solution, to the invertebrate's diet or foodstuff. Alternatively, the inoculum can be injected into the invertebrate. The volume of inoculum and the site of its injection obviously depends upon the species and the age of the invertebrate being inoculated. After an appropriate incubation period, the virus from infected invertebrates may be harvested (see Section III,A,2) or the intact invertebrate may be stored frozen.

Invertebrate cell lines have been inoculated with infected, insect-derived virus. There are several sources of inocula and these include tissue extracts, invertebrate hemolymph, purified virus, and viral DNA (Bellet and Mercer, 1964; Granados and Naughton, 1975; Kelly and Tinsley, 1974a,b; Knudson and Tinsley, 1974; Goodwin *et al.*, 1970; Granados *et al.*, 1974; Ignoffo *et al.*, 1971; Vail *et al.*, 1973; Webb *et al.*, 1974). Granados (1976) has recently reviewed the methods that have been employed to infect invertebrate cell lines and his discussion of the topic is recommended reading. Isolates of pathogenic-invertebrate viruses usually consist of invertebrate-tissue extracts, purified virus, or infected cell culture fluids.

2. Purification

Procedures for producing arbovirus-infected mouse brain suspensions have been described in detail by Casals (1967) in Volume III of this series. Arbovirus-infected cell culture fluids are prepared by lysing the cells by sonication or cycles of freezing and thawing and clarifying the suspension by centrifugation at 5000–10,000 g for 15–20 minutes. The supernatant fluid represents the virus. Further virus purification by two-phase separation, differential centrifugation, and density-gradient centrifugation may be achieved following methodologies described in Volume II of this series (Maramorosch and Koprowski, 1967). Japanese encephalitis virus grown in baby hamster kidney cell line (BHK-21) and *A. albopictus* cell line, for example, has been purified from clarified, infected cell culture fluids by polyethylene glycol 6000 precipitation and density-gradient centrifugation (Igarashi *et al.*, 1973b). Infected cell cultures are clarified by centrifugation for 15 minutes at 8500 *g*. The supernatant is adjusted to 0.5 *M* NaCl and 8% (w/v) polyethylene glycol 6000, and it is allowed to stand overnight at 4 °C. The suspension is centrifuged for 15 minutes at 8500 *g* and the supernatant is decanted. The pellet is resuspended in 1/20 of the original supernatant volume in STE(0.1 *M* NaCl, 0.01 *M* Tris-HCl, and 0.001 *M* EDTA, pH 7.6) and the suspension is centrifuged for 10 minutes at 700 *g*. The supernatant can then be analyzed by density-gradient centrifugation. The analogous procedure of McSharry and Benzinger (1970) has been used in this laboratory for the concentration of several rhabdoviruses from infected cell culture fluids (D. H. Clarke, personal communication). The fluid is adjusted to 0.5 *M* NaCl and 6% (w/v) polyethylene glycol 6000, and treated similarly as described above. Lower centrifugation *g* forces, however, have been used with good recoveries of infectivity and hemagglutinating activity of the virus.

The pathogenic-invertebrate viruses are also amenable to similar purification protocols. In general, the viruses have been purified from infected invertebrate tissues by differential and density-gradient centrifugation. Recently, baculoviruses have been similarly purified from infected cell cultures (Knudson and Tinsley, 1974; Summers and Volkman, 1976; Henderson *et al.*, 1974).

3. Storage

Ward (1968) has discussed methods for storage and preservation of animal viruses in Volume IV of this series and these methods will not be reiterated here.

4. Viral Antigen Preparation from Cell Cultures

Viral antigens for use in complement-fixation and hemagglutination tests have been produced in invertebrate cell lines. Complement-fixing antigens have been detected in *A. albopictus* cell cultures infected with dengue viruses, types 1–4 (Pavri and Gosh, 1969), and also with West Nile virus (Ajello *et al.*, 1975). In the latter method, monolayer cell cultures of *A. albopictus* growing in the Mitsuhashi–Maramorosch medium were inoculated with serial dilutions of West Nile virus (strain Egypt 101, 11th mouse brain passage with a titer of 10^{10} PFU/ml of a 10% infant mouse brain suspension) in VERO cell cultures using 0.75% bovine albumin in 0.05 *M* phosphate-buffered saline, pH 7.2, as the diluent. The cultures were incubated at 30 °C and the fluids were tested at 4-, 7-, and 11-day intervals postinoculation (p.i.) for the presence of complement-fixing antigen employing the complement-fixation test described by Casals (1967) in Volume III of this series (Section C,2). The culture fluids were used as the antigen source without further manipulation such as freezing and thawing cycles or clarification by centrifugation. Complement-fixation titers varying from 1:32 to 1:128, depending upon the strength of the antiserum, may be expected 7 and 11 days p.i. in cell cultures inoculated with dilutions up to 10^{-8} of the virus. This complement-fixing antigen may be stored in sealed ampoules at −65 °C and the antigen is specific.

Hemagglutinating antigen of eastern equine encephalomyelitis virus (2nd mouse brain passage) has been produced in 8–10-day-old primary cultures of *A. aegypti* embryos (Peleg, 1968a). Cultures were inoculated with 10^3 $LD_{50}/0.03$ ml of the virus and incubated. Culture fluids were removed from two to three cultures at intervals and they were pooled. An aliquot was titrated in mice (Section III,B,1,a) and the remainder was extracted with an equal volume of Genetron 133 (trifluorotrichloroethane). This mixture was shaken for 5 minutes and the phases were separated by centrifugation for 5 minutes at 3000 rpm. The upper phase was removed, and it was assayed for the presence of hemagglutinating antigen (Casals, 1967; Clarke and Casals, 1958; Section III,C,3). Provided the virus inoculum had a titer of at least 6×10^6 $LD_{50}/0.03$ ml, hemagglutinating antigen titers ranging from 1/8 to 1/128 had been detected. Although this procedure was described for primary cultures, it should be directly applicable to established cell lines. Recent work suggests that some flaviviruses such as Kunjin do not yield hemagglutinin when grown in *A. albopictus* even though high titers (10^8 PFU/ml) of the virus are recovered (E. G. Westaway, personal communication). Thus, the use of invertebrate cell lines for the production of hemagglutinin for all arboviruses may not be feasible and vertebrate cell lines may be the practical alternative system.

B. Infectivity Assays

The quantitative assay of the pathobiological activity of a virus preparation relies on the need of a system in which a detectable response is observed with dosage. Both *in vivo* and *in vitro* systems are commonly employed in assessments of viral infectivity. *In vivo* systems, usually whole organisms, are used for end point dilution assessments where the calculated titer represents a statistical, biological unit. Enumerative assessments, such as plaque assays, give titers which represent one infectious virus particle per biological unit. Both dose-response methods have their statistical validity in the conceptual constraints of Poisson distribution. Bryan (1957) has discussed the principles inherent in quantitative studies on animal viruses and his discussion represents an excellent source for a clear description of the theoretical basis of infectivity assays. The choice of the method for virus infectivity assays depends upon the particular investigation, the laboratory facilities, and the nature of the response of the biological test system.

1. *End Point Dilution Methods*

A virus suspension, a diluent, and a biological test system are required for end point dilution assays of viral infectivity. The virus suspension may be clarified extracts of infected tissues or cells or purified preparations of virus. The diluent employed must be compatible with the virus and the biological test system. A diluent frequently employed for arboviruses, for example, consists of 0.75% (w/v) bovine plasma albumin (Fraction V, Armour) in 0.05 M phosphate-buffered saline, pH 7.2 (Dick and Taylor, 1949), and another common diluent is cell culture medium. Serial dilutions of the virus suspension are made in volumes appropriate for the test in test tubes. The tubes may be placed in an ice-water bath during the test to minimize any potential loss of virus titer due to thermal inactivation. Aliquots of the dilutions are inoculated into the biological test system which are subsequently maintained or incubated. The test subjects are observed or examined for different types of reactions such as lethality or cytopathic effect (CPE), as compared to the diluent-inoculated controls. The frequency of the response is recorded and end point titers, such as lethal dose (LD_{50}), infectious dose (ID_{50}), or tissue culture infectious dose, ($TCID_{50}$), are calculated by the methods of either Reed and Muench (1948) or Kärber (1931). The titer represents the dilution of the original virus suspension that will induce the observed response in 50% of the biological test units.

a. In Vivo. The preferable susceptible host or biological test systems for arboviruses is the mouse (either newborn, suckling, or adult, depending upon the particular arbovirus). The test material is usually inoculated directly into the target organ, such as intracerebrally, in 0.02 ml or 0.03 ml amounts. The animals are observed for 14–21 days for their response following the inoculation of each of the serial dilutions of the test material into batches of either 8 newborn or 6 adult mice, and the LD_{50} end point titer is calculated. For simple qualitative monitoring for the presence or absence of virus, an undiluted or a 10^{-2} diluted sample may be inoculated into the test animal.

The susceptible host for pathogenic-invertebrate viruses is entirely dependent upon the nature of the viral agent. Nevertheless, the procedure is essentially analogous to that used for arboviruses with the exception that many pathogenic-invertebrate viruses are infective for the host *per os*. The polyhedra of baculoviruses, cytoplasmic polyhedrosis viruses, and entomopoxviruses are particularly amenable to this approach. Although the procedure of infecting the host may seem simplified, it is not without problems. Vail (1975) has recently discussed some of the pitfalls associated with standardization and quantification of these viruses.

b. *In Vitro.* The BHK-21 cell line (Macpherson and Stoker, 1962) has excelled with regard to rapid cellular growth, low maintenance cost, and sensitivity to arboviruses (Sellers, 1963; Karabatsos and Buckley, 1967; Miles and Austin, 1963). Although CPE has been described with some flaviviruses in *Aedes* cell lines (Singh, 1972; Varma *et al.*, 1974a) (Section III,E), the vertebrate cell lines are usually used for *in vitro* assays of arboviruses. The methods used in this laboratory are described briefly. Stock cultures are grown in media consisting of 80% (v/v) Eagle's basal medium (BME) (Eagle, 1955) prepared with Hanks' balanced salt solution (HBSS) (Hanks and Wallace, 1949), 10% (v/v) tryptose phosphate broth (29.5 gm/liter), and 10% (v/v) fetal bovine serum. They are transferred every 3 to 4 days by a 1:14 split. Stationary tube cultures are prepared by seeding 5×10^4 cells/ml in the same medium, and they are used in 2 to 3 days. The maintenance conditions for the stationary tube cultures are pretested before virus inoculation. The outgrowth medium is discarded, and the cultures are inoculated with 0.1 ml of the respective diluent and 0.9 ml of the maintenance medium consisting of 97% (v/v) BME, 3% (v/v) fetal bovine serum, and 2 M Trizma base (Sigma Chemical Company). The latter supplement varies in amount from 0.05 ml, 0.1 ml, 0.15 ml, and 0.2 ml per 100 ml of maintenance medium. Cultures are incubated at 36°C for 15 minutes in a maintenance medium with a Trizma concentration that will render the pH of incubated cultures to be from 7.2 to 7.4. A pH of from 7.2 to 7.4 appears to be optimal for maintenance of the fibroblast BHK-21 cell line, and it is also optimal for growth of arboviruses in this particular cell culture system. The serially diluted test materials are inoculated in 0.1 ml aliquots (usually at least 3 replicate tubes per dilution) directly into the tube cultures, and they are fed 0.9 ml of appropriate maintenance medium. Inoculated cultures are incubated at 36°C. They are examined at daily intervals for development of CPE, and the culture fluids are changed at 3- to 4-day intervals. The tube cultures are routinely observed for 21 days following inoculation and the $TCID_{50}$ titer is calculated.

CPE in the *A. albopictus* cell line has also been observed in this laboratory with West Nile virus (strain Egypt 101). Confluent monolayers in French square bottles (160 ml) of *A. albopictus* cells are dispersed mechanically with a combination of a rubber policeman and vigorous pipetting. Stationary tube cultures are prepared by dispensing 0.5 ml/tube of a cell suspension representing a 1:20 split of the stock culture. Confluency is reached in the stationary tube cultures within 3 to 7 days incubation at 28 °C, depending on the number of cells present in the stock cultures. *Aedes albopictus* cells are maintained prior to virus inoculation by replacing the growth medium with Mitsuhashi–Maramorosch medium in which the concentration of fetal bovine serum is reduced from 20% to 3% (v/v), and 0.9 ml of maintenance medium is added per stationary tube culture. West Nile virus stock is serially diluted and each dilution is inoculated into 3 tubes (0.1 ml/tube). Control tube cultures are inoculated with 0.1 ml of diluent, 0.75% (w/v) bovine albumin in 0.05 M phosphate-buffered saline, pH 7.2. All cultures are incubated at 35 °C p.i., and they are examined for development of CPE for up to 25 days p.i., using a light microscope at a magnification of 100 ×. The titration end point is calculated and expressed as $TCID_{50}$/ml. The titration end point is usually reached 5 days p.i.

Granados (1976) has recently reviewed the literature on the topic of titration of insect pathogenic viruses. His discussion represents a good account of recent progress in this area. Baculoviridae has received considerable research activity such that there have been several reports on macro- and micromethods for end point titration of these viruses (Faulkner and Henderson, 1972; Knudson and Tinsley, 1974; Vaughn and Faulkner, 1963; Vaughn and Stanley, 1970). A number of laboratories are now using the recently described micromethod for end point dilution titrations of baculoviruses (Knudson, 1975; Brown and Faulkner, 1975). The method used in this laboratory will now be described.

Microtest plates (number 3034, Falcon Plastics), containing 60 wells per plate with a working volume of 0.015 ml/well, are used. Each well is seeded with 0.005 ml of a cell suspension containing 3×10^5 viable cells/ml (1500 viable cells/well) and 0.01 ml of the appropriate serial dilution of virus is added to each well. The microtest plate is incubated at 27 °C inside sealed plastic containers lined with moist filter paper which helps to maintain a high level of humidity. After 4 to 6 days, the wells are microscopically examined for infected cells displaying the CPE characteristic of baculovirus infections, i.e., inclusions or polyhedra in the nucleus of the cells (Harrap, 1973). This system has proved satisfactory for cell lines of *S. frugiperda*, *S. littoralis*, and *T. ni* (Knudson, 1975; Brown and Faulkner, 1975). Recently, this system has also been applied to the assay of *Oryctes* virus in *S. frugiperda* and *A. albopictus* cell lines (Kelly, 1976).

A tube method for end point dilution titration of *Chilo* iridescent virus (type 6, Kelly and Tinsley, 1970) in the *T. ni* cell line has been described by McIntosh and Kimura (1974), and recently, a bioassay method for mosquito iridescent virus (type 3, Kelly and Tinsley, 1970) in an *A. aegypti* cell line has also been reported by Webb and co-workers (1975). Paschke and Webb (1976) have recently reviewed literature on the problems associated with the assay of iridescent viruses in invertebrate cell cultures.

2. *Enumerative Methods*

The enumerative unit, unlike the LD_{50} unit, is not a statistical unit of virus activity, but it is a real measure of the biological activity of the virus. There are several methods for the assessment of virus activity and these are described. These methods all rely on an *in vitro* test system and on the appearance or detectability of a virus-induced change in the test system.

a. Plaque Assay. The plaque assay (Dulbecco, 1952) of animal viruses has been introduced to this series in Volume III by Cooper (1967). His discussion represents an excellent source of information on methods, interpretations, and potential pitfalls associated with the technique. Ideally, one virus particle induces the formation of a plaque or an eruption in a confluent monolayer of cells. These plaques may be counted and plaque-forming units (PFU) calculated per milliliter of original virus suspension.

i. Arboviruses. Arbovirus plaquing in two Simian kidney cell lines, VERO (African green monkey kidney, *Ceropthicus aethiops*), and LLC-MK$_2$ (*Macacus rhesus*), has been a routine procedure in this laboratory for many years (Buckley, 1974, 1975; Halstead, 1970; Karabatsos, 1969; Stim, 1969; Buckley and Casals, 1970). The methods presently used are now described.

Roux bottle cultures of both cell lines are grown in 90% (v/v) Eagle's minimal essential medium (MEM) (Eagle, 1959) in Earle's balanced salt solution (EBSS) (Earle, 1943), and 10% (v/v) fetal bovine serum. The cultures are subcultured at 2-week intervals at 1:4 or 1:5 split ratios. The cells are dispersed with equal volumes (10 ml) of 0.25% (w/v) trypsin and versene (1:5000) and resuspended in MEM in HBSS. For plaque assays, the cells from 2-week-old Roux bottle are suspended in 320 ml of growth medium prepared with HBSS and 8 ml aliquots are dispensed per 2-oz. flint glass prescription bottles. The plaque bottles are incubated at 36 °C for 4 days until the cells form a confluent monolayer. In the event of marked acidity in the plaque bottles, 0.1 ml of 7.5% (w/v) sodium bicarbonate may be added to individual bottles. On the day of inoculation, the growth medium is decanted and the monolayers are inoculated with 0.2 ml of serial dilutions of test material. An adsorption period of 1 hour at 36 °C is routinely used and two types of agar overlay are used depending on the virus being assayed. These agar overlays are described below.

Dengue agar overlay (Halstead, 1970) in LLC-MK$_2$ is used for plaque assay of dengue viruses, types 1–4. Two percent (w/v) ionagar No. 2S (Oxoid) in water is prepared by autoclaving the solution 15 minutes at 121 °C and cooling it to 56 °C in a water bath. The 2X nutrient portion is prepared and contains 20 ml fetal bovine serum (heat inactivated at 56 °C for 30 minutes), 20 ml 10X BME (in HBSS and without phenol red, NaHCO$_3$, and L-glutamine), 2 ml 200 mM L-glutamine, 2.5 ml 7.5% (w/v) NaHCO$_3$ (pH 8.5), 1 ml neutral red (1:200), 4 ml penicillin (10,000 Units/ml) and streptomycin (10,000 μg/ml) solution, 0.5 ml 100X amphotericin B (250 μg/ml), and 50 ml sterile distilled H_2O. The pH of the 2X nutrient protion should be 8.0 and it is prewarmed to 37 °C. Equal volumes of 2X nutrient and 2% agar are mixed and 8 ml of the mixture are added to an inoculated, confluent plaque bottle.

Diethylaminoethyl- (DEAE) dextran agar overlay is used for plaquing other arboviruses in VERO or LLC-MK$_2$ cells (Karabatsos, 1969). The 2% ionagar No. 2S is prepared as described above. The 2X nutrient portion contains 18 ml 10X EBSS (without NaHCO$_3$ and phenol red), 3 ml 10% (w/v) lactalbumin hydrolysate, 1.2 ml 5% (w/v) yeast extract, 3.6 ml fetal bovine serum (heat inactivated 56 °C for 30 minutes), 3 ml neutral red (1:1000), 0.9 ml 2% (w/v) DEAE-dextran, 1.8 ml penicillin-streptomycin solution, 1.8 ml amphotericin B, and H_2O up to 90 ml. The 2X nutrient portion is prewarmed to 37 °C and equal volumes of the 2X nutrient portion and the 2% agar are mixed. The inoculated plaque bottles are overlayed with 8 ml of the freshly prepared overlay. Five milliliters instead of 8 ml of agar overlay will give better defined plaques for Crimean hemorrhagic fever–Congo group viruses. Overlayed cultures are incubated inverted at 35°–36°C for up to 14 days, and thereafter, at ambient temperatures up to 21 days p.i. Plaques are counted at intervals and PFU per milliliter are calculated.

ii. Plaque assay in Aedes cell lines. Some arboviruses produce plaques in the *A. albopictus* cell line and the method of Yunker and Cory (1975) is described. *Aedes albopictus* cells are propagated in lactalbumin hydrolysate growth medium in Hanks' balanced salt solution (HLH) (Grand Island Bidogical Co.) containing the following ingredients per 1000 ml HLH: 100 ml fetal bovine serum (inactivated at 56 °C for 60 minutes), 1 gm bovine plasma albumin, 10^5 Units of penicillin G (sodium salt), 100 mg of streptomycin sulfate, and 40 mg of neomycin sulfate. The pH is adjusted to 6.8 and the complete medium is sterilized by positive pressure filtration. Stock cultures are grown as monolayers in 250 ml plastic flasks (Falcon Plastics) in 12 ml of medium at 27 °C, and they are subcultivated at 7- to 10-day intervals. Monolayer cultures are prepared for virus inoculation by replacing the growth medium of 4 confluent stock culture flasks with fresh growth medium and removing the cells from the surface of the vessel. The cell suspensions are pooled in a 100 ml flask; it is adjusted to a cell density of 2 to 3 $\times$ 10^5 cells/ml with growth medium; and 5 ml are seeded into 30 ml plastic flasks. The cultures are incubated at 27 °C for 2 to 7 days and used when confluent. Serial dilutions of the virus are prepared in growth medium. The medium is decanted from the cell monolayer and 0.2 ml of each virus dilution is added flask. The caps are replaced and the inoculum in distributed over the monolayer by tilting of the flask. After 1 hour at ambient temperature on a rocking platform, the flasks are overlayed with 4 ml of the primary overlay which contains a mixture of 3 parts growth medium held at 44 °C and 1 part 2% (w/v) agarose (Seakem) in Hanks' balanced salt solution. The caps are replaced, the flasks are brought to a horizontal position, and the first overlay is allowed to solidify for at least 15 minutes. Cultures are incubated at 35°–37 °C for 5 to 7 days, whereupon 2 ml of a second overlay medium is introduced into each flask. The second overlay medium is identical to the first overlay, except that it contains neutral red (1:5000, final concentration). Plaques are counted after 1 to 5 days of further incubation in the dark at 35°–37 °C, and the PFU titer is calculated.

iii. Pathogenic-Invertebrate viruses. Hink and Vail (1973) reported the first plaque assay procedure for a pathogenic-invertebrate virus in an invertebrate cell line. The baculovirus, *Autographa californica* nuclear polyhedrosis virus produced plaques, i.e., foci of infection in the *T. ni* cell line when overlayed with a growth medium (TNM–FH) containing 0.6% (w/v) methylcellulose (4000 centipoise). The procedure of Hink and Vail (1973) is described. Plastic petri dishes (60 × 15 mm) are seeded with $1.2 × 10^6$ *T. ni* cells in 1 ml and 2 ml of TNM–FH, medium is added. The dishes are rotated to ensure a uniform cell suspension and are incubated at 28 °C for 2 hours, allowing attachment of the cells to the surface of the dish. The growth medium is decanted and 0.35 ml of serial dilutions of test virus material are inoculated per dish. After a 1 hour absorption period, the inoculum is decanted and 3 ml of the methylcellulose-supplemented TNM–FH medium is added. The petri dishes are incubated at 28 °C until scored. The cells can be stained using neutral red (0.01% w/v) in Hanks' balanced salt solution to which glucose (14 gm/liter) has been added to render the solution isotonic for the cells. The methylcellulose–TNM–FH overlay is decanted and 3 ml of neutral red solution is added. After 1–2 hours incubation at 28 °C, the plaques can be scored with the aid of a stereoscopic microscope. The plaques appear pale pink or unstained, while the uninfected cells are stained red. Methylcellulose was used because Hink and Vail (1973) reported little or no success with agar, agarose, starch, and gelatin overlays. They also reported that methylcellulose functioned in their plaque assay system by keeping the cells attached as they grew and by limiting the lateral spread of the virus in the petri dish. It should be pointed out, however, that a 0.6% (w/v) methylcellulose solution is not solid, and the petri dishes cannot be inverted for incubation. Although the overlay did appear to limit the lateral spread of this virus in this system such that plaques were observed, the low viscosity of the methylcellulose may not prevent extensive lateral transmission of other viruses, in particular, some of the smaller icosahedral viruses. Since Ackers and Steere (1967), in Volume II of this series, have reported that the effective pore radius for solid 1% (w/w) agar or agarose gels is approximately 120 nm, a fluid methylcellulose overlay could allow considerable diffusion of these viruses. Knudson and Tinsley (1974) subsequently reported that plaques or foci of infection of *S. frugiperda* nuclear polyhedrosis virus could be produced without the use of methylcellulose in the medium in the *S. frugiperda* cell line. Their evidence suggested that the appearance of plaques was dependent upon initial cell concentration at the time of seeding. Further, Volkman and Summers (1975) have reported the use of a modified formulation of the plaque assay procedure of Hink and Vail (1973) in which the concentration of methylcellulose was increased from 0.6% (w/v) to 0.9% (w/v).

The plaque assay of pathogenic-invertebrate viruses is an active research area. The need for solid agar or agarose techniques for plaque assay should provide the impetus for the development of such methods in the near future.

b. Infectious Center Assay. The units enumerated in the infectious center assay are single, intact, infected cells which are cocultivated with susceptible cells which will exhibit a CPE (Weller, 1953; Weller *et al.*, 1958). They induce plaque formation when plated in serial dilutions onto susceptible cell monolayer cultures maintained under an overlay (Rapp and Benyesh-Melnick, 1963). The infectivity titers are expressed as the percentage of infected cells present in 1 ml of a cell suspension, the viable cell count of which has been determined prior to plating, and the percentages of infected cells may vary from 100% to fractions of 1%.

The method used in this laboratory for *Aedes* cell lines infected with two Rhabdoviridae (Wildy, 1971; Knudson, 1973) is described. Infected monolayers are incubated at 30 °C in either replicate stationary tube cultures, 2-oz. or 3-oz. prescription bottles are harvested, and the pooled cell suspension is pipetted vigorously in order to obtain a single cell suspension. The cells are washed 3 times in 10 ml of Rinaldini's salt solution, resuspended in 1 ml of Mitsuhashi–Maramorosch medium supplemented with 20% (v/v) of heat-inactivated fetal bovine serum, and counted in a hemacytometer in the presence of 0.5% (w/v) trypan blue. Then 0.2 ml aliquots of serial dilutions of the cell suspension are plated on VERO or LLC-MK$_2$ cells. After an adsorption period of 60 minutes at 36 °C, the cultures are overlayed and incubated as described above (Section III,B,2,a). The number of infectious centers, i.e., plaques, in VERO of LLC-MK$_2$ cells represents the number of infected cells in the sample. If the plating of a cell suspension containing 10^7 cells/ml results in a titer of $10^{5.6}$ PFU/ml, then the ratio of infected to uninfected cells is 1/25, i.e., 4% of cells are infected (Libiková and Buckley, 1971).

c. Fluorescent Antibody Technique. When applying the fluorescent antibody technique (Coons and Kaplan, 1950; Coons *et al.*, 1942) as an enumerative method, the enumerated units are the individual cells showing specific, virus antigen localization and/or accumulation. The infectivity titer is expressed, therefore, as the percentage of cells exhibiting specific immunofluorescence. The percentages of fluorescent cells may also vary from 100% to fractions of 1%. The technical details of the method have been discussed extensively in Volume III of this series by Casals (1967). The indirect fluorescent antibody technique (Weller and Coons, 1954) or "sandwich" technique (Nairn, 1969), is a method of choice for several reasons. The indirect method is from 4- to 12-fold more sensitive than the direct method (Coons, 1956; Pressman et al., 1958). Also, only one conjugate is used per species.

The methods of Buckley and Clarke (1970) and Libiková and Buckley (1971) are described. Cell cultures, either vertebrate or invertebrate, are prepared in either Leighton tubes containing 11 × 22 mm cover slips, or Lab-Tek tissue culture chamber slides (Lab-Tek Products, Miles Laboratories), and the cultures are infected. At the appropriate interval, the infected cultures are rinsed with a salt solution, phosphate-buffered saline for vertebrate cultures and Rinaldini's salt solution for invertebrate cultures, and fixed in three changes of cold (−20°C) acetone. The specimens are air dried, immersed briefly in phosphate-buffered saline, pH 7.8, and drained. They are overlayed with either immune or nonimmune unconjugated mouse sera which have been diluted 1:10. They are placed in a humid chamber for 30 minutes and then washed for 10 minutes with three changes of phosphate-buffered saline. The specimens are drained and allowed to react for 30 minutes with antimouse serum to which fluorescein isothiocyanate (FITC) (Riggs et al., 1958) has been conjugated. The FITC-antimouse conjugate is prepared by the method described by Goldstein and colleagues (1961), and it is subjected to DEAE-cellulose chromatography to eliminate both under- and overconjugated antibody molecules. A dilution of FITC-antimouse serum which contains 1 mg conjugated protein/ml is routinely used in the test. After the reaction, the specimens are again washed in three changes of phosphate-buffered saline and then mounted in glycerol:phosphate-buffered saline (90:10% v/v). The specimens are examined by fluorescence microscopy. Ten microscopic fields are routinely evaluated and the number of fluorescent cells as well as the total number of cells per microscopic field are enumerated. The proportion of fluorescent cells to nonfluorescent cells determines the infectivity titer, which is expressed as the percentage of fluorescent cells. One specificity test consists of the overlaying of infected preparations with nonimmune, unconjugated mouse serum, diluted 1:10, prior to reaction with the FITC-antimouse conjugate. The percentage of fluorescent cells should be nil in the control series. The advantage of the indirect FA method over the direct method lies in the fact that one conjugate can localize numerous arboviruses, provided the homologous immune sera are available.

  d. *Fluorescent Focus Assay.* Fluorescent antibody techniques were first used for enumerative purposes with vertebrate cells infected with West Nile virus (Noyes, 1955). Subsequent studies in other systems have shown that the results of fluorescent focus assays agree well with plaque assays and they are consistent with the hypothesis that one infectious particle is sufficient to initiate infection (Igarashi and Mantani, 1974; Rapp *et al.*, 1959).

  In this laboratory, cells are cultured on Lab-Tek tissue culture chamber slides (Lab-Tek Products, Miles Laboratories) at a cell density of 10^6 cells/ ml. The cultures are infected as described for plaque assays. After a final washing of the cell monolayer, an overlay containing 2% (w/v) Sephadex G-200 (L. Schneider, 1973) in a nutrient medium appropriate for the cell line is added. Earle's balanced salt solution base, which is buffered with *N*-Tris(hydroxymethyl)methyl-2-aminoethanesulfonic acid (TES) to pH 7.2–7.5, is nutrient medium for vertebrate cells. The nutrient medium for *A. albopictus* cells is the Mitsuhashi–Maramorosch medium with 20% (v/v) fetal bovine serum, and it does not require an additional buffer system. At an appropriate time interval postinoculation, the slides are prepared for fluorescence microscopy by removing the Sephadex overlay with a phosphate-buffered saline wash and proceeding with direct or indirect techniques as described above. The number of infectious units of a virus suspension is determined by microscopically counting the number of immunofluorescent foci in the infected culture. A focus reduction neutralization test can also be done using this system. These techniques as used in this laboratory have been described in detail for rabies virus and the reader should consult Smith *et al.* (1977) for a complete treatment of this topic.

  C. Serologic Tests

  As pointed out (Section II,E), some arboviruses which replicate in one or the other invertebrate cell systems are capable of inducing a persistent infection. It is, therefore, important to monitor the known infecting agent, and a variety of serologic tests are available. The appropriateness of a given serologic test depends, to some extent, upon the properties of the individual arbovirus being studied.

  1. *Immunofluorescence*

  The details have been described (Section III,B,2,c). The indirect fluorescent antibody method (Weller and Coons, 1954) is used primarily because of its greater sensitivity (Coons, 1956; Pressman *et al.*, 1958). Antigen of the known infecting agent is localized by specific immunofluorescence. It is, therefore, reidentified with the aid of a homologous immune serum and the corresponding conjugate.

2. Complement Fixation

Cell culture fluids obtained from infected *A. albopictus* cell cultures represent a source of specific complement-fixing antigen (Section III,A,4). The infected culture fluid is the antigen when reidentifying the known infecting agent.

3. Hemagglutination and Hemagglutination Inhibition

The techniques pertaining to hemagglutination (HA) and hemagglutination inhibition (HI) (Clarke and Casals, 1958) have been described by Casals (1967) in Volume III of this series and they are applicable as such to invertebrate tissue culture systems. However, it is important to note that culture fluids from uninfected *A. albopictus* cell cultures agglutinated goose red blood cells (Ghose and Bhat, 1971). This HA activity has been found only in extracellular fluid and not in intracellular material. Careful control studies, therefore, are indicated to distinguish between nonspecific and specific HA activity when applying HA and HI techniques for reidentification of infecting agents.

4. End Point Dilution Neutralization Test

In the end point serologic tests or neutralization tests, virus test materials are diluted serially. The dilutions are mixed in equal parts with undiluted, heat-inactivated (56°C for 30 minutes) immune or nonimmune serum. The mixtures are incubated in a 37°C water bath for 60 minutes, and then inoculated into a susceptible *in vivo* or *in vitro* indicator system (Section III,B). The neutralization index (NI) is the difference in the infectivity titers calculated for the two series, one containing the immune serum and the other containing the nonimmune serum. The NI is usually expressed as LD_{50} or $TCID_{50}$. For some of the flaviviruses, neutralization tests are carried out in invertebrate tissue culture systems.

5. Plaque Reduction Neutralization Test

When reidentifying the infecting agent by the plaque reduction neutralization test, two methods may be used and these are discussed in detail by Casals (1967) in Volume III of this series. The two methods include serial virus dilutions with indiluted serum or serial serum dilutions with constant amounts of virus. Prior to tests, sera are inactivated for 30 minutes at 56°C. Virus–serum mixtures are incubated in a 37°C water bath for 60 minutes, and then inoculated into VERO or LLC-MK$_2$ plaque bottles as described (Section III,B,2,a). Serologic results are expressed as the neutralization index (NI) of the PFU titer, or as the 50% (80% or 90%) plaque reduction end point of the serum, i.e., the highest dilution of serum which reduces the plaque count by 50% (80% or 90%). Choice of the method is dependent upon the particular experiment (see discussion of Casals, 1967, in Volume III of this series), but the undiluted serum method involves the fewest technical difficulties.

D. OPTIMAL CONDITIONS FOR VIRUS REPLICATION

It is probably naive to assume that in nature viruses seek suitable hosts which optimize their replication potential. It is more realistic and admittedly teleological to assume that viruses seek conditions which best ensure their survival and/or continuance in nature. Undoubtedly, there will be situations in which viral replicative events will be optimal and situations in which they will be suboptimal. The latter could be envisaged when optimal replication would cause extinction of the host. Yet, virologists frequently attempt to study viral replicative events under optimized conditions. The study of virus replication in cell culture, a host system which is amenable to controls, represents the most rewarding avenue of study. Nevertheless, the plaguing problem, "How much carry-over is there from the *in vitro* system to the *in vivo* system?" presents itself. *In vitro* studies allow a defining of parameters that are important in virus replication and/or continuance, and more importantly, they provide a standard to which other data are comparable. A few of these parameters that must be considered in *in vitro* studies on virus replication are discussed here.

1. *Tissue Specificity and Susceptibility*

Tables II–X represent a compilation of data on the *in vitro* specificity and susceptibility of arboviruses and pathogenic-invertebrate viruses for invertebrate cell lines. It may be premature to speculate on the significance of these data, but some degree of specificity does, in fact, exist. Parameters that may be important for arbovirus specificity may include the innate differences between the invertebrate cell lines, the heterogeneity of this group of viruses with regard to their vectors in nature, and physical properties such as the presence or absence of a virion envelope. Enveloped arboviruses are sensitive to the action of sodium deoxycholate (Casals, 1971; Theiler, 1957; Borden *et al.*, 1971), while unenveloped arboviruses, such as orbiviruses, are relatively resistant. The unenveloped tick-borne and phlebotomine orbiviruses replicate in the *A. albopictus* cell line, while the enveloped tick-borne flaviviruses, Louping ill, Langat, and Kadam (Mugo and Shope, 1972), as well as the unclassified tick-borne virus, Quaranfil, do not. Varma and colleagues (1974b) have demonstrated that Louping ill, Langat and Quaranfil viruses grow in a tick cell line. A flavivirus, West Nile, has been isolated from mosquitoes and from ticks, *Argus reflexus hermanni*, in Egypt

(Theiler and Downs, 1973), and it replicates in both mosquito and tick cell lines. Similarly, Ganjam virus, grouped as a tick-borne virus, has been isolated from mosquitoes (Theiler and Downs, 1973), and it replicates in the *A. albopictus* cell line. The *in vitro* susceptibility of a given arbovirus seems to correlate well with its *in vivo* host, i.e., a host in which the virus replicates and not necessarily one by which the virus is mechanically transmitted. Specificity seems to reside at the taxonomic level of the order, but the data are sparse and the present comments can only be considered speculative.

What little data there are on pathogenic-invertebrate virus *in vitro* specificity suggest that cytoplasmic polyhedrosis viruses and Iridoviridae lack specificity because they replicate in dipteran and lepidopteran cell lines. Many Baculoviridae, on the other hand, display a remarkable specificity which apparently resides at the generic level. There are exceptions to this generic rule, since the nuclear polyhedrosis viruses of *Autographa californica* and *T. ni* replicate in lepidopteran cell lines from several generic groups (Table IX).

2. Media and Cell Condition

The population doubling time of a cell line is probably the simplest measurable parameter that gives some indication as to the condition or fitness of the cells in the culture. If cell numbers are assessed with time at a specified temperature, a growth curve for cell population can be constructed and population doubling times may be calculated. Any or all subsequent culture alterations may then be compared with this defined norm or standard. For example, the adaptation of a cell line to a new medium formulation may be followed by this type of measurement. The reason for the concern for condition of the cell line is that its state may also influence the replication of a given virus. Igarashi and co-workers (1973c) adapted the *A. albopictus* cell line to Eagle's minimal essential medium supplemented with 10% (v/v) bovine serum. They found that nonessential amino acids such as serine and, to a lesser extent, proline were required for optimal cell growth. In contrast, when these cultures were inoculated with a flavivirus, Japanese encephalitis virus, the authors found that proline and, to a lesser extent, glycine were necessary for optimal virus replication. Thus, the media and cell condition should be considered when attempting to optimize conditions for virus replication.

3. Temperature

a. Cell Lines. In contrast to vertebrate cell lines whose optimal growth temperature is 37°C, invertebrate cell lines such as dipteran and lepidopteran grow optimally at 28°C.

b. Viruses. Viruses are often defined as obligate intracellular parasites and as such their optimal replication is directly dependent upon optimal cellular function. It is not surprising that optimal replication of Togaviridae in invertebrate and vertebrate cell lines was directly correlated with the optimal growth temperature of the host cell system (Davey *et al.*, 1973). Knudson and Tinsley (1974) reported a similar finding for the replication of a baculovirus in the *S. frugiperda* cell line. Therefore, the choice of temperature which will optimize virus replication in a cell culture is likely to be one that also optimizes cellular functions.

4. *Virion to Infectious Unit Ratio and Multiplicity of Infection*

The virus particle to infectious unit ratio, or absolute efficiency of plating, may be determined by a variety of methods. Electron microscopy is commonly employed in this type of determination because it allows a direct, visual enumeration of intact virions. Microscopic methods for determining the parameter have been described in Volume III of this series by Horne (1967). This parameter may vary from unity to 10,000 for animal viruses (Fenner *et al.*, 1974) and Knudson and Tinsley (1974) have reported that a baculovirus grown in and purified from a lepidopteran cell culture exhibited a ratio of approximately 250 virus particles/PFU. The low efficiencies of some animal viruses are indicative of noninfectious particles in virus suspensions. This parameter reflects either the extent of inactivation of viral infectivity occurring during manipulative purification procedures, or the production of genetically defective virus particles, i.e., lacking a complete functioning genome during the course of virus replication in a given host system.

Multiplicity of infection (m.o.i.) is the ratio of infectious virions introduced into an *in vitro* system divided by the number of cells present in the system. This ratio is important because it is indicative of the degree of synchrony or asynchrony in the initiation of virus infections. The validity of this parameter relies upon the constraints of the Poisson distribution of particles. The Poisson distribution, for example, predicts that a m.o.i. of three will statistically ensure that 95% of the cells will encounter an infectious unit, while 5% will not. Fifteen percent will encounter a single infectious unit, while the remaining 80% should statistically receive 2 or more infectious units. When high m.o.i., such as greater than 100 PFU/cell, are used to initiate virus infections, it is advisable to ensure that the virus stock does not contain defective particles which may interfere with the viral replicative events (Section IV). Arbovirus inocula often consist of infected mouse brain virus stocks which are diluted 10^{-2} to 10^{-3} in an attempt to minimize effects of defective virions.

5. *Adsorption and Penetration*

There are many factors influencing the efficiency of virus attachment with the cell membrane and the subsequent penetration of the virus to the site of its replication. Factors, such as ionic strength, pH, and temperature, are important and the reader is encouraged to consult a recent review on the topic by Lonberg-Holm and Philipson (1974) in which the interaction between viruses and cells is discussed in detail.

The length of the adsorption period for animal viruses varies from report to report, but 60 minutes at an appropriate temperature is a common interval employed. Artsob and Spence (1974) have demonstrated that the inclusion of 0.1% (w/v) DEAE-dextran in the diluent for a rhabdovirus, vesicular stomatitis virus, results in an increased number of infected *A. albopictus* cells. The mode of action of the polymer is not completely understood, but it may effect increased virus adsorption by cells by electrostatic interaction (Allison and Valentine, 1960) or by reversible damage to the cell membrane (Ryser, 1967).

6. *Growth Curves or Multiplication Cycles*

Growth curves relay graphic information on the kinetics of viral replication by monitoring the amount of infectious progeny with time. Characteristically, the multiplication cycle is divided into three distinct phases, lag (including eclipse and latent periods), exponential, and stationary phase. The duration of the phases is influenced by the host cell system, the virus, the incubation temperature, the sampling interval, and m.o.i. Thus, these parameters must be stated in assessments of the viral multiplication cycle. The frequency of sampling is also important because with more data points, the profile of the cycle approaches the situation of continuous monitoring. The host cells may be grown as stationary cultures or as spinning suspension cultures. Infectivity assays are performed on samples which reflect extracellular virus, intracellular virus, and total virus produced during the sampling interval.

Growth curves in invertebrate cell lines have been reported for arboviruses. Buckley and colleagues (1975) reported comparative growth curves for small and large plaque variants of Chikungunya virus in two vertebrate cell lines (36°C) and *Aedes* cell lines (29° ± 1 °C). CPE was seen only in the vertebrate cell lines, whereas noncytocidal infections accompanied by virus shedding over a period of several days was observed in the *Aedes* cell lines. There are also examples of baculovirus growth curves (Knudson and Tinsley, 1974; Raghow and Grace, 1974; Volkman and Summers, 1975; Potter *et al.*, 1976).

Cells of invertebrate cell lines are, in general, either loosely attached to the surface of the culture vessel or suspended, unattached, in the medium. Thus, steps must be taken to avoid the loss of infected cells during washing and sampling intervals. If systems yielding high virus titers are sought, then the passage history of the virus may also be important. Igarashi and co-workers (1973a) demonstrated that Japanese encephalitis virus strains with passage histories in primary monkey kidney or BHK-21 cells produced higher titers in the *A. albopictus* cell line than did strains which were passaged in suckling mouse brains. Passage effects on virus stocks have also been reported for baculoviruses (Brown and Faulkner, 1975; Faulkner and Henderson, 1972; Hink and Strauss, 1976b; Knudson and Harrap, 1976; Ignoffo *et al.*, 1971; MacKinnon *et al.*, 1974; Potter *et al.*, 1976).

E. Cytopathology

Enders (1959) described three factors which he felt were responsible for the increased importance of cell culture systems in the study of animal viruses. One of these was that "many viruses as they multiply produce degenerative changes in cultured cells which can be easily distinguished." It was significant that the use of the whole, living animals was not necessary for assays of virus infectivity and for serological identifications of the agent. In this section, methods and strategies for the study of the effects that arboviruses and pathogenic-invertebrate viruses have on cells in culture are discussed.

1. *Gross Cytological Changes*

Gross cytological changes in cell lines infected with virus means that a macroscopically visible change in the cells or cell monolayers is detectable. The formation of plaques under agar or agarose overlays is the obvious example. Although vertebrate cell lines infected with arboviruses commonly exhibit plaque formation under overlays, this situation is not universally true for invertebrate cell lines. Suitor (1969) reported the formation of plaques in *A. albopictus* cells infected with Japanese B encephalitis virus. CPE was subsequently described by Cory and Yunker (1972) in *A. albopictus* cells infected with a rhabdovirus, vesicular stomatitis virus (Indiana), and several flaviviruses, including West Nile, Japanese B encephalitis, yellow fever (17D strain), and dengue (types 1–4). The number of arboviruses plaquing in the *A. albopictus* cell line was later increased by Yunker and Cory (1975) to 124 virus strains belonging to 21 serological groups. Nearly all the mosquito-borne flaviviruses as well as 30 strains from Bunyaviridae plaqued in the *Aedes* cell line. Alphaviruses, with the exception of Chikungunya, and members of the California serogroup of the Bunyaviridae failed to induce gross cytological changes in the *Aedes* cell lines. A prolonged incubation, up to 12 days, and a reduced incubation temperature, 35 °C, was necessary for the plaquing of four members of the Bunyamwera group of viruses.

Gross cytological changes have been observed in invertebrate cell lines infected with pathogenic-invertebrate viruses. Specifically, the plaque formation of baculoviruses in lepidopteran cell lines has been reported (Hink and Vail, 1973; Knudson and Tinsley, 1974).

2. Light Microscopy of Cytopathic Effects

Singh and Paul (1968a,b) observed that certain mosquito-borne flaviviruses produced a CPE in *A. albopictus* cells. The CPE was characterized by the necrosis of individual cells, by the development of syncytial masses, and by the increased numbers of giant multinucleated cells. In time, "healthy" cells reappear and the infected cultures seem to recover from the infection, but they remain persistently infected (Section V). West Nile virus and Japanese B encephalitis virus produced CPE in two newly established cell lines, *Aedes malayensis* and *A. pseudoscutellaris*, while dengue virus (type 2) produced CPE only in the *A. pseudoscutellaris* cell line (Varma *et al.*, 1974a). When mosquito cells were grown on a plastic substrate, virus-induced syncytia were more pronounced. A similar phenomenon has been observed in *A. albopictus* cells infected with dengue virus (type 2) (Suitor and Paul, 1969). Different sublines of *A. albopictus* in combination with strains of arboviruses with varying passage histories seem to behave differently in different laboratories. West Nile virus, for example, failed to induce CPE in *A. albopictus* cells (Murray and Morahan, 1973), and dengue virus (type 2) similarly did not produce a CPE (Sinarachatanant and Olson, 1973). There are no clear explanations accounting for these observations, but "environmental conditions (may) prove to be of importance for yield of virus and to some degree for establishing a cytopathic effect" (Mussgay *et al.*, 1975). Further, arbovirus-infected *A. albopictus* cells may only show CPE "when stressed in a particular way" (Dalgarno and Davey, 1973).

In this laboratory, CPE in *A. albopictus* cells has been observed with West Nile virus (Egypt 101 strain). Cultures were prepared and infected as described previously (Section III,B,1,b). Cytopathic effects began 48 hours p.i. and consisted initially of multinucleated giant cells located mainly at the periphery of the cell sheet. Subsequently, fusion of the small cells occurred, resulting in formation of syncytial areas. These areas and multinucleated giant cells became necrotic and detached from glass by 3 to 4 days p.i. In addition, cells with elongated processes appeared, and pycnosis of groups of cells or single cells was prominent. On day 5 p.i., cultures inoculated with 10^{-2}–10^{-9} dilutions of virus showed clear-cut CPE which subsequently reached grades $3+$ (50% of cells detached) and 3 to $4+$ (75–90% of cells detached) in those cultures inoculated with terminal dilutions. Cultures inoculated with diluent failed to display either multinucleated giant cells or fusion of cells. The development of CPE was specifically inhibited in cultures inoculated with a mixture of West Nile virus and a homologous immune mouse ascitic fluid.

Granados (1976) has recently reviewed the cytopathic effects in invertebrate cell lines produced by pathogenic-invertebrate viruses. The majority of the research activity has been directed toward Baculoviridae, specifically, nuclear polyhedrosis viruses, as reflected by the literature. Baculoviridae, cytoplasmic polyhedrosis viruses, and entomopoxviruses are occluded viruses which in general induce a clear distinguishable CPE in invertebrate cell lines. The inclusion body which develops is characteristic of infections by these viruses (Harrap. 1973). There is comparatively little information on CPE produced in invertebrate cell lines by nonoccluded viruses. Kelly and Tinsley (1974a) reported that iridescent virus (types 2 and 6) replication in *A. aegypti* and *A. eucalypti* cells resulted in a contraction in cell size and the detachment of cells from the surface of the culture vessel. The CPE, however, was not proportional to the input virus, and the authors suggested that the observed effect was cytotoxic as opposed to cytopathic. Replication of *Oryctes rhinoceros* virus, a nonoccluded baculolike agent in the *S. frugiperda* and the *A. albopictus* cell lines, resulted in "cell fusion and/or nuclear proliferation" in the latter cell line and "disorganized" cells with no cell growth in the *S. frugiperda* cell line (Kelly, 1976).

Thus virus CPE in invertebrate cell lines should be distinguishable from mock-infected controls. Its expression should be proportional with the m.o.i. which is indicative of CPE and not simply a cytotoxic effect. Virus-specific antisera should neutralize CPE at limiting dilutions of the virus.

3. *Immunofluorescence Microscopy*

Methods for immunofluorescence microscopy have been described (Section III,C) and the topic is mentioned here only to emphasize its usefulness as a technique for the early detection of virus effects upon cells. The technique will undoubtedly be beneficial as an early diagnosis procedure for either demonstrations of virus antigen prior to CPE or identification of unknown viral agents.

4. *Electron Microscopy*

Methods for the preparation of invertebrate cell lines for electron microscopic studies are essentially the same as those used for vertebrate cell lines. The chapter of Morgan and Rose (1967) and Granboulan (1967) in Volume III of this series describe methods for thin sectioning of materials and for autoradiography, respectively.

The buffers commonly used in the fixing and washing of cells for embedding may not be isotonic for some invertebrate cell lines. This may present one troublesome area which must be considered prior to the microscopy study. Knudson and Harrap (1976), for example, used Trager's B medium (Trager, 1935) as the solvent for fixative agents and for washes in electron microscopy study of a baculovirus replication in the *S. frugiperda* cell line.

5. Time-Course Studies on Cytopathology

Time-course studies on the cytopathology of virus replication follows the ordered sequence of cytopathic and virion morphogenic events in time. One prerequisite for a study of this nature is to have sufficient numbers of cells exhibiting similar effects at the same time. This constraint cannot be met *in vivo*, and hence, this type of study requires an *in vitro* system, such as cell cultures. Although it is impossible to infect the cells of an *in vivo* system synchronously, it is possible, in some instances, to determine a plausible sequence of events occurring during the course of virus replication. The validity of the sequence and its place in time, nevertheless, must be confirmed in the *in vitro* system. The ability to initiate synchronous infections in an *in vitro* system, therefore, represents a distinct advantage of the system over *in vivo* systems. It must be emphasized that the majority of reported time-course studies on viral replication are synchronously initiated infections, and this is not to be confused with synchronous infections. The latter implies that the same stage of the virus replication is taking place concurrently in all cells of the culture. A synchronous infection would require the infection of all cells while all cells were in the same stage of their growth cycle. Since virus replication may be dependent upon the stage of cell cycle, synchronous infections require a synchronously growing cell population. Thus, infections are synchronously initiated in an asynchronous cell population. Before a synchronous infection can be assured, the cell growth cycle must be known and controlled, and the cycle's effect on the virus replication must be assessed.

For the synchronous initiation of virus infections, a m.o.i. that statistically ensures that all the cells are infected must be employed. Normally, a m.o.i. of 25 to 30 should suffice because, even with three washes, the inoculum would not be diluted to the point of asynchrony in initiation of the infection. In electron microscopic studies where the early events of adsorption, penetration, and viral uncoating are to be followed, the m.o.i. must be increased to 100 or more to ensure visualization of these events.

IV. Viral Interference

Viral interference has been defined as a state induced by an interfering virus which results in resistance of the cells to superinfection or to infection by a challenge virus (Schlesinger, 1959; Fenner *et al.*, 1974). Four cases of viral interference are distinguishable and they are discussed in relation to arboviruses and pathogenic-invertebrate viruses in invertebrate cell lines.

A. VIRUS-ATTACHMENT INTERFERENCE

This case of interference is typified by the alteration or destruction of virus receptors on the cell surface by the interfering virus. The specific attachment and subsequent penetration of the challenge virus, therefore, is prevented (Fenner *et al.*, 1974). At present, there are no examples of this type of interference for either group of invertebrate-associated animal viruses.

B. HOMOLOGOUS INTERFERENCE

Homologous interference or defective particle interference refers to an intracellular interference which functions against homologous or homotypic virus challenge. The mechanism of this type of interference is not completely understood, but it is generally regarded to be mediated via the competitive action of defective virus particles in the inoculum for substrates necessary for virus replication (Fenner *et al.*, 1974). Defective virus particles may be mechanically produced by UV-inactivation, or they may be generated for certain viruses under conditions of high multiplicity or undiluted passage. The latter was initially described for influenza virus (von Magnus, 1954). Influenza and vesicular stomatitis viruses both produce large amounts of antigenically homogeneous defective-interfering (DI) virus particles upon serial undiluted passage, and this type of interference has been denoted as autointerference, a special case of homologous interference (Fenner *et al.*, 1974). DI particles have been observed in many animal virus systems (Huang, 1973) and their possible role in disease processes has also been examined (Huang and Baltimore, 1970). The vertebrate host cell also influences the extent of DI particle production as reported for arboviruses such as vesicular stomatitis virus (Huang and Baltimore, 1970), Semliki Forest virus (Levin *et al.*, 1973), and West Nile virus (Darnell and Koprowski, 1974). Observed variations in susceptibility to interference by DI of vesicular stomatitis virus have been shown to be host cell dependent (Huang and Baltimore, 1970; Perrault and Holland, 1972).

Although serial undiluted passage of Sindbis virus in vertebrate cells, BHK-21, or chick embryo fibroblasts generates DI particles (Guild and Stollar, 1975; Shenk and Stollar, 1972; Schlesinger *et al.*, 1972), passage of the virus in the *A. albopictus* cell line, however, did not yield DI particles even after 20 to 30 undiluted passages (Stollar *et al.*, 1975). Characteristically, the production of Sindbis DI particles in cell cultures may be detected by monitoring the titer of infectious virus produced with passage or examining the species of viral RNA produced during the course of the infection (Shenk and Stollar, 1972). The Sindbis DI particles also exhibit a slightly greater density when compared to the infectious particles, i.e., 1.22 gm/ml compared to 1.20 gm/ml (Shenk and Stollar, 1973). When both titer and viral RNA were monitored in Sindbis passages in *A. albopictus* cells, evidence for DI production was not obtained. Similarly plaque-purified Semliki Forest virus, another alphavirus, was used to generate DI particles in BHK cells by serial undiluted passage. The presence of DI particles in the inoculum altered the pattern of RNA synthesis in BHK and VERO cells, but the RNA pattern obserbed in *A. albopictus* cells was analogous to that seen when either BHK or VERO are infected with standard infectious virus (Eaton, 1975). Thus, the *A. albopictus* cell line would appear to be restrictive for DI particle generation for alphaviruses. Although homologous interference is demonstrable in vertebrate cell lines, it has not been shown in invertebrate cell lines. However, homologous interference has been demonstrated in Sindbis persistently infected cultures and this is discussed elsewhere (Section V).

C. HETEROLOGOUS INTERFERENCE

When challenge virus is different from virus inducing the state of interference, the interference is called heterologous interference and, sometimes, intrinsic interference (Fenner *et al.*, 1974). Classically, the interference is induced by infectious virus and it is not sensitive to actinomycin D. Other examples of heterologous interference exist and these have been reviewed (Fenner *et al.*, 1974).

Attempts to demonstrate heterologous interference in invertebrate cell lines have been made using persistently infected invertebrate cultures (Section V), and negative results have been reported (Peleg, 1972; Kascsak and Lyons, 1974; Libiková and Buckley, 1971; Stollar and Shenk, 1972).

D. INTERFERON-MEDIATED INTERFERENCE

Interferons are a heterogeneous class of inducible cellular proteins which exhibit species specificity in their ability to interfere with virus replication in homologous cells (Isaacs and Lindemann, 1957; Wagner, 1963). The antiviral protein(s) is inducible by a variety of agents but commonly it is activated by foreign nucleic acids. Interferons are of cellular origin and they are not cytotoxic (Fenner *et al.*, 1974). Detailed methods for the study of interferons in animal virus cell systems are presented in Volume IV of this series (Wagner et al., 1968) and they are not discussed here.

Although arboviruses are interferon inducers in vertebrate cell systems, attempts to demonstrate interferon production in Aedes cell lines have resulted in conflicting accounts. Reports on the failure to demonstrate interferon have been published (Peleg, 1969; Davey and Dalgarno, 1974; Kascsak and Lyons, 1974; Libiková and Buckley, 1971; Murray and Morahan, 1973; Stollar and Shenk, 1972), and there are two accounts of interferon induction in Aedes cell lines (Enzmann, 1973; Bergold and Ramirez, 1972). Yet, these latter accounts remain to be confirmed. Enzmann (1973) reported an attempt to demonstrate interferon in Sindbis persistently infected cultures. Although the inducible substance was actinomycin D sensitive, heterologous interference was not attempted.

Interference has not been demonstrated for pathogenic-invertebrate viruses in invertebrate cell lines.

V. Persistent Infections

A persistent infection in cell culture is an example of a chronic infection. A chronic infection is clinically characterized by the demonstrability of virus, but disease is either absent or associated with "immunopathological disturbances" (Fenner et al., 1974). The latter constraint has little relevance to persistently infected cell lines, but the "shedding" of virus by the culture is a key factor in determining whether the infection is persistent. Persistent infections of cell cultures are characterized by several parameters. A percentage or all of the cells continue to grow and divide, and thus, subcultivation of the cells is possible. Virus or virus-specified functions can be continuously detected. Although there exist a number of virus–cell persistent interactions, they may be placed into one of three definable categories and these have been defined previously (Fenner et al., 1974). The tumor viruses induce one class of persistent infection in cell culture by the integration of viral genome into cellular genome. This class of persistent infections has been well documented. A second class, steady-state infection, is characterized by the continuous release of virus, albeit at a rate slower than a primary infection, from the majority of infected cells and by the observation that the infection of culture cannot be "cured" by viral-specific antibody. Numerous examples of this type of infection in vertebrate cells have been described and reviewed (Fenner et al., 1974). The last class, carrier culture, can be distinguished from the other two classes because this state is often induced by the deliberate or primary infection of the cell culture. Only some of the cells are infected, and thus, uninfected cells can be recovered from the persistently infected culture by cell cloning techniques. Evidence thus far would indicate that all arbovirus-initiated persistent infections of invertebrate cell lines are of the carrier culture type (Buckley, 1976).

A. DEVELOPMENT OF CARRIER CULTURES

Singh's *A. albopictus* cell line has been the favored invertebrate cell system for the establishment of arbovirus carrier cultures. One disadvantage of the cell system, however, is the ease with which the carrier culture state can be initiated, and thus, the line can become contaminated with other arboviruses (Section II,E). Optimal conditions for the maintenance of the carrier culture may be variable and these factors may have to be determined empirically (Banerjee and Singh, 1968).

Mosquito-borne arbovirus carrier cultures are developed by the primary infection of *A. albopictus* cultures. Uninoculated and infected cell cultures are subcultured at a split ratio of 1:8 to 1:16 7 days post-primary-infection (Section II,C). The cultures are incubated at 30°C until confluent and, thereafter, at ambient temperature (20°–25°C). The titer of the fluid phase and of the cell washing is determined by inoculation into the appropriate *in vivo* or *in vitro* vertebrate system. The infecting virus is also reidentified by mouse or plaque reduction neutralization tests. The persistently infected culture is subcultured at weekly intervals at the aforementioned split ratios.

Procedures for the establishment of tick-borne and phlebotomine orbiviruses, and lyssaviruses (Shope, 1975), such as kotonkan (Kemp et al., 1973), Obodhiang, and Mokola (Kemp et al., 1972; Shope et al., 1970) are essentially the same as described above. The only modifications are to extend the transfer interval to 2 weeks with one medium change, to use split ratios of 1:2 to 1:4 for subculturing, and to incubate the cultures at 30°C immediately after subculture and at 35°C prior to subsequent subculture (Libikova and Buckley, 1971).

B. DETECTION AND ASSAY

Since the carrier culture does not exhibit a cytopathic effect, other methods must be employed to detect the presence of virus. Immunofluorescence techniques are commonly employed for the detection of viral antigens in the carrier culture (Section III,E). Although the persistent infection is noncytocidal in the invertebrate cells, CPE can be observed in vertebrate cell lines. The vertebrate cell systems, therefore, represent an additional detection system (Section III,B). The use of the infectious center assay is a further refinement which permits assessments to be made of the percentage of infected cells in the carrier culture (Section III,B,2).

C. Nature of the Carrier Culture State

Carrier cultures have been differentiated into three types (Fenner *et al.*, 1974): "1. Most of the cells in the culture are genetically resistant to the carried virus but the infection is perpetuated in a minority of susceptible cells. 2. The cells are genetically susceptible to the carried virus but the transfer of virus from cell to cell is limited by antiviral factors in the medium. 3. The cells are genetically susceptible, but most cells are made temporarily refractory by interfering factors produced within the carrier culture." These distinctions are not meant to imply that the carrier culture state is one of the above since alternative situations may yet be demonstrated. Furthermore, it is also conceivable that the carrier culture state may result from a combination of the above or may evolve through the various possibilities.

The genetic resistance or susceptibility of the cells in culture may be determined by cell cloning techniques (Section II,C) and by the subsequent establishment of persistently infected or uninfected cloned cultures. Given that the cells are genetically susceptible, the second type may be confirmed by washing the carrier culture and monitoring the percentage of infected cells, i.e., a dilution effect upon the antiviral factor in the medium. Proof of the third type relies upon the techniques used in the study of viral interference (Section IV).

Some of the characteristics of arbovirus persistent infections in invertebrate cell lines, particularly for the *A. albopictus* cell line, have been reviewed recently (Buckley, 1976). The question of which type of carrier culture arboviruses produce is not readily resolvable, but a summary of evidence, albeit simplistic, may be indicative of the type. In general, the majority of the cells appear to be genetically susceptible, as demonstrated by primary infection rates, and therefore, the type 1 carrier culture can be eliminated. The type 2 carrier culture is also not particularly attractive because the percentage of infected cells decreases with passage after primary infection. Type 3, temporarily refractory cells due to viral interference, seems the choice by default. The four cases of viral interference (Section IV) have not been completely investigated in regard to arbovirus persistent infections in invertebrate cells. Viral attachment interference has not been demonstrated. Persistently infected cultures exhibit homologous interference, but Stollar and colleagues (1975) have demonstrated that DI particles produced in vertebrate cells have no effect in invertebrate cells. Further, Sindbis virus isolated from Sindbis persistently infected *A. albopictus* cells is temperature sensitive (Shenk *et al.*, 1974). Heterologous interference in persistently infected cell cultures is not demonstrable, and interferon-mediated interference has not been satisfactorily shown.

Clearly, further investigation is warranted because " the systems under study are complex and are influenced by a number of factors such as temperature, viral interference, relative multiplicity of different viral genotypes, and perhaps the host cell" (Shenk *et al.*, 1974). These studies may be particularly relevant to a frequent host of arboviruses, the mosquito. Initially the gut cells are infected and "cured," and subsequently neuronal and salivary tissues are infected. The latter tissues appear to remain persistently infected with little or no deleterious effect to the insect (Marshall, 1973).

There is little evidence to suggest that pathogenic-invertebrate viruses induce persistent infections in invertebrate cell lines. Mitsuhashi (1967) described a persistent infection of a hemocyte cell line of *Chilo suppressalis* induced by the accidental introduction of *Chilo* iridescent virus (type 6, Kelly and Tinsley, 1970) into the cultures. The presented evidence suggested that the persistent infection was an example of the carrier culture state. Unfortunately, the cell line has been lost (Hink, 1976) and thus, it remains as an unconfirmed observation. Yet, the report is of interest because more recent studies on iridescent virus replication in invertebrate cell lines have revealed several curious features. This area has been reviewed recently (Paschke and Webb, 1976) and the problem of recovering infectious virus from cell lines is discussed. Although some of the mosquito iridescent viruses do produce progeny in cell cultures, it is not infectious for other cell cultures or insects. Further, if infected cultures are permitted to grow, the cells "recover" from the infection and appear "normal." The cell-derived progeny are enveloped in contrast to the insect-derived virus which is unenveloped. Therefore, it has been suggested that the envelopment of the viruses renders them incapable of entering other cells (Paschke and Webb, 1976). Thus, a form of viral-attachment interference may be operating in this virus–cell system. The nature of these observations, however, requires further data before they can be resolved.

VI. Viral Attenuation

The occurrence of marked genetic heterogeneity of wild-type virus stocks is a common phenomenon which has been recognized, for example, for flaviviruses (Mayer and Kozuch, 1969; Smorodintsev *et al.*, 1969). Wild-type virus stocks may be cloned by plaque-picking techniques and the virulence of the cloned virus for vertebrate hosts is one marker that is sometimes determined. Avirulent virus strains are useful for potential vaccines and thus, measures which enhance the probability of the isolation of such strains are important. Viral attenuation is an example of such measures. In this context, viral attenuation refers to a reduction in virulence of a virus for a given vertebrate host. The attenuation process is a selective one and a few experimental approaches are discussed.

Baseline or control data on wild-type virus are prerequisite before attempting to attenuate a particular arbovirus. Attenuation has been accomplished for some members of the Togaviridae by manipulations such as serial subculture of intact, infected cells (Banerjee and Singh, 1969) and serial passage of virus derived from cell culture fluids in uninfected cell cultures (Peleg, 1971). Japanese encephalitis virus variants have been induced by ethyl methanesulfonate (an alkylating agent characterized by low cytotoxicity) treatment of infected *A. albopictus* cell cultures (K. Banerjee, personal communication), and this procedure is now described.

The *A. albopictus* cell line is grown as a spinning suspension culture using the normal growth medium. Ten milliliters of cell suspension containing approximately 5×10^6 cells/ml are removed from a 7-day-old spinner culture. The cells are pelleted for 10 minutes at 1600 rpm and washed four times with Rinaldini's salt solution. The cells are resuspended into 100 ml of growth medium, placed in a spinner culture, and infected with wild-type virus at a m.o.i. of 0.1 LD_{50}/cell. Then 48 hours p.i., 20 ml of the infected cell suspension are removed and washed four times with Rinaldini's salt solution by centrifugation. The cells are resuspended in 100 ml of growth medium and placed into a spinner flask. Ethyl methanesulfonate, 0.2 ml, is added to the culture for two consecutive days and on the third and fourth day following the addition of the mutagen, mutants are selected. A 5 ml aliquot of the infected cell suspension is frozen and thawed twice and centrifuged at 10,000 rpm for 30 minutes. After clarification, the supernatant is plaque-assayed in VERO cells by inoculating monolayers with serial dilutions with 6 replicate cultures per dilution. The overlaid cultures are incubated at 37 °C, 30 °C, or 25 °C and plaques are picked which increase in size at temperatures other than 37 °C. They are transferred to stationary tube cultures of *A. albopictus* cells and a virus stock is grown. The virus clones are tested for virulence in mice by inoculating mice of different age groups intracerebrally and intraperitoneally. The animals are observed for sickness and/or death for 14 consecutive days. Mice recovering from transient illness or those appearing healthy at 21–23 days are challenged with the wild-type stock of virus by the intraperitoneal route of inoculation. The challenge virus dose should kill all control mice by the intracerebral route of inoculation and thus, approximately $10^{5.0}$–$10^{5.6}$ LD_{50}/mouse of virus should be used. Blood from surviving groups of mice is pooled and tested for neutralizing antibody by the plaque reduction method (Section III,C,5). Variant strains have been obtained by this method. Some were apathogenic for mice when inoculated intraperitoneally and proved to be immunogenic. One of the advantages of chemical mutagensis as compared to selection of clones either by transfers of intact, infected cells or by virus passages in fresh cell cultures is that the chemical method is economical with regard to time and material.

VII. Conclusion

Methods for studying DNA, RNA, and protein synthesis during the course of virus replication in invertebrate cell lines have not been described in this chapter. Perhaps a discussion of such methodologies may seem germane, but it was omitted for two reasons. Although the bulk of the studies thus far reported for arbovirus replication have used the vertebrate cell system (Mussgay *et al.*, 1975), there is little experimental data available on this topic for arbovirus or pathogenic-invertebrate virus replication in invertebrate cell lines. Second, a discussion of biochemical methods is a subject area whose scope precludes any detailed treatment in this chapter.

One aim of this chapter has been to point out the differences and similarities between invertebrate cell systems and vertebrate cell systems. Although there are differences, it is hoped that this contribution has emphasized that there are also similarities between the systems such that methods developed for the study of animal viruses in vertebrate cell systems should be adaptable, perhaps with minor modifications, to invertebrate cell systems. Thus, the other contributions in this series should provide the background for studies on animal virus replication in invertebrate cell lines.

ACKNOWLEDGMENTS

We wish to thank Drs. K. Banerjee, U. M. Bhat, D. H. Clarke, R. H. Goodwin, T. Lescott, J. F. Longworth, A. H. McIntosh, K. Mifune, H. G. Miltenburger, T. Motohashi, J. Peleg, P. D. Scotti, K. R. P. Singh, A. Smith, G. H. Tignor, M. G. R. Varma, S. R. Webb, and E. G. Westaway for kindly allowing us to include some of their unpublished data in this chapter. We are grateful to Dr. R. E. Shope for his reading of the manuscript and for his thoughtful suggestions, and to Dr. J. Casals for his helpful comments. The secretarial staff of the Yale Arbovirus Research Unit, particularly Miss C. Mousch, is also acknowledged for their competent assistance in the preparation of the manuscript.

REFERENCES

Ackers, G. K., and Steere, R. L. (1967). *In* "Methods in Virology" (K. Maramorosch and H. Koprowski, eds.), Vol. 2, pp. 325–365. Academic Press, New York.

Ajello, C., Gresiková, M., Buckley, S. M., and Casals, J. (1975). *Acta Virol. (Engl. Ed.)* **19**, 441–442.

Allison, A. C., and Valentine, R. C. (1960). *Biochem. Biophys. Acta* **40**, 400–410.

Armstrong, D. (1973). *In* "Contamination in Tissue Culture" (J. Fogh, ed.), pp. 51–64. Academic Press, New York.

Artsob, H., and Spence, L. (1974). *Can. J. Microbiol.* **20**, 329–336.

Bailey, L., Newman, J. F. E., and Porterfield, J. S. (1975). *J. Gen. Virol.* **26**, 15–20.

Banerjee, K., and Singh, K. R. P. (1968). *Indian J. Med. Res.* **56**, 812–814.

Banerjee, K., and Singh, K. R. P. (1969). *Indian J. Med. Res.* **57**, 1003–1005.

Barile, M. F. (1973a). *In* "Contamination in Tissue Culture" (J. Fogh, ed.), pp. 131–172. Academic Press, New York.

Barile, M. F. (1973b). *In* "Tissue Culture: Methods and Applications" (P. F. Kruse, Jr. and M. K. Patterson, Jr., eds.), pp. 729–735. Academic Press, New York.

Bayne, C. J. (1976). *In* "Invertebrate Tissue Culture: Research Applications" (K. Maramorosch, ed.), pp. 61–74. Academic Press, New York.

Bellet, A. J. D., and Mercer, E. H. (1964). *Virology* **24**, 645–653.

Berge, T. O. (1975). "International Catalogue of Arboviruses Including Certain Other Viruses of Vertebrates," 2nd ed., DHEW Publ. No. (CDC) 75–8301. U.S. Dept. of Health, Education and Welfare.

Bergold, G. H., and Ramirez, N. (1972). *Monogr. Virol.* **6**, 56–59.

Bhat, U. K. M., and Singh, K. R. P. (1970). *Curr. Sci.* **39**, 388–390.

Borden, E. C., Shope, R. E., and Murphy, F. A. (1971). *J. Gen. Virol.* **13**, 261–271.

Brooks, M. A., and Kurtti, T. J. (1971). *Annu. Rev. Entomol.* **16**, 27–52.

Brown, A., and Officer, J. E. (1968). *In* "Methods in Virology" (K. Maramorosch and H. Koprowski, eds.), Vol. 4, pp. 531–564. Academic Press, New York.

Brown, M., and Faulkner, P. (1975). *J. Invertebr. Pathol.* **26**, 251–257.

Bryan, W. R. (1957). *Ann. N.Y. Acad. Sci.* **69**, 698–728.

Buckley, S. M. (1969). *Proc. Soc. Exp. Biol. Med.* **131**, 625–630.

Buckley, S. M. (1971a). *Curr. Top. Microbiol. Immunol.* **55**, 133–137.

Buckley, S. M. (1971b). *Trans. R. Soc. Trop. Med. Hyg.* **65**, 535–536.

Buckley, S. M. (1971c). *In* "Proceedings of the International Symposium on Tick-borne Arboviruses" (M. Gresiková, ed.), pp. 43–52. Slovak Acad. Sci., Bratislava.

Buckley, S. M. (1972). *J. Med. Entomol.* **9**, 167–169.

Buckley, S. M. (1973). *Appl. Microbiol.* **25**, 695–696.

Buckley, S. M. (1974). *Proc. Soc. Exp. Biol. Med.* **146**, 594–600.

Buckley, S. M. (1975). *Ann. N.Y. Acad. Sci.* **266**, 241–250.

Buckley, S. M. (1976). *In* "Invertebrate Tissue Culture: Research Applications" (K. Maramorosch, ed.), pp. 201–232. Academic Press, New York.

Buckley, S. M., and Casals, J. (1970). *Am. J. Trop. Med. Hyg.* **19**, 680–691.

Buckley, S. M., and Clarke, D. H. (1970). *Proc. Soc. Exp. Biol. Med.* **135**, 533–539.

Buckley, S. M., and Tignor, G. H. (1975). *J. Clin. Microbiol.* **1**, 241–242.

Buckley, S. M., Casals, J., and Downs, W. G. (1970). *Nature (London)* **227**, 174.

Buckley, S. M. Singh, K. R. P., and Bhat, U. K. M. (1975). *Acta Virol. (Engl. Ed.)* **19**, 10–18.

Casals, J. (1967). *In* "Methods in Virology" (K. Maramorosch and H. Koprowski, eds.), Vol. 3, pp. 113–198. Academic Press, New York.

Casals, J. (1971). *In* "Comparative Virology" (K. Maramorosch and E. Kurstak, eds.), pp. 307–333. Academic Press, New York.

Casals, J., and Buckley, S. M. (1973). *Dengue Newslett. Pan Am. Health Organ.* **2**, 6–7.

Cassell, E. A. (1965). *J. Appl. Microbiol.* **13**, 293–296.

Chiu, R. J., and Black, L. M. (1967). *Nature (London)* **215**, 1076–1078.

Clarke, D. H., and Casals, J. (1958). *Am. J. Trop. Med. Hyg.* **7**, 561–573.

Converse, J. L., and Nagle, S. C. (1967). *J. Virol.* **1**, 1096–1097.

Coons, A. H. (1956). *Int. Rev. Cytol.* **5**, 1–23.

Coons, A. H., and Kaplan, M. H. (1950). *J. Exp. Med.* **91**, 1–13.

Coons, A. H., Creech, H. J., Jones, R. N., and Berliner, E. (1942). *J. Immunol.* **45**, 159–170.

Cooper, J. E. K. (1973). *In* "Tissue Culture: Methods and Applications" (P. F. Kruse, Jr. and M. K. Patterson, Jr., eds.), pp. 266–269. Academic Press, New York.

Cooper, P. D. (1967). *In* "Methods in Virology" (K. Maramorosch and H. Koprowski, eds.), Vol. 3, pp. 243–311. Academic Press, New York.

Coriell, L. L. (1973). *In* "Contamination in Tissue Culture" (J. Fogh, ed.) pp. 29–49. Academic Press, New York.

Cory, J., and Yunker, C. E. (1972). *Acta Virol. (Engl. Ed.)* **16**, 90.

Cunningham, A., Buckley, S. M., Casals, J., and Webb, S. R. (1975). *J. Gen. Virol.* **27**, 97–100.

Dalgarno, K., and Davey, M. W. (1973). *In* "Viruses and Invertebrates" (A. J. Gibbs, ed.), pp. 245–270. North-Holland Publ., Amsterdam.

Danielová, V. (1973). *Acta Virol. (Engl. Ed.)* **17**, 249–252.

Darnell, M. B., and Koprowski, H. (1974). *J. Infect. Dis.* **129**, 248–256.

Davey, M. W., and Dalgarno, L. (1974). *J. Gen. Virol.* **24**, 1–11.

Davey, M. W., Dennett, D. P., and Dalgarno, L. (1973). *J. Gen. Virol.* **20**, 225–232.

Day, M. F., and Grace. T. D. C. (1959). *Annu. Rev. Entomol.* **4**, 17–38.

Demal, J., and Leloup, A. M. (1972). *In* "Invertebrate Tissue Culture" (C. Vago, ed.), Vol. 2, pp. 3–39. Academic Press, New York.

Dick, G. W. A., and Taylor, R. M. (1949). *J. Immunol.* **62**, 311–317.

Dulbecco, R. (1952). *Proc. Nat. Acad. Sci. U.S.A.* **38**, 747–752.

Eagle, H. (1955). *Proc. Soc. Exp. Biol. Med.* **89**, 362–364.

Eagle, H. (1959). *Science* **130**, 432–437.

Earle, W. R. (1943). *J. Nat. Cancer Inst.* **4**, 167–212.

Eaton, B. T. (1975). *Virology* **68**, 534–538.

Enders, J. F. (1959). *In* "Viral and Rickettsial Infections of Man" (T. M. Rivers and F. L. Horsfall, eds.), 3rd ed., pp. 209–229. Lippincott, Philadelphia, Pennsylvania.

Enzmann, P. J. (1973). *Arch. Gesamte Virusforsch.* **41**, 382–389.

Faulkner, P., and Henderson, J. F. (1972). *Virology* **50**, 920–924.

Fedoroff, S. (1967). *J. Nat. Cancer Inst.* **38**, 607–611.

Fenner, F. (1976). *Intervirology* **7**, 1–115.

Fenner, F., McAuslan, B. R., Mims, C. A., Sambrook, J., and White, D. O. (1974). "The Biology of Animal Viruses," 2nd ed. Academic Press, New York.

Fogh, J., ed. (1973). "Contamination in Tissue Culture." Academic Press, New York.

Gardiner, G. R., and Stockdale, H. (1975). *J. Invertebr. Pathol.* **25**, 363–370.

Gartler, S. M., and Farber, R. A. (1973). *In* "Tissue Culture: Methods and Applications" (P. F. Kruse, Jr. and M. K. Patterson, Jr., eds), pp. 797–804. Academic Press, New York.

Ghose, S. N., and Bhat, U. K. M. (1971). *Curr. Sci.* **40**, 354–355.

Giannotti, J., and Vago, C. (1971). *Physiol. Veg.* **9**, 541–553.

Gibbs, A. J., ed. (1973). "Viruses and Invertebrates." North-Holland Publ., Amsterdam.

Goldstein, G., Slizys, I. S., and Chase, M. W. (1961). *J. Exp. Med.* **114**, 89–110.

Gomot, L. (1972). *In* "Invertebrate Tissue Culture" (C. Vago, ed.), Vol. 2, pp. 41–136. Academic Press, New York.

Goodwin, R. H. (1975). *In Vitro* **11**, 369–378.

Goodwin, R. H., Vaughn, J. L., Adams, J. R., and Louloudes, S. J. (1970). *J. Invertebr. Pathol.* **16**, 284–288.

Goodwin, R. H., Vaughn, J. L., Adams, J. R., and Louloudes, S. J. (1973). *Misc. Publ. Entomol. Soc. Am.* **9**, 66–72.

Grace, T. D. C. (1962). *Nature* (London) **195**, 788–789.

Grace, T. D. C. (1968). *Exp. Cell Res.* **52**, 451–458.

Granados, R. R. (1976). *Adv. Virus Res.* **20**, 189–236.

Granados, R. R., and Naughton, M. (1975). *Intervirology* **5**, 62–68.

Granados, R. R., McCarthy, W. J., and Naughton, M. (1974). *Virology* **59**, 584–586.

Granboulan, N. (1967). *In* "Methods in Virology" (K. Maramorosch and H. Koprowski, eds.), Vol. 3, pp. 617–637. Academic Press, New York.

Greene, A. E., and Charney, J. (1971). *Curr. Top. Microbiol. Immunol.* **55**, 51–61.

Greene, A. E., and Charney, J. (1973). *In* "Tissue Culture: Methods and Applications" (P. F. Kruse, Jr. and M. K. Patterson, Jr., eds.), pp. 753–758. Academic Press, New York.

Greene, A. E., Charney, J., Nichols, W. W., and Coriell, L. L. (1972). *In Vitro* **7**, 313–322.

Guild, G. M., and Stollar, V. (1975). *Virology* **67**, 24–41.

Halstead, S. B. (1970). *Yale J. Biol. Med.* **42**, 350–362.

Ham, R. G. (1973). *In* "Tissue Culture: Methods and Applications" (P. F. Kruse, Jr. and M. K. Patterson, Jr., eds.), pp. 254–261. Academic Press, New York.

Hanks, J. H., and Wallace, R. E. (1949). *Proc. Soc. Exp. Biol. Med.* **71**, 196–200.

Hann, W. D., and Clarke, R. B. (1971). *Curr. Top. Microbiol. Immunol.* **55**, 149–150.

Hannoun, C., and Echalier, G. (1971). *Curr. Top. Microbiol. Immunol.* **55**, 221–230.

Hansen, E. L. (1976). *In* "Invertebrate Tissue Culture: Research Applications" (K. Maramorosch, ed.), pp. 75–99. Academic Press, New York.

Harrap, K. A. (1973). *In* "Viruses and Invertebrates" (A. J. Gibbs, ed.), pp. 271–299. North-Holland Publ., Amsterdam.

Hayflick, L. (1973). *In* "Tissue Culture Methods and Applications" (P. F. Kruse, Jr. and M. K. Patterson, Jr., eds.), pp. 822–827. Academic Press, New York.

Henderson, J. F., Faulkner, P., and MacKinnon, E. A. (1974). *J. Gen. Virol.* **22**, 143–146.

Hink, W. F. (1970). *Nature (London)* **226**, 466–467.

Hink, W. F. (1972a). *Adv. Appl. Microbiol.* **15**, 157–214.

Hink, W. F. (1972b). *In* "Invertebrate Tissue Culture" (C. Vago, ed.), Vol. 2, pp. 363–387. Academic Press, New York.

Hink, W. F. (1976). *In* "Invertebrate Tissue Culture: Research Applications" (K. Maramorosch, ed.), pp. 319–369. Academic Press, New York.

Hink, W. F., and Ellis, B. J. (1971). *Curr. Top. Microbiol. Immunol.* **55**, 19–28.

Hink, W. F., and Strauss, E. (1976a). *In* "Invertebrate Tissue Culture: Applications in Medicine, Biology, and Agriculture" (E. Kurstak and K. Maramorosch, eds.), pp. 297–300. Academic Press, New York.

Hink, W. F., and Strauss, E. (1976b). *J. Invertebr. Pathol.* **27**, 49–55.

Hink, F., and Vail, P. (1973). *J. Invertebr. Pathol.* **22**, 168–174.

Hirumi, H. (1976). *In* "Invertebrate Tissue Culture: Research Applications" (K. Maramorosch, ed.), pp. 233–268. Academic Press, New York.

Horne, R. W. (1967). *In* "Methods in Virology" (K. Maramorosch and H. Koprowski, eds.), Vol. 3, pp. 521–574. Academic Press, New York.

Hsu, S. H. (1971). *Curr. Top. Microbiol. Immunol.* **55**, 140–148.

Hsu, S. H., Wang, B. T., Huang, M. H., Wong, W. J., and Cross, J. H. (1975). *Am. J. Trop. Med. Hyg.* **24**, 881–888.

Huang, A. S. (1973). *Annu. Rev. Microbiol.* **27**, 101–117.

Huang, A. S., and Baltimore, D. (1970). *Nature (London)* **226**, 325–327.

Hukuhara, T., and Hashimoto, Y. (1967). *J. Invertebr. Pathol.* **9**, 278–281.

Igarashi, A., and Mantani, M. (1974). *Biken J.* **17**, 87–93.

Igarashi, A., Sasao, F., Wungkobkiat, S., and Fukai, K. (1973a). *Biken J.* **16**, 17–23.

Igarashi, A., Fukuoka, T., Sasao, F., Surimarat, S., and Fukai, K. (1973b). *Biken J.* **16**, 67–73.

Igarashi, A., Sasao, F., and Fukai, K. (1973c). *Biken J.* **16**, 95–101.

Ignoffo, C. M., Shapiro, M., and Hink, W. F. (1971). *J. Invertebr. Pathol.* **18**, 131–134.

Isaacs, A., and Lindemann, J. (1957). *Proc. R. Soc. London* **147**, 258–267.
Kärber, G. (1931). *Naunyn-Schmiedebergs Arch. Exp. Pathol. Pharmakol.* **162**, 480–483.
Karabatsos, N. (1969). *Am. J. Trop. Med. Hyg.* **18**, 803–810.
Karabatsos, N., and Buckley, S. M. (1967). *Am. J. Trop. Med. Hyg.* **16**, 99–105.
Kascsak, R. J., and Lyons, M. J. (1974). *Arch. Gesamte Virusforsch.* **44**, 1–6.
Kawarabata, T., and Hayashi, Y. (1972). *J. Invertebr. Pathol.* **19**, 414–415.
Kelly, D. C. (1976). *Virology* **69**, 596–606.
Kelly, D. C., and Tinsley, T. W. (1970). *J. Invertebr. Pathol.* **16**, 470–472.
Kelly, D. C., and Tinsley, T. W. (1974a). *Microbios* **9**, 75–93.
Kelly, D. C., and Tinsley, T. W. (1974b). *J. Invertebr. Pathol.* **24**, 169–178.
Kemp, G. E., Causey, O. R., Moore, D. L., Odeola, A., and Fabiyi, A. (1972). *Am. J. Trop. Med. Hyg.* **21**, 356–359.
Kemp, G. E., Lee, V. H., Moore, D. L., Shope, R. E., Causey, O. R., and Murphy, F. A. (1973). *Am. J. Epidemiol.* **98**, 43–49.
Kenny, G. E. (1973). *In* "Contamination in Tissue Culture" (J. Fogh, ed.), pp. 107–129. Academic Press, New York.
Kitamura, S. (1974). *Abstr. Conf. Invertebr. Tissue Cult.: Appl. Fund. Res., 1974* pp. 14–15.
Knudson, D. L. (1973). *J. Gen. Virol.* **20**, Suppl., 105–130.
Knudson, D. L. (1975). D.Phil. Thesis, University of Oxford, United Kingdom.
Knudson, D. L., and Harrap, K. A. (1976). *J. Virol.* **17**, 254–268.
Knudson, D. L., and Tinsley, T. W. (1974). *J. Virol.* **14**, 934–944.
Kurtti, T. J. (1976). *In* "Invertebrate Tissue Culture: Research Applications" (K. Maramorosch, ed.), pp. 39–56. Academic Press.
Kurtti, T. J., and Brooks, M. A. (1971). *Curr. Top. Microbiol. Immunol.* **55**, 204–208.
Landureau, J. C. (1966). *Exp. Cell Res.* **41**, 545–556.
Levin, J. G., Ramseur, J. M., and Grimley, P. M. (1973). *J. Virol.* **12**, 1401–1406.
Libiková H., and Buckley, S. M. (1971). *Acta Virol. (Engl. Ed.)* **15**, 393–403.
Lonberg-Holm, K., and Philipson, L. (1974). *Monogr. Virol.* **9**, 1–148.
Lwoff, A., Dulbecco, R., Vogt, M., and Lwoff, M. (1955). *Virology* **1**, 128–139.
McIntosh, A. H. (1976). *In* "Invertebrate Tissue Culture: Research Applications" (K. Maramorosch, ed.), pp. 3–12. Academic Press, New York.
McIntosh, A. H., and Kimura, M. (1974). *Intervirology* **4**, 257–267.
McIntosh, A. H., and Rechtoris, C. (1974). *In Vitro* **10**, 1–5.
MacKinnon, E. A., Henderson, J. F., Stoltz, D. B., and Faulkner, P. (1974). *J. Ultrastruct. Res.* **49**, 419–435.
Macpherson, I. (1969). *In* "Fundamental Techniques in Virology" (K. Habel and N. P. Salzman, eds.), Vol. 1, pp. 17–20. Academic Press, New York.
Macpherson, I. (1973a). *In* "Tissue Culture Methods and Applications" (P. F. Kruse, Jr. and M. K. Patterson, Jr. eds.), pp. 241–244. Academic Press, New York.
Macpherson, I. (1973b). *In* "Tissue Culture: Methods and Applications" (P. F. Kruse, Jr. and M. K. Patterson, Jr., eds.), pp. 276–280. Academic Press, New York.
Macpherson, I., and Stoker, M. (1962). *Virology* **16**, 147–151.
McSharry, J., and Benzinger, R. (1970). *Virology* **40**, 745–779.
Malková, D., and Marhoul, Z. (1973). *Proc. Int. Colloq. Invertebr. Tissue Cult. 3rd, 1971* pp. 267–274.
Maramorosch, K., ed. (1976). "Invertebrate Tissue Culture: Research Applications." Academic Press, New York.
Maramorosch, K., and Koprowski, H., eds. (1967). "Methods in Virology," Vol. 2, Academic Press, New York.
Marhoul, Z. (1973a). *Proc. Int. Colloq. Invertebr. Tissue Cult. 3rd, 1971* pp. 275–290.

Marhoul, Z. (1973b). *Acta Virol. (Engl. Ed.)* **17**, 507–509.
Marks, E. P. (1973). *In* "Tissue Culture: Methods and Applications" (P. F. Kruse, Jr. and M. K. Patterson, Jr., eds.), pp. 153–156. Academic Press, New York.
Marshall, I. D. (1973). *In* "Viruses and Invertebrates" (A. J. Gibbs, ed.), pp. 406–427. North-Holland Publ., Amsterdam.
Mayer, V., and Kozuch, O. (1969). *Acta Virol. (Engl. Ed.)* **13**, 469–482.
Mettler, N. E., and Buckley, S. M. (1973). *Proc. Int. Colloq. Invertebr. Tissue Cult. 3rd, 1971* pp. 255–265.
Meynadier, G. (1971). *In* "Invertebrate Tissue Culture" (C. Vago, ed.), Vol. 1, pp. 141–167. Academic Press, New York.
Miles, J. A. R., and Austin, F. J. (1963). *Aust. J. Sci.* **25**, 466–467.
Mirchamsy, H., Hazrati, A., Bahrami, S., and Shafyi, A. (1970). *J. Vet. Res.* **31**, 1755–1761.
Mitsuhashi, J. (1967). *Nature (London)* **215**, 863–864.
Mitsuhashi, J., and Maramorosch, K. (1964). *Contrib. Boyce Thompson Inst.* **22**, 435–460.
Morgan, C., and Rose, H. M. (1967). *In* "Methods in Virology" (K. Maramorosch and H. Koprowski, eds.), Vol. 3, pp. 575–616. Academic Press, New York.
Mugo, W. N., and Shope, R. E. (1972). *Trans R. Soc. Trop. Med. Hyg.* **66**, 300–304.
Murray, A. M., and Morahan, P. S. (1973). *Proc. Soc. Exp. Biol.* **142**, 11–15.
Mussgay, M., Enzmann, P. J., Horzinck, M. C., and Weiland, E. (1975). *Prog. Med. Virol.* **19**, 257–323.
Nairn, R. C. (1969). "Fluorescent Protein Tracing," 3rd ed. Williams & Wilkins, Baltimore, Maryland.
Nichols, W. W., Bradt, C., and Bowne, W. (1971). *Curr. Top. Microbiol. Immunol.* **55**, 61–69.
Noyes, W. F. (1955). *J. Exp. Med.* **102**, 243–248.
Ohanessian, A. (1971). *Curr. Top. Microbiol. Immunol.* **55**, 230–233.
Paschke, J. D., and Webb, S. R. (1976). *In* "Invertebrate Tissue Culture: Research Applications" (K. Maramorosch, ed.), pp. 269–293. Academic Press, New York.
Paul, J. (1961). "Cell and Tissue Culture," 2nd ed. Williams & Wilkins, Baltimore, Maryland.
Pavri, K. M., and Ghosh, S. N. (1969). *Bull. W.H.O.* **40**, 984–986.
Peleg, J. (1968a). *Am. J. Trop. Med. Hyg.* **17**, 219–223.
Peleg, J. (1968b). *Virology* **35**, 617–619.
Peleg, J. (1969). *J. Gen. Virol.* **5**, 463–471.
Peleg, J. (1971). *Curr. Top. Microbiol. Immunol.* **55**, 155–161.
Peleg, J. (1972). *Arch. Gesamte Virusforsch.* **37**, 54–61.
Perrault, J., and Holland, J. (1972). *Virology* **50**, 148–158.
Phillips, H. J. (1973). *In* "Tissue Culture: Methods and Applications" (P. F. Kruse, Jr. and M. K. Patterson, Jr., eds.), pp. 406–408. Academic Press, New York.
Potter, K. N., Faulkner, P., and MacKinnon, E. A. (1976). *J. Virol.* **18**, 1040–1050.
Pressman, D., Yagi, Y., and Hiramoto, R. (1958). *Int. Arch. Allergy Appl. Immunol.* **12**, 125–136.
Printz, P. (1973). *Adv. Virus Res.* **18**, 143–157.
Raghow, R. S., and Grace, T. D. C. (1974). *J. Ultrastruct. Res.* **47**, 384–399.
Raghow, R. S., Davey, M. W., and Dalgarno, L. (1973a). *Arch. Gesamte Virusforsch.* **43**, 165–168.
Raghow, R. S., Grace, T. D. C., Filshie, B. K., Bartley, W., and Dalgarno, L. (1973b). *J. Gen. Virol.* **21**, 109–122.
Rapp, F., and Benyesh-Melnick, M. (1963). *Science* **141**, 433–434.
Rapp, F., Seligman, J., Jaross, L. B., and Gordon, I. (1959). *Proc. Soc. Exp. Biol. Med.* **101**, 289–294.
Reed, L. J., and Muench, H. (1938). *Am. J. Hyg.* **27**, 493–497.

Rehaček, J. (1968). *Acta Virol.* (*Engl. Ed.*) **12**, 241–246.
Rehaček, J., and Kozuch, O. (1969). *Acta Virol.* (*Engl. Ed.*) **13**, 253.
Riggs, J. L., Seiwald, R. J., Burckhalter, J. H., Downs, C. M., and Metcalf, T. G. (1958). *Am. J. Pathol.* **34**, 1081–1097.
Rinaldini, L. M. (1954). *Nature* (*London*) **173**, 1134–1135.
Ryser, H. J. P. (1967). *J. Cell Biol.* **32**, 737–750.
Sanders, F. K. (1973). *In* "Contamination in Tissue Culture" (J. Fogh, ed.), pp. 243–256. Academic Press, New York.
Sanford, K. K. (1973). *In* "Tissue Culture: Methods and Applications" (P. F. Kruse, Jr. and M. K. Patterson, Jr., eds.), pp. 237–241. Academic Press, New York.
Sanford, K. K., Earle, W. R., and Likely, G. D. (1948). *J. Nat. Cancer Inst.* **9**, 229–246.
Schlesinger, R. W. (1959). *In* "The Viruses" (F. M. Burnet and W. M. Stanley, eds.), Vol. 3, pp. 157–194. Academic Press, New York.
Schlesinger, S., Schlesinger, M., and Burge, B. W. (1972). *Virology* **48**, 615–617.
Schneider, I. (1969). *J. Cell Biol.* **42**, 603–606.
Schneider, I. (1973a). *In* "Tissue Culture: Methods and Applications" (P. F. Kruse, Jr. and M. K. Patterson, Jr., eds.). pp. 150–152. Academic Press, New York.
Schneider, I. (1973b). *In* "Tissue Culture: Methods and Applications" (P. F. Kruse, Jr. and M. K. Patterson, Jr., eds.), pp. 788–700. Academic Press, New York.
Schneider, L. (1973). *In* "Laboratory Techniques in Rabies" (M. M. Kaplan and H. Koprowski, eds.), 3rd ed., pp. 339–342. World Health Organ., Geneva.
Sellers, R. F. (1963). *Trans. R. Soc. Trop. Med. Hyg.* **57**, 433–437.
Shannon, J. R., and Macy, M. L. (1973b). *In* "Tissue Culture: Methods and Applications" (P. F. Kruse, Jr. and M. K. Patterson, Jr., eds.), pp. 712–718. Academic Press, New York.
Shannon, J. R., and Macy, M. L. (1973b). *In* "Tissue Culture: Methods and Applications" (P. F. Kruse, Jr. and M. K. Patterson, Jr., eds.), pp. 804–807. Academic Press, New York.
Shenk, T. E., and Stollar, V. (1972). *Biochem. Biophys. Res. Commun.* **49**, 60–67.
Shenk, T. E., and Stollar, V. (1973). *Virology* **53**, 162–173.
Shenk, T. E., Koshelnyk, K. A., and Stollar, V. (1974). *J. Virol.* **13**, 439–447.
Shope, R. E. (1975). *In* "The Natural History of Rabies" (G. M. Baer, ed.), pp. 141–152. Academic Press, New York.
Shope, R. E., Murphy, F. A., Harrison, A. K., Causey, O. R., Kemp, G. E., Simpson, D. I. H., and Moore, D. L. (1970). *J. Virol.* **6**, 690–692.
Sinarachatanant, P., and Olson, L. C. (1973). *J. Virol.* **12**, 275–283.
Singh, K. R. P. (1971). *Curr. Top. Microbiol. Immunol.* **55**, 127–133.
Singh, K. R. P. (1972). *Adv. Virus Res.* **17**, 187–206.
Singh, K. R. P., and Paul, S. D. (1968a). *Curr. Sci.* **37**, 65–67.
Singh, K. R. P., and Paul, S. D. (1968b). *Indian J. Med. Res.* **56**, 815–820.
Singh, K. R. P., and Paul, S. D. (1969). *Bull. W.H.O.* **40**, 982–983.
Singh, K. R. P., Goverdhand, M. K., and Bhat, U. K. M. (1973). *Proc. Int. Colloq. Invertebr. Tissue Cult., 3rd, 1971* pp. 291–298.
Smith, A., Tignor, G. H., Mifune, K., and Motohashi, T. (1977). *Intervirology* (in press).
Smith, C. N., ed. (1966). "Insect Colonization and Mass Production." Academic Press, New York.
Smorodintsev, A. A., Dubov, A. V., Ilynko, V. I., and Platonov, V. G. (1969). *J. Hyg.* **67**, 13–21.
Sohi, S. S. (1971). *Can. J. Zool.* **49**, 1355–1358.
Sohi, S. S. (1973). *Proc. Int. Colloq. Invertebr. Tissue Cult. 3rd, 1971* pp. 75–92.
Sohi, S. S., and Cunningham, J. C. (1972). *J. Invertebr. Pathol.* **19**, 51–61.
Stanley, M. S. M., and Vaughn, J. L. (1967). *J. Insect Physiol.* **13**, 1613–1617.

Stevens, T. M. (1970). *Proc. Soc. Exp. Biol. Med.* **134**, 356–361.
Stim, T. B. (1969). *J. Gen. Virol.* **5**, 329–338.
Stollar, V., and Shenk, T. E. (1972). *J. Virol.* **11**, 592–595.
Stollar, V., Shenk, T. E., Koo, R., Igarashi, A., and Schlesinger, W. R. (1975). *Ann. N. Y. Acad. Sci.* **266**, 214–231.
Stulberg, C. S. (1973). *In* "Contamination in Tissue Culture" (J. Fogh, ed.), pp. 1–27. Academic Press, New York.
Suitor, E. C., Jr. (1969). *J. Gen. Virol.* **5**, 545–546.
Suitor, E. C., Jr., and Paul, F. J. (1969). *Virology* **38**, 482–485.
Suitor, E. C., Jr., Chang, L. L., and Liu, H. H. (1966). *Exp. Cell Res.* **44**, 572–578.
Summers, M. D., and Volkman, L. E. (1976). *J. Virol.* **17**, 962–972.
Sweet, B. H., and Unthank, H. D. (1971). *Curr. Top. Microbiol. Immunol.* **55**, 150–154.
Theiler, M. (1957). *Proc. Soc. Exp. Biol. Med.* **96**, 380–382.
Theiler, M., and Downs, W. G. (1973). "The Arthropod-Borne Viruses of Vertebrates". Yale Univ. Press, New Haven, Connecticut.
Trager, W. (1935). *J. Exp. Med.* **61**, 501–513.
Vago, C. (1967). *In* "Methods in Virology" (K. Maramorosch and H. Koprowski, eds.), Vol. 1, pp. 567–602. Academic Press, New York.
Vago, C. (1970). *Rev. Pathol. Comp. Med. Exp.* **7**, 259–268.
Vago, C., ed. (1971). "Invertebrate Tissue Culture," Vol. 1. Academic Press, New York.
Vago, C., ed. (1972). "Invertebrate Tissue Culture," Vol. 2. Academic Press, New York.
Vail, P. V. (1975). *In* "Baculoviruses for Insect Pest Control: Safety Considerations" (M. Summers *et al.*, eds.), pp. 44–46. Am. Soc. Microbiol., Washington, D. C.
Vail, P. V., Jay, D. L., and Hink, W. F. (1973). *J. Invertebr. Pathol.* **22**, 231–237.
Varma, M. G. R., and Pudney, M. (1969). *J. Med. Entomol.* **6**, 432–439.
Varma, M. G. R., Pudney, M., and Leake, C. J. (1974a). *Trans. R. Soc. Trop. Med. Hyg.* **68**, 374–382.
Varma, M. G. R., Pudney, M., and Leake, C. J. (1974b). *J. Med. Entomol.* **11**, 698–706.
Varma, M. G. R., Pudney, M., Leake, C. J., and Peralta, P. H. (1976). *Intervirology* **6**, 50–56.
Vaughn, J. L. (1968). *Curr. Top. Microbiol. Immunol.* **42**, 108–128.
Vaughn, J. L. (1971). *In* "Invertebrate Tissue Culture" (C. Vago, ed.), Vol. 1, pp. 3–40. Academic Press, New York.
Vaughn, J. L., and Faulkner, P. (1963). *Virology* **20**, 484–489.
Vaughn, J. L., and Stanley, M. S. M. (1970). *J. Invertebr. Pathol.* **16**, 357–362.
Volkman, L. E., and Summers, M. D. (1975). *J. Virol.* **16**, 1630–1637.
Volkman, L. E., and Summers, M. D. (1976). *In* "Invertebrate Tissue Culture: Applications in Medicine, Biology, and Agriculture" (E. Kurstak and K. Maramorosch, eds.), pp. 289–296. Academic Press, New York.
von Magnus, P. (1954). *Adv. Virus Res.* **2**, 59–79.
Wagner, R. R. (1963). *Annu. Rev. Microbiol.* **17**, 285–296.
Wagner, R. R., Levy, A. H., and Smith, T. J. (1968). *In* "Methods in Virology" (K. Maramorosch and H. Koprowski, eds.), Vol. 4, pp. 1–52. Academic Press, New York.
Ward, T. G. (1968). *In* "Methods in Virology" (K. Maramorosch and H. Koprowski, eds.), Vol. 4, pp. 481–489. Academic Press, New York.
Webb, S. R., Paschke, J. D., Wagner, G. W., and Campbell, W. R. (1974). *J. Invertebr. Pathol.* **23**, 255–258.
Webb, S. R., Paschke, J. D., Wagner, G. W., and Campbell, W. R. (1975). *J. Invertebr. Pathol.* **26**, 205–212.
Weiss, E. (1971). *Curr. Top. Microbiol. Immunol.* **55**, 1–288.

Weller, T. H. (1953). *Proc. Soc. Exp. Biol. Med.* **83**, 340–346.
Weller, T. H., and Coons, A. H. (1954). *Proc. Soc. Exp. Biol. Med.* **86**, 789–794.
Weller, T. H., Wilton, H. M., and Bell, E. J. (1958). *J. Exp. Med.* **108**, 843–868.
White, G. F. (1931). *J. Parasitol.* **18**, 133.
Whitney, E., and Deibel, R. (1971). *Curr. Top. Microbiol. Immunol.* **55**, 138–139.
Wildy, P. (1971). *Monogr. Virol.* **5**, 1–81.
Willis, N. G., and Campbell, J. B. (1973). *Proc. Int. Colloq. Invertebr. Tissue Cult. 3rd, 1971*
    pp. 347–366.
Yang, Y. J., Stoltz, D. B., and Prevec, L. (1969). *J. Gen. Virol.* **5**, 473–483.
Yunker, C. E. (1971). *Curr. Top. Microbiol. Immunol.* **55**, 113–126.
Yunker, C. E., and Cory, J. (1968). *Am. J. Trop. Med. Hyg.* **17**, 889–893.
Yunker, C. E., and Cory, J. (1969). *J. Virol.* **3**, 631–632.
Yunker, C. E., and Cory, J. (1975). *Appl. Microbiol.* **29**, 81–89.
Yunker, C. E., Vaughn, J. L., and Cory, J. (1967). *Science* **155**, 1565–1566.

11 Techniques of Invertebrate Tissue Culture for the Study of Plant Viruses

D. V. R. Reddy

I. Introduction

Viruses causing plant disease have played a significant role in the discovery of viruses. Unfortunately, the virus–plant host system later proved to be considerably less attractive than bacterial or vertebrate virus systems for fundamental research. This was primarily because the inoculation of plant viruses to plants or to walled plant cells in cultures suffered from many limitations for precise assays and for investigations at the cellular and subcellular level. Advancements are now being made in obtaining plant protoplast cultures, yet technology is still not available to maintain them continuously and they therefore have obvious disadvantages when compared to continuous cell lines.

During a period of controversy over whether any plant viruses multiplied in their insect vectors, experiments were conducted on this question with several pathogens then regarded as viruses. Some of them were later shown to be nonviral. In retrospect it is clear that the first plant pathogen proven to be a virus and proven to multiply in its insect vector was wound tumor virus (WTV) (Black and Brakke, 1952; Brakke *et al.*, 1953, 1954). The idea of using cultured insect vector cells for plant virus study, as cultured vertebrate cells were being used for the study of vertebrate viruses, must have occurred in many laboratories after it was believed that certain plant viruses do multiply in their vectors. Because of the state of the science of growing insect cells in cultures at that time and because most of the vectors involved weighed about 1 mg, a number of attempts to culture their cells probably ended in several laboratories as unpublished failures. As in any other tissue culture systems, techniques and culture media had to be developed which would support a particular insect cell line. The first significant progress toward this goal came with the establishment of *Antheraea eucalypti* Scott cell cultures by Grace (1962) and the successful growth of insect vector cells in primary cultures by Hirumi and Maramorosch (1964a,b,c) and Mitsuhashi and Maramorosch (1964). These discoveries led to a renewed attack on the problem in Black's laboratory (Chiu *et al.*, 1966) and ultimately Chiu and Black (1967, 1969) achieved success in obtaining continuous cell cultures of agallian leafhopper vectors. Using the WTV system in the newly developed *Agallia constricta* Van Duzee cell lines, they showed that virus extracted from infected plant tissues could initiate infection in insect monolayers. It is therefore apparent that continuous cultures of vectors of plant viruses can be of use in the investigations of several aspects of plant virus research.

The primary objective of this chapter is to describe the various methods that have been used to establish leafhopper cell lines and the methods available to infect them with plant viruses. A brief account of methods available for obtaining primary aphid cell cultures will also be described. A large number of recent reviews dealing with homoptera cell cultures and their application to the study of plant viruses have been published (Black, 1969; Hirumi, 1971; Hirumi and Maramorosch, 1971; Maramorosch, 1976; Mitsuhashi, 1972; Vago, 1967).

II. Establishment of Primary Cultures

Attempts to establish cell lines from tissues and organs derived from nymphs or adult leafhoppers have not been successful. All the established leafhopper cell lines have been derived from embryonic tissues.

A. COLLECTION OF EGGS

Hirumi and Maramorosch (1964b) have demonstrated that 6- or 7-day-old eggs were a better source for obtaining embryos than older or younger eggs. Six-day-old eggs were distinguished by a small pigmented eye spot, which migrated to the anterior region (blastokinetic movement) during the subsequent 2 or 3 days (Fig. 1). At the same time the spot increased in size and density of color. The eggs harvested between the 7th and 9th day were most suitable for obtaining embryos. Normally the eye spot could be clearly visualized only after it migrated to the anterior region.

To obtain the eggs of *Macrosteles fascifrons* Stal., Hirumi and Maramorosch (1964a) recommended transferring approximately 200 male and female adult insects to rye (*Secale cereale* L.) plants maintained at 25°C with 16 hours of light per day at 6600 lumens/m² from standard fluorescent tubes. The insects were transferred every day. Eggs of the appropriate age were dissected from the plants.

In order to obtain the embryos of *Nephotettix cincticeps* Uhler, mated females were confined to rice reedlings (*Oryza sativa* L.) in glass tubes for 2 days and then transferred to a fresh set of rice seedlings (Mitsuhashi, 1965).

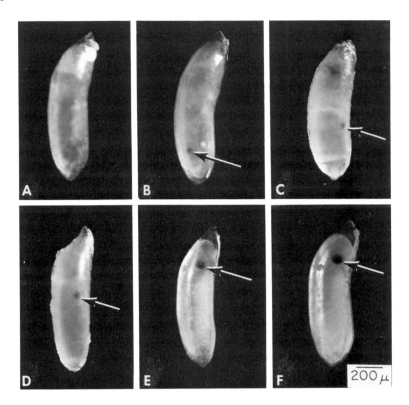

FIG. 1. Developmental stages of the embryo of *Macrosteles fascifrons* (×30). (A) The 5-day-old egg. The eye spot is not visible. (B–D) Seven- or 8-day-old eggs. Note the movement of the pigmented eye spot from the posterior to the anterior region of the egg. Normally the eye spot when it reaches the stage in (D) is clearly visible and the eggs at this stage are most suitable for obtaining embryos. (E) and (F) represent 10- and 11-day-old eggs, respectively. Note the increase in size and density of color of the eye spot. The eggs in stages (E) and (F) are not suitable for establishing primary cultures. [From Hirumi and Maramorosch (1964b). By courtesy of *Science*. Copyright 1964 by the American Association for the Advancement of Science.]

For obtaining the eggs of *A. constricta* and *Agalliopsis novella* Say, 50 or more adult female insects, drawn from a mixed colony, were transferred to small crimson clover (*Trifolium incarnatum* L.) plants with 2 or 3 petioles. Insects were removed after a day and the plants were kept at 27 °C with 16 hours of light per day. Eggs were dissected from the petioles 7 days after oviposition, in the case of *A. constricta*, and 6 days in the case of *A. novella* (Liu and Black, 1976). Petioles were disinfected in 70% ethanol for 1 minute, followed by at least 6 rinses in sterile distilled water before the eggs were removed from them (Chiu *et al.*, 1966). Hirumi and Maramorosch

(1971) used rye plants instead of clover for obtaining the eggs of *A. constricta*.

Mitsuhashi (1970) devised a cage in order to collect the eggs of some planthoppers and leafhoppers. The insects were transferred to a cage with a medium container compartment. Stretched parafilm was used to separate the liquid container from the cage. The insects were kept at 25 °C under 16 hours of light per day. Insects laid their eggs through the parafilm, some of which dropped into the media while others hung onto the parafilm singly or in clusters. Five percent and 10% sucrose solutions were found to be suitable for the oviposition of several planthopper and leafhopper species (Mitsuhashi and Koyama, 1975). *Nephotettix cincticeps*, *Balclutha viridis* Matsumura, and other leafhopper species did not oviposit in sucrose solutions. The embryogenesis progressed quite normally when the eggs were left in distilled water for long periods.

B. SURFACE DISINFECTION OF EGGS

It is important to choose suitable sterilizing agents because embryos of many species of leafhoppers are very sensitive to certain ones. Treatment with mercuric chloride should be used cautiously (Mitsuhashi, 1972). It is convenient to transfer the eggs to depression slides of 1.0 ml capacity for surface sterilization (Fig. 2). Although several methods have been described in the literature, treatments in 70% ethanol for 1 minute (Mitsuhashi, 1965;

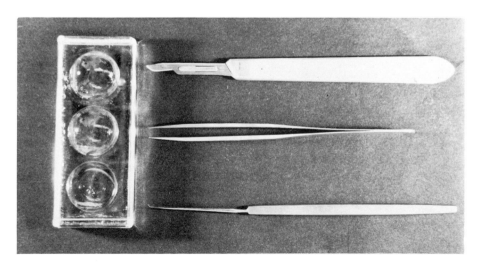

FIG. 2. Instruments required for dissecting the embryos. Note that the needle is bent at the tip.

Hirumi and Maramorosch, 1968) appeared to be the most desirable one. Following treatment in ethanol, the eggs were rinsed in the same slide by replacing ethanol with 3 to 4 changes of sterile distilled water.

C. DISSECTION OF EMBRYOS

Eggs rinsed in sterile water were washed in Tyrode's salt solution (to prepare 1 liter, 8.0 gm of NaCl, 0.2 gm of KCl, 0.2 gm of $CaCl_2$, 0.1 gm of $MgCl_2 \cdot 6H_2O$, 0.05 gm of $NaH_2PO_4 \cdot H_2O$, and 1.0 gm of glucose were added) containing 0.5% of bovine serum albumin (Chiu *et al.*, 1966). All dissections were done under a dissection microscope at 16 × magnification. The various steps for dissecting the embryos and the required instruments are presented in Figs. 2 and 3. Holding the posterior end of the egg with a fine forceps, an incision was made at the anterior end with a surgical blade (#15) attached to a scalpel. The embryo was gently squeezed out of the eggshell using pressure at the posterior end while the anterior end was held with a fine needle. The eggshell and yolk were removed. At least 35 embryos were collected in each well of the depression slide. Hirumi and Maramorosch (1964a) used Earle's balanced salt solution and Mitsuhashi and Maramorosch (1964) employed Rinaldini's salt solution (Rinaldini, 1954) for dissecting the embryos.

D. TRYPSINIZATION

Hirumi and Maramorosch (1964a) recommended treatment of the tissue, after being cut into small fragments, with 0.02% tyrpsin prepared in Earle's balanced salt solution at 26 °C for 10 minutes. Mitsuhashi and Maramorosch (1964), Chiu *et al.* (1966), and Chiu and Black (1967) recommended treatment with 0.25% trypsin (Difco trypsin, 1:250) prepared in Rinaldini's citrate salt solution (Rinaldini, 1959) at room temperature for 15 minutes. Trypsinization beyond 30 minutes normally rendered the tissue unsuitable for culturing. The initial growth of cells was improved as a result of trypsin treatment of the embryos (Chiu *et al.*, 1966), although it was not absolutely essential (Mitsuhashi and Maramorosch, 1964; Chiu and Black, 1967; Tokumitsu and Maramorosch, 1967).

Following tyrpsinization, the embryos should be handled more carefully since they tend to adhere to each other or to surface with which they come in contact. The trypsinization process was stopped by washing several times with culture medium. Alternate low-speed centrifugation of the tissues (at 1000 rpm for 5 minutes) in conical 5 ml tubes and resuspension in culture medium with a Pasteur pipette was commonly employed to accomplish the washing of tissues. This procedure resulted in the loss of some tissue frag-

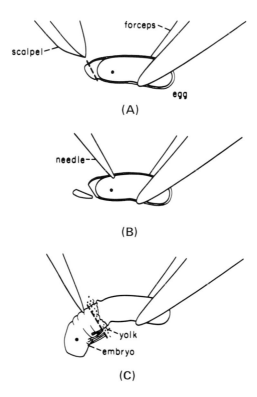

FIG. 3. Dissection of leafhopper embryos. (A) Holding the posterior end of the egg with a fine forceps, an incision is made with a surgical blade (#15) attached to a scalpel. Dotted line represents the area for making the incision. (B and C) Squeezing of the embryos from the egg shell, using pressure at the posterior end while holding the anterior end with a fine needle. Part of the embryo, marked with a dotted line (in C), is used for starting the primary cultures.

ments because of their adherance to glass surface. Liu and Black (Section III,H) devised a simple method to sediment the tissues to the bottom of the well of the depression glass slide, in which they were dissected. Thus the loss of tissue during low-speed sedimentation and resuspension was eliminated.

E. CULTURE VESSELS

Sitting drop cultures, prepared in culture flasks containing approximately 0.2 ml of the medium with trypsinized tissue fragments derived from about 10 embryos, was employed by Hirumi and Maramorosch (1964a). The sitting drop culture technique was also employed in vessels prepared from a glass ring (10 mm in height) and sealed on top and bottom using 25 mm cover glasses (Mitsuhashi and Maramorosch, 1964; Mitsuhashi, 1965; Tokumitsu and Maramorosch, 1967; Hirumi and Maramorosch, 1968). Trypsinized tissue fragments derived from at least 35 embryos, in a small drop of the medium (approximately 0.1 ml), were transferred to a Falcon tissue culture dish (35 × 10 mm) (Falcon Plastics, Los Angeles, California) and placed in a sealable glass dish. From 8 to 10 drops of the medium were added to the glass dish to maintain the desired vapor pressure and then sealed with silicone grease (Fig. 4) (Liu and Black, 1976).

F. CULTURE MEDIUM

The medium employed by Chiu and Black (1967, 1969) (Table IV) was used for setting up primary cultures and for growing established cell lines. Although the medium formulated by Martínez-López and Black (1974) (Table V) was shown to be superior for growing established agallian leafhopper cell lines, it was unsatisfactory for establishing primary cell cultures (Liu, 1975). Thus caution should be exercised in setting up primary cultures using media developed for continuous cell cultures. The medium of Mitsuhashi and Maramorosch (1964) was used successfully to establish primary cultures. According to Chiu and Black (1967), this medium was not suitable for established cell lines. In addition there are no reports on the successful use of this medium to grow continuous leafhopper cell cultures.

FIG. 4. Vessel for growing primary culture. (A) 50 mm diameter sealable glass petri dish. (B) Small drops of the growth medium (8–10) to maintain the desired vapor pressure. (C) 35 × 10 mm Falcon tissue culture dish and (D) trypsinized embryo fragments in a small drop of the medium.

The compositions of the media employed for growing primary cultures are given in Tables I–IV. The medium developed by Chiu and Black (Table IV) was successfully used to establish cultures of *A. constricta*, *A. novella*, *Agallia quadripunctata* Provancher, *Aceratagallia sanguinolenta* Provancher (Chiu and Black, 1967, 1969; Liu and Black, 1976), and *Colladonus montanus* Van Duzee (Richardson and Jensen, 1971). The medium is prepared by mixing

TABLE I

CULTURE MEDIUM FOR GROWING LEAFHOPPER PRIMARY
CELL CULTURES[a]

Components	Concentration (ml)
Morgan's synthetic medium Tc199[b]	100
Modified Vago's medium BM22[c]	100
Fetal bovine serum	30
Penicillin–streptomycin mixture[d]	5

[a] From Hirumi and Maramorosch (1964).
[b] From Microbiological Associates, Bethesda, Maryland.
[c] 1 liter of medium contains: 1.2 gm $NaH_2PO_4 \cdot H_2O$; 3.0 gm $MgCl_2 \cdot 6H_2O$; 4.0 gm $MgSO_4 \cdot 7H_2O$; 3.0 gm KCl; 1.0 gm $CaCl_2 \cdot 2H_2O$; 0.7 gm glucose; 0.4 gm sucrose; 0.4 gm fructose; 10.0 gm lactalbumin hydrolysate; and 5.0 ml 0.5% phenol red.
[d] Microbiological Associates Inc., Bethesda, Maryland.

TABLE II

CULTURE MEDIUM FOR GROWING LEAFHOPPER PRIMARY
CELL CULTURES[a]

Components	Concentration (gm/liter)
$CaCl_2 \cdot 2H_2O$	0.2
KCl	0.2
$MgCl_2 \cdot 6H_2O$	0.1
NaCl	7.0
$NaHCO_3$	0.12
$NaH_2PO_4 \cdot H_2O$	0.2
D-Glucose	4.0
Lactalbumin hydrolysate	6.5
Yeastolate	5.0
Fetal bovine serum	200 ml
Penicillin–streptomycin mixture[b]	10,000 (IU)
pH	6.5

[a] From Mitsuhashi and Maramorosch (1964).
[b] Microbiological Associates Inc., Bethesda, Maryland.

each of the following components, prepared in glass distilled water, in the following manner:

0.4 gm $CaCl \cdot 2H_2O$ in 7 ml
1.85 gm $MgSO_4 \cdot 7H_2O$ in 7 ml
0.8 gm KCl, 0.3 gm KH_2PO_4, and 1.05 gm NaCl in 15 ml
0.35 gm $NaHCO_3$ in 7 ml
4 gm of dextrose in 40 ml
6.50 gm lactalbumin hydrolysate in 200 ml
5.0 gm yeastolate in 100 ml
200 ml fetal bovine serum
100,000 Units of penicillin G, 0.1 gm streptomycin, and 0.05 gm neomycin in 20 ml

Fetal bovine serum is brought quickly to 56 °C with stirring, held at that temperature for 30 minutes and then cooled. Lactalbumin hydrolysate is prepared fresh each time by adding 6.5 gm to 200 ml of glass distilled water and heated to 50 °C until the solution is complete. When both fetal bovine serum and lactalbumin hydrolysate are cooled, all components are added to 401.5 ml of distilled water in 1.5 liter Erlenmeyer flask in the order listed above, while stirring vigorously.

The medium is sterilized by pressure filtration in previously autoclaved 142 mm Millipore filters in which 0.8, 0.45, and 0.22 μm filters are arranged serially with progressively larger sizes located on top and each filter

TABLE III

CULTURE MEDIUM FOR GROWING LEAFHOPPER PRIMARY
CELL CULTURES[a]

Components	Concentration (gm/liter)
NaCl	2.8
KCl	0.08
$CaCl \cdot H_2O$	0.08
$MgCl_2 \cdot 6H_2O$	0.04
NaH_2PO_4	0.08
$NaHCO_3$	0.05
D-Glucose	1.60
Lactalbumin hydrolysate	5.20
Bactopeptone	5.20
Yeastolate	2.00
Dihydrostreptomycin sulfate	0.05
Medium 199	200 ml
Fetal bovine serum	200 ml

[a] From Mitsuhashi (1965, 1972).

TABLE IV

CULTURE MEDIUM FOR GROWING LEAFHOPPER PRIMARY CELL
CULTURES AND ESTABLISHED CELL CULTURES[a]

Components	Concentration (gm/liter)
$CaCl_2 \cdot 2H_2O$	0.4
KCl	0.8
KH_2PO_4	0.3
$MgSO_4 \cdot 7H_2O$	1.85
NaCl	1.05
$NaHCO_3$	0.35
Dextrose	4.00
Lactalbumin hydrolysate	6.50
Yeastolate	5.00
Fetal bovine serum	200 ml
Neomycin	0.05
Penicillin G	100,000 (IU)
Streptomycin	0.10
pH	~7.0

[a] From Chiu and Black (1969).

separated from one another with dacron separators. A prefilter is used on top of the 0.8 μm filter.

Mitsuhashi's (1972) (Table III) medium was successfully employed to culture embryonic tissues of *M. fascifrons*, *A. constricta*, *Dalbulus elimatus* Ball, *N. cincticeps*, *Nephotettix apicalis* Motschulsky, and *Inazuma dorsalis* Motschulsky. The medium developed by Hirumi and Maramorosch (Table I) was suitable for establishing primary cultures of *M. fascifrons* (Hirumi and Maramorosch, 1964a,b,c). Attempts to obtain *M. fascifrons* primary cultures in Chiu and Black's medium were not successful whereas Hirumi and Maramorosch's medium yielded primary cultures of *M. fascifrons* (Liu and Black, 1976).

G. GROWTH OF PRIMARY CELL CULTURES

A few hours after cultivation, a number of dipolar single cells became attached to the surface of the culture flasks. They increased 10-fold during the next 2 days, yet they did not remain viable for long periods. Different cell types (largely epithelial and fibroblastic types) appeared 24–48 hours after cultivation. These increased in number to form cell sheets (Fig. 5), surrounding the original tissue explant (Hirumi and Maramorosch, 1964a,b,c, 1968; Mitsuhashi and Maramorosch, 1964; Mitsuhashi, 1965; Chiu *et al.*, 1966; Chiu and Black, 1967). The detailed morphology and behavior of other cell

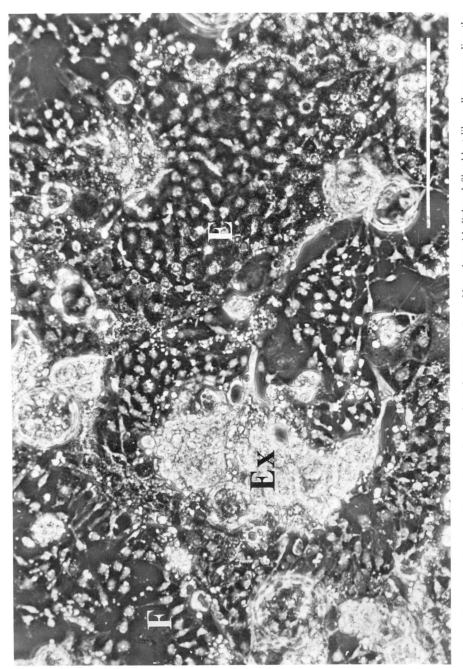

FIG. 5. Primary cell cultures from the embryos of *Nephotettix cincticeps*. Note the epithelial and fibroblast-like cells surrounding the original tissue explant. Ex, Explant; E, Epithelial-like cells; F, Fibroblast-like cells. Bar represents 200 μm. (Courtesy of Dr. J. Mitsuhashi.)

types observed were described by Hirumi and Maramorosch (1964c), Mitsuhashi and Maramorosch (1964), and Mitsuhashi (1965).

H. ESTABLISHMENT OF CELL LINES

Chiu and Black (1967) established the first continuous cell line from leafhoppers. Following their successful results, Richardson and Jensen (1971) and McBeath and Chen (1973) established cell lines of *C. montanus* and *D. elimatus*. A culture of *M. fascifrons* which had been obtained by Hirumi and Speyer but had not been fully developed into a continuous line was established into continuous culture by Liu and Black (1976) using Hirumi and Maramorosch's medium (Table I). Attempts to obtain the continuous cell lines from the primary cultures of *M. fascifrons* by growing them in Martínez-López and Black's (1974) medium were not successful (Liu and Black, 1976).

The procedure originally adopted by Chiu and Black (1967) yielded an *A. constricta* cell line in about 4 months subsequent to the establishment of the primary cultures. Methods to obtain cell lines have been constantly improved in Black's laboratory and the following procedure devised by Liu and Black (1976) yielded cell lines of *A. constricta* and *A. novella* in 10 weeks after the embryonic tissues were explanted. Embryos were dissected in glass depression slides (Fig. 2) from at least 35 eggs in Tyrode's salt solution (Section II, C) and were treated for 15 minutes with 0.25% trypsin prepared in Rinaldini's (1959) solution. A gentle swirling with a needle resulted in the sedimentation of the tissues to the bottom of the depression slide. This facilitated repeated washing with the medium in the well of a depression slide. After washing 3 to 4 times in Chiu and Black's medium (Table IV) embryo fragments suspended in the same medium were transferred to a small Falcon plastic tissue culture (35 × 10 mm) dish (Fig. 4) and sealed in a glass dish. Cell growth started in 2–3 days, and the medium was changed after 7 days. After 2 weeks of maintenance, Chiu and Black's medium was replaced with Martínez-López and Black's medium (Table V). Four weeks after explantation, the cells were detached by treating with 0.05 mg/ml of papain in a diluent described by Martínez-López and Black (Section III,A). The cells from tissue were dispersed in the same area of the plastic dish. The papain solution was replaced by the Martínez-López and Black's medium. Papain treatment was repeated weekly until the entire surface of the culture dish was covered with cells. The entire process took approximately 8 weeks. The cells were treated with papain at a concentration of 0.01 mg/ml, as described by Martínez-López and Black (1974) (Section III,A), and transferred to 60 × 15 mm disposable Falcon plastic dishes. In about 2 weeks the cells multiplied to cover the entire area of

TABLE V

CULTURE MEDIUM FOR GROWING LEAFHOPPER
ESTABLISHED CELL CULTURES[a]

Components	Concentration (gm/liter)
$CaCl_2 \cdot 2H_2O$	0.36
$MgSO_4 \cdot 7H_2O$	1.2
KH_2PO_4	0.27
KCl	1.6
NaCl	1.0
$NaHCO_3$	0.9
Dextrose	9.0
Lactalbumin hydrolysate	10.0
Yeastolate	9.0
Fetal bovine serum	100 ml
L-Histidine HCl (monohydrate)	5.09
L-Histidine (free base)	3.21
Penicillin G	100,000 (IU)
Streptomycin	0.1
Neomycin	0.05˙
pH	6.43
Osmotic pressure	360 ± 5 mOsm

[a] From Martínez-López and Black (1974).

the dish and the cell population became adequate for transfer to one 25 cm² Falcon plastic flask for continuous subculturing.

The medium of Martínez-López and Black is prepared by mixing each component, prepared in double distilled water, in the following manner:

1. Histidines are first dissolved in about 300 ml of water. Then KH_2PO_4 and $NaHCO_3$ are added.
2. To another 300 ml of water $CaCl_2$, $MgSO_4$, KCl, NaCl, dextrose, antibiotics, and yeastolate are added and dissolved.
3. Solutions in steps 1 and 2 are then mixed.
4. Fetal bovine serum is heated at 56°C for 30 minutes and added to the salt mixture of step 3.
5. Lactalbumin hydrolysate is prepared by adding 100 ml water to weighed amount of the powder and the mixture is warmed to 50°C until the solution is completed. Then it is added to the mixture in step 3.

The volume is made up to 1 liter with distilled water.

It is convenient to prepare antibiotics as follows: 1,000,000 IU penicillin G is dissolved in 10 ml H_2O, 1 gm streptomycin in 2 ml H_2O, and 0.5 gm neomycin is 5 ml H_2O. For each liter of medium 1 ml of penicillin solution, 0.2 ml of streptomycin solution, and 0.50 ml of neomycin solution are added.

The procedure for sterilization by using Millipore filters is as previously described.

Primary cultures of certain cell lines can be established with relative ease, e.g., *A. constricta* and *A. novella*. In certain situations establishment of a continuous cell line for various reasons may be difficult to achieve, e.g., *M. fascifrons* (H. Hirumi and G. Speyer, unpublished; H. Y. Liu and L. M. Black, unpublished). In addition, production of primary cell cultures has not immediately led to the establishment of continuous cultures, e.g., *N. cincticeps*, despite concerted efforts to this end (J. Mitsuhashi, personal communication; I. Kimura, personal communication). It is not currently clear precisely why some primary cell cultures are more conducive to establishment into continuous cultures than others.

III. Maintenance and Growth of Established Leafhopper Cell Lines

A. MAINTENANCE OF CELL LINES

For maintenance and experimental purposes cultures were normally grown in Falcon 25 cm² disposable plastic flasks and incubated at 28°C. After the cells were fully confluent, subcultures were usually made every 4 to 5 days.

Chiu and Black (1967) employed trypsinization for the dissociation and detachment of cells from the flasks. The cells were washed once with 2 ml of Rinaldini's solution (1959) at pH 7.5 and replaced with 2 ml of Rinaldini's solution containing 0.05% trypsin. After 5 to 6 minutes incubation at room temperature the tryptic action was ended by dilution with 2 ml of growth medium. The cells were flushed from the surface of the flask using a Pasteur pipette. They were then transferred to a 15 ml conical tube which was capped and centrifuged at about 1000 rpm for 2 minutes in a clinical centrifuge. The supernatant above the sedimented cells was removed and replaced with a volume of fresh medium, depending on the cell density and the number of flasks to be seeded. Each flask was usually seeded with 2 ml of resuspended cells and then 2 ml of fresh medium was added. Special care was taken to obtain an even distribution of seeded cells on the growing surface of the flask. Stirring the cell suspension, before seeding, by using a mixer (Vortex-Genie, Model K 550 G) for at least 5 times at its maximum setting for 2 to 3 seconds each time assured uniform dispersion of cells. In addition gentle mixing of cells in the flask before laying them on a level rack in the incubator was essential (Martínez-López, 1973).

Martínez-López and Black (1974) improved the transfer procedure by using papain prepared in a new diluent of the same pH as that optimal for the growth of *A. constricta* cells.* The diluent was prepared by mixing 3.2 gm KCl, 2.0 gm NaCl, 0.69 gm K_2HPO_4, 24.36 gm dextrose, 0.38 gm disodium dihydrogen ethylene diaminetetraacetate dihydrate (EDTA), 0.88 gm cysteine, 2.39 gm histidine HCl (monohydrate), 5.21 gm histidine free base, and the same concentration of antibiotics as in the medium (Table V). The pH of the solution was 6.4 with an osmolality of 360 mOsm. The monolayers were first washed with the diluent and then treated with 2 ml of the same solution (per 25 cm² culture flask) containing papain at a concentration of 0.01 mg/ml. The cells were normally treated for 5 minutes in papain and the action of the enzyme was stopped by adding 2 ml of the Martínez-López and Black's (1974) medium (Table V). The cells were subsequently handled as previously described.

B. CONDITIONS FOR GROWTH OF CELL LINES

Martínez-López and Black (1974) undertook extensive tests on various parameters essential for the growth of AC20 cells derived from *A. constricta*. Chiu and Black's medium (Table IV) which was commonly used for culturing various agallian leafhopper cell lines did not always support good cell growth. Since this medium was developed by trial and error, it was essential to formulate a medium to allow for optimal conditions of cell growth. Martínez-López and Black (1974) determined various conditions optimal for the growth of the cells. Their new procedure for enumeration of cell counts, as described in Section V,C, was the major technique for dermining the optimal concentration of each of the medium components.

The best growth of AC20 cell line was obtained at 28 °C, pH 6.43, osmotic pressure 360 mOsm, and component concentrations per liter of 0.36 gm $CaCl_2 \cdot 2H_2O$, 1.2 gm $MgSO_4 \cdot 7H_2O$, 0.27 gm KH_2PO_4, 1.6 gm KCl, 1.0 gm NaCl, 0.9 gm $NaHCO_3$, 9.0 gm dextrose, 10.0 gm lactalbumin hydrolysate, 8.0 gm yeast hydrolysate, 4.8 gm histidine–HCl (monohydrate), 3.45 gm histidine free base, and 100 ml of fetal bovine serum (heat-treated at 56 °C for 30 minutes). In addition, Martínez-López and Black emphasized the importance of testing different commercial batches of certain components such as fetal bovine serum, lactalbumin hydrolysate, and histidine. Incorporating the various components at their optimal concentration, Martínez-López and Black (1974) formulated a new medium for the growth of AC20 cell lines. This medium not only resulted in shorter doubling times, but noticeable increases of cells in mitosis, an improved cell appearance, and

* Martínez-López and Black reported crystalline papain as suitable for this purpose. Recent experiments by H. Y. Liu (personal communication) indicate that technical grade papain, which is considerably less expensive than the crystalline grade papain, can be used with equally good results.

significant reduction or virtual elimination of cell debris. The details of the preparation of this medium are given in Table V.

The pH of the medium was able to be adjusted with HCl or NaOH to meet the requirements of different cell lines. The histidine concentration was able to be varied so as to obtain a pH of 6.30, which was optimal for the growth of *A. sanguinolenta* (AS2) cells. Moreover, it was possible to alter the osmolality of the medium to suit the requirements of different cell lines. Although Martínez-López and Black's medium provided excellent growth for agallian leafhopper cell lines, it was necessary to test batches of different components of the medium to obtain the optimal growth of cells.

McIntosh *et al.* (1973) have successfully adapted *A. constricta* cell to a mammalian cell medium consisting of equal parts of medium 199 with Hanks' salt and Melnick's monkey kidney medium "A," both without sodium bicarbonate. Inactivated fetal bovine serum to a final concentration of 10%, 50 Units of penicillin, 50 μg of streptomycin per milliliter of medium, and 1 ml of 200 m*M* glutamine per 100 ml of medium were added. The final pH of the medium was 6.9 and osmotic pressure was 288 mOsm. At least 6 months adaptation was necessary to achieve the normal cell growth of AC20 cell line.

McBeath and Chen (1975) have succeeded in growing *D. elimatus* continuous cell culture in a medium prepared by mixing commercially available tissue culture media. It contained 84% Schneider's revised Drosophila medium, 6% medium CMRL-1066 (1X) without L-glutamine, and 10% fetal bovine serum. Its pH was 6.9 and its osmolality was 345 mOsm.

Liu and Black (1976) have been successful in growing *A. constricta, A. novella, A. Sanguinolenta* and *M. fascifrons* cells in a solution prepared by mixing commercially available media (Table VI). Although complete data on the ability of various cell lines to withstand several subculturings and the data on doubling times are not available, the general appearance of cells is equivalent to those maintained in Martínez-López and Black's medium. *Agallia constricta* cell lines (AC20) can be maintained for 16 subculturings without any noticeable change in appearance. There is no need for any adaptation before growing the cells in this medium. Since the components of this medium are available commercially, and the medium can be used for widely different cell lines, it appears to be an excellent choice for maintaining leafhopper cell lines. The medium reported by McBeath and Chen (1975) is not suitable for the growth of any of the above cell lines (H. Y. Liu, personal communication).

Many leafhopper species contain intracellular symbionts (Buchner, 1953; Mitsuhashi and Kono, 1975). These can also be observed in primary cell cultures (Mitsuhashi and Maramorosch, 1964; Mitsuhashi and Kono, 1975). In addition organisms such as mycoplasma may also be introduced into the

TABLE VI

Culture Medium for Growing Leafhopper Established Cell Cultures[a]

Components	Concentration (ml)
Schneider's *Drosophila* medium (revised)[b]	100
Medium 199-10X, with Hanks' salts (with glutamine, without sodium bicarbonate)	10
CMRL 1066 (without glutamine)[b]	5
Fetal bovine serum[c]	30
Antibiotics[d]	1
Histidine-buffer solution[e]	90
pH	6.35–6.40

[a] From Liu and Black (1976).

[b] Grand Island Biological Company, Grand Island, New York.

[c] Microbiological Associates, Bethesda, Maryland. Heated at 56 °C for 30 minutes, passed through 0.45 μm Millipore filter, and stored at 4 °C.

[d] Potassium penicillin G 1,000,000 IU, streptomycin sulfate 1 gm, and neomycin sulfate 0.5 gm, dissolved in 20 ml distilled water, stored at 4 °C.

[e] L-Histidine-HCl (monohydrate) 0.8 gm, and L-histidine (free base) 1.0 gm, in double distilled water.

cell cultures through a variety of sources: serum, enzymes, isotopes, and aerosols (Barile *et al.*, 1973).

The presence of mycoplasma may not be immediately apparent. Therefore, it is extremely important to ascertain periodically whether the cell cultures are contaminated by such organisms.

Highly convenient procedures for screening the cell lines for the presence of covert contaminants, such as mycoplasma, are available (Brown *et al.*, 1974). These procedures involve the use of the scanning electron microscope for examining monolayers grown on cover slips. The method is relatively simple and much more rapid and reliable than other methods of detection. The cells grown on 15 mm glass cover slips, without antibiotics in the growth medium (Section VI), are fixed in 2.5% glutaraldehyde and 0.5% osmium tetroxide prepared in 0.1 M cacodylate buffer of pH 7.4 at O °C. The cells are dehydrated in a graded series of ethanols, critical-point dried using either Freon or liquid CO_2, and giratory coated with heavy metal. The cover slips are then examined under a scanning electron microscope. By employing this procedure *A. constricta* (AC20) cell line maintained in Black's laboratory was found to be free of contamination (R. MacLeod, personal communication). A simple procedure for detecting mycoplasma contamination by employing Giemsa staining on vertebrate cells is also available (Harnett *et al.*, 1974). The effectiveness of this procedure for detecting mycoplasmas in insect cells has not been reported.

In view of the recent publication of Hirumi *et al.* (1976) it is also important to examine the cell lines periodically for possible contamination with viruses. In this regard Hirumi and co-workers observed the presence of 5 different viruslike particles in cultures of *Aedes albopictus.* Transmission electron microscopic observations of cells in this laboratory have shown the leafhopper cell lines to be free of extraneous viruses.

Besides being susceptible to contamination with microorganisms an established cell line can also be contaminated with cells from a different species. It is therefore extremely important to maintain and handle the established insect lines under conditions which would not possibly allow their mixing. In this laboratory the different cell lines are always subcultured separately, with every precaution taken not to cross-contaminate the established lines. Should mixing of the cell lines occur it is currently impossible to distinguish the cells of one line from those of the other (see Section IV).

C. MORPHOLOGY OF CELL LINES

Extensive studies on morphological characteristics of leafhopper cell lines have not been published. *Agallia constricta* (AC20) cell lines are mainly composed of epithelial-like cells, differing in their size and shape, intermingled with a few giant cells (Chiu and Black, 1967) (Fig. 6). Some cell lines contain distinctly slender cells with compact cytoplasm (Chiu and Black, 1967; Hirumi, 1973). Generally the smaller cells contain high electron-dense cytoplasmic matrices with an irregularly shaped nucleus. The larger cells consist of low electron-dense cytoplasmic matrices with a spherical nucleus (Hirumi, 1973).

IV. Identification of Cell Lines

Although karyotyping of cultured cells is widely employed for identification and for detection of interspecies contamination of one vertebrate cell line with another, its potential has not been fully explored with cultured leafhopper cells. So far only *A. constricta* (AC20) cell line was karyotyped by Hirumi (1973), who employed a modified Onuki's squash technique. Before karyotyping the cells were grown in the presence of colcemid (Grand Island Biological Co., Grand Island, New York) at a concentration of 0.2 μg/ml of growth medium. After 18 hours of exposure to Colcemid, the cells were rinsed in Earle's balanced salt solution, collected by low-speed centrifugation, and resuspended in a hypotonic solution consisting of 1 part growth medium and 3 parts distilled water. After centrifugation the cell pellets were fixed in 50% acetic acid for 20 minutes. Following the

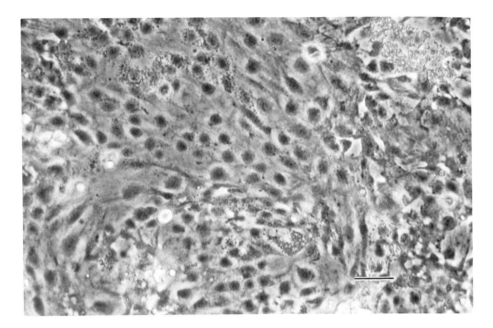

FIG. 6. *Agallia constricta* (AC20) cell line. Composed of mostly epithelial-type cells (×480). Bar represents 100 μm. (Courtesy of Dr. H. Y. Liu.)

squash preparation, the cells were stained in 1% acetic orcein in 50% acetic acid. The spermatocytes and oocytes from male and female adult insects, undergoing meiosis, as well as stained cultivated cells were examined. The normal chromosome complement of *A. constricta* males was found to be $2n = 21(2A + X)$ and in females $2n = 22(2A + XX)$. The chromosome number of cultivated cells varied considerably from 24 to 122 and appeared in tetraploid or octaploid ranges, which represented two frequent chromosome numbers. Thus the results indicated that the AC20 cell line contained a mixed populations of cells with $4n$ and $8n$ chromosome numbers. When 50-day-old primary cultures of *N. cincticeps* were karyotyped, the majority of the cells were found to be diploid (Mitsuhashi, 1966). Although experimental evidence is lacking, primary cultures of the AC20 line probably consisted of diploid stem cells. During several subculturings shifts in chromosome numbers have occurred.

Another useful criterion for distinguishing cell lines of leafhopper vectors is their relative susceptibility to one or more vector-transmitted viruses. Although established cell cultures were shown to be heterogeneous (Hirumi, 1973) with respect to cell types, there was consistency in their susceptibility to a particular viral infection. All *A. constricta* cell lines established so far support excellent WTV multiplication and the susceptibility of two other leafhopper cell lines, *A. sanguinolenta* and *A. novella* was found to be different (Chiu and Black, 1969; H. T. Hsu and L. M. Black, unpublished; H. Y. Liu and L. M. Black, unpublished). It was interesting to note that *D. elimatus* (DE) cells were completely resistant to infection by WTV and CYDV whereas a small proportion of cells took infection when inoculated with SYDV (H. T. Hsu and L. M. Black, unpublished).

Because of the heterogeneity of leafhopper cell lines it is likely that karyotyping may not yield conclusive results on the identity of each cell line. Since infectivity measurements are dependent on several factors, it is not advisable to use them alone for a critical identification of a cell line. More conclusive evidence on the identity of a cell line could probably be obtained by biochemical characterization, such as characterization of isoenzymes, and by serological methods as applied to distinguish mosquito cell lines (Greene and Charney, 1973).

V. Cell Enumeration

Techniques involving a hemocytometer or an electronic cell counter are employed for the enumeration of cultured leafhopper cells. Both methods involve removal of cells from culture substrates. Enumeration of the cells in 10 squares on a diagonal line of a graded ocular micrometer in 8 different microscopic fields was used by Mitsuhashi *et al.* (1970). Martínez-López and Black (1973) reported a new method for indexing cell growth by the use of a reticule in the eyepiece of an inverted phase contrast microscope.

A. HEMOCYTOMETER

The counting chamber of a hemocytometer was divided into sections of calibrated area (ruled area) and depth. The chambers and cover glass had to be carefully cleaned free of dust, lint, or grease. Cells were collected by low-speed centrifugation and resuspended in culture medium at various dilutions. The chambers of the hemocytometer were filled immediately with cell suspension and immediately examined under one of the higher objective lenses of a microscope. Care was taken to add sufficient cell suspension to load the chamber without overflow. If air bubbles were present in the chamber, or the cells were not evenly distributed, the chamber was cleaned and refilled. After allowing a few minutes for the cells to settle the counts were recorded. In the Neubauer-type chamber with 0.1 mm depth, 10 of the 1-mm squares were counted. For ease and accuracy of counting the cell chambers were filled with cell suspensions containing approximately 30 to 45 cells/mm (Sanford *et al.*, 1951). In order to avoid confusion in counting cells located on the lines of each square, all cells present on the upper and left lines of the square were counted as if they were inside the square and those cells located on the lower and right lines of the square were neglected. If the number of cells counted in 10 1-mm squares was 350, and since the depth of the chamber was 0.1 mm, the sample volume was 10^{-3} ml and the sample contained 350,000 cells/ml. The counts of cells per milliliter when multiplied by the dilution factor yielded the total number of cells in the undiluted sample.

Although counting of cells with a hemocytometer can be easily performed, there are several possible sources of error. These include inaccuracies in preparing dilutions, inadequate dispersion of cells, loss of cells during the process of their dispersion, and incorrect filling of the chamber. Berkson et al. (1939) and Sanford et al. (1951) estimated counting errors up to 10%.

B. Electronic Coulter Counter

Employing a Coulter counter, Model B (Coulter Electronics, Inc., Hialeah, Florida), one can automatically count several thousand cells with precision and reproducibility. The instrument functions on the basis of impedance of conductivity due to the temporary passage of cells through the narrow aperture separating two platinum electrodes, one located inside the tube with cells in it and the other outside the tube. These charges are electrically amplified and transmitted to a scaling circuit with adjustable upper and lower threshold levels. The signals are visualized on an oscilloscope and recorded in a decade counter.

Martínez-López and Black (1971) reported the following procedure for counting leafhopper cells in the Coulter counter, Model B. Cells removed from flasks were collected by low-speed centrifugation and resuspended in growth medium. Two milliliters of cell suspension were diluted in 12 ml of 0.9% NaCl, previously passed through a Millipore filter of 0.45 μm pore diameter. Samples were stirred and could be left at 0 °C until the counts were made. Countings were made under the following conditions (Martínez-López, 1973):

Aperture tube	100 μm
Manometer size	500 μl
1/Amplification	setting 1
1/Aperture current	setting 2
Matching switch	setting 64H
Gain control	setting 52
Lower threshold control	setting 10
Upper threshold control	setting 100

Calibration of the instrument was achieved with cell suspensions initially counted in a hemocytometer. Once the instrument was calibrated the cell counts obtained from the Coulter counter represented the absolute number of cells. At least 3 counts were made on each sample and the average value was adjusted using the correction chart for the 100 μm aperture. Some of the errors inherent in counts made by a hemocytometer are also present in the counts made in Coulter counters. These include errors introduced during suspension of cells, aggregation of cells, dilution of cell suspensions, and because of the presence of cell debris, and cell aggregates.

C. METHOD OF INDEXING CELLS *in Situ*

In order to avoid the pitfalls inherent in counts by either the hemocytometer or the electron Coulter counter, Martínez-López and Black (1973) devised a method for counting cells as monolayers. Cell nuclei were enumerated employing a reticule inserted in the eyepiece of an inverted phase contrast microscope. The number of cell nuclei that coincided with the preset points in the reticule were counted. The counts were then used as an index of the number of cells in a given flask. Furthermore, the absolute number in a given flask was obtained from calculations based on area of the bottom of the flask, the average value for the square of the diameter of the nucleus, and an average count of nuclei which fell on the reticule point per field.

The following equation was derived for *A. constricta* cell line (AC20) grown in a 25 cm^2 culture flask and examined under an inverted phase contrast Zeiss microscope equipped with a pH 2 Plan 40/0.60 objective and a KPl 10× eye-piece housing the reticule. The reticule contained 25 points distributed uniformly over a large circle. The number of cells per flask was computed by using the formula:

$$65.8 \times A/B \times 10^6$$

where A was the average number of nuclei counted on the reticule points for the unknown and B was the square of the average diameter, in micrometers, of the nuclei of the unknown. The number of microscopic fields and their positions were determined as described by Martínez-López and Black (1973), assuming the cells were evenly distributed. Fields were determined on a random basis if the cells were not evenly distributed. In the latter instances more counts per flask were necessary.

This technique of cell counting has many advantages over the hemocytometer and Coulter counter. The number of flasks required for accurate cell counts and the time required is greatly reduced. Since the technique did not require the removal of cells, errors inherent in cell removal and manipulation are reduced. More details of the technique and its application to leaf-hopper cell monolayers appear elsewhere (Martínez-López and Black, 1977).

VI. *Preparation of Cells For Inoculation*

We have found that actively growing cells were the best source for seeding cell monolayers on cover slips. Cells from aged cultures tended to clump and formed multiple layers. Moreover, when younger cells were seeded on cover slips, they grew more vigorously and supported better viral replication.

The assay system for plant viruses in vector cell monolayers is based primarily on immunofluorescent staining. This necessitated the use of cover slip cultures for infectivity measurements. Glass cover slips of 15 mm diameter were cleaned in acetone using an ultrasonic cleaner. The acetone was changed at least 3 times after an interval of 5–10 minutes of ultrasonication and were rinsed in fresh acetone before they were air dried on a sheet of lint-free paper. Twenty to 30 cover slips were wrapped in aluminum foil and sterilized in an oven. During the entire cleaning process cover slips were handled with fine-tipped forceps.

Cells were harvested from 2- to 4-day-old cultures after they were confluent, using papain as described (Section III,A), were collected by low-speed centrifugation (750–1000 rpm for 2 minutes) in a conical centrifuge tube, and were washed once with the growth medium. The cells were resuspended in fresh medium to obtain the proper cell density for seeding. The optimum cell density for seeding on cover slips was about 1 to 2 $\times$ 10^6 cells/ml and 0.1 ml of this cell suspension (1 to 2 $\times$ 10^5 cells) was added to each cover slip. Prior to adding the cell suspension, cover slips (normally 3) were securely positioned in a sealable glass dish (50 mm diameter) with a drop of growth medium (Fig. 7). It was important to keep the individual cover slips from touching one another in order to prevent spreading of the cell suspension and overlapping of one cover slip by another. After seeding of cells on cover slips, the dishes were sealed, using a nontoxic silicon lubricant (high vacuum grease, Dow Corning Corporation, Midland, Michigan), and incubated at 28 °C. If growth was normal, the cells were confluent in 14–20 hours. If the monolayers required more than 2 days to become confluent, this indicated that they were not under optimal conditions and they were therefore discarded.

When the cells were seeded by the procedure described above, the cell density was not uniform. The edge of the cover slip ($\sim$3 mm wide ring around the margin) contained a lower number of cells (Kimura and Black, 1971). However, counts of infected cells located in the central 8 mm area

FIG. 7. Preparation of cells for inoculation, showing three 15 mm cover slips securely positioned in a sealable glass petri dish (50 mm diameter) with a drop of growth medium. Each cover slip is seeded with 1 to 2 × 10⁵ cells suspended in 0.1 ml of the growth medium, sealed in a petri dish, and incubated at 28 °C. If growth is normal the cells are confluent in 14–20 hours and are ready for inoculation.

provided precise infectivity measurements (Kimura and Black, 1972). If an even distribution of cells was desired throughout the entire area of the cover slip, the entire dish containing the cover slip was flooded with a cell suspension. In this case cell density was adjusted to give 1 to 2 × 10⁵ cells per cover slip. The dishes were carefully transferred to an incubator and incubated at 28°–29°C. After overnight incubation, cover slips with confluent monolayers were transferred aseptically to another dish (H. T. Hsu, unpublished).

VII. Preparation of Inoculum

Extracts obtained from infected plants, viruliferous insects, and infected cultures have been used as inocula. The inocula were subjected to special treatments to eliminate contamination or they were diluted to a degree at which contamination posed no problem for inoculation. It is important to prepare the inocula in a suitable buffer solution in order to preserve infectivity. In addition, other factors regulating the infectivity are discussed below.

A. FROM PLANTS

Infected material was thoroughly rinsed in distilled water and cut into small pieces of approximately 1 gm weight. The tissue was disinfected in 70% ethanol for 1/2 to 1 minute and thoroughly rinsed in sterile distilled water. Before maceration the tissue was finally rinsed in the same buffer as that used for making up the dilutions of the extract. The tissue was triturated, normally in 10 times its weight, in a buffer solution, abbreviated as His–Mg buffer, containing 0.1 M histidine and 0.01 M MgCl$_2$ (prepared by mixing 0.1 M L-histidine free base containing 0.01 M MgCl$_2$, and 0.1 M L-histidine hydrochloride containing 0.01 M MgCl$_2$ in such proportion as to yield the desired pH), using a sterile pestle and mortar (Kimura and Black, 1971). After clarifying in sterile centrifuge tubes at 6000 rpm for 5 to 10 minutes at 4 °C in a Sorvall super-speed centrifuge, supernatant fractions were collected aseptically and further dilutions were made in His–Mg solution at optimal pH for inoculation.

B. FROM VIRULIFEROUS INSECTS

In order to facilitate inoculation of cell monolayers with concentrated extracts at 10^{-2} or lower dilutions, pretreatment of insects was necessary. Attempts to eliminate contamination by using Millipore filters of pore size 8.0 μm or smaller diameter resulted in loss of infectivity (Reddy and Black, 1972). Extracts from insects treated with 1% mercuric chloride were occasionally toxic to monolayers. Treatments with antibiotics alone did not always yield sterile extracts. The following treatment of insects reported Reddy and Black (1972) effectively controlled contamination without detectable loss of infectivity. Insects were first maintained at −13 °C for about an hour and then placed on a piece of sterile cheesecloth fastened by a rubber band to the rim of a beaker. Insects were washed first with sterile water and then with 50% ethanol, followed immediately by a wash with a mixture containing 100 μg/ml of each of the following antibiotics: Fungizone[®], Actidione[®], and griseofulvin. Then the insect samples were washed several times with sterile water before being used for extraction. Treated insect samples were allowed to dry and macerated in 32 times their weight (weight of each insect was assumed to be 1 mg) of 0.1 M His–Mg buffer, pH 7.0, in a prechilled mortar. They were then clarified in a Sorvall super-speed centrifuge at 6000 rpm in the SS-34 rotor for 5 minutes. The dilution of such a preparation was assumed to be $10^{-1.5}$. After suitably diluting the extracts in His–Mg buffer of appropriate pH, they were employed for inoculating monolayers of vector cells.

C. FROM INFECTED VECTOR CELL MONOLAYERS

Harvested cells were collected by low-speed centrifugation in Corex 15 ml capacity tubes employing a Sorvall super-speed centrifuge, washed once in His–Mg buffer, pH 7.0, and resuspended in the same buffer. For 0.1 ml of packed cell suspension, 0.9 ml of buffer was added and sonified at 0 °C for 15 seconds at an output of 40–50 W, employing a Sonifier Cell Disruptor (Model W185D, Heat Systems-Ultrasonic Inc., Plainview, New York). The extract was then clarified by centrifugation at 6000 rpm for 10 minutes. Supernatants were used for inoculations (Hsu and Black, 1974).

D. PURIFIED VIRUS PREPARATION AS INOCULUM

Viral zones were drawn from sucrose gradients aseptically by inserting a bent sterile needle from the top of the gradient or by puncturing the bottom of the tube. When diluted appropriately, traces of sugar present in the inoculum did not interfere. Purified virus inocula were normally free of interfering microbial contamination.

If it was necessary to remove sugar from purified virus, dialysis was preferred to high-speed sedimentation and resuspension of the virus.

Aliquots of purified virus inocula were stored in sterile ampules (1.2 ml capacity, Wheaton Glass Co., Millville, New Jersey) and hermetically sealed. The inocula were then quick-frozen in a mixture of Dry Ice and ethanol and stored at −80 °C. No loss in infectivity was noticeable in inocula stored in His–Mg buffer of optimum pH during several years of storage at −80 °C (Reddy and Black, 1973; Hsu and Black, 1973).

VIII. Conditions for Inoculation

A. SOLUTIONS FOR PREPARING INOCULA

The ideal solution for diluting inocula should stabilize the virus and should be nontoxic to the cell monolayers during the inoculation procedure.

Chiu and Black (1969) used growth media with reduced fetal bovine serum or 0.1 M glycine containing 0.01 M $MgCl_2$, pH 7.0. Among several solutions tested for infecting cell monolayers with WTV from crude extracts of sweet clover (*Melilotus officinalis*) tumors, Kimura and Black (1971) found His–Mg buffer better than any other for inoculating *A. constricta* cells with WTV. His–Mg buffer was also found to be an ideal solution for inoculating *A. constricta* cells with WTV derived from insect extracts (Reddy and Black,

1972), with purified virus preparations (Reddy and Black, 1973), and for inoculations with PYDV (Hsu ånd Black, 1973).

B. EFFECT OF IONS

Mg^+ ion (supplied as $MgCl_2$) was important for inoculating WTV on vector cell monolayers. The effects of $MgSO_4$ and $CaCl_2$ proved to be almost the same as those of $MgCl_2$. Na^+ and K^+ ions, Al^{3+} ion supplied by $AlCl_3$, DEAE-dextran, and dextran sulfate either reduced or completely eliminated WTV infectivity (Kimura and Black, 1971).

C. EFFECT OF pH

The dramatic effect of the pH of the inoculum, prepared in His–Mg buffer, was evident when crude extracts were used for preparing inocula. Kimura and Black (1971) showed that the optimum pH value for inoculating cell monolayers with WTV from crude tumor extracts, prepared in His–Mg buffer, was 6.0, and that below pH 5.0 and above pH 6.6 the infectivity of WTV decreased rapidly. On the other hand, when crude viruliferous insect extracts were used for inoculation, the optimal pH of His–Mg buffer was between 5.0 and 5.8 (Reddy and Black, 1972). When purified WTV was used as inoculum, no detectable variation in infectivity was noticeable when the pH of His–Mg buffer was varied between 5.0 and 7.0 (D. V. R. Reddy unpublished). Thus it is likely that the effect of pH is mediated through host material present in the inocula.

The pH was extremely important for inoculating the two vector-specific isolates of PYDV. The optimal pH for inoculation of purified SYDV preparations was 5.9 whereas it was 5.3 for inoculating CYDV (Hsu and Black, 1973). By performing inoculations on two different cell lines, Hsu and Black (1973) concluded that the effect of pH was mediated through the virus rather than the host cell.

D. PROCEDURE FOR INOCULATION

Culture medium was removed from the cover slips and the monolayers were washed twice with His–Mg buffer of the same pH as that employed for diluting the virus inoculum. It is important to remove as much as possible of the buffer after the second wash. Then 0.05 to 0.1 ml of inoculum was added to each cover slip. Each dilution of inoculum was added to at least 2 or preferably 3 cover slips. Petri dishes were sealed again and incubated at 28 °C. At the end of the inoculation period, the inoculum was removed, the cover slips were washed three times in the growth medium, and the cells were then overlaid with about 0.1 ml of growth medium. Inoculated cell monolayers were normally incubated for about 40 hours at 28°C.

E. EFFECT OF INOCULATION PERIOD ON INFECTIVITY

The optimal inoculation period may be affected by the concentration of the virus and the volume of the inoculum. At concentrations of $8.3 \times 10^{7.5}$, $8.3 \times 10^{7.0}$, and $8.3 \times 10^{6.5}$ virions per milliliter of crude tumor extract diluted in His–Mg buffer, the most efficient inoculation periods were 1.5, 2, and 3 hours, respectively (Kimura and Black, 1971). With extracts from insects diluted in His–Mg buffer to 10^{-2} and higher dilutions, the most suitable inoculation period was 3 hours (Reddy and Black, 1972). A 45–60-minute inoculation period was satisfactory for obtaining maximum infections with SYDV or CYDV (Hsu and Black, 1973).

F. MAINTENANCE AFTER INOCULATION

The culture medium used for incubating the inoculated cells contained penicillin G, neomycin, and streptomycin. In most cases bacterial or fungal contaminations were not encountered in inoculated cultures. Occasionally inoculations performed with high concentrations of extracts from viruliferous insects gave rise to bacterial contamination. Under these conditions inoculated cultures were incubated in the growth medium supplemented with 500 μg/ml of kanomycin sulfate, and the use of kanomycin had no effect on WTV multiplication (Reddy and Black, 1972).

It is desirable to observe the inoculated cells under an inverted phase contrast microscope on the day following inoculation to ascertain that the cultures are not contaminated and the inoculated cells are in satisfactory condition.

IX. Detection of Infection

It is important to include uninoculated cover slip monolayers to serve as controls. This would not only aid in distinguishing infected cells from healthy ones but also serve as a routine check to ascertain that the cultures are indeed uninfected before inoculation.

A. STAINING BY FLUORESCENT ANTIBODIES

Procedures described below are suitable for labeling the antibodies of WTV and PYDV and for immunofluorescent staining of inoculated monolayer cultures of leafhopper cells. For more comprehensive information, the reader is referred to the review by Spendlove (1967). As "there is no single procedure that is best for all investigators or investigations" (Spendlove, 1967), it is advisable to investigate the necessary modifications to suit the specific virus–host systems.

1. *Preparation of Fluorescein-Labeled Antibodies*

It was essential to obtain high-titered antisera for the conjugation of antibodies with fluorescein isothiocyanate (FITC) (Sinha and Reddy, 1964). Gamma globulins were first isolated from the antisera and then conjugated with FITC. The following procedure, described by Liu (1970), is essentially similar to those described by Campbell *et al.* (1964) and Spendlove (1967). One part of saturated $(NH_4)_2SO_4$ solution was added drop by drop to two parts of the serum, while stirring in a beaker at room temperature. The pH of the mixture was adjusted to 7.8 with 2 N NaOH. The mixture was stirred for 2–3 hours at 4°C and the precipitate was collected by centrifuging for 30 minutes in a clinical centrifuge. The supernatant was discarded and the precipitate was dissolved in normal saline (0.15 M NaCl) to restore the original volume of the serum. The serum globulins were precipitated twice in ammonium sulfate employing the same procedure as described above. After the third precipitation the γ-globulins were dissolved in saline to yield one-half of the initial serum volume.

Ammonium sulfate was removed by dialysis against saline at 4°C. The saline was changed several times until no sulfate was detected when a few drops of $BaCl_2$ were added to the dialysate. The globulins were removed from the dialysis bag and centrifuged at 3000 rpm at 4°C for 30 minutes in a Sorvall super-speed centrifuge in order to remove insoluble material formed during dialysis. The solution was then diluted with phosphate-buffered saline (PBS, 0.01 M sodium phosphate, 0.15 M NaCl, pH 7.5) and the protein concentration was determined by measuring the optical density at 280 nm in a spectrophotometer. The protein concentration was calculated from the formula:

protein concentration (mg/ml) = optical density × dilution factor/1.8,

where 1.8 = extinction coefficient of rabbit γ-globulin (McGuigan and Eisen, 1968).

From 15 to 25 mg of crystalline FITC (from Baltimore Biological Laboratory, Baltimore, Maryland), dissolved in 0.6 ml of 0.1 M Na_2HPO_4, was added to 1.0 ml of serum globulins diluted to contain 1 mg/ml protein concentration, while the solution was constantly stirred. More 0.1 M Na_2HPO_4 was added to make up the volume to 1.8 ml. The pH of the solution was adjusted to 9.5 with 0.04 N NaOH (approximately 0.2 ml is required). The solution was then wrapped in tinfoil in order to protect it from exposure to strong light. Conjugation was allowed to take place for 30 minutes at room temperature without stirring. The conjugate was then passed through a Sephadex G-50 column (2 × 20 cm) equilibrated with PBS. The column was then flushed with PBS and the visible band of labeled globulins which clearly separate was collected. The conjugate was then diluted with PBS and tested in inoculated and healthy cell monolayers grown on cover slips. The maximal dilution that gave good staining of infected cells but no nonspecific staining of uninfected cells was determined. Diluted serum (in PBS) was distributed in 1.0 ml aliquots and frozen at −20°C for future use.

2. Fluorescent Antibody Staining of Cell Monolayers

The following procedure was originally described by Chiu and Black (1969) and slight modifications introduced by Reddy and Black (1972) were incorporated. The monolayers were washed at least 4 times in PBS, after the growth medium was removed with a Pasteur pipette. Addition of adequate PBS to cover each individual cover slip was preferred to flooding the glass dish with PBS. Flooding sometimes resulted in the overlapping of cover slips one over the other and led to the removal of cell monolayers. Thorough washing in PBS was important in obtaining a clear background after fluorescent antibody staining. Then the monolayers were air dried, fixed in ice-cold acetone for 3–5 minutes, and agian air dried. Fixation in acetone was performed by carefully adding enough acetone to the glass dish to immerse the cover slips. If fluorescent antibody staining could not be performed immediately fixed specimens were stored at −25°C for at least a week. The monolayers were washed in PBS and stained with fluorescein-labeled antibody at 37°C for at least 60 minutes in a moist chamber. The stained monolayers were washed 5 times in PBS, the excess PBS was drained off, and the bottom side of each cover slip was wiped clean with lint-free absorbent tissue. The white styrene metabolic inhibition test trays are convenient for handling the cover slips during the staining and subsequent washing in PBS. The details of these procedures are lucidly described by Spendlove (1967, pp. 503–506). Cover slips were then mounted on a regular microscopic slide, with the cell monolayers facing downward in PBS containing 60% glycerol. The slides were examined under a UV fluorescence microscope. A dry condenser is preferable to oil immersion when transmitted light was used. The mounted cover slips can be stored at 4°C for at least 2 weeks before examination.

3. Quantitation of the Infection

Infected cells were distinguished from the surrounding healthy cells by the presence of bright apple-green fluorescence either in the cytoplasm as observed in WTV infection (Fig. 8) (Chiu and Black, 1967) or in the nucleus as noted in the case of infections with PYDV (Fig. 9) (Chiu et al., 1970).

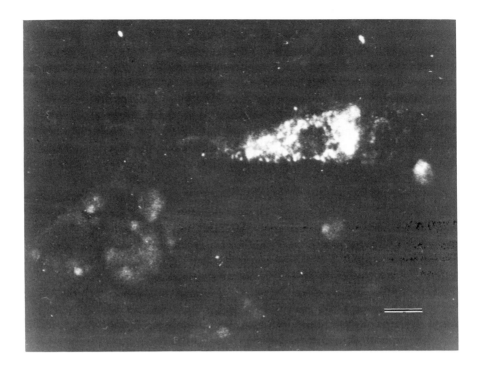

FIG. 8. *Agallia constricta* (AC20) cells inoculated with wound tumor virus. The unstained nucleus (dark circle) is surrounded by granular staining in the cytoplasm. Occasionally, stained antigen in the cytoplasm, above or below the nucleus, gives the false impression of specific staining in the nucleus (×400). Bar represents 50 μm. [From Chiu and Black (1967). By courtesy of *Nature* (*London*).]

When the incubation period was restricted to about 40 hours after inoculation of monolayers of cells, WTV infection was either limited to the primarily infected cell or a small group of cells, termed an "infection focus" (Kimura and Black, 1971). In the case of infection with PYDV, there was little, if any, evidence of intercellular spread of the virus (Hsu and Black, 1973).

Chiu and Black (1969) counted fluorescent cells either in entire monolayers or in randomly selected microscopic fields. Each of the latter represented an area of 0.075 mm^2 when 500× magnification was used. Since this area represented 1/400 of the total area covered by the confluent cell monolayers, the total number of infected cells per cover slip could be calculated by multiplying the average number of infected cells per field by 400. Two factors introduced error in counting the total number of infected cells: (1) virus released from primary infected cells may infect surrounding healthy cells, and these cells were scored along with those infected from the original inoculum; and (2) if the virus being assayed produced no cytopathic effects

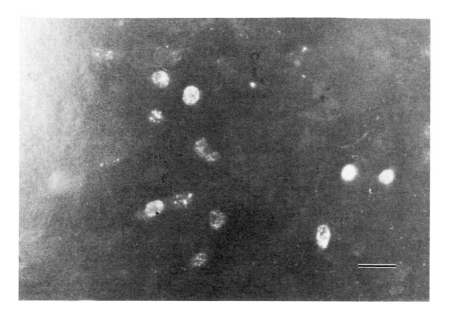

FIG. 9. *Agallia constricta* (AC20) cells inoculated with potato yellow dwarf virus. Note the intense staining in the nucleus and the lack of staining in the cytoplasm (×400). Bar represents 100 μm. [From Chiu *et al.* (1970).]

on the susceptible host cells, the infected cells may divide during the incubation period. Although counting the total number of infected cells may yield overestimates of absolute cell-infecting units, the method was successfully used to determine the relationship between infected cell counts and relative virus concentrations (Chiu and Black, 1969).

Kimura and Black (1971, 1972) found that it was sometimes more convenient and accurate to count each infective focus to measure the amount of infection, rather than to count the numbers of infected cells, as described above. Focus counts were made in both horizontal and vertical diametral zones on each monolayer, according to Spendlove (1967) and Liu (1969). Each diametral zone contained 25 microscopic fields at a magnification of 200×. The distribution of foci in the diametral zone was uniform except in 3 or 4 microscopic fields (1.8–2.4 mm diameter) near the margin of the cover slips. For more precise estimates of WTV infectivity, foci in only 13 central fields of each diametral zone were counted (Kimura and Black, 1972, p. 552).

B. ELECTRON MICROSCOPY

Although it is time consuming and less sensitive than fluorescent antibody staining, electron microscopy of infected cells is vital to ascertain the subcellular sites of viral multiplication and accumulation. Cells are processed for electron microscopy as intact monolayers or after dissociation according to the procedure of Chiu *et al.* (1970) and Hirumi (1973).

When cells were embedded as a monolayer, heat-stabilized carbon was deposited on cover slips before they were seeded with infected cells. The culture medium was removed from the monolayers and replaced with 3% glutaraldehyde in Cunningham's saline, pH 6.5. Until the final stage of dehydration, the cover slips were placed on crushed ice in order to keep them at 0 °C. After 45 minutes of fixation in the glutaraldehyde the mono-layers were gently rinsed with Cunningham's saline and flooded with cold 2% osmium tetroxide in Cunningham's saline. Postfixation in osmium tetroxide was allowed to proceed for 60 minutes, and cells were then rinsed with dis-tilled water and rapidly dehydrated through a graded series of ethanols.

After an overnight period of infusion with "Epon" (Epoxy resin–propy-lene oxide mixture), the cell monolayers were carefully peeled off the cover slip and trasferred as sheets to Epon in Beem capsules, or when the cells were grown on heat-stabilized carbon coated cover slips, a Beem capsule filled with Epon was inverted over the tissues. In the latter case, after the resin had polymerized, the cover slip was easily separated from the resin, leaving the monolayer clearly visible at the surface of the epoxy resin block. The block was then trimmed for ultramicrotomy.

Alternatively the infected cells were harvested from culture flasks, col-lected after centrifugation at 1000 rpm for 2–3 minutes in a clinical centri-fuge, resuspended in PBS, and placed in an ice bath. An equal volume of 6% glutaraldehyde in PBS was added, mixed with the cell suspension, and left at 0 °C for 60 minutes. Postfixation in osmium tetroxide was carried out as de-scribed above. The cells were collected again by centrifugation and resus-pended in a small amount of distilled water. Dehydration was carried out in a graded series of ethanols, allowing 10-minute intervals between each step. The cells were collected by centrifugation, resuspended in absolute ethanol, and washed once more in absolute ethanol. The cells were resuspended in an ethanol–propylene oxide mixture (1:1 proportion), centrifuged, and rinsed twice with propylene oxide before being transferred to a 1:1 propylene oxide–resin mixture. After overnight infusion with propylene oxide–resin mixture, the cells were centrifuged and most of the supernatant was removed, leaving a small amount of resin mixture covering the cell surface. The cell suspension was gently mixed and transferred with a Pasteur pipette to Beem capsules filled with epoxy. Beem capsules were then centrifuged in a clinical centrifuge to pack the cells at the tip of the capsule and the capsules were placed at 60 °C in an oven for 36 hours to accelerate the polymerization (H. T. Hsu, unpublished).

Hirumi (1973), after rinsing the cells grown in 25 cm² flasks in Earle's balanced salt solution, fixed the cells for 30 minutes at 4 °C with 3% glutaraldehyde prepared in 0.1 M sodium cacodylate buffer, pH 7.2, and then removed them from the flasks. The cells were rinsed with the buffer for 30 minutes, collected by low-speed centrifugation, and postfixed in 2% osmium tetroxide prepared in the same buffer with 5% sucrose, for 30 minutes at 4 °C. After brief rinsing with the buffer solution, cell pellets were embedded in 1.5% agar, the agar blocks were cut into small pieces, and were dehydrated by treatment with cold-graded ethanol series and propylene oxide. The specimens were embedded in epoxy resin.

In all situations the ultrathin sections were poststained with uranylacetate and lead citrate (Chiu *et al.*, 1970; Hirumi, 1973). Sections deposited on uncoated 300 mesh copper grids were stained in 2% uranyl acetate, pH 2.5, for 2 hours at 50 °C, followed by 7-minute poststaining in Reynold's lead citrate (Chiu *et al.*, 1970).

X. Storage of Leafhopper Cell Lines

A. SHORT-TERM CELL STORAGE

Cell cultures grown in 25 cm² flasks were stored for short intervals at 9 °C by the procedure described by Schlegel and Black (1969). Before storage, cultures were grown until nearly confluent, and the medium was replaced with 5 ml of fresh medium. After removal from storage at 9 °C, the medium was again replaced, the cultures were kept at 30 °C overnight, and are then subcultured. Medium changes were made every 3–4 months during storage. *Agallia constricta* cells could be stored for nearly 1 year, and *A. sanguinolenta* cultures for a maximum of 20 weeks (in Chiu and Black's medium) under these conditions (Liu, 1969). A comparison of longevity of cell cultures stored at 9 °C and 12°–13 °C indicated that storage at 12°–13 °C was more desirable (Liu, 1969), and cell monolayers could be kept at 13 °C for nearly 2 years after the culture medium (5 ml of Martínez-López and Black's medium for 25 cm² flasks) was replaced every 3 to 4 months (Martínez-López, 1973). Under similar conditions, it was also possible to store WTV-infected cell monolayers for nearly 2 years. *Aceratagallia sangiunolenta* and *A. novella* cells could be stored at 13 °C for nearly 1 year. Martínez-López (1973) attributed the longer viability to the improved condition of the cells in the new medium used.

B. LONG-TERM CELL STORAGE

Cells of *A. constricta* and *A. sanguinolenta* in the presence of dimethylsulfoxide (DMSO) or glycerol could be preserved under liquid nitrogen

($-190\,°C$). Fully confluent cell monolayers were harvested, collected by low-speed centrifugation, and then suspended in culture medium containing 5–10% DMSO or 10% glycerol to give a cell density of about 7 to 8 × 10⁶ cells/ml. Then 0.6 ml of cell suspension was hermetically sealed and labeled in each 1.2 ml ampule (Wheaton Glass Co., Millville, New Jersey). Freezing was carried out at a rate of 1 °C or less per minute until the temperature reached $-40\,°C$. The ampules were then stored at $-80\,°C$ or were placed under liquid nitrogen (Hsu and Black, 1969).

Thawing was accomplished by placing the ampules in a water bath at 30 °C, and gently agitating them until no ice remained. The suspension in each ampule was then rinsed into 1 ml of culture medium and diluted to a final volume of 10 ml by increments of 1 ml of medium per minute. The cells were sedimented by centrifugation, resuspended in 4–5 ml of fresh medium, and were centrifuged again. The pellet of cells was then suspended in 4 ml of fresh medium for seeding in a 25 cm² culture flask.

After demonstrating that cultured leafhopper cells could be stored at cryogenic temperature, several cell lines were stored at either $-80\,°C$ or $-190\,°C$. Leafhopper cell lines could be kept viable at $-80\,°C$ for about 9 months (Hsu, 1971). However, cells stored in liquid nitrogen remained viable for at least 7 years, the longest period so far tested.

XI. Transportation of Cell Lines

Cells stored at $-80\,°C$ by the procedure described by Hsu and Black (1969) were suitable for shipping, provided the cells were kept frozen at $-80\,°C$ or lower. Fully confluent monolayers in culture flasks could also be shipped by completely filling the flasks with the growth medium, securely tightening the screw cap covers, except for a few air bubbles, and wrapping caps and neck with parafilm. However, early attempts to ship cells to South America by this method were not successful (Anderson and Black, 1975).

Anderson and Black (1975) have devised a valise for carrying cell cultures. The valise is made of fabricated 16 gauge (1.6 mm) aluminum sheets and insulated with styrofoam. It can hold eight 25 cm² flasks. Two plastic jars filled with ice and water keep the temperature in the neighborhood of 14 °C for about 24 hours. It is more convenient to use plastic jars with ice and water for this purpose than plastic freezer packs. The valise is equipped with a readily observable thermometer. Flasks with confluent cell monolayers are filled with 4 ml of fresh medium before carrying. By using this box, cells have been successfully transported to Taiwan, Japan, and South America.

XII. Availability of Established Cell Cultures

Cell lines from 7 species of leafhoppers have been established so far. These were derived from vectors of not only plant pathogenic viruses, but also other plant pathogens such as spiroplasmas, mycoplasmas, and the clover club leaf organism. Different laboratories which have leafhopper cell lines in culture or stored at cryogenic temperature are given in Table VII.

XIII. Aphid Cell Cultures

Although attempts to establish cell lines from aphids have not been successful, primary cultures from aphids have been successfully inoculated with viruses. Tokumitsu and Maramorosch (1966) first attempted to culture aphid cells. They have reported the survival of some cells of the aphid, *Acyrthosiphum pisum* Harris, in Grace's Lepidoptera medium (1962) lacking calcium pantothenate. Peters and Black (1970) were successful in preparing primary cultures of ovarian and embryonic tissues from the aphid *Hyperomyzus lactucae*. Ovaries and embryos were removed from surface sterilized apterous adults and cut into small parts with steel knives. The tissue fragments were treated with 0.1% Pronase prepared in a salt solution containing 6.45 gm NaCl, 0.35 gm KCl, 0.94 gm $CaCl_2 \cdot 2H_2O$, 0.86 gm $MgSO_4 \cdot 7H_2O$, 1.5 gm tricine, 1 gm galactose, 1 gm trehalose, 0.1 gm streptomycin, 0.05 gm neomycin, and 0.0025 gm Fungizone per liter. The treatment with Pronase was continued 5–8 minutes and the cell suspension was centrifuged at 1000 rpm for 5 minutes and washed in a culture medium containing, per liter, 13 gm sucrose, 0.4 gm fructose, 0.7 gm glucose, 0.5 gm KH_2PO_4, 1.1 gm KCl, 0.5 gm $CaCl_2$, 1.14 gm $MgCl_2 \cdot 6H_2O$, 1.39 gm $MgSO_4 \cdot 7H_2O$, 200 ml of fetal bovine serum, 100,000 Units penicillin, 0.1 gm streptomycin, 0.05 gm neomycin, 0.0025 gm Fungizone, and amino acids and organic acids to the same proportion as in Grace medium (1962). The suspension of cells was taken in a small drop of the culture medium, seeded on cover slips, and incubated at room temperature in sealed glass petri dishes. Within 1–2 days, the cells adhered to the glass surface. Large numbers of cells migrating from the tissue explant were observed. Cells could be maintained in the medium for a period of 10 days, and following this interval the cells gradually degenerated. No cell multiplication was observed in these cultures (Peters and Black, 1970).

Purified sowthistle yellow vein virus was sterilized by passing through Millipore filters with an average pore diameter of 0.3 μm, and the cover slip

TABLE VII

ESTABLISHED LEAFHOPPER CELL LINES AND LABORATORIES MAINTAINING THEM

Cell line	Established by	Laboratories[a,b]
Agallia constricta		
AC13[c]	Chiu and Black (1967, 1969)	None
AC15[c]	Chiu and Black (1967, 1969)	None
AC16	Chiu and Black (1967, 1969)	A
AC17[c]	Chiu and Black (1967, 1969)	None
AC18[c]	Chiu and Black (1967, 1969)	None
AC20	Chiu and Black (1967, 1969)	A,B,C,D,E,G
AC21	Chiu and Black (1967, 1969)	A
AC24	Liu and Black (1976)	A
Agallia quadripunctuta[c]	Chiu and Black (1967)	None
Agalliopsis novella		
AN3[c]	Windsor (1972)	None
AN4	Liu and Black (1976)	A,G
AN5	Liu and Black (1976)	A,G
Aceratagallia sanguinolenta		
AS1	Chiu and Black (1969)	A
AS2	Chiu and Black (1969)	A,G
Colladonus montanus[c]	Richardson and Jensen (1971)	None
Dalbulus elimatus	McBeath and Chen (1975)	F
Macrosteles fascifrons	Liu, Hirumi, and Black	A

[a] Maintaining the cell lines at the present time.

[b] (A) Dr. L. M. Black, Department of Genetics and Development, University of Illinois, Urbana, Illinois; (B) Dr. R. J. Chiu, Plant Industry Division, Joint Commission on Rural Reconstruction, Taipei, Taiwan, China; (C) Dr. I. Kimura, Institute for Plant Virus Research, Japan; (D) Dr. G. Martínez-López, Institute Colombianao Agropecuario, Bogota, Colombia, South America; (E) Dr. A. McIntosh, Institute of Microbiology, Rutgers University, New Brunswick, New Jersey; (F) Dr. T. A. Chen, Department of Plant Pathology, Rutgers University, New Brunswick, New Jersey; (G) Cell Culture Department, American Type Culture Collection, Rockville, Maryland.

[c] No longer available.

cell cultures were inoculated with the virus for 2 hours. Presence of viral antigens in the cells was detected by fluorescent antibody staining. After 2 days of incubation, several specifically stained cells could be observed (Peters and Black, 1970).

Hirumi (1971) succeeded in growing *A. pisum* cells from embryonic tissues in a culture medium, consisting of 10 parts of Schneider's medium, 10 parts of Vago's B.M.22 medium (see Table I), and 3 parts of fetal bovine serum with antibiotics. The majority of the cells in aphid primary cultures were fibroblastlike cells and large multinucleate cells (Hirumi, 1971; Peters and Black, 1970). Among several media that they tried Matisova and Valenta (1975) reported Hirumi's medium as the most suitable one for culturing *A.*

pisum cells, and in addition, they reported that the osmolality of the medium was a major factor influencing cell growth. They reported survival of cells for nearly 3 weeks, but attempts to subculture the cells failed. Peters (1975) succeeded in obtaining primary cultures of the aphids *Spodoptera frugiperda* and *Adoxophyes orana* after developing a new medium. Currently there are no reports on the successful establishment of aphid cell cultures. (See Note Added in Proof, p. 434.)

XIV. Conclusions

Considerable progress has been made during the past decade in the development of plant virus–vector cell monolayer systems. These have already been of substantial significance in the field of plant virus research, particularly in supplying an extremely sensitive assay system. They have not yet been fully exploited for thorough basic penetrating studies of the mode of replication of the plant viruses that they support, yet progressive improvements in techniques may eventually lead to an understanding of the mode of replication of several of these plant viruses. Certain plant viruses are capable of multiplying in nonvector cell lines. The efforts involved in obtaining continuous cultures of all vector cells might be saved by attempting to initiate infection of the virus in previously established cell lines.

One would be remiss not to mention the several inherent difficulties that may be encountered with plant virus–vector cell line systems. The cycle of virus replication in the insect monolayer is considerably slower, with the appearance of viral progeny antigen being evident seldom earlier than 24 hours postinoculation. Associated with the slower replicative cycle the incorporation of radioactive isotopes into viral intermediates is slow. This can be particularly troublesome in experiments in which antagonists of protein or nucleic acid synthesis, such as actinomycin D and cycloheximide, are to be used to examine the mode of replication of viruses. Frequently these antagonists show overt cytopathic effects on the insect cells before adequate incorporation of isotopes into viral intermediates occurs. Several of these disadvantages may be related to the temperature at which the cells are grown, which is considerably lower than the optimum temperature for growing vertebrate cell lines.

The viruses that have been thus far grown successfully in vector cell lines show no signs of cytopathogenesis. This, of course, need not always be related to the absolute amount of virus production by the supporting cell, yet in the insect monolayer system yields of virus proved to be lower than those routinely obtained from classic vertebrate virus systems. These aspects of replication of plant viruses in vector cell monolayer systems make complete analysis of their mode of replication considerably more challenging, but may lead in general to an understanding of the basis of the noncytopathic relationship of viruses in cells.

It is desirable for laboratories successful in obtaining continuous monolayers of cells to store them in liquid nitrogen and to deposit their cell lines with organization similar to ATCC in order to assure their supply to scientists in need of them. Close coordination and free flow of information between various laboratories working on cultivation of cell lines is of paramount importance to aid them in their efforts to establish cell lines and infect them with viruses.

Although attempts to obtain established cell lines from aphids have not been successful so far, such cell lines could be employed for infections with viruses propagative in the aphid vector and could provide an excellent means of assaying these viruses. Media for culturing aphid or leafhopper cells, developed by mixing commercially available culture media, appear to be a logical choice for starting new vector cell lines or for maintaining established cell lines. The medium developed by Hirumi (1971) for culturing aphid cells and Liu and Black's medium (Table VI) for maintaining leafhopper cell lines can be prepared by mixing commercially available culture media. Besides, efforts should be made to develop a defined medium for growing leafhopper cell lines to achieve rapid progress in studying the molecular biology of vector-transmitted viruses.

ACKNOWLEDGMENT

I gratefully acknowledge the many valuable discussions with Dr. Roderick MacLeod during the preparation of this chapter. I also thank my colleagues, Drs. H. T. Hsu, H. Y. Liu, and I. Kimura for providing extremely useful information which is incorporated in various sections of this work. I thank Drs. Hirumi and J. Mitsuhashi for providing Figs. 1 and 5 and for their helpful suggestions. I express my appreciation to Miss Connie Bowton for her help in the preparation of the manuscript.

REFERENCES

Anderson, J., and Black, L. M. (1975). *In Vitro* **11**, 322.
Barile, M. F., Hopps, H. E., Grabowski, M. W., Riggs, D. B., Del Giudice, R. A. (1973). *Ann. N. Y. Acad. Sci.* **225**, 251.
Berkson, J., Magath, T. B., and Hurn, M. (1939). *Am. J. Physiol.* **128**, 309.
Black, L. M. (1969). *Annu. Rev. Phytopathol.* **7**, 73.
Black, L. M., and Brakke, M. K. (1952). *Phytopathology* **42**, 269.
Brakke, M. K., Maramorosch, K., and Black, L. M. (1953). *Phytopathology* **43**, 387.
Brakke, M. K., Vatter, A. E., and Black, L. M. (1954). *Brookhaven Symp. Biol.* **6**, 137.
Brown, S., Teplitz, M., and Revel, J. (1974). *Proc. Natl. Acad. Sci. U.S.A.* **71**, 464.
Buchner, P. (1953). "Endosymbiosen der Tiere mit Pflanzlichen Mikroorganismen." Birkhaeuser, Basel.

Campbell, D. H., Garvey, J. S., Cremer, N. E., and Sussdorf, D. H. (1964). *In* "Methods in Immunology," p. 263. Benjamin, New York.
Chiu, R. J., and Black, L. M. (1967). *Nature (London)* **215**, 1076.
Chiu, R. J., Black, L. M. (1969). *Virology* **37**, 667.
Chiu, R. J., Reddy, D. V. R., and Black, L. M. (1966). *Virology* **30**, 562.
Chiu, R. J., Liu, H. Y., MacLeod, R., and Black, L. M. (1970). *Virology* **40**, 387.
Grace, T. D. C. (1962). *Nature (London)* **195**, 788.
Greene, A. E., and Charney, J. (1973). *In* "Tissue Culture: Methods and Applications" (P. F. Kruse, Jr. and M. K. Patterson, Jr., eds.), p. 750. Academic Press New York.
Harnett, G. B., Phillips, P. A., and MacKay-Scollay, E. M. (1974). *J. Clin. Pathol.* **27**, 70.
Hirumi, H. (1971). *Curr. Top. Microbiol. Immunol.* p. 171.
Hirumi, H. (1973). *Proc. Int. Colloq. Invertebr. Tissue Cult., 3rd, 1971* p. 233.
Hirumi, H., and Maramorosch, K. (1964a). *Exp. Cell Res.* **36**, 625.
Hirumi, H., and Maramorosch, K. (1964b). *Science* **144**, 1465.
Hirumi, H., and Maramorosch, K. (1964c). *Contrib. Boyce Thompson Inst.* **22**, 343.
Hirumi, H., and Maramorosch, K. (1968). *Proc. Int. Colloq. Invertebr. Tissue Cult., 2nd, 1967* p. 203.
Hirumi, H., and Maramorosch, K. (1971). *In* "Invertebrate Tissue Culture" (C. Vago, ed.), Vol. 1, p. 304. Academic Press, New York.
Hirumi, H., Hirumi, K., Speyer, G., Yunker, C. E., Thomas, L. A., Cory, J., and Sweet, B. H. (1976). *In Vitro* **12**, 83.
Hsu, H. T. (1971). Ph.D. Thesis, University of Illinois, Urbana.
Hsu, H. T., and Black, L. M. (1969). *Phytopathology* **59**, 1032 (abstr.).
Hsu, H. T., and Black, L. M. (1973). *Virology* **52**, 187.
Hsu, H. T., and Black, L. M. (1974). *Virology* **59**, 331.
Kimura, I., and Black, L. M. (1971). *Virology* **46**, 266.
Kimura, I., and Black, L. M. (1972). *Virology* **49**, 549.
Liu, H. Y. (1969). Ph.D. Thesis, University of Illinois, Urbana.
Liu, H. Y. (1970). *Plant Prot.* **24**, 170.
Liu, H. Y. (1975). *Proc. Int. Congr. Virol., 3rd, 1975* Abstracts, p. 88.
Liu, H. Y., and Black, L. M. (1976). *Proc. Am. Phytopathol. Soc.* Vol. 3.
McBeath, J. H., and Chen, T. A. (1973) *Proc. Int. Congr. Phytopathology, 1973* Abstracts, p. 278.
McBeath, J. H., and Chen, T. A. (1975). *Proc. Int. Congr. Virology, 3rd, 1975* Abstracts, p. 87.
McGuigan, J. E., and Eisen, H. N. (1968). *Biochemistry* **7**, 1919.
McIntosh, A. H., Maramorosch, K., and Rechtoris, C. (1973). *In Vitro* **8**, 375.
Maramorosch, K. (1976). *In* "U.S. Japan Conference on Invertebrate Tissue Culture: Research Applications" (K. Maramorosch, ed.). Academic Press, New York (in press).
Martínez-López, G. (1973). Ph.D. Thesis, University of Illinois, Urbana.
Martínez-López, G., and Black, L. M. (1971) *Phytopathology* **61**, 902 (abstr.).
Martínez-López, G., and Black, L. M. (1973). *In Vitro* **9**, 1.
Martínez-López, G., and Black, L. M. (1974). *Phytopathology* **64**, 1040.
Martínez-López, G., and Black, L. M. (1977). *Methods Cell Biol.* **15** (in press).
Matisova, J., and Valenta, V. (1975). *Proc. Int. Congr. Virol., 3rd, 1975* Abstracts, p. 89.
Mitsuhashi, J. (1965). *Jpn. J. Appl. Entomol. Zool.* **9**,107.
Mitsuhashi, J. (1966). *Appl. Entomol. Zool.* **1**, 103.
Mitsuhashi, J. (1970). *Appl. Entomol. Zool.* **5**, 47.
Mitsuhashi, J. (1972). *In* "Invertebrate Tissue Culture" (C. Vago, ed.), Vol. 2, p. 343. Academic Press, New York.

Mitsuhashi, J., and Kono, Y. (1975). *Appl. Entomol. Zool.* **10**, 1.

Mitsuhashi, J., and Koyama, K. (1975). *Appl. Entomol. Zool.* **10**, 123.

Mitsuhashi, J., and Maramorosch, K. (1964). *Contrib. Boyce Thompson Inst.* **22**, 435.

Mitsuhashi, J., Grace, T. D. C., and Waterhouse, D. F. (1970). *Entomol. Exp. Appl.* **13**, 327.

Peters, D. (1975). *Proc. Int. Congr. Virol. 3rd, 1975.* Abstracts, p. 88.

Peters, D., and Black, L. M. (1970). *Virology* **40**, 847.

Reddy, D. V. R., and Black, L. M. (1972). *Virology* **50**, 412.

Reddy, D. V. R., and Black, L. M. (1973). *Virology* **54**, 150.

Richardson, J., and Jensen, D. D. (1971). *Ann. Entomol. Soc.* **64**, 722.

Rinaldini, L. M. (1954). *Nature (London)* **173**, 1134.

Rinaldini, L. M. (1959). *Exp. Cell Res.* **16**, 477.

Sanford, K. K., Earle, W. R., Evans, V. J., Waltz, H. K., and Shannon, J. E. (1951). *J. Natl. Cancer Inst.* **11**, 773.

Schlegel, D., and Black, L. M. (1969). *Annu. Rev. Phytopathol.* **7**, 84.

Sinha, R. C., and Reddy, D. V. R. (1964). *Virology* **24**, 626.

Spendlove, R. S. (1967). *In* "Methods in Virology" (K. Maramorosch and H. Koprowski, eds.), Vol. 3, p. 485. Academic Press, New York.

Tokumitsu, T., and Maramorosch, K. (1966). *Exp. Cell Res.* **44**, 652.

Tokumitsu, T., and Maramorosch, K. (1967). *J. Cell Biol.* **34**, 677.

Vago, C. (1967). *In* "Methods in Virology" (K. Maramorosch and H. Koprowski, eds.) Vol. 1, p. 567. Academic Press, New York.

Windsor, I. (1972). Ph.D. Thesis, University of Illinois, Urbana.

NOTE ADDED IN PROOF

Adam and Sander (1976) (*Virology* **70**, 502) reported establishment of primary cultures of *Myzus persicae* and their successful inoculation with pea enation mosaic virus (PEMV). Embryonic tissues and ovaries were employed to obtain cell suspensions of *M. persicae*. A new culture medium was developed by the authors. Nearly 95% of the quasi-cell mono-layers of cells obtained were viable for 4 to 5 days after preparation. Following this interval the cells gradually degenerated and only few cells were observed to be viable up to 14 days. Cells were inoculated, 2 days after preparation, with purified PEMV. Infected cells could be detected 38 hours after inoculation employing FITC-labeled antibodies using the direct staining method.

12 *Use of Protoplasts for Plant Virus Studies*

S. Sarkar

I. Introduction

Protoplasts offer some significant advantages over a whole organism or an organ thereof for studies on viruses. Practically 100% of the protoplasts can be infected with a virus almost synchronously, so that a fairly homogeneous population of virus-synthesizing units can be analyzed as a function of time. Not only in virology, but for physiological and genetical studies the great

435

importance of single cells and protoplasts had been recognized very early and, since the end of the last century, many attempts have been made to isolate them in a viable state. Some of our valuable data stem from mechanically dissociated cells and protoplasts. The historical background of this research has been reviewed in recent years by Cocking (1972) and Zaitlin and Beachy (1974), among others, and some early achievements have also been documented (Kohlenbach, 1959, 1966; Melchers, 1975).

The term "protoplast" has not been used very consistently (see Cocking, 1972). In this contribution the contents of a single cell after removal of the cell wall will be called a protoplast, without being very critical about an eventual retention of a last trace of the cell wall material, say, after enzymic digestion. Under certain conditions it is possible to obtain the contents of a cell as more than one unit. Such subprotoplasts, particularly when isolated from fungi and algae, often lack a nucleus and do not "survive." We will be restricted here to the use of protoplasts of higher plants for studies on plant viruses. Some representative experimental procedures will be cited for practical work. The technique is undergoing rapid modifications to suit specific host–virus combinations and hardly two laboratories seem to follow exactly the same procedure.

The scope of the use of protoplasts in plant virology and some of the major achievements in this field have been discussed and reviewed recently by Takebe (1975). As already mentioned by Zaitlin and Beachy (1974), the studies of Takebe and his associates (Takebe *et al.*, 1968; Takebe and Otsuki, 1969; Aoki and Takebe, 1969) have ushered in a new era in plant virology. The enzymic procedure for the isolation of protoplasts has become the method of choice (Cocking, 1960; see also Ruesink, 1971).

II. Materials

For quick and efficient isolation of plant protoplasts and their infectivity with viruses, certain particular reagents and equipment have proved to be most useful. Since they are not available from every dealer, the name of a manufacturer or at least of one supplier will be cited. There are now a few booklets containing information specific for the isolation and infection of protoplasts. Some examples are: (1) "Enzymatic Isolation of Living Cells and Their Protoplasts from Higher Plants." Kanematsu-Gosho Inc., New York, New York; (2) "Plant Tissue Culture Methods," by O. L. Gamborg and L. R. Wetter. National Research Council of Canada, Saskatoon, Canada; (3) "The Use of Protoplasts from Fungi and Higher Plants as Genetic Systems. A Practical Handbook," by E. C. Cocking and J. F. Peberdy, Department of Botany, University of Nottingham, Nottingham,

England; and (4) "From Single Cells to Plants," by E. Thomas and M. R. Davey. Wykeham Publications, London, 1975.

A. REAGENTS

Reagents used are the following:

Cellulase–"Onozuka R–10," and macerozyme R–10 Kinki Yakult Company Ltd., Nishinomiya, Japan, supplied by All Japan Biochemicals Ltd., Nishinomiya, Osaka, Japan.

Driselase (Kyowa Hakko Kogyo Company Ltd., Ohtemachi, Chiyoda-Ku, Tokyo, Japan).

Meicelase (Meiji Seika Kaisha Ltd., Kyobashi, Chuo-Ku, Tokyo, Japan).

Pectinol fest and Rhozyme HP 150 (Roehm and Haas Company, Darmstadt, Germany, or Rohm and Haas Company Ltd., Philadelphia, Pennsylvania).

MES buffer (2-N-morpholino-ethane sulfonic acid) and pectinase (Sigma Chemical Company, St. Louis, Missouri).

Helicase (L'Industrie Biologique Francais, Gennevillies, France).

Potassium dextran sulfate (MW over 500) (Meito Sangyo Company, Meitosangyo, Nagoya, Japan). Poly-L-ornithine (MW 130,000) (Pilot Chemicals Inc., Watertown, Massachusetts; also available from Sigma London Chemicals Company Ltd., Surrey, England).

Domestos (Lever Brothers Ltd., London, England).

Carbanecillin (Beecham Research Laboratories, Brentford, England).

Aureomycin, gramicidine, and other antibiotics are available from various suppliers of fine chemicals.

B. EQUIPMENT

It is not essential to have a sterile room for working with protoplasts. Laminar airflow transfer cabinets have proved to be efficient and are now available in many countries. Names of some suppliers are South London Electrical Equipment, London; John Bass Ltd., Southampton, England; Ralph E. Benner Ltd., Ontario, Canada; and CEAG, Dortmund, West Germany.

The term "protoplast pipette" is used here for a volumetric pipette whose capacity is about 15 ml. Fitted with a cotton plug and a rubber ball it has been found to be convenient for pipetting of protoplasts during the washing and transferring steps. It is basically similar to a Pasteur pipette of large dimensions, with a gradually tapering end. A length of 30–35 cm is useful and the opening at the tip should be broad enough (diameter about 3 mm) to allow a uniform flow of protoplasts without injury. These pipettes are used in Japan under the name "Kommagome" pipettes.

Enzyme solutions that are unable to be sterilized by autoclaving are filtered through membrane filters using a positive pressure from a compressed air cylinder. The membrane is held in place in any suitable filtration unit. A convenient unit (Model 16510 with 110 silicone rings) whose parts can be detached, cleaned, and sterilized by autoclaving before use, is supplied by Sartorius-Membranfilter GmbH, Göttingen, West Germany. Membrane filters, syringes, and adapters can be obtained from the same suppliers or from others, such as the Millipore Corp., Bedford, Massachusetts, whose agencies are distributed over the world. Disposable membrane filter units are supplied by, among others, Canadian Laboratory Supplies, Winnipeg, Canada. Stainless steel mesh is supplied by The W. S. Tyler Company of Canada Ltd., Ontario, Canada, and nylon screens are available from B. and S. H. Thompson Company Ltd., Montreal, Quebec, Canada. Heat sterilizable nets can also be purchased from Vereinigte Seidenwebereien AG, Krefeld, West Germany. Of a large variety of products "Monodur-Polyester tissues" of 800 and 140 μm mesh are recommended.

Observation of protoplasts *in situ*, i.e., within a flask or a petri dish, is facilitated by use of an inverted microscope, e.g., the Diavert, made by Ernst Leitz GmbH, Wetzlar, West Germany.

III. Preparation of Protoplasts

A. CULTURE AND PRETREATMENT OF PLANTS

Probably the most important single factor determining the success in isolation of viable protoplasts is the condition of the starting material, i.e., the plant or the tissue. The species and age of the plant and the environmental conditions during growth need particular attention. Reproducible results are obtained only when these are kept as constant as possible (Motoyashi *et al.*, 1973; Otsuki *et al.*, 1974; Watts *et al.*, 1974; Harrison *et al.*, 1975; and others). For the isolation of mesophyll protoplasts from leaves a general rule is to keep the plants under rapid growth with sufficient nutrients and to collect the leaves that have reached or will soon attain their full size. However, older leaves have been found to be more convenient with certain plants, as *Brassica* and potato, and even with tobacco under certain conditions. Plants grown in an environmental chamber may be more suitable for attaining reproducible results. During winter months a few hours of additional illumination may be essential. It is advisable to limit the supply of water to the plants for several hours before sampling (Takebe *et al.*,

1968). Under such a condition the leaves lose their turgidity to some extent and a better yield of protoplasts can be achieved.

B. Sequential or Two-Step Procedure

The standard method for the isolation of tobacco mesophyll protoplasts is described in detail. It is based mainly on the investigations of Takebe and co-workers (Takebe et al., 1968; Takebe and Otsuki, 1969). According to this procedure the cells are first released by a digestion of the middle lamella with the help of polygalacturonase. Then the suspension of cells is subjected to the action of cellulase. The liberation and stability of the protoplasts are facilitated by plasmolysis caused by a solution of mannitol of suitable molarity.

The steps in the procedure are as follows:

1. Dissolve 0.5 gm macerozyme R-10 and 0.3 gm potassium dextran sulfate in 100 ml of 0.7 M mannitol in distilled water. Adjust the pH to between 5.5 and 5.8 with 0.1 N KOH or 0.1 N HCl. Sterilize by filtration through a Millipore filter (pore size between 0.2 and 0.4 μm). It has been found that for tobacco protoplasts the molarity of mannitol can be changed with advantage from 0.7 to 0.6 M throughout the treatments (I. Takebe, personal communication).

2. Dissolve 0.5 gm cellulase in 100 ml of 0.7 M mannitol in distilled water. Adjust the pH to between 5.2 and 5.5 as above. Clarify the solution by a short centrifugation at about 5000 g (or by filtration) and then sterilize by filtration through a membrane filter (pore size about 0.4 μm). A second filtration through a finer filter (0.2–0.1 μm) is recommended. Depending on the purity of the enzymes available, the concentrations of macerozyme and cellulase can be reduced to half or even less. Sometimes the potassium dextran sulfate has also been used at a lower concentration or even totally eliminated.

3. Collect tobacco leaves in the proper stage of growth (see above) and wash them gently in a soap solution. If the plants have been raised in a clean greenhouse or environmental chamber, the leaves can be washed with sterile tap water only. Sterilize the surface of the leaves by dipping them for 5 seconds in 50–60% ethanol and then in 5–10% Domestos bleach (20–30 minutes) or in a sodium hypochlorite solution for half a minute. Commercially available solutions can be diluted 15- to 20-fold to give a final concentration of about 100 gm active Cl per liter. Transfer the leaves quickly to about 1 or 2 liters of sterile distilled water. From this stage all operations should be carried out under sterile conditions (in a sterile room or under a laminar airflow hood). Change the water three times at intervals of about 2 minutes each to remove hypochlorite. No significant injury

should be visible on the leaf surface. The concentration of hypochlorite and the duration of treatment should be standardized for the type of leaf used in a preliminary test. The ethanol treatment may be harmful to soft delicate leaves. Ethanol can be replaced by a dilute (1 drop per 10 ml) solution of a detergent like Tween 80.

4. Take out the leaves with sterile forceps and place them on a double sheet of blotting paper (previously sterilized in a drying oven for 2 hours at 120 °C). Remove excess water by pressing gently with another sheet of blotting paper. The leaves become slightly flaccid due to loss of water by transpiration. This takes about half an hour or longer.

5. Using forceps and sterile gloves, remove the lower epidermis of the leaves as far as possible. Pierce the tip of the forceps through the epidermis near the junction of a side vein with the midrib and peel off the epidermis. This needs some practice. However, it is not possible to peel off the whole epidermis from many kinds of leaves. In these cases the leaf can be cut into narrow strips (2 mm or less) to expose the tissue to the enzymes. As another alternative, brushing the epidermis with added carborundum has been reported to give good results (Beier and Bruening, 1975) but an appreciable amount of carborundum is usually retained in the protoplast preparations even after several washings. This difficulty has been averted by avoiding carborundum and using only a 1.5 cm wide nylon brush with bristles 1.5 cm long (Shepard, 1975). The leaf should be stroked with the brush until the surface appears shiny green.

6. After removal of the epidermis cut the leaf in small pieces of about 1 to 2 cm square (preferably avoiding the midrib) and place about 2 gm (fresh weight) of leaf pieces in 20 ml of the maceration medium in a sterile conical flask (capacity 100 ml) plugged with cotton and covered preferably with a double layer of aluminum foil. The leaf pieces can be infiltrated with the macerozyme solution by applying a low vacuum for 2 minutes and releasing the vacuum slowly. Alternatively, the pieces can be collected first in 0.7 M mannitol, shaken for a few minutes, and, after decanting most of the mannitol solution, suspended in the maceration medium.

7. Shake on a reciprocal shaker at 25 °C (about 80–120 excursions per minute, stroke about 4.5 cm). Remove the solution with a protoplast pipette or by decanting after about 15 minutes. This suspension usually contains a high proportion of broken cells.

8. Add 20 ml of fresh maceration medium and continue shaking as above. At intervals of about 30 minutes check the presence of liberated cells under the microscope by collecting a drop with a sterile pipette. Filter through a nylon gauge (or a fine wire netting) and preserve the filtrate. Treat the rest of the softened leaf pieces again with macerozyme and collect a second or third batch of cells in the same way. Maceration is generally complete

in less than 2 hours. Prolonged incubation with macerozyme should be avoided.

9. Allow the isolated cells to settle down slowly in centrifuge tubes or collect them by low speed centrifugation (about 50–100 g, generally below 800 rpm in a bench centrifuge with swing-out tube holders) for 2 minutes. Remove most of the supernatant with the help of a protoplast pipette and suspend the cells in 0.7 M mannitol containing 0.1 mM CaCl$_2$. This washing medium (WM) should have a pH around 5.5. Centrifuge again as above, remove the supernatant, and suspend the cells in the cellulase solution. Place the suspension in a 100 ml conical flask as before and shake gently at about 30°C (about 50 excursions per minute, stroke about 4.5 cm).

10. Check for digestion of the cell wall and conversion of the cells to perfectly round protoplasts by removing a small sample from time to time. Depending on the type of cells, the digestion may have to be continued for 2 hours or longer (Figs. 1 and 2).

11. Filter through a nylon gauge of 100 μm pore mesh. Centrifuge filtrate at 50 g for 2 minutes. Remove the supernatant with a protoplast pipette. Wash the loosely packed protoplasts twice with WM, each time collecting the protoplasts by low-speed centrifugation as before. Suspend them in WM or in the desired inoculation medium.

Another two-step procedure, where inorganic salts replace mannitol, has been described by Meyer (1974). After removing the epidermis, the pieces of leaf are preplasmolyzed in a solution containing 13% mannitol (about 0.71 M). Maceration is done for 1 hour without shaking in 0.02% (w/v) pectinol fest (Roehm and Haas, Darmstadt, West Germany) in a saline plasmolyticum (SP), containing 2.5% w/v KC1 and 1% w/v MgSO4· 7H$_2$O. The pH is adjusted to 5.2. After washing in SP the softened leaf pieces are incubated in a mixture of 0.02% w/v Pectinol fest and 0.4% cellulase–Onozuka R-10 in SP at room temperature. The flasks are shaken gently every 20 or 30 minutes and the protoplasts are collected after 90 minutes by filtration and washing in SP.

Instead of peeling the epidermis, Schilde-Rentschler (1972) digested the cutin of the epidermal cells with β-glucuronidase from *Escherichia coli* (type II of Sigma Chemical Company, St. Louis, Missouri) and pectin-glycosidase from *Aspergillus* (T. Schuchardt, Munich, West Germany). The *E. coli* enzyme was dissolved at a concentration of 5 mg/ml in 0.067 M phosphate buffer of pH 7.0 with or without 0.7 M mannitol. Pectin-glycosidase was dissolved (10 mg/ml) either in 0.067 M phosphate buffer of pH 5.0 containing 0.7 M mannitol or in phosphate-free 0.7 M mannitol of pH 5.0 (adjusted with 0.1 N HCl after solution of the enzyme). Any turbidity of the enzyme solution was removed by centrifugation at 12,000 g for 10

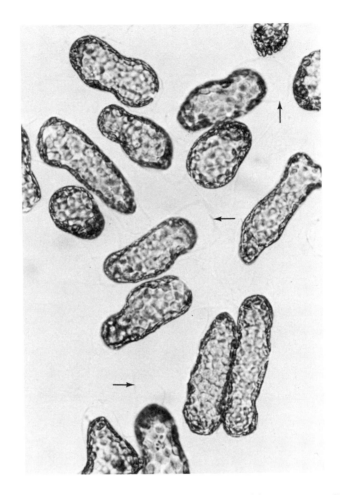

Fig. 1. Palisade cells of tobacco, isolated by treatment with macerozyme. Note that at several places the cell wall is visible, as the contents are partially plasmolyzed in 0.6 M mannitol. ×350. (Photograph by courtesy of I. Takebe.)

minutes and sterilized by filtration through a membrane filter of pore size 0.2 μm (No. 11307 of Sartorius, Göttingen, West Germany). An amount of 40 ml of each enzyme solution for 5 gm of leaf pieces (size about 4 cm²) was used. The suspension was shaken at 30 °C on a reciprocal shaker (120 excursions per minute, stroke 4.5 cm) for 1 hour. The liquid was decanted, and the softened leaf pieces were washed twice with 0.7 M mannitol. Cells and protoplasts were then isolated using macerozyme and cellulase according to the procedure of Takebe and his associates as previously described.

FIG. 2. Protoplasts of palisade cells of tobacco in 0.6 M mannitol after treatment of isolated cells with cellulase. $\times 350$. (Photograph by courtesy of I. Takebe.)

C. MIXED-ENZYME METHOD OR ONE-STEP PROCEDURE

Several methods have been reported for a direct isolation of protoplasts with a mixture of enzymes. The following procedure is based on the data of Kassanis and White (1974).

1. Peel the lower epidermis of tobacco leaves with fine jeweler's forceps. Cut pieces of 3–6 cm square. Float them, peeled side down, on 30 ml of a mixture of macerozyme (0.3–0.4%), cellulase-Onozuka R-10 (0.6–1.2%) and D-mannitol (13.2% w/v) poured in large petri dishes (diameter about

15 cm). Incubate in the dark at 25°C for about 17 hours, conveniently overnight. After incubation swirl the petri dishes gently, filter the proto-plasts through muslin, centrifuge them down at 35 g for 3 minutes and wash them in 13.2% (w/v) mannitol. Before inoculation wash the protoplasts at 100 g for 2 minutes.

Due to its simplicity and the high yield of protoplasts the above mixed-enzyme method is very useful. However, the protoplasts often fuse spon-taneously to give rise to many multinucleate products of variable sizes. This difficulty can be overcome to a large extent by initially using a low

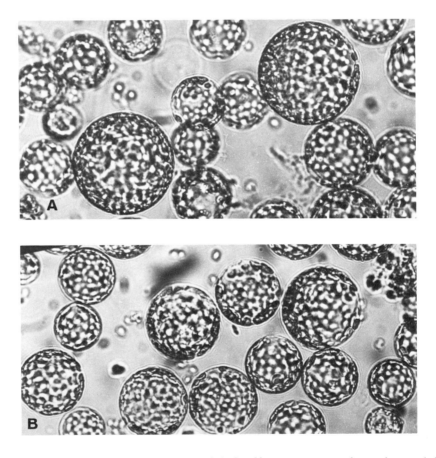

FIG. 3. (A) Protoplasts of *Brassica perviridis* isolated by a one-step procedure and suspended in 0.35 M mannitol. Through a spontaneous fusion of protoplasts large units are produced. (B) Mesophyll protoplasts of the same plant isolated by a modification of the one-step pro-cedure. The enzyme mixture initially contained only about 1/10 the concentration of cellulase, as in (A), and then the protoplasts were incubated for an hour in cellulase alone. ×400.

concentration of cellulase in the mixture (S. Sarkar, unpublished). Pieces of peeled tobacco leaves are treated as above with a mixture of 0.3% macerozyme and 0.06% cellulase overnight at room temperature (about 24 °C). The protoplasts obtained by this method may retain a part of their cellulose wall. The washed protoplasts are then incubated further with a fresh solution of cellulase only (about 0.5–1.0% w/v, depending on the specific activity of the enzyme sample) at 30 °C for an hour or longer with gentle shaking (Fig. 3). In some cases it is advisable to soften only the pieces of leaf with pectinase alone and then isolate the protoplasts with cellulase (Watts *et al.*, 1974).

2. Good results have also been obtained by using a mixture of Driselase (0.25%) and cellulase (about 0.3%) in 0.7 M mannitol at pH 5.8–5.9 (W. Keller, unpublished). Narrow strips of tobacco leaves are incubated in this mixture at 25 °–30 °C for 2 hours with gentle shaking on a reciprocal shaker. Preplasmolysis for 10 minutes in 0.7 M mannitol before the addition of the enzyme mixture is recommended.

D. SOME COMMENTS ON THE PREPARATION AND HANDLING OF PROTOPLASTS

Isolated protoplasts are subject to mechanical injury. Therefore, pipetting and resuspension of pellets must be done slowly and carefully. The centrifugal force for sedimenting the protoplasts should not exceed 100 g.

Mannitol has proved to be the most suitable plasmolyticum for plant protoplasts, although other substances such as sucrose, sorbitol, KCl, etc., might be good under specific conditions. The molarity of mannitol has to be adjusted to the type of cells chosen. Tobacco protoplasts are handled best in 0.6–0.7 M mannitol, those of cowpea (*Vigna unguiculata*) and Chinese cabbage (*Brassica sinensis*) in 0.6 M (Hibi *et al.*, 1975; Renaudin *et al.*, 1975). Those of potato, tomato, and *Physalis* species keep well in 0.5 M mannitol, whereas for *Brassica perviridis* the molarity may have to be reduced to 0.4 or even 0.35 (S. Sarkar and R. Kaus, unpublished).

Mannitol-containing washing medium may be enriched with $10^{-4} M$ CaCl$_2$ (Takebe *et al.*, 1968) and buffered with a low concentration of citrate or phosphate. Ca^{2+} can be partially replaced by Mg^{2+} (Kassanis and White, 1974). However, an unbuffered solution of mannitol is good for several kinds of protoplasts. Probably the proteins on the surface of protoplasts and some ions released from protoplasts contribute a sufficient buffering and protecting property to the suspending solution. In this respect the ratio between the amount of plant tissue and the volume of the macerating medium may be of some importance. However, the activity of the enzymes may also be significantly reduced by a rapid change in the H^{+}-ion concentration of an unbuffered medium. For a stabilization of the pH of the medium during enzymic digestion of tobacco leaves, Shepard (1975) found it better to use MES buffer (2-*N*-morpholinoethane sulfonic acid) at a concentration of 0.01 *M* and a pH of 5.6 than citrate or phosphate.

The tobacco variety *Xanthi*, commonly used in Japan, is not the same as *Xanthi* nc. It is a systemic host for tobacco mosaic virus (TMV) and is comparable in this respect to the variety samsun (Melchers *et al.*, 1940) of *Nicotiana tabacum* L., whereas the *Xanthi* nc. (*N. tabacum* L. cultivar *Xanthi necroticum*: Takahashi, 1956) produces local lesions with all strains and mutants of TMV when cultivated at temperatures below 30 °C. Interestingly, however, isolated protoplasts of *Xanthi* nc, when inoculated with TMV, do not undergo necrosis and produce as much virus as those of the samsun variety. Obviously, the necrotic or hypersensitive reaction of *Xanthi* nc, caused by the *N* gene, is expressed only when the infected cells are surrounded by uninfected ones in a tissue (Otsuki *et al.*, 1972b).

It is not always possible to judge the quality of protoplasts simply by observation under the microscope. Apparently very good preparations may be poor with respect to their ability to take up and multiply a virus. On the other hand, it has been found that protoplasts that did not look very "nice" at first, did so after several hours in a nutrient medium and multiplied a virus efficiently.

IV. Inoculation of Protoplasts

A. Virus Nucleoprotein as the Infecting Agent

The medium in which protoplasts are suspended is of great importance for a successful inoculation with virus and also with viral nucleic acids. Media and inoculation schedules have been described for several host–virus combinations but only a few standard methods will be presented here. With other viruses some minor modifications might be needed for optimal efficiency.

For the tobacco protoplast–TMV system the following procedure has been standardized and described in detail by Takebe and Otsuki (1969):

a. Suspend a freshly prepared pellet of tobacco protoplasts in a medium containing 0.8 *M* mannitol, 0.2 m*M* KH$_2$PO$_4$, 1 m*M* KNO$_3$, 0.1 m*M* MgSO$_4$, 0.1 m*M* CaCl$_2$, 1 μ*M* KI, 0.01 μ*M* CuSO$_4$, 1 μg/ml 6-benzyladenine, 300 μg/ml cephaloridine, and 10 μg/ml rimocidin to give a concentration of 1 to 4 $\times$ 10^6 protoplasts/ml. The medium should be sterilized by autoclaving without the last two components, cooled, and then, after adding the two antibiotics, adjusted with KOH to pH 5.4.

b. Dissolve purified TMV in 0.02 M potassium citrate buffer of pH 5.0 containing 0.8 M mannitol and poly-L-ornithine. The concentration of TMV and the polycation should be 2 μg/ml each. Preincubate the virus with the polycation at 25 °C for 10 minutes and mix with an equal volume of suspension of protoplasts. Mix and keep for 20 minutes at 25 °C with occasional swirling. The time of inoculation may be prolonged to 1 hour, but most of the protoplasts get infected during the first 10 minutes. Collect the protoplasts by low-speed centrifugation, wash them two or three times with WM (0.8 M mannitol with $10^{-4} M$ CaCl$_2$), suspend them in the incubation medium (see Section V,A) to a density of 1 to 4 $\times$ 10⁵ protoplasts/ml. Incubate each 10 ml of the suspension in conical flasks (capacity preferably 100 ml) for the desired period. In later experiments it has been found that the mannitol concentration can be reduced to 0.7 or even 0.6 M throughout.

The above procedure can be called an "indirect" method (Takebe et al., 1968) in contrast to a "direct" method, where the inoculation medium containing the infecting agent is used directly to suspend the protoplasts in a pellet (Motoyashi et al., 1974a; Kubo et al., 1974; Sarkar et al., 1974).

The following procedure has been recommended by Kubo et al. (1974) for the infection of tobacco protoplasts with the tobacco rattle virus (TRV): Preincubate a mixture of virus and poly-L-ornithine at concentrations of 4 and 1 μg/ml, respectively, in 0.025 M phosphate buffer, pH 6.0, containing 0.7 M mannitol for 10 minutes at 25 °C. Suspend the pelleted protoplasts directly in this inoculum, keep for 10 minutes at 25 °C with occasional gentle shaking and wash twice with 0.7 M mannitol containing $10^{-4} M$ CaCl$_2$. Resuspend in nutrient media and incubate at 22 °C under continuous illumination (3700 lux).

The following procedure for inoculation of tobacco protoplasts with potato virus X (PVX) was adopted by Shalla and Petersen (1973): Suspend tobacco protoplasts (about 0.25 ml packed volume) in 3 ml of a solution containing 0.9 M mannitol, 13 mM CaCl$_2$, 7 mM ZnSO$_4$, 1.3 mM MgSO$_4$, and 10 μg/ml poly-L-ornithine. Mix with 0.1 mg PVX suspended in 1 ml distilled water. Keep for 30 minutes. Wash the protoplasts twice with 0.8 M mannitol and suspend in incubation medium. Otsuki et al. (1974) inoculated tobacco protoplasts at 25 °C by the indirect method using a mixture of PVX and poly-L-ornithine at final concentrations of 5 μg/ml and 3 μg/ml, respectively. The medium containing 0.7 M mannitol was buffered to pH 5.8 with 0.02 M potassium phosphate.

For cowpea chlorotic mottle virus (CCMV) mix a suspension of tobacco protoplasts at a density of 3 to 10 $\times$ 10⁵/ml in 0.7 M mannitol with an equal volume of a preincubated (10 minutes at 25 °C) mixture of CCMV and poly-L-ornithine, so that the final concentrations of CCMV and polycation during inoculation will be 0.5 μg/ml and 1 μg/ml, respectively (Motoyashi *et al.*, 1973). Preincubation is done in CBM, i.e., in 0.02 M potassium citrate buffer of pH 5.2 containing 0.7 M mannitol.

For cowpea mosaic virus (Hibi *et al.*, 1975) use protoplasts isolated from fully expanded primary leaves of 9- to 11-day-old seedlings of cowpea (*Vigna unguiculata* L.). Proceed as described for CCMV above but maintain the mannitol concentration at 0.6 M. At the time of inoculation the concentrations of protoplasts, virus, and poly-L-ornithine should be about 2.5 × 10^5/ml, 5 μg/ml, and 0.5 μg/ml, respectively.

B. Viral Nucleic Acid as the Infecting Agent

There are two significant advantages of using viral nucleic acid for inoculation over viral nucleoprotein: the residual inoculum, not taken up by the protoplasts, can be removed more efficiently by washing, and the production of viral antigen in the cells starts from zero.

In their first successful inoculation of a high percentage of tobacco protoplasts with TMV–RNA, Aoki and Takebe (1969) reduced a chance inactivation of the nucleic acid by ribonucleases before penetration of the protoplasts by performing the inoculation at 0°C. The procedure was as follows: After isolation, washing, and pelleting, tobacco protoplasts were suspended at a density of 1 to 3 × 10^6/ml in 0.9 M mannitol containing 13.3 mM $CaCl_2$, 6.6 mM $ZnSO_4$, 1.3 mM $MgSO_4$, and either 26.6 μg/ml protamine sulfate or 6.6 μg/ml poly-L-ornithine. Three volumes of protoplast suspension were mixed with 1 volume of an aqueous solution of TMV–RNA at 2–6 mg/ml. The mixture was kept at 0°C for 20 minutes. Then it was diluted with 2 or 3 volumes of wash medium (0.8 or 0.7 M mannitol, 10 mM $CaCl_2$, and 1 mM $MgSO_4$). It was washed 2 or 3 times with the wash medium, and each time the protoplasts were collected by centrifugation at about 100 g for 1 minute. About 7% of the protoplasts were infected by this method, as determined by the fluorescent antibody technique (see below).

For inoculation of tobacco protoplasts with RNA of cowpea chlorotic mottle virus, the procedure described by Motoyashi *et al.* (1973) should be followed: Dilute CCMV–RNA to a concentration of 12 μg/ml in CBM (0.02 M potassium citrate buffer of pH 5.2 containing 0.7 M mannitol) and prepare a solution of poly-L-ornithine (10 μg/ml) in CBM. In a 100 ml centrifuge tube mix 6 ml CBM with 2 ml poly-L-ornithine solution and add 2 ml of the RNA solution. Shake for 2 minutes at 25°C. Suspend a pellet of protoplasts in 10 ml of 0.7 M mannitol and add the suspension to the preincubated mixture of RNA and poly-L-ornithine. Shake with 80 excursions per minute at 25°C for 10 minutes. At the time of inoculation the density of protoplasts should be 1.5 to 5.0 × 10^5/ml and the concentration of CCMV–RNA and poly-L-ornithine preferably 1.2 μg/ml and 1.0 μg/ml, respectively.

As in the previous procedure with TMV–RNA, only about 7% of the protoplasts can be infected by this method, but a much lower concentration of RNA is required. Moreover, the inoculation is done at room temperature.

Sarkar, et al. (1974) achieved a very high efficiency of infection with TMV–RNA by use of an alkaline buffer. It had been observed earlier (Sarkar, 1963) that the infectivity of TMV–RNA for tobacco leaves can be increased very efficiently by use of a buffer solution of pH 8.7–8.8. Moreover, Keller and Melchers (1973) showed that tobacco protoplasts can be induced to fuse with one another if suspended in a buffer of high pH containing Ca^{2+} ions. Since protoplasts can withstand a pH of even 11 for some time, an alkaline inoculation medium was developed on the basis of the above results. If the protoplasts have been isolated in 0.7 M mannitol (by a two-step or a one-step procedure) they can be infected with TMV–RNA in the following way: Prepare an inoculation medium (IM_1) containing 0.1 M glycine, 0.8 M mannitol, 0.1 M sucrose, and 0.1 M NaCl. Recent experiments indicate that the addition of NaCl is not essential. Adjust pH to 9.0 or 9.1 with 1 N KOH. Divide into portions of about 40 ml each and sterilize by autoclaving once. Before use, mix 5 ml of IM_1 with 25 μl of a 1 M $MgCl_2$ solution and 20 μl of TMV–RNA solution (concentration about 1 mg/ml in 0.1 M glycine-KOH buffer, pH 8.8). Filter the RNA solution through a membrane filter (e.g., Millipore) using a Swiny adapter (see Section II, B) fitted to a syringe. Use the sterile filtrate to resuspend freshly pelleted protoplasts in a 30 ml centrifuge tube. Inoculation by this direct method is very efficient at room temperature, although the temperature can be raised to even 30 °C. Swirl or shake gently to give a uniform suspension. After 15 minutes, during which the contents of the tube should be gently swirled occasionally, add about 20 ml of WM (0.7 M mannitol with $10^{-4} M$ $CaCl_2$, pH about 5.5) and collect the protoplasts by centrifugation at 50–100 g for 1–2 minutes. Wash the protoplasts a second and third time with WM and suspend them in the nutrient inoculation medium.

Another inoculation medium of high pH, in which the mannitol is replaced by KCl and $MgCl_2$, has also been found to be very effective (Sarkar et al., 1974). This medium (IM_2) contains 2.5% w/v KCl, 1.0% w/v $MgCl_2$· $6H_2O$, and 0.1 M glycine. The pH is adjusted to 9.0 with 1 N KOH. The viral RNA is dissolved in this buffer. The rest of the inoculation procedure is similar to that described above and is carried out at a temperature between 20° and 30 °C. This method is recommended for protoplasts prepared in a saline plasmolyticum (Meyer's method, see Section III, B, 2). Unless the protoplasts are of very good quality, they do not withstand a rapid transfer from mannitol- to salt-containing media, and vice versa. When the pH of the inoculation medium is near 9.0, it is unnecessary and even harmful to add poly-L-ornithine.

C. ROLE OF POLYCATIONS AND OTHER COMPONENTS IN THE INOCULATION MEDIUM

Instead of being taken up, virus particles are expected to be repelled by plant protoplasts, since both of them usually carry an overall negative charge. The charge on the virus particles was probably neutralized or reversed by a preincubation with poly-L-ornithine (Takebe and Otsuki, 1969), resulting in an efficient inoculation of protoplasts. Probably the same mechanism is responsible for the increased uptake of proteins by mammalian cells in the presence of polycations (Ryser, 1968). In general the infectivity of all viruses that are negatively charged is enhanced in the presence of polycations, whereas no polycation is needed for infection with positively charged (in commonly used acidic inoculation media) viruses such as pea enation mosaic virus (PEMV) (Motoyoshi and Hull, 1974) and brome mosaic virus (Motoyoshi et al., 1974b). However, the situation is more complex, since even with these viruses a stimulation of infection has been achieved with poly-L-ornithine. Besides poly-L-ornithine other polycations, such as poly-L-histidine, poly-L-lysine, and poly-L-arginine show similar effects, though with less efficiency. As mentioned earlier, no polycation is required for inoculation with viral RNA at high pH. The exact role of polycations is not yet understood (see Takebe, 1975).

The ionic composition of the inoculation medium has a profound effect on the stability and properties of the protoplast membrane and more work is needed to determine the optimal conditions for inoculation. In contrast to citrate-containing buffers that had been used with success from the very beginning (Takebe and Otsuki, 1969), a phosphate-containing one has been found to be more efficient for inoculation with TRV (Kubo et al., 1974; Harrison et al., 1975). Sarkar et al. (1974) and other workers have used inoculation media that have a slightly higher osmolarity than the medium used to wash the protoplasts just before inoculation. How much this practice is really of advantage needs further investigation.

V. Culture of Protoplasts

A. CULTURE MEDIA

The composition of an incubation medium for tobacco protoplasts (Takebe et al., 1968) is given below: KH_2PO_4, 0.2 mM; KNO_3, 1 mM; $MgSO_4$, 0.1 mM; $CaCl_2$, 0.1 mM; KI, 1 μM; $CuSO_4$, 0.01 μM; 6-benzyladenine, 1 μg/ml; cephaloridine, 300 μg/ml; rimocidin, 10 μg/ml; and mannitol, 0.7 or 0.6 M. The pH is adjusted to 5.4 after autoclaving and addition of the last two components. Cephaloridine and rimocidin were added to prevent the growth of contaminating bacteria and fungi, respectively. Later on Aoki and Takebe (1969) modified the medium by using 10 mM CaCl$_2$, 1 mM MgSO$_4$, omitting benzyladenine, and adding 1 μg/ml 2,4-dichlorophenoxyacetic acid.

On the basis of the data of Murashige and Skoog (1962) Nagata and Takebe (1971) standardized another medium suitable for tobacco protoplasts. One liter of final medium contains the following amounts of salts and organic constituents: NH$_4$NO$_3$, 0.825 gm; KNO$_3$, 0.95 gm; CaCl$_2$·2H$_2$O, 0.22 gm; MgSO$_4$·7H$_2$O, 1.233 gm; KH$_2$PO$_4$, 0.68 gm; Na$_2$EDTA, 37.3 mg; FeSO$_4$·7H$_2$O, 27.8 mg; H$_3$BO$_3$, 6.2 mg; MnSO$_4$·4H$_2$O, 22.3 mg; ZnSO$_4$·4H$_2$O, 8.6 mg; KI 0.83 mg; Na$_2$MoO$_4$·2H$_2$O, 0.25 mg; CuSO$_4$·5H$_2$O, 25 μg; CoSO$_4$·7H$_2$O, 30 μg; sucrose, 10 gm; *meso*-inositol, 100 mg; thiamine-HCl, 1 mg; 1-naphthaleneacetic acid (NAA), 3 mg; 6-benzyl-aminopurine (6-BAP), 1 mg; and D-mannitol, 0.7 M. Adjust pH to 5.8 with KOH or HCl. Autoclave at 120°C for 15 minutes. This medium, in contrast to that of Aoki and Takebe (1969), contains sucrose and is somewhat more susceptible to contamination by microorganisms.

For protoplasts from a plant other than tobacco the concentration of mannitol and the pH have to be adjusted. It may be necessary to try some other culture medium. Composition of several media can be obtained from the booklets previously mentioned (Section II) and from original publications.

To prevent the growth of microorganisms in the cultures one can add antibiotics such as Aureomycin®, cephaloridine, Rimocidin®, Gentamycin, carbanecillin, or Mycostatin®, at concentrations around 10–100 μg/ml. However, in order to be sure about their noninterference with the problem under investigation, preliminary control experiments with and without the antibiotics should be performed. For a continued culture of protoplasts and induction of cell division, the media should be supplemented with auxin and kinetin (3 μg/ml of NAA and 1 μg/ml of benzyladenine).

B. Conditions for Incubation

Protoplasts in suspension are generally incubated in conical flasks or petri dishes in diffuse light (about 3000 lux) at a temperature between 25° and 28°C. As with cultures of bacteria, the protoplasts can be subjected to other temperature and light conditions and the air can be enriched with CO$_2$, O$_2$, or other gases, if desired. However, light intensities over 5000 lux are harmful to many kinds of protoplasts. It is not necessary to shake the flasks or to keep them in slow movement for virus multiplication but occasional

swirling and redistribution of the protoplasts (say, every 12 hours) is recommended. As a standard procedure 10 ml of a suspension of protoplasts
containing 1 to 5 × 10^5 protoplasts/ml should be poured in each 100 ml
conical flask.

Division of cells leading to the formation of calluses is clear proof of
an initially "living" state of the protoplasts. However, in the absence of
auxins and kinetin mesophyll protoplasts seldom divide, and many experiments on the viral infection of protoplasts are terminated within 2–3 days.
If a good virus multiplication takes place within the period of incubation,
it is obvious that the protoplasts were in an actively metabolizing state,
at least for virus multiplication.

VI. Detection and Estimation of Virus in Protoplasts

A. ASSAY OF INFECTIVITY

Whether a virus has increased in the protoplasts after a desired period
of incubation can be determined by breaking up the protoplasts and testing
for infectivity on a sensitive host. As a routine procedure, collect the protoplasts by low-speed centrifugation and test the pellet and the supernatant
for infectivity. Any virus present in the supernatant by a chance breakage
of protoplasts or by a probable active liberation of virions by the protoplasts
can be assayed on a host plant. However, the supernatant should be diluted
at least 5-fold in order to reduce the concentration of mannitol or other
plasmolyticum that interferes with the bioassay. Suspend the protoplast
pellet in a small volume of suitable buffer solution and break the protoplasts
(which may have a more or less rigid wall) by homogenization either in a
suitable homogenizer or in a mortar with pestle. In the latter case, if the
sample contains a stable virus like TMV, freeze and thaw once to facilitate
the liberation of the virus. Prepare a series of dilutions and assay by standard
methods (Kleczkowski, 1968).

In all cases where the virus under investigation has a local lesion host,
the number of lesions produced by the extract from protoplasts is tested in
a series of dilutions. For TMV, a proportionality between the concentration
of virus and the number of lesions per *Xanthi* nc leaf is maintained, so long
as the lesion number does not exceed 300 per half-leaf (preferably below 150
per half-leaf). From a comparison with the lesions produced by a standard
dilution of a purified virus preparation the concentration of virus per
milliliter of extract can be estimated. If the protoplast pellet and the supernatant are tested separately, the amount of virus per milliliter of the original
protoplast suspension in the culture medium can be calculated by adding
the values together, paying due attention to the dilution factors (Takebe

et al., 1968; Otsuki *et al.*, 1972b; Sarkar *et al.*, 1974; and others). In the absence of a local lesion host, the dilution point giving a 50% infection of a systemic host can be determined using a sufficient number of replications.

The average amount of virus produced per protoplast can be calculated by dividing the quantity of virus by the number of protoplasts in a sample. The number of protoplasts per unit volume can be counted directly under a microscope using a hemocytometer.

Besides bioassay, the quantity of virus can also be determined by some physicochemical or serological means. The serological method has the advantage that it can be applied to crude extracts or partially purified preparations of virus. If the particle weight of the virus is known, the number of virions per protoplast can also be estimated. Under optimal conditions as many as 10^6 or 10^7 virus particles can be produced in a single protoplast. However, the conclusion drawn would need a correction, depending on the proportion of infected and uninfected protoplasts in a sample. The infected protoplasts can be identified by a fluorescent antibody technique, as described below. Only in cases of protoplasts of Chinese cabbage, infected with turnip yellow mosaic virus (TYMV) chloroplasts within the cells aggregate to form so-called "polyplasts" (Renaudin *et al.*, 1975) and the individual chloroplasts undergo a characteristic change in shape under strong illumination (Matthews and Sarkar, 1976). The proportion of TYMV-infected protoplasts can thus be determined by a direct observation under the light microscope.

B. Fluorescent Antibody Technique

Prepare antiserum to the virus in a rabbit following one of the methods described earlier (Matthews, 1967). Purify the globulins partially from an excess of albumins by a fractional precipitation with ammonium sulfate. Dissolve sufficient $(NH_4)_2SO_4$ of analytical grade in PBS7.4 (phosphate-buffered saline, pH 7.4, prepared by dissolving 8 gm NaCl, 0.2 gm KCl, 1.5 gm Na_2HPO_4, and 0.2 gm KH_2PO_4 in 1 liter of distilled water, according to the procedure of Kawamura, 1969) to give a saturated solution. Cool and allow to equilibrate with an excess of undissolved $(NH_4)_2SO_4$ at $0°-4°C$. Add an equal volume of this saturated solution dropwise to antiserum which had previously been diluted 1:1 with PBS7.4 in the cold. After 30–60 minutes at $0°-4°C$ collect the sediment by centrifugation at about 8000 g, dissolve in PBS7.4, and precipitate again with about 3/4 volume of saturated $(NH_4)_2SO_4$ solution. Collect the sediment by centrifugation after standing in the cold, redissolve in PBS7.4, and treat with 1/2 volume of $(NH_4)_2SO_4$ solution. Dissolve the final sediment in PBS7.4 and dialyze exhaustively against the same buffer at $0°-4°C$ until the buffer outside the dialysis bag becomes free of sulfate (tested with a solution of $BaCl_2$). Dissolve the partially purified immunoglobulin in 0.02 M Na$_2$CO$_3$–NaHCO$_3$ buffer, pH 9.8, to give a protein concentration of nearly 1% and dialyze against 10 volumes of 0.01% fluorescein-isothiocyanate in the above buffer in the cold for 24 hours (Clark and Shepard, 1963). Prepare a column (1.8 × 23 cm) of Sephadex G-25, pour the dialysate slowly, and elute with the above-mentioned buffer. The conjugated product moves ahead of the free dye. Collect the peak fractions together and preserve frozen at 0°–4°C. To avoid nonspecific staining it is recommended to adsorb the preparation with acetone-extracted powder of healthy tobacco leaves. The rest of the procedure is taken from the data of Otsuki and Takebe (1969b, 1973).

Wash protoplasts in 0.7 M mannitol. Place a drop of thick protoplast suspension in 0.7 M mannitol on a glass slide which has been smeared with Meyer's albumen (25 ml eggwhite + 25 ml glycerine + 0.5 gm sodium salicylate). Spread the drop by blowing on it and then dry the protoplasts in a stream of warm air. Dehydrate the protoplasts by dipping the slide in acetone or 95% ethanol for 30 minutes at room temperature (Otsuki *et al.*, 1972a,b). Wash for 2 hours in PBS7.0 (0.01 M phosphate buffer, pH 7.0, containing 0.85% NaCl), keeping the buffer in slow movement with, say, a magnetic stirrer. Change the buffer 3 or 4 times. Cover the area of the slide

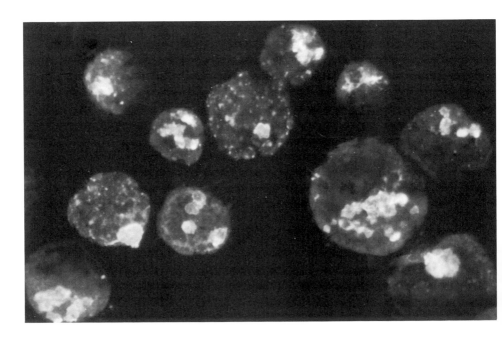

FIG. 4. Fluorescence of tobacco palisade protoplasts 24 hours after inoculation with tobacco mosaic virus. The viral antigen combines with FITC-conjugated antibody. ×700. (Photograph by courtesy of I. Takebe.)

containing the attached protoplasts with sufficient FITC-conjugated purified antiserum and incubate for 2 hours at 36 °C in a humid chamber. Wash with PBS7.0, mount in PBS7.0 containing 10% glycerine, and observe in UV light. Otsuki and Takebe used a Zeiss fluorescent microscope with Osram HBO 200 mercury vapor lamp and suitable filters (BG 3 exciter filter and barrier filters No. 44, 47, and 65). The typical appearance of some TMV-infected tobacco protoplasts is shown in Fig. 4.

C. Other Methods

A direct estimation of virus by spectrophotometry is possible only after a high degree of purification of the virions. Dependable results can be obtained in some cases by measuring the optical density of the virus band after a centrifugation of the crude extract in a density gradient (Motoyashi et al., 1974a). The amount of virus produced can also be determined by labeling with radioactive precursors or by quantitative electron microscopy. However, all methods should be judged against proper controls. The residual infectivity due to virus particles adsorbed to the surface of protoplasts can be quite significant, even after washing, if the virus is stable and if a high concentration of virus had been used for inoculation. However, this error is minimized to a great extent whenever free nucleic acid can be used as inoculum, as mentioned earlier.

Isolated protoplasts are now being used to study the mode of entry of virions, the time course of multiplication of viruses, the specific organelles involved in virus multiplication, the precursors and accessory products of virus biosynthesis, the existence and induction of resistance to virus infection, and the ultrastructural changes within the protoplasts (see, for example, Takebe, 1975; Coutts et al., 1972). The procedures are essentially the same as those applied to various other kinds of cells and tissues, such as autoradiography, electrophoresis, electron microscopy, and the standard genetic and microbiological methods.

Acknowledgments

My sincere thanks are due to Dr. G. Melchers, Director, Max-Planck-Institut für Biologie, Tübingen, Germany for his support and encouragement. I am also very grateful to Dr. I. Takebe, Director, Institute for Plant Virus Research, Aoba-cho, Chiba, Japan for some of the data and photographs.

References

Aoki, S., and Takebe, I. (1969). *Virology* **39**, 439.
Beier, H., and Bruening, G. (1975). *Virology* **64**, 272.
Clark, H. F., and Shepard, C. C. (1963). *Virology* **20**, 642.

Cocking, E. C. (1960). *Nature (London)* **187**, 962.

Cocking, E. C. (1972). *Annu. Rev. Plant Physiol.* **23**, 29.

Coutts, R. H. A., Cocking, E. C., and Kassanis, B. (1972). *J. Gen. Virol.* **17**, 289.

Harrison, B. D., Kubo, S., and Robinson, D. J. (1975). *Proc. Int. Congr. Virol., 3rd, 1975* Abstract, p. 205.

Hibi, T., Rezelman, G., and van Kammen, A. (1975). *Virology* **64**, 308.

Kassanis, B., and White, R. F. (1974). *J. Gen. Virol.* **24**, 447.

Kawamura, A., Jr. (1969). "Fluorescent Antibody Techniques and Their Applications." Univ. of Tokyo Press, Tokyo.

Keller, W. A., and Melchers, G. (1973). *Z. Naturforsch. Teil C* **28**, 737.

Kleczkowski, A. (1968). *In* "Methods in Virology" (K. Maramorosch and H. Koprowski, eds.), Vol. 4, p. 615. Academic Press, New York.

Kohlenbach, H. W. (1959). *Naturwissenschaften* **46**, 116.

Kohlenbach, H. W. (1966). *Z. Pflanzenphysiol.* **55**, 142.

Kubo, S., Harrison, B. D., and Robinson, D. J. (1974). *Intervirology* **3**, 382.

Matthews, R. E. F. (1967). *In* "Methods in Virology" (K. Maramorosch and H. Koprowski, eds.), Vol. 3, p. 199. Academic Press, New York.

Matthews, R. E. F., and Sarkar, S. (1976). *J. Gen. Virol.* **33**, 435.

Melchers, G. (1975). *Ber. Dtsch. Bot. Ges.* **88**, 473.

Melchers, G., Schramm, G., Trurnit, H., and Friedrich-Freksa, H. (1940). *Biol. Zentralbl.* **60**, 524.

Meyer, Y. (1974). *Protoplasma* **81**, 363.

Motoyashi, F., and Hull, R. (1974), *J. Gen. Virol.* **24**, 89.

Motoyashi, F., Bancroft, J. B., Watts, J. W., and Burgess, J. (1973). *J. Gen. Virol.* **20**, 177.

Motoyashi, F., Bancroft, J. B., and Watts, J. W. (1974a). *J. Gen. Virol.* **25**, 31.

Motoyashi, F., Watts, J. W., and Bancroft, J. B. (1974b). *J. Gen. Virol.* **25**, 245.

Murashige, T., and Skoog, F. (1962). *Physiol. Plant.* **15**, 473.

Nagata, T., and Takebe, I. (1971). *Planta* **99**, 12.

Otsuki, Y., and Takebe, I. (1969a). *Plant Cell Physiol.* **10**, 917.

Otsuki, Y., and Takebe, I. (1969b). *Virology* **38**, 497.

Otsuki, Y., and Takebe, I. (1973). *Virology* **52**, 433.

Otsuki, Y., Takebe, I., Honda, Y., and Matsui, C. (1972a). *Virology* **49**, 188.

Otsuki, Y., Shimomura, T., and Takebe, I. (1972b). *Virology* **50**, 45.

Otsuki, Y., Takebe, I., Honda, Y., Kajita, S., and Matsui, C. (1974). *J. Gen. Virol.* **22**, 375.

Renaudin, J., Bové, J. M., Otsuki, Y., and Takebe, I. (1975). *Mol. Gen. Genet.* **141**, 59.

Ruesink, A. W. (1971). *In* "Methods in Enzymology" (A. San Pietro, ed.), Vol. 23, Part A, p. 197. Academic Press, New York.

Ryser, H. J. P. (1968). *Science* **159**, 390.

Sarkar, S. (1963). *Virology* **20**, 185.

Sarkar, S., Upadhya, M. D., and Melchers, G. (1974). *Mol. Gen. Genet.* **135**, 1.

Schilde-Rentschler, L. (1972). *Z. Naturforsch, Teil B* **27**, 208.

Shalla, T. A., and Petersen, L. J. (1973). *Phytopathology* **63**, 1125.

Shepard, J. F. (1975). *Virology* **66**, 492.

Takahashi, W. N. (1956). *Phytopathology* **46**, 654.

Takebe, I. (1975). *Annu. Rev. Phytopathol.* **13**, 105.

Takebe, I., and Otsuki, Y. (1969). *Proc. Natl. Acad. Sci. U.S.A.* **64**, 843.

Takebe, I., Otsuki, Y., and Aoki, S. (1968). *Plant Cell Physiol.* **9**, 115.

Watts, J. W., Motoyashi, F., and King, J. (1974). *Ann. Bot. (London)* [N.S.] **38**, 667.

Zaitlin, M., and Beachy, R. N. (1974). *Adv. Virus Res.* **19**, 1.

13 Nucleic Acid Hybridization Technology and Detection of Proviral Genomes

Eng-Shang Huang and Joseph S. Pagano

I. Introduction

There has been a flood of techniques developed in the last few years to define sequence homology between species of nucleic acids. The power and versatility of these techniques are nowhere better displayed than in virus systems, especially those that have features of integration, proviral states, or other forms of latency of the viral genome or portions of it. Viral genomes, because of their uniqueness and limited complexity, lend themselves especially well to varied hybridization approaches, but there is no doubt that more complex genomes, including mammalian DNA, will become accessible to such analytic techniques; such efforts have already been initiated.

We have chosen to consider several techniques of nucleic acid hybridization in some detail rather than to catalog all the available methods. The prin-

ciples and general framework are similar in the many different systems that
have been studied. Our focus is on hybridization systems that have proven
of value in DNA virus systems. The same techniques can be used in RNA
virus systems. However, they involve special problems, and we do not take
them up in detail.

The purpose of this Chapter, the intent of which is practical rather than
theoretical, is to deal with methods for the labeling, detection, quantitation,
and determination of homology of viral DNA and RNA. We do this in the
context of presentations of specific methods while indicating some of the
practical theory behind the methods. We illustrate their uses and examine
problems that are likely to be encountered in the laboratory and in interpretation of results. We also take up other selected methods of characterization of DNA that can fill an accessory role in hybridization approaches.

The topics included are preparation and radiolabeling of nucleic acids
used as probes, DNA–DNA and RNA–DNA hybridization with immobilized nucleic acids, cytohybridization, and methods of kinetic analyses of
DNA–DNA and RNA–DNA hybridization in solution. Finally, we present
a limited discussion of the use of restriction endonuclease technology especially as it applies to blot-transfer techniques of nucleic acid hybridization
and to the use of defined fragments of DNA as special probes. We finish with
a consideration of heteroduplex technology, including R-loop technique.
The electron microscopic approaches are not only sensitive and direct but
they often decisively complement the other approaches.

II. General Utility

Hybridization techniques have become fashionable, and they are sometimes used when it is unnecessary. When should these techniques be used?
What is required for their use? The prime application of these techniques
is for the detection of viral genomes that cannot be disclosed by tests for
infectivity. In principle, there is no more sensitive technique than simple
determination of infectivity. However, if the viral genome is defective and
unable to replicate, if the viral genome is integrated in whole or in part into
the cellular chromosome, or if the viral genome is present as an episome,
then hybridization techniques are indispensible. Systems, especially in the
RNA tumor virus field which require helper viruses for expression of infectivity, also call for nucleic hybridization technology. *In vitro* systems for
synthesis of viral DNA or RNA at some stage require identification and
quantitation of the products by specific hybridization. Nucleic acid hybridization provides the only accessible method for the identification of messenger RNA. These methods are particularly valuable for decisive identification

of viruses or viral genetic material. They provide the ultimate methods for the determination of homology between viruses both in kind and in degree. They make it possible to determine the percentage of the genome or transcripts of the genome present in cells or tissue. The technique of cytohybridization actually makes possible localization of viral genetic material to specific cell type and even intracellular site. For sensitivity and quantitation of nucleic acids hybridization approaches are the methods nonpareil.

There are, however, requirements that must be met before these techniques can be employed. First, the identity of the nucleic acid being sought must be known or suspected. There is no way that one can detect a specific genome in whole or in part without the use of the appropriate viral probe unless through an accident of partial homology. Second, the nucleic acid used as probe must be purified and radiolabeled to a high degree of specific activity. The general limitation on the minimum amount of DNA that can be detected in liquid systems at present depends upon the complexity of the genome. However, it is already within the realm of possibility to improve upon this level of sensitivity up to 10-fold. This can be done by using probes of restricted fragments of nucleic acid that might be present in the cell system under analysis. Cytohybridization is a less sensitive technique at present, but it is advantageous for certain purposes. The overriding power of these techniques is the certainty with which it is possible to uncover the presence of specific nucleotide sequences. Viral genetic material that defies detection by any other approach can be defined with precision by the proper use of the technology we are about to describe.

III. Basic Methods

A. Preparation and Radiolabeling of Probes

The purity and specific radioactivity of the nucleic acids used as probes are crucial to the conduct of successful nucleic acid hybridization. The representative nature or fidelity of the probe if copied sequences are being used is also important. In some cases a suitable probe can be achieved by direct incorporation of radiolabeled uridine or thymidine during replication of the virus in cell culture. However, with many viruses there is insufficient incorporation. In general nucleic acids with specific radioactivity in excess of 10^6 cpm/μg, preferably about 10^7 cpm/μg, are needed. If labeling *in vivo* does not reach this level of radioactivity, then labeling during *in vitro* synthetic reactions is needed. None of the current *in vitro* methods of labeling yields completely representative probes; label may be unevenly distributed in the repair synthesis methods, or some sequences may not be copied into the probe.

These shortcomings do not obviate the value of such probes for most purposes, but they do make it mandatory to carry out appropriate reconstruction or saturation analyses. Methods of labeling nucleic acids by direct or indirect iodination should in principle avoid the deficiencies of the other methods in the construction of representative probes. However, iodination introduces its own problems; to date this approach has not been as useful as had been hoped.

In general, if labeling of nucleic acids can be accomplished in infected cell cultures, then this is the preferable method. However, labeling of DNA *in vitro* by repair synthesis is a versatile technique that gives most satisfactory results. In situations in which the nucleic acid that serves as template is available only in minute amounts, the synthesis of complementary nucleic acids, either DNA or RNA, can provide excellent probes and at the same time augment the amount available.

The purity of the nucleic acid used as probe is the most critical element in specific hybridization. Virus harvested from extracellular fluids rather than from infected cells is the better source of probe nucleic acid, particularly if methods involving complementary RNA (cRNA) or complementary DNA (cDNA) are employed. Virus harvested from cells carries with it contaminating cellular DNA. Even virus from extracellular fluids is contaminated with host–cell DNA which is not always eliminated by DNase treatment because of protection by extraneous proteins. The first step, then, is to procure virus which is highly purified by any method that conserves the integrity of the virion so that treatment with DNase will not cause fragmentation of the encapsidated genome. Next, the extracted viral DNA should be rigorously purified. The genome should emerge from purification largely unfragmented so that it can be separated from cellular nucleic acids on the basis of physical properties such as size, as well as density or supercoiled state. Physical homogenity is the aim. Analysis of the isolated components of nucleic acid both by optical and radioisotope counting methods is desirable since contaminating cellular DNA may not have been labeled. The final test of purity of the probe is a complete set of analyses of the behavior of the probe during hybridization in the presence of heterologous as well as homologous nucleic acids. These tests should include DNA from uninfected host–cell material. There is always some degree of nonspecific background hybridization, but the level should be low, and it should be consistent for each probe prepared.

The probes include DNA radiolabeled either *in vivo* or *in vitro*, cRNA, and cDNA. The isotopes of most utility are ^{3}H, ^{32}P, and ^{125}I. Methods of iodination are discussed last in this section.

1. DNA Labeled in Vivo

Some viral DNAs can be labeled to high specific activity during replication in cell cultures, but other viral DNAs incorporate radiolabeled thymidine much less efficiently. Viruses such as herpes simplex which induce virus-specific thymidine kinase can be labeled to relatively high specific activities, as shown by Davis and Kingsbury (1976). In contrast, human cyto-megaloviruses (CMV) which appear not to induce novel thymidine kinases (J. Estes and E.-S. Huang, unpublished data) cannot be labeled well *in vivo*. The efficiency of incorporation of radiolabeled thymidine into viral DNA works best if viral DNA is replicated through salvage pathways rather than by *de novo* synthesis. In the case of Epstein–Barr virus, (EBV), incorporation of tritiated thymidine into viral DNA is also inefficient, and it is difficult to obtain EBV DNA with specific radioactivities of more than 10^5 cpm/μg. This is generally inadequate for hybridization techniques. With this virus the situation is complicated further by the very low yields of virus per unit of culture fluid, and it is impractical to add enough radioisotope to the large volumes of culture medium needed to produce sufficient virus. If, however, viral DNA can be radiolabeled *in vivo*, then this is usually considered the method of choice because of the uniform radiolabeling that results.

Herpes simplex virus (HSV) types 1 and 2 induce their own specific thymidine kinase in virus-infected cells. The salvage pathway as well as *de novo* viral DNA synthesis is utilized in the presence of viral replication. Tritiated thymidine of high specific activity (60 Ci/mM) in aqueous sterile H$_2$O is added to HSV type 1-infected cells at a concentration of 50 μCi to 100 μCi per milliliter immediately after virus adsorption. The most economical and efficient procedure is to grow the cells in Wheaton roller bottles; 10–15 milliliters of culture media is enough to cover a monolayer of 5×10^7– 1×10^8 cells in the bottle. Twenty-four hours after infection HSV can be purified from the cytoplasmic fraction of the infected cells; 48–72 hours after infection the virus can be purified from extracellular fluid. The [^{3}H] HSV DNA is purified from virions, as previously described (Huang *et al.*, 1973). The specific activities obtained for HSV 1 and HSV 2 are $1–2 \times 10^7$ and $2–5 \times 10^6$ cpm/μg; respectively (Davis and Kingsbury, 1976; E.-S. Huang, unpublished data).

2. *DNA Labeled in Vitro*

The most widely used method for labeling of viral DNA *in vitro* is nick translation or repair synthesis with DNA polymerase I (Aposhian and Kornberg, 1962). In this technique, purified viral DNA is first nicked under carefully controlled conditions with pancreatic DNase I and then subjected to repair synthesis in the presence of radiolabeled [^{3}H]TTP or [^{125}I]dCTP (Shaw *et al.*, 1975). DNase digestion has to be done carefully so as to avoid undue hydrolysis. The nicks introduced are assumed to be distributed at random through the genome. Since repair synthesis is initiated at the nicks, the introduction of radiolabel should also be at random sites along the genome. Under optimal conditions, up to 35% of ^{32}P murine CMV genome can be replaced in either strand by newly synthesized [³H]DNA (E.-S. Huang, unpublished results). The conditions, especially the temperature in which repair synthesis is conducted, are crucial in order to prevent redundant and branched replication in which the newly synthesized strand segments are themselves copied. Such structures behave anomalously during digestion with S1 single-strand specific nuclease. In any case, viral DNA with specific activities between 10^6–10^7 cpm/μg can be produced with this method (Nonoyama and Pagano, 1972; Frenkel et al., 1972; Huang et al., 1973).

DNA polymerase I is used to prepare tritium-labeled viral DNA in vitro, and was purified according to Jovin et al., (1969), as described previously (Huang et al., 1973).

The details of the in vitro synthesis reaction were originally described by Nonoyama and Pagano (1972). Two micrograms of purified EBV DNA are incubated with 5×10^{-2} μg of DNase I for 10–20 minutes at 37 °C in 0.45 ml of 70 mM potassium phosphate, pH 7.4, 1 mM 2-mercaptoethanol, and 7 mM MgCl$_2$. DNase is inactivated by heating the mixture for 10 minutes at 70 °C. The nicked DNA is then labeled in a 0.5 ml reaction which contains: 70 mM potassium phosphate, pH 7.4, 1 mM 2-mercaptoethanol, 7 mM MgCl$_2$, 0.1 mM each of dATP, dCTP, and dGTP, 250 μCi of [³H] TTP (20 Ci/mM), and 2 units of DNA polymerase I. The mixture is incubated at 18 °C until incorporation of [³H] TMP (measured by counting TCA precipitates of aliquots taken during the course of the reaction) has reached a plateau (approximately 5 hours); then Sarkosyl and neutralized EDTA are added to 1% and 10 mM, respectively, and the mixture is passed through G-50 Sephadex equilibrated with 10 mM Tris-HCl, pH 7.4, neutralized 1 mM EDTA, and 0.1% Sarkosyl. The TCA-precipitable radioactivity in the fractions eluted from the column are pooled and mixed with an equal volume of water-saturated phenol. Following centrifugation, the DNA in the aqueous phase is precipitated overnight with salt-saturated ethanol at −20 °C. The specific activity of the DNA is estimated before G-50 Sephadex chromatography from the DNA concentration and the TCA-precipitable radioactivity in an aliquot of the reaction mixture.

3. Messenger RNA

Viral mRNA can be labeled in infected cells at various times after infection with [³H]uridine or ortho ³²P or in vitro by direct iodination. The specific activity of mRNA is dependent on the amount of radioactive material used. In general, for short-term labeling in tissue culture, larger quantities of isotope (up to 100 μCi/ml) can be used; for long-term labeling the quantity should be decreased to 3–10 μg/ml in order to avoid radiation damage. The times selected for labeling depend on the purpose of the experiment.

Messenger RNA of high specific activity can be obtained by iodination of mRNA after purification with ^{125}I by the thallium chloride technique. However, the labeled product is suitable only for hybridization purposes and not for size characterization; this is due to fragmentation of the iodinated product. Specific activities of viral mRNA up to $5–10 \times 10^7$ or even higher can be obtained, but undesired effects are encountered with specific activities greater than 5×10^8 cpm/μg. The ideal specific activity for consistent results and low background is between 2×10^7 and 1×10^8 cpm/μg (our unpublished results). The method for iodinating viral mRNA is described in Section III, A,6.

a. Preparation of Viral mRNA. Several features affecting extraction of mRNA, both viral and cellular, should be mentioned. (i) Most but not all cellular and viral mRNA contain poly(A) stretches at the 3′ OH end of the RNA molecule (Edmonds and Caramela, 1969; Edmonds and Kopp, 1971; Sheldon *et al.*, 1972; Hadjivassilions and Brawerman, 1966; Kates, 1970; Phillipson *et al.*, 1971). At neutral pH and cold temperatures, the poly(A)-containing mRNA is retained in the nonaqueous phase during phenol extraction (Georgiev and Mantieva, 1962) from which it can be recovered by reextraction with a slightly alkaline buffer (pH 9) (Brawerman *et al.*, 1963). (ii) Retention of poly(A)-containing RNA in the nonaqueous phase at neutral pH is caused by interaction between the poly(A) and the denatured ribosomal proteins. Tris is protonated at neutral pH; a solution of 0.1 *M* Tris has a concentration of monovalent cations equal to 0.1 *M* K^+. This monovalent cation will promote the interaction of poly(A) and protein. This kind of interaction can be overcome by extraction at slightly alkaline pH; at pH 9 Tris is in an almost nonionized condition (Edmonds and Caramela, 1969). (iii) Phenol extraction at alkaline pH is very effective for mRNA extraction, but at low temperature DNA is extracted together with mRNA. Although the cell DNA can be removed by digestion with pancreatic DNase, contamination of the DNase preparation with RNase causes fragmentation of the RNA; consequentially there are almost always problems in recovery of mRNA. The hot phenol–SDS method (Girard, 1967) for mRNA extraction facilitates recovery of mRNA; this method is as efficient as phenol extraction at pH 9, and it does not trap mRNA in the nonaqueous phase. Also DNA is excluded by retention in the nonaqueous phase.

Extraction of polysomal RNA or RNA from the cytoplasmic fraction is best carried out in an ice bath. The sample in 0.05–0.1 *M* Tris-HCl, pH 9, and 0.5% SDS is extracted three times with equal volumes of water-saturated redistilled phenol (80% phenol). The mixture of sample and phenol is vigorously stirred for 5 minutes and then centrifuged at 5,000 rpm in the Sorvall HB-4 or SS-34 rotor (10,000–12,000 *g*) for 5 minutes. The upper aqueous phase containing mRNA and rRNA is saved. The nonaqueous phase (phenol phase and interphase) is reextracted with a suitable volume of 0.1 M Tris-HCl, pH 9, and 0.5% SDS to recover the RNA trapped in the nonaqueous phase. The aqueous supernatant fractions of each extract are pooled, and RNA is precipitated with 2.5 volumes of ethyl alcohol in the presence of 0.1 M NaCl at -20 °C overnight. The precipitate is collected by centrifugation in the HB-4 or SS-34 rotor at 10,000 rpm for 30 minutes and washed twice with 70–90% alcohol containing 0.1 M NaCl to remove residual phenol. The precipitate can be dissolved in any desired buffer and stored at -20 °C, but preferably at -70 °C.

For the extraction of RNA from nuclei or whole cells, the hot phenol–SDS method is recommended to avoid entrapment of RNA in cellular DNA (Girard, 1967; Brawerman, 1974). The extraction procedures are identical to those used for polysomal or cytoplasmic RNA except that at least the first cycle of SDS–phenol extraction is carried out in a 60 °C water bath. After vigorous extraction at 60 °C for 5 minutes, the sample is chilled quickly in an ice bath. Under these conditions RNA is released to the aqueous phase while cell DNA remains in the nonaqueous phenol phase. Separation of aqueous and nonaqueous phases by centrifugation and alcohol precipitation follow.

b. Preparation of Poly(A)-Containing mRNA. Poly(A)-containing mRNA has several properties useful for extraction. (i) Due to complementary base pairing it can absorb to poly(U)-Sepharose or oligo(dT) cellulose columns at high ionic strengths. (ii) The poly(A) segment of mRNA is able to bind to nitrocellulose filters in a manner similar to denatured single-stranded DNA in the presence of 0.5 M KCl and low levels of actinomycin; the size of the poly(A) stretch should be greater than 50 nucleotides (Lee *et al.*, 1971). This RNA does not adsorb to filters in the presence of detergent, and it can be removed from nitrocellulose filters with 0.5% SDS in 0.1 M Tris-HCl, pH 9. (iii) Poly(A) also binds to denatured protein in the presence of sufficiently high concentrations of monovalent cations (Brawerman, 1974).

Based on these properties, mRNA containing poly(A) can be isolated by poly(U)-Sepharose, poly(U) glass filters and oligo(dT) cellulose, and by B-6 Millipore filters.

Poly(U)-Sepharose (Adesnik *et al.*, 1972) and oligo(dT)-cellulose (Aviv and Leder, 1972) are commercially available. Poly(U) glass filters can be prepared by coupling poly(U) to glass fiber filters by means of UV irradiation. The method has been described (Sheldon *et al.*, 1972; Adesnik *et al.*, 1972). In brief, 0.1 ml of poly(U) stock solution (2 mg/ml sterilized distilled water) is applied to Whatman GF/C filters. After drying at room temperature, the filters are dried in a vacuum at 37 °C and then UV-irradiated on both sides for 5 minutes with a 15–30 W GE germicidal lamp (distance, 20 cm). The unadsorbed poly(U) is rinsed away with sterile distilled water. The filter is dried at 37 °C and kept at 4 °C.

Before hybridization of mRNA to the poly(U) glass filters, the filters are first washed with distilled water by Millipore filtration in individual holders. The number of filters used is totally dependent on the amount of RNA expected; several layers of filters can be used. The RNA solution without DNase digestion in SDS buffer (1% SDS, 0.001 M EDTA, 0.1 M NaCl, 0.01 M Tris-HCl, pH 7.4) is slowly passed through the poly(U) glass filter at a flow rate of 1–2 ml per minute at room temperature. After washing with SDS buffer, the poly(U)-containing mRNA is eluted from the filter with low ionic strength 0.1 × SDS buffer. The elution can be done by crushing the filter in 0.01 × SDS buffer and removing the glass fiber by centrifugation. The poly (A)-containing RNA is precipitated by alcohol.

The method for isolation of poly(A)-containing mRNA by poly(U)-Sepharose 4B is described in detail in the Pharmacia poly(U)-Sepharose 4B Bulletin. In brief, the gel is first swollen in 1 M NaCl (in 0.05 M Tris-HCl, pH 7.5) for 5 minutes and washed with 0.1 M NaCl in 0.05 M Tris-HCl, pH 7.4. The column is then equilibrated with a concentrated salt buffer (CSB; 0.7 M NaCl, 0.05 M Tris-HCl, 0.01 M EDTA, and 25% formamide, final pH 7.5). The RNA sample in detergent solution (1% lauroyl-sarcosine, 0.03 M EDTA) is diluted 5 times with CSB and applied to the column. After sample application the column is washed with CSB buffer, and the poly(A)-containing mRNA is recovered with eluting buffer (EB; 0.01 M potassium phosphate, 0.01 M EDTA, 0.2% lauroyl-sarcosine, 90% formamide, final pH 7.5).

The procedure for isolation of poly(A)-containing mRNA with oligo(dT) cellulose columns is described by Aviv and Leder (1972). The RNA solution is applied to the oligo(dT) column in the presence of 0.5 M KCl (in 0.01 M Tris-HCl, pH 7.5), and the mRNA is eluted from the column with 0.01 M Tris-HCl, pH 7.5.

Although up to 60 μg of poly(A)-containing RNA can adsorb to a single nitrocellulose Millipore filter in the presence of 0.5 M KCl, the technique in which Millipore filters are employed is not popular (Brawerman *et al.*, 1972). This is due to extensive contamination of isolated mRNA with rRNA. The detailed procedure is described by Brawerman *et al.* (1972). The RNA in KCl buffer (0.5 M KCl, 1 mM MgCl$_2$, 0.01 M Tris-HCl, pH 7.6) free of detergent and at a concentration less than 300 μg/ml (Brawerman, 1974) is slowly passed through a Millipore nitrocellulose filter (0.5 ml or less per minute). The filter is washed with KCl buffer and then kept in 0.5–1.0 ml of elution buffer (containing 0.5% SDS, 0.1 M Tris-HCl, pH 9) in ice for about 30 minutes to elute the poly(A)-containing mRNA from the filters.

4. *Complementary RNA: Practical Theory*

In cases in which very small amounts of nucleic acid are available to serve as probes or the specific radioactivity achieved by direct incorporation of label into the probe nucleic acid is too low, then the use of cRNA may be of great value. If cellular DNA that contains hybridizable sequences is denatured and affixed to membrane filters, then self-annealing of DNA is prevented. Instead the DNA is accessible for hybridization with complementary sequences in the probe nucleic acid, in this case, cRNA. Complementary RNA, preferably in excess, is applied to duplicate filters containing the immobilized DNA to be tested and hybridized under appropriate ionic conditions (0.3–0.9 M NaCl) at 66 °C for a prolonged period, such as 16–20 hours. Usually about 20–50 μg of DNA are affixed to each filter; this quantity is confirmed by a diphenylamine reaction after hybridization has been carried out. In the neighborhood of 100,000 counts of the cRNA (specific activity, 1×10^7 cpm/μg) are usually applied to each filter. Extensive rinsing and treatment with RNase are important steps in the procedure to eliminate nonspecifically adherent cRNA. RNA–DNA hybrid duplexes are resistant to RNase. Hybridization of cRNA to homologous sequences of DNA is directly proportional to the concentration of homologous DNA until all of the sequences present in the probe and in the homologous DNA have been hybridized, in which case saturation conditions are met.

Procedure. This procedure was used for the preparation of EBV cRNA (Nonoyama and Pagano, 1971). Highly purified *Escherichia coli* DNA-dependent RNA polymerase was prepared according to Burgess (1969). The reaction mixture contained 0.04 M Tris, pH 7.9, 0.01 M MgCl$_2$, 0.1 mM dithiothreitol, 0.15 M KCl, 0.15 mg/ml bovine serum albumin, 0.15 mM GTP, ATP, and CTP, 0.15 mM [^{3}H]UTP (17–30 Ci/mM), 16 units of enzyme (400 units per milligram protein) and 2 μg EBV DNA in 0.25 ml. The mixture was kept at 37 °C for 2–5 hours when the reaction reached a plateau, the DNA was then digested with DNase (RNase-free batch, Worthington Biochemical, Freehold, N.J.) at the concentration of 20 μg/ml for 20 minutes at 37 °C. Yeast RNA (1 mg) and SDS (0.5% final concentration) were added, and the mixture was chromatographed through a Sephadex G-50 column (1.1 $\times$ 30 cm) in Tris-EDTA (TE; 0.05 M Tris-HCl, pH 7.4; 0.001 M EDTA) with 0.1% SDS. The first peak was collected, treated with water-saturated phenol, and dialyzed against TE with 0.01% SDS at 4 °C for 36 hours with three changes of the buffer. The incorporation efficiency in these conditions was 3 to 6% of the added [^{3}H]UTP. From the specific radioactivity of the [^{3}H]UTP and the guanine-cytosine content of EBV DNA, the product was expected to have 10^7 cpm/μg.

When the synthesized cRNA was analyzed by velocity sedimentation on sucrose gradients, radioactivity was distributed from 16 S to 4 S, with an obvious peak at the 12 S to 16 S region. As the incubation time was so long, some degradation of the RNA could be expected by a trace of RNase in the reaction mixture.

5. *In Vitro Synthesis of DNA Transcripts (cDNA) of RNA Virus Genomes*

The largest components of RNA tumor virus genomes are 60–70 S RNA (Gillespie *et al.*, 1975; Robinson *et al.*, 1965). At an early stage of investigation, there were indications that fragmentation of the viral genomes occurred upon heat treatment; the fragments produced are heterogeneous in size (Duesberg, 1970). In the case of avian tumor viruses, this 60–70 S RNA consists predominantly of two 35 S RNA subunits with approximate molecular weights of 3×10^6 (Duesberg and Vogt, 1973). These data imply that the 60–70 S DNA is diploid and composed of two identical subunits.

Transcription of RNA tumor virus genomes *in vitro* by viral RNA-dependent DNA polymerase generates chiefly short DNA transcripts ranging up to 2×10^5 MW in size (Rokutanda *et al.*, 1970; Temin and Baltimore, 1972; Duesberg and Canaani, 1970). Duesberg and Canaani demonstrated that DNA made *in vitro* with Rous sarcoma virus (RSV) renders most of viral RNA resistant to RNase digestion; Duesberg and Canaani (1970) and Garapin *et al.* (1973) suggested that the synthesized DNA is complementary to the entire viral genome.

Recently the difficulty in obtaining full-length viral DNA transcripts *in vitro* has been overcome by the finding of the optimal triton X-100 concentration necessary for expression of viral DNA synthesis. With optimal triton concentration (0.0225%, v/v) about 1–10% of the transcribed DNA has a size between $2.5–3.7 \times 10^6$ MW which is complementary to the total viral RNA (Junghaus *et al.*, 1975).

Radioactive DNA transcripts of an RNA tumor virus were used by Gelb *et al.* (1971) to detect the viral DNA in mammalian cells in a murine leukemia virus system (MuLV). The virus DNA was made in reaction mixtures containing 0.05 M Tris-HCl, pH 7.8, 0.002 M dithiothreitol, 0.06 M NaCl, 5×10^{-4} M deoxyribonucleotide triphosphate (specific activity of dXTP was 161 mCi/mM), 0.014% Triton X-100, and 100 μg of virus protein per milliliter. The reaction mixture was incubated at 37°C for 18 hours. The double-stranded DNA was then extracted by a phenol–SDS method, and the viral RNA was digested by pancreatic RNase at 10 μg/ml for 2 hours at room temperature. The details are given in Gelb *et al.* (1971). The specific activity of ^{3}H-labeled murine leukemia virus DNA was around 2.5×10^5 cpm/μg. Reassociation kinetics analysis of this *in vitro* [^{3}H]MuLV DNA showed a Cot_{50} of $2.8–3.4 \times 10^3$ mol sec/liter, equivalent to a genome complexity (molecular weight) of $5.0–6.1 \times 10^6$ (Gelb *et al.*, 1971).

The procedure and the reaction mixture used by Junghaus *et al.* (1975) for *in vitro* synthesis of full-length DNA transcripts of RSV RNA are briefly described for reference. Rous sarcoma virus equivalent to 1.25 mg of viral protein was added to a reaction mixture containing 10^{-4} M dATP, dGTP;

10^{-5} M dTTP, dCTP; 0.002 M magnesium acetate, 0.03 M dithiothreitol; 0.1 M Tris-HCl, pH 7.6; 0.825 μM [³H]TTP (20 Ci/mM); and 0.0225% TritonX-100. The total volume of the reaction mixture was 4 ml. The reaction was carried out at 41 °C for 18 hours and was terminated by the addition of EDTA to 0.015 M, SDS to 1%, NaCl to 0.1 M, and β-mercapto-ethanol to 2%; 80 μg of denatured salmon sperm DNA was added as carrier. The product DNA was extracted with phenol, and viral RNA was removed by digestion with pancreatic RNase (Gelb et al., 1971). The specific activity of the viral DNA synthesized was around 2 × 10⁵ cpm/μg.

6. Iodination Methods: Direct and Indirect Labeling of DNA and RNA

There are several methods for labeling of nucleic acids in vitro. These include (i) tritium exchange (Borenfreund et al., 1959), (ii) methylation of DNA by tritium-labeled dimethyl sulfate (Smith et al., 1967), and (iii) iodination of nucleic acids in the presence of thallic chloride (Commerford, 1971). The specific activity achieved by tritium exchange and methylation does not meet the requirements for a sensitive probe in either membrane hybridization or reassociation kinetics analysis.

The technique for direct labeling of nucleic acid with radioactive iodine in the presence of thallic chloride at low pH was first established by Commerford (1971). The method yields nucleic acid of very high specific activity and offers economical and simple processing. The ¹²⁵I is incorporated into DNA through a covalent bond as 5-iodocytosine (Commerford, 1971). There are two ways to label DNA or RNA with ¹²⁵I: directly and indirectly. The direct iodination technique has recently been modified for labeling of 5 S RNA, mRNA, or denatured DNA (Prensky et al., 1973; Tereba and McCarthy, 1973; Scherberg and Refetoff, 1974). The technique for labeling DNA and RNA is essentially the same except that the DNA should be maintained in the single-stranded state and should be kept from self-annealing during iodination.

Several important considerations should be kept in mind for high efficiency of labeling. (i) The specific activity of the nucleic acid is dependent on the amount and molar ratio of iodine to cytosine used in the iodination system. For reproducible specific activities, a constant molar ratio of iodine to cytosine should be used (optimal ratio about 1). (ii) The purity of thallium chloride (K & K Chemical Co., Plainview, N. Y.) plays an important role in the efficiency of iodination. Poor iodination will result if a chelating agent is present; the amount of thallium ion should be in excess of any chelating agent present. A molar ratio of TlCl₃ to iodine of 6 to 1 is popularly used (Commerford, 1971). A fresh preparation of thallium chloride solution at pH 5 is recommended. (iii) The pH of the reaction mixture is critical; it should be 4.5–5.0. (iv) The efficiency of iodination of double-stranded DNA is low.

The DNA should be denatured first by boiling in low ionic strength buffer or distilled water. The ionic strength of the iodination buffer should be maintained below 0.1 M Na$^+$ in order to prevent reannealing at the iodination step.

An example of a procedure for labeling human CMV DNA is as follows: Human CMV DNA is dialyzed against distilled water, denatured by boiling for 10 minutes, and rapidly chilled in an ice bath. The DNA solution is then adjusted to pH by adding sodium acetate–acetic acid buffer (final concentration, 0.01 M sodium acetate). For iodination of single-stranded DNA or RNA, when the denaturation step is not needed, the sample is simply subjected to dialysis against 0.1 M acetate buffer, pH 5. The reaction mixture (Commerford, 1971), assembled at 0°C, contains in a total volume of 100 μl, 0.1 M (or 0.01 M for denatured DNA) acetate buffer (pH 5); 2 μg denatured CMV DNA; 0.25 mM KI in chemical equilibrium with the desired amount of ^{125}I, and 1.5 mM TlCl$_3$ (added last in sodium acetate buffer, pH 5). The reaction is mixed and heated to 60°C in a water bath for 15 minutes and then chilled in ice. The specific activity can be controlled by the length of the reaction time. Twenty-five microliters of freshly prepared 0.1 M Na$_2$SO$_3$ are added to reduce the excess amount of thallium chloride; β-mercaptoethanol at a final concentration of 30 mM is recommended in place of Na$_2$SO$_3$. The noncovalently bound ^{125}I can be removed by competition with cold sodium iodide in order to reduce the nonspecific background radioactivity after this stage. The pH of the reaction mixture is then raised to pH 9 by adding 0.05 ml of 1 M ammonium acetate–0.5 M ammonium hydroxide solution. The mixture is then heated at 60°C for 15 minutes to disassociate the unstable intermediate products (Commerford, 1971). At the end of the reaction the mixture is chromatographed on a Sephadex G-50 or hydroxyapatite column to recover the DNA or RNA (for details, see Commerford, 1971; Prensky *et al.*, 1973).

Iodination of nucleic acid by the direct method incurs treatment with oxidizing and reducing reagents at high temperature and at high and low pH. Also, native DNA cannot be labeled efficiently. Indirect methods of labeling viral DNA and cRNA are available that avoid exposure of DNA to high temperature, unfavorable pH, and oxidizing or reducing conditions. DNA can be recovered in the native state after iodination. The indirect method involves first synthesis of 5-^{125}iodo-dCTP or 5-^{125}iodo-CTP and then use of these reagents as precursors for repair synthesis of viral DNA *in vitro* and synthesis of cRNA.

The method for iodination of dCTP and CTP is essentially the same as the method described above for iodinating viral DNA and RNA, except that after iodination the [^{125}I]dCTP and CTP are recovered by DEAE-cellulose chromatography. The iodination reaction mixture is diluted to a salt concentration of less than 0.01 M and loaded on a DEAE-cellulose column (Whatman DE52, 0.9 by 22 cm) that has been prewashed with 0.3 column volume of 1 M triethylamine carbonate, pH 8.0, and pre-equilibrated with 0.01 M triethylamine carbonate, pH 8. The triethylamine carbonate is prepared by passing CO_2 into a mixture of 170 ml water and 30 ml redistilled triethylamine in ice until a single phase of triethylamine carbonate forms. The pH of the solution is adjusted by adding CO_2 to pH 8. After loading, the column is washed with 0.01 M triethylamine carbonate, and then the sample is eluted with a linear gradient (0.01–0.5 M) of triethylamine carbonate, pH 8. The peak fractions containing 5-iodo-dCTP or 5-iodo-CTP are pooled and lyophilized to dryness. The final product is dissolved in H_2O and stored at $-20°C$.

Synthesis of ^{125}I-labeled viral DNA or cRNA are essentially as described for ^3H viral DNA and $[^3H]cRNA$ earlier (Sections III,A,2 and 4). The details are given by Shaw *et al.* (1975).

B. DNA–DNA Hybridization on Nitrocellulose Filters

The classic method of hybridization is with probe DNA to denatured DNA affixed to membrane filters. The DNA probe can be radiolabeled either *in vivo* or *in vitro*. This method represents a direct approach to hybridization, but it presents a number of problems that have curtailed its use with the availability of new methods. Direct DNA–DNA hybridization requires relatively large amounts of probe DNA in contrast to kinetic methods of hybridization which are conducted in liquid media (Section III,F). As noted above, the difficulty of attaining sufficiently high specific radioactivity can be overcome with radiolabeling of the DNA *in vitro*. The amounts usually required for repeated hybridization tests mean that the labeled DNA must be prepared more frequently than if it is to be used in other hybridization systems. Another problem with DNA–DNA hybridization is self-annealing of the DNA probe. Ideally, DNA–DNA hybridization utilizes separated complementary strands as probe. These in turn have special uses and shortcomings. Also, it is not as easy to eliminate unhybridized DNA as cRNA; in cRNA–DNA hybridization RNase can be used since the RNA in hybrid molecules is not digested by the enzyme. Nevertheless, DNA–DNA hybridization has given important leads (Fujinaga and Green, 1966; zur Hausen and Schulte-Holthausen, 1970) and will probably continue to prove useful in the future. The method for DNA–DNA hybridization is similar to RNA–DNA hybridization except that in DNA–DNA hybridization S1 or *Neurospora crassa* single-stranded nuclease is used instead of RNase to digest nonhybridized single-stranded DNA.

C. RNA–DNA HYBRIDIZATION ON NITROCELLULOSE FILTERS

The principles and procedures of RNA–DNA hybridization were described by Gillespie and Spiegelman (1965). The denatured DNA is first immobilized on nitrocellulose filters and then annealed with radioactive virus-specific RNA. There are two ways to denature the DNA for immobilization: one is by boiling in low salt solution and rapid chilling, and the other is by denaturation in alkali. An example of membrane-filter hybridization is shown in Fig. 1.

Fifty micrograms of HEp-2 DNA with graded amounts of CMV DNA in 2 ml of 0.1 × SSC in 1 mM EDTA were denatured in 0.5 N NaOH for 2 hours at 37 °C. The DNA solution was neutralized with 1.1 N HCl in 0.2 M Tris in an ice bath and then adjusted to 6 × SSC. The solution was slowly passed through a Bact-T-flex type 6 (Schleicher and Schuell Co., Keene, N. H.) nitrocellulose membrane filter (prewashed and soaked in 6 × SSC). After filtration washing, the filters were dried at room temperature for at least 3 hours or overnight and baked in a vacuum oven at 80 °C for 3 hours. The filters can be stored in the refrigerator or at room temperature for several months without loss of hybridization capacity.

The DNA filters were then immersed in 1 ml of 6 × SSC containing CMV [^{3}H] cRNA (1.5 × 10^5 cpm; specific activity, 1 × 10^7 cpm/μg), 1 mg of yeast RNA, and 0.1% SDS. The hybridization was carried out at 66 °C (20°–25 °C below T_m) for 20 hours. Hybridization can be conducted at lower temperature (40 °C) in the presence of 50% formamide for 40 hours; the reaction mixture contained 1.5 × 10^5 cpm [^{3}H] CMV cRNA, 1 mg yeast RNA in 50% formamide, 0.9 M NaCl, 0.1% SDS, and 0.01 M Tris-HCl, pH 7.9. After hybridization the filters were extensively washed with 2 × SSC. The unhybridized [^{3}H] cRNA was removed by digestion with 40 μg/ml of RNase (preheated at 80 °C for 10 minutes to inactivate contaminating DNase) at 37 °C for 30 minutes. The filters were then washed by filtration on both sides with 2 × SSC, dried, and counted.

1. Reconstruction Curve

Figure 1A and B show reconstruction curves of CMV and EBV DNA hybridization, respectively. These data provide the basis for calculation of the number of viral genome equivalents detected by hybridization. These curves were constructed from the amount of radioactive CMV cRNA or EBV cRNA that hybridized to graduated amounts of CMV DNA or EBV DNA in the presence of 50 μg of HEp-2 or calf thymus DNA. If the molecular weight of CMV DNA and diploid human cell DNA are 10^8 and 4 × 10^{12}, respectively, then 0.1 μg of CMV DNA in 50 μg of human tissue DNA

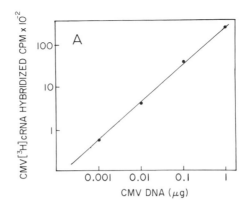

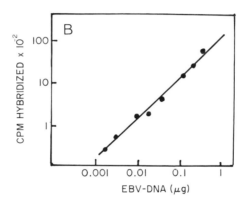

FIG. 1. (A) Hybridization of CMV cRNA to CMV DNA. Graded amounts of CMV DNA were mixed with 50 μg of HEp-2 DNA and denatured in 0.5 N NaOH for 2 hours at 37 °C. The DNA solution was neutralized with 1.1 N HCl in 0.2 M Tris, adjusted to 6 × SSC, and immobilized on filters. The input cRNA for each filter was 1.5×10^5 cpm (specific activity was about 10^7 cpm/μg). The amount of DNA retained on the filter, determined by a diphenylamine test after hybridization, remained constant. The background hybridized counts for 50 μg of HEp-2 (130 cpm) were subtracted from each value. (From Huang *et al.*, 1973). (B) Calculation of numbers of EBV genomes. EBV DNA in the quantities indicated was mixed with 50 μg HeLa cell DNA and denatured in alkali. DNA–RNA hybridization was carried out according to Gillespie and Spiegelman (1965) except that SDS was present at the final concentration of 0.1% during hybridization. The DNA concentrations on the filters were determined by the diphenylamine test after hybridization, and the hybridized counts were expressed per 50 μg DNA. The input cRNA was 75,000 cpm per filter. The hybridized value for 50 μg of HeLa DNA alone was subtracted. (From Nonoyama and Pagano, 1971).

TABLE I

DNA–RNA HYBRIDIZATION TESTS[a,b]

DNA on filter	cRNA hybridized (cpm/50 μg DNA)	Estimated number of genome equivalents per cell
HR-1 Burkitt lymphoma	12,131	680
32°C for 10 days	17,392	810
IF negative[c]	596	32
Raji Burkitt lymphoma		
IF negative	1126	65
Chromosomes[d]	1099	62
EBV infected[e]	22,617	1170
6410 myelogenous leukemia[f]	803	45
F-265 normal patient[g]	1650	100
NC-37 human healthy donor[g]	1343	80
HeLa human carcinoma	152	< 2
HEp-2 human carcinoma	126	< 2

[a] From Nonoyama and Pagano (1971).

[b] The conditions for hybridization were the same as in Fig. 1B. All DNA was prepared by treatment with SDS and Pronase followed by phenol extraction; 150 cpm, the hybridized value for HeLa cell DNA, was subtracted to estimate viral genome number equivalents.

[c] HR-1 cell line that no longer sheds EBV.

[d] Raji cells were arrested at metaphase by 0.05 μg/ml of Colcemid, and chromosomes were prepared according to Maio and Schildkraut (1967).

[e] Raji cells were infected with EBV, and the DNA was extracted after 48 hours of infection.

[f] Human lymphocyte line obtained from Dr. W. Henle.

[g] Human lymphocyte lines obtained from the John L. Smith Memorial for Cancer Research, Pfizer Co.

is equivalent to 80 CMV genomes per cell. The counts of cRNA hybridized to 0.1 μg of DNA in 3882 cpm, which is equal to 48 cpm/genome/cell. By this calibration, we should be able to detect as few as two viral genome equivalents per cell.

Typical results of cRNA–DNA hybridization applied to the detection of EBV genomes in human tissue are shown in Table I. P3HR1 cells (virus-producing Burkitt-lymphoma cell line) showed 680 EBV genome equivalents per cell. When the cells were incubated at 32°C for 10 days, the number of genomes increased. A P3HR1 cell line that was no longer positive for EBV capsid antigens by immunofluorescence (IF) test was available, and DNA from these cells was extracted and tested. The number of genome equivalents was drastically reduced to 32 per cell, but virus-specific DNA was still present.

Raji cells, a line of Burkitt lymphoma that does not produce EBV or viral antigens, had been reported to contain EBV DNA (zur Hausen and Schulte-Holthausen, 1970). This observation was confirmed by cRNA–

DNA hybridization, but about ten times more genomes per cell were found. zur Hausen and Schulte-Holthausen (1970) assumed in the calculations from the DNA–DNA hybridization tests that all of the DNA on a filter was hybridized at the saturation level of added [³H]DNA, which probably led to the underestimates.

2. *Messenger RNA–DNA Hybridization*

The RNA–DNA filter hybridization techniques can be used to detect and also to select for virus-specific RNA. Hybridization is carried out with immobilized purified viral DNA on the filter. The amount of viral DNA used is totally dependent on the purpose of the experiment. Reaction conditions, including 50% formamide at low temperature, are more favorable to select for viral mRNA with minimum breakdown of the macromolecules. The radioactive mRNA hybridized to viral DNA on the filter can be eluted by elevation of the formamide concentration in low salt and precipitation with alcohol in the cold. Messenger RNA–DNA hybridization on filters is still the most popular technique to examine the size of viral message in various gradients.

3. *Multisampled DNA Concentration Curves (cRNA–DNA Hybridization on Filters)*

Complementary RNA–DNA hybridization is customarily carried out with a single concentration of the DNA to be analyzed. It is feasible, however, to conduct an analysis in special cases with multiple concentrations of the test-cell DNA. This procedural variation affords some increase in sensitivity when relatively small amounts of hybridizable DNA are present. In the illustration presented, there is approximately a 2- to 3-fold increase in sensitivity. In this example, tissue from a Burkitt's lymphoma without detectable EBV DNA by the conventional cRNA–DNA hybridization was analyzed more closely. Different concentrations of control-cell DNA (HEp 2) and DNA from the lymphoma (FM) were fixed onto nitrocellulose filters and exposed to the same amount of cRNA (Fig. 2). Although the background hybridization increased with higher concentrations of DNA, hybridization for HEp-2 DNA and the degree of hybridization for the lymphoma did not differ significantly with concentrations of up to 200 μg of cellular DNA. The results indicated clearly that there was less than one EBV genome equivalent per cell.

4. *Competition Hybridization*

This is a useful technique to compare quantitatively and qualitatively species of viral RNA as well as to define sequence of gene transcription. There are two methods of approach: (i) simultaneous hybridization competi-

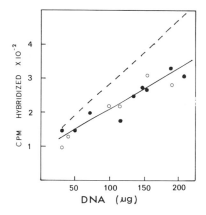

FIG. 2. Comparison of hybridization efficiency between HEp-2 DNA and DNA from the single Burkitt's lymphoma without detectable EBV DNA. Each amount of HEp-2 DNA or the lymphoma (FM 1143) DNA was fixed on a nitrocellulose membrane filter and hybridized with 1.5×10^5 cpm of cRNA specific to EBV DNA. (O—O) FM DNA; (●—●) HEp–2 DNA; (———) one genome per cell. (From Nonoyama et al., 1973.)

tion in which hybridization experiments are performed by incubation of DNA-containing filters with a constant saturation amount of radioactive RNA and increasing amounts of unlabeled competitive RNA; and (ii) sequential competition in which hybridization experiments are carried out sequentially by preincubating the viral DNA filter with increasing amounts of unlabeled competitive RNA for 12 hours and then adding a saturating amount of radioactive RNA. The use of saturating amounts of radioactive indicator RNA in competition experiments is essential. Saturation curves determined with the same batch of DNA filters and radioactive RNA are also essential. Representative examples of the application of hybridization competition can be found in Oxman et al. (1971) and Levine et al. (1973) in an adenovirus 2-SV40 hybrid system and in Huang et al. (1972) in an SV40 system.

D. RNA–DNA Hybridization in Liquid Medium

RNA–DNA hybridization can be performed in liquid without immobilization of DNA on solid support. This technique was introduced by Nygaard and Hale (1964). The handicap implicit in the technique is self-annealing of the DNA in liquid which prevents the hybridization of RNA to the homologous DNA if the ratio of RNA to its homologous DNA sequences is too low. The technique is at its best when separated strands of DNA can be used, for example, in SV40 and adenovirus systems (Sharp et al., 1974; Flint *et al.*, 1975, 1976a,b). Self-annealing of DNA is obviated, and the RNA–DNA duplexes formed as the result of hybridization can be treated with RNase to remove unhybridized radioactive RNA and precipitated with TCA to quantitate the RNase-resistant counts in the hybrid. Alternatively, hydroxyapatite chromatography can be used to separate the duplexes from single strands of DNA and RNA.

E. Complementary RNA–DNA Cytohybridization *in Situ*

In the variety of cRNA–DNA hybridization known as cytohybridization *in situ*, hybridization of the cRNA is carried out in fixed cells rather than to extracted cellular DNA (Gall and Pardue, 1971; Jones and Corneo, 1971; Huang *et al.*, 1973; Pagano and Huang, 1974). The great advantage of this technique is its ability to localize virus-specific DNA according to cell type or intracellular location by autoradiography and observation with light microscopy. Essentially similar techniques have been used to localize virus-specific RNA with radiolabeled DNA probes. Potentially it should be possible to extend these methods by the use of high resolution techniques that have been perfected for direct autoradiographic approaches.

General Description of Method. Cells in suspension, exfoliated cells, or tissue sections containing virus-specific DNA sequences are fixed on a slide or cover slip and exposed under controlled conditions to alkali or heat so that the cellular DNA is partially denatured *in situ*. The crucial aspect of this step is avoidance of excessive damage to the cell structure. It is then possible to use either tritiated or iodinated (Shaw *et al.*, 1975) cRNA hybridization directly to the cell preparation or tissue section; this is done under standard buffer conditions for essentially the same period of time used to conduct hybridization on membrane filters. The steps that follow include repeated rinsing and treatment with RNase to remove the nonspecifically adherent cRNA. Autoradiography follows.

Cytohybridization is not as sensitive as cRNA–DNA hybridization on membrane filters in which the DNA from many cells is pooled for hybridization. For example, with EBV DNA sequences in the nonvirus-producing Raji lymphoblastoid cell line, the 60 genome equivalents per cell that these cells contain is well above the limit of detectability by hybridization on membrane filters, i.e., two genome equivalents per cell. By cytohybridization with the same cRNA probes, however, only about 10–15 grains can be seen after a 4-week exposure time (Pagano and Huang, 1974). The number of grains is slightly higher than the number found in the background. If, however, most of the viral genetic material were confined to a few cells, then such cells would stand out, and the test would become relatively sensitive. It is not quite a quantitative test for determination of the amount of viral genome present, but it is quantitative in the sense that the percentage of cells bearing homologous DNA can be determined. As indicated before, the test has the special advantage of permitting localization of the viral DNA, both with respect to cell type and subcellular site.

This procedure has been used with considerable success in studies of infection with SV40, EBV, human CMV, and adenoviruses. Recent methods that employ iodinated cRNA (Shaw et al., 1975) and scintillation fluor (Huang et al., 1976b, and described below) greatly reduce the exposure time required for in situ cytohybridization. Although with ^{125}I the grain size on the autoradiograms is larger, the results by light microscopy are still acceptable.

This procedure for cytohybridization in situ is adapted from that of Gall and Pardue (1971) and Jones (1970) with modification (Huang et al., 1973).

Cells or nuclei are pelleted by centrifugation at 1000 rpm (about 100 g) for 10 minutes and resuspended in hypotonic solution (0.1 × SSC or Hanks' solution) for 20 minutes at 37 °C; this step is omitted for examination of cytoplasmic viral DNA or RNA. The pellet is resuspended and pelleted at the same centrifugal force and then fixed with freshly prepared ice-chilled ethanol (3 parts) and acetic acid (1 part) for 15 minutes. The cells and nuclei are spun off and washed once with the same fixative. The cells are gently dispersed in fixative with 0.2 ml of the fixative left after centrifugation at low speed. The cell suspension is then spread onto a clean precooled slide and dried in air or quickly in a flame.

In tissue obtained at autopsy or biopsy, tissue blocks are embedded in Ames O.C.T. compound and sliced into sections 0.6–10 μm in thickness and then applied to the slide and fixed with freshly prepared ice-chilled fixative for 15 minutes. After fixation, the slides are dipped in 90% alcohol and absolute alcohol to remove the residual acetic acid, and then dipped into 0.4% agarose at 60 °C and air dried to form a thin agarose layer on the slide to prevent detachment of cells during alkalization and hybridization.

The denaturation of DNA is carried out by alkalinizing the specimens in 0.07 N NaOH for 3 minutes; an alternate method of treatment with 0.2 N HCl for 20 minutes has been used by McDougall et al. (1972). The slides are then washed extensively with 70% alcohol and absolute alcohol and air dried.

One-tenth of a milliliter of $[^3H]$ (or ^{125}I) cRNA (3 × 10^5 cpm/0.1 ml with 1 mg yeast RNA, 0.1% SDS, and 6 × SSC) is applied to each slide and covered with a cover slip to prevent evaporation. The specific activity of the $[^3H]$RNA is about 1 × 10^7 cpm/μg and of the $[^{125}I]$RNA about 4 × 10^7 cpm/μg (Shaw et al., 1975). The hybridization is carried out in a moist chamber at 66 °C for 20 hours. The slides are rinsed with 2 × SSC four times and treated with 40 μg/ml pancreatic RNase for 30 minutes at 37 °C. After

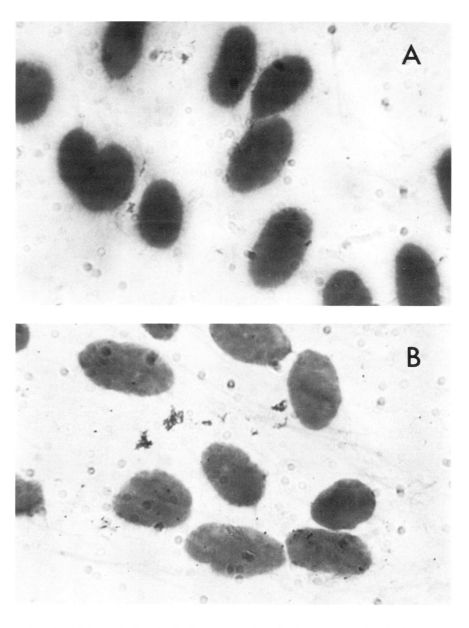

FIG. 3. Cytohybridization *in situ* for the detection of cytomegalovirus DNA: time course of appearance of viral DNA in infected cells. WI-38 cells on cover slips were infected at a high multiplicity (1–2 PFU/cell). The nuclei were released by hypotonic treatment and fixed. The DNA was denatured with 0.07 N NaOH for 3 minutes. After washing and fixation cRNA–DNA hybridization was carried out with an input of 5×10^5 cpm CMV cRNA in 0.1 ml per cover

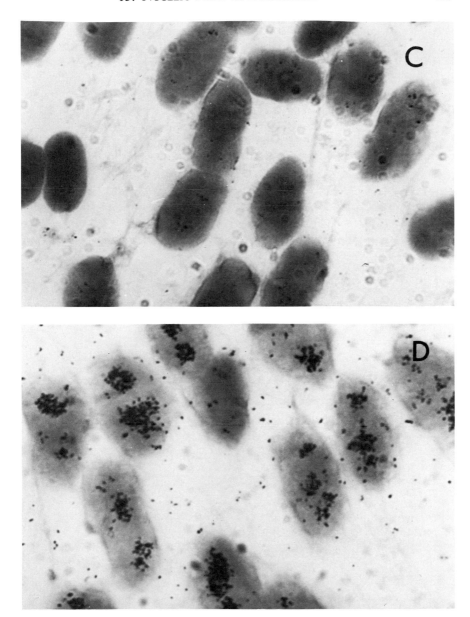

slip. After 22 hours of hybridization at 66°C the preparations were rinsed and treated with RNase and then extensively washed with 2 × SSC; autoradiography was carried out with NTB 2 emulsion. (A) Uninfected control cells. (B) Seven hours after infection. (C) Twenty-four hours after infection. (D) Thirty-two hours after infection. (From Pagano and Huang, 1974.)

extensive washing again with 2 × SSC, the slides are sequentially dehydrated with 70%, 95%, and absolute alcohol and dried.

Kodak Nuclear Track NTB 2 is used for autoradiography. The exposure time for tritiated labeled material ranges from 4 weeks to 2 months, but for ^{125}I material the time is overnight to one week; the time depends mainly on the specific radioactivity of viral cRNA and also on the viral DNA content of the cells. Recently we have been able to shorten the exposure time of ^{3}H-labeled material from 4 weeks to 1 or 2 days by dipping the emulsion-covered slide (after complete drying) in liquid scintillation fluid— dioxane and PPO (1% w/v) and POPOP (0.02%). The slides are developed in Kodak D-19 developer for 3 minutes, gently rinsed with water for 30 seconds, and then fixed in Kodak rapid fixer for 2–3 minutes. After rinsing in water and drying in air, the slides are stained with Giemsa for 30 minutes.

Figure 3 shows an example of *in situ* RNA–DNA cytohybridization used to observe the time course of viral DNA synthesis in human CMV-infected cells. Under the conditions of the experiment, viral DNA synthesis commenced in these cells 20–24 hours after infection and reached its peak 70 hours after infection (Huang *et al.*, 1973). The first indication of hybridization, found 7 hours after infection, is almost certainly the input virus. However, within 24 hours after infection, replicated CMV DNA is found in two acrocentric areas rather than diffusely scattered through the nuclei. Only nuclei are shown since hypotonic treatment was used, but if preparations of whole cells were used, then it would be possible to show the transition of viral genetic material from intranuclear to cytoplasmic sites.

Figure 4 shows an example of cRNA–DNA cytohybridization *in situ* with CMV[^{3}H]cRNA applied to kidney tissue (in collaboration with Dr. G. Nankervis). The tissue was from autopsied kidney of a congenitally infected infant with CMV infection. This photograph does not represent the incorporation of [^{3}H]thymidine in the DNA, but rather the hybridization of CMV[^{3}H]cRNA to the viral DNA in the cells. The grains which deposit heavily in cuboidal epithelial cells in the collection tubules represent the localization of viral DNA. As shown before (Huang *et al.*, 1976b), CMV viral structural antigen was also detected in these cells by the ACIF test.

The application of cytohybridization techniques to EBV and SV40 systems is shown in Figs. 5 and 6.

Cytohybridization is also potentially useful for the detection and localization of virus-specific mRNA and has been used for this purpose by McDougall *et al.* (1972) and Dunn *et al.* (1973). For the detection of virus-specific mRNA, it would be necessary to use radioactively labeled pure viral DNA. There are two problems: (1) Viral DNA of high specific activity is required; viral DNA of sufficiently high specific radioactivity is now available from *in vitro* labeling techniques (either tritiated or [^{125}I]DNA). (2) An enzymic

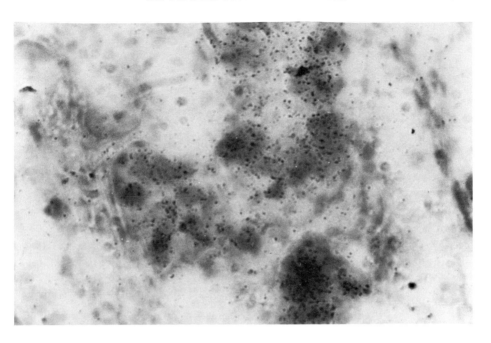

FIG. 4. Complementary RNA–DNA cytohybridization *in situ* with CMV[³H] cRNA applied to kidney. A tissue block from the kidney of a congenitally CMV-infected infant was embedded in Ames O.C.T. compound. The tissue was sliced into sections 6 μm thick, transferred to slides, and fixed with freshly prepared fixative. After dehydration the section was exposed to 0.07 *N* NaOH for 3 minutes and dehydrated with 70% and 95% ethyl alcohol. The cytohybridization was carried out as described in the text. (From Huang *et al.*, 1976b.)

means of hydrolyzing residual labeled DNA, equivalent to RNase treatment, without destruction of duplex molecules is necessary. For this purpose single-strand-specific nuclease, such as S1 enzyme from *Aspergillus oryzae*, or the nuclease with similar properties from *Neurospora crassa* have proved to be suitable.

F. DNA–DNA RENATURATION KINETICS ANALYSIS*

1. *Practical Theory*

In aqueous solution single-stranded complementary segments of DNA will reassociate and form double-stranded hydrogen-bonded structures (Britten *et al.*, 1974; Britten and Kohne, 1968). Reassociation is generally

* Modified from Shaw and Pagano (1976).

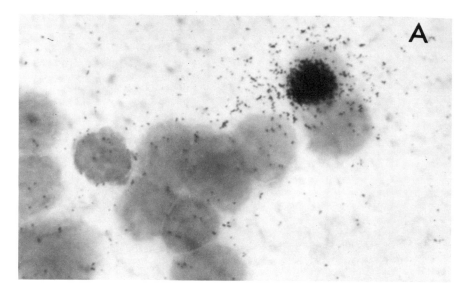

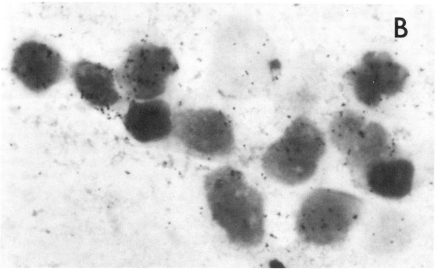

FIG. 5. Cytohybridization with EBV cRNA to the virus-producing (HR-1) line and a non-virus-producing line (Raji) or Burkitt's lymphoma cells. The HR-1 or Raji cells were washed and treated with 0.1 × SSC for 20 minutes at 37 °C. The cells were then fixed and applied to the slide for cytohybridization. The amount of EBV [^{3}H] cRNA applied to each slide was 5×10^5 cpm in 0.1 ml 6 × SSC with 1 mg of yeast RNA and 0.1% SDS. (A) The virus-producing cell line HR-1 (exposure time 2 weeks). A heavy deposition of grains was concentrated in the virus-producing cells. This indicates the heterogeneity of HR-1 cells. (B) The nonvirus-producing cell line Raji (exposure time 4 weeks). A scattering of grains higher than background was found in the nuclei of Raji cells; Raji cells contain 60 EBV genomes per cell, located in the chromosomes, and distributed uniformly among the cells. (From Pagano and Huang, 1974.)

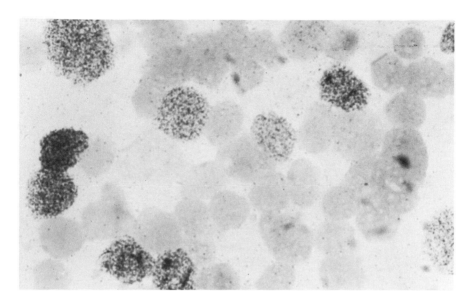

FIG. 6. SV40-transformed WI-38 human fibroblasts—WI-38Val3. The cells grown on a cover slip were hypotonically treated and fixed with fresh fixative as described in the text. SV40 cRNA of specific activity 1×10^7 cpm/μg was used for cytohybridization, and the exposure time was 3 weeks. There is a striking difference in the distribution of the hybridizable SV40 DNA in the nuclei. WI-38 cells are partially permissive for the replication of SV40. The phenomenon of active SV40 DNA replication in these transformed cells may be similar to that for EBV in the HR-1 cell system ($\times$1000). (From Pagano and Huang, 1974.)

carried out in dilute solution so that it is possible to follow the kinetics which are typically those of a second-order rate reaction. The kinetics are significantly influenced by temperature and ionic milieu. The rate-limiting step is an in-register collision at one or more sites along pairs of single-stranded complementary segments of DNA (Wetmur and Davidson, 1968). The equation for an ideal second-order reaction is $C/Co = 1/(1 + KCot)$, where C is the concentration of a single-stranded DNA, Co is the total DNA concentration, K is the reassociation rate constant, and t is time (Britten et al., 1974). The units for C and Co are moles of nucleotides per liter; t is in seconds. The reassociation rate constant K depends on the incubation conditions and complexity of the DNA, and it is the reciprocal of the $Cot_{1/2}$ value, defined later. The rate of reassociation depends on DNA complexity and concentration, length of the single-stranded segments, viscosity, temperature, and salt concentration. When these variables are standardized, the reaction rate becomes directly dependent on DNA concentration.

Since all of the DNA in the reaction mixture is initially denatured and, therefore, present as single strands at the beginning of the reaction, the extent of reassociation is determined at any time during incubation by measurement of the amount of double-stranded DNA formed. This determination requires discrimination of single- and double-stranded DNA. Hydroxyapatite is used extensively for this purpose because of its capacity to retain double-stranded DNA relatively exclusively at low (0.14 M potassium phosphate), but not at high (0.4 M) salt concentrations (Kohne and Britten, 1971). Another method for the discrimination of double-and single-stranded DNA is by digestion of single-stranded DNA remaining in the reaction mixture by a single-strand specific nuclease, S1 enzyme (Ando, 1966), and nuclease from *Neurospora crassa*. Ideally, digestion with such nucleases leaves only double-stranded DNA precipitable from the reaction mixture. If the DNA has been labeled with a radioactive precursor, then the amount of duplex DNA formed can be inferred by scintillation spectrometry.

The results of a reassociation experiment may be expressed in terms of Co/C versus time if a linear representation of the data is desirable, or as the percentage of double-stranded DNA formed versus time of incubation. In the latter case, it is convenient to plot the time scale as the product (Cot) of total nucleic acid concentration, Co (moles of nucleotides per liter), and time t (seconds). Plots of data in terms of Cot are useful for estimating the completion of the reaction and for comparing DNAs from various sources. When different DNAs are compared, it is usually at their $Cot_{1/2}$ value, the value obtained when 50% of the single-stranded radioactive probe DNA has reassociated.

2. *Factors Influencing the Rate of Reassociation*

Several such factors have been described in detail by Wetmur and Davidson (1968). The main points can be summarized as follows:

(a) Effect of temperature on the rate of reassociation. The rate of reassociation has a bell-shaped dependence on temperature with a region of maximum reassociation around $T_m - 25$ °C. The reassociation rate reaches a broad plateau from 15 °C to 30 °C below T_m; it drops significantly when the reaction temperature is lower than 30 °C below T_m or higher than 15 °C below T_m.

(b) Effect of DNA complexity on the rate of reassociation. The rate of reassociation is inversely proportional to the complexity or number of base pairs (or molecular weight) in the nonrepeating DNA complement of the viral or cellular genome. The DNA should be sheared to a homogeneous size, 300–400 nucleotide pairs.

(c) Effect of pH. Within the pH range of 5–9, the rate of reassociation in 0.4 M sodium ion is essentially independent of pH. There is a remarkable decrease in the rate of reassociation when the pH is above this range.

(d) Effect of ionic strength and viscosity. The rate of reassociation is dependent on the ionic strength below 0.4 M Na$^+$; there is almost a 7-fold increase in the rate of reassociation when 0.4 M Na$^+$ is used as compared with 0.15 M. The rate is almost independent of salt concentration when ionic strength is above 0.4 M. The differences in the rate of reassociation at 0.4 M and 1.0 M is 2-fold.

Since the rate of reassociation is proportional to the frequency of mechanical collision of the two complementary DNA segments, the rate of reassociation decreases with increasing solvent viscosity. Because of this, the viscosity should be properly controlled.

3. Uses and Procedure

Reassociation curves can establish whether DNA contains repeated sequences (a rapidly reassociating fraction), as is characteristic of eukaryotic DNA, or whether it is unique DNA (DNA for which repeated sequences are not recognized). One of the most useful applications of reassociation experiments has been in the area of viral oncology where it has been possible to detect small numbers of copies of viral genomes in the DNA isolated from cells suspected of carrying viral information, but in which biologically active virus is not demonstrable. It is possible to detect as little as 0.1–0.2 viral genome per cell in some instances. The sensitivity of the assay depends on the use of viral probes which have been labeled to high specific radioactivity.

In a typical reassociation experiment, purified viral DNA, labeled during virus replication or by *in vitro* techniques, is mixed with a large excess of unlabeled test (cellular) DNA; DNA with no detectable sequence homology to the labeled probe would be selected as a substitute for the test DNA in a control reaction mixture. The DNAs are sheared to uniform size, denatured, and then the mixtures are adjusted to the proper incubation conditions. At various times during incubation aliquots are removed from the mixtures to determine the amount of double-stranded DNA formed. At the end of the incubation period, the reassociation rates are compared. If the test DNA lacks viral DNA sequences, then the reassociation rates of the test and control reactions would be the same. The rates would be the same because the concentration of viral DNA in each mixture is identical. If unlabeled viral DNA sequences are present in the test reaction, then the total viral DNA concentration would be greater than that of the control reaction. This would increase the rate of reassociation of the labeled DNA in the test reaction over the labeled DNA in the control reaction and would indicate that viral DNA sequences are present in the test DNA.

Outlined below is a technique which utilizes DNA–DNA reassociation kinetics to detect the presence of EBV DNA in the DNA of two B-lymphoblastoid cell lines; Raji, a nonvirus-producing cell line containing 50–60 copies of viral DNA per cell, and P3HR-1, an EBV-productive cell line harboring 800–1000 copies of viral DNA per cell.

Stock solutions of Raji, P3HRl, calf thymus, and ³H-labeled EBV DNA free of protein and RNA are sonicated. The DNA is concentrated to 10 mg/ml (Raji, P3HRl, and calf thymus DNA) or 2×10^{-2} μg/50 μl (EBV DNA) in neutralized 2.5 mM EDTA, and dialyzed extensively against the same solution. The solutions are stored at $-20\,°C$ until used. To demonstrate the effect of DNA concentration on the rate of reassociation, four reassociation mixtures are assembled at room temperature, as shown in Table II: a control containing calf thymus DNA, two concentrations of Raji-cell DNA, and one concentration of P3HR1 DNA. Because viscosity affects the rate of reassociation, all mixtures are adjusted to the same final DNA concentration by adding sonicated calf thymus DNA.

All reaction components are added except NaCl. The mixtures are heated for 10 minutes in a boiling water bath and quick-chilled in ice water. NaCl is then added, and each mixture is divided into 9 or 10 fractions by withdrawal of 0.1 ml of the mixture into the center of a 100 μl pipette. The pipettes are sealed by flame and are placed in a water-filled tube which is immersed in a $67°–70\,°C$ water bath. At various times during incubation, fractions are removed and frozen at $-20\,°C$.

At the end of the incubation period, each sample is added to 1.9 ml of S1 digestion buffer; then each mixture is split into two 1-ml fractions, one to be digested with S1-nuclease, the other to serve as a control. Five to ten microliters of S1-nuclease are added to one tube of each set. After 2–3 hours at 40$°C$, each sample is chilled to 0$°C$ and mixed with 0.2 ml of ice cold 100% TCA to precipitate undigested DNA. The precipitates are collected on Millipore filters by suction and washed twice with cold 5% TCA.

TABLE II

REASSOCIATION KINETICS ANALYSIS OF EBV DNA[a,b]

	[³H]EBV DNA	Calf thymus DNA	Raji DNA	P3HRl DNA	Reassociation buffer	H₂O	6 M NaCl[c]
Control	0.05	0.20	—	—	0.1	0.40	0.25
Raji	0.05	0.16	0.04	—	0.1	0.40	0.25
Raji	0.05	0.12	0.08	—	0.1	0.40	0.25
P3HRl	0.05	—	—	0.20	0.1	0.40	0.25

[a] From Shaw and Pagano (1976).
[b] Figures in milliliters.
[c] Added after heat denaturation.

The filters are dried at room temperature and counted in a toluene-based scintillation fluid.

To determine the fraction of double-stranded DNA present in each 100 μl sample, the ratio of cpm of the digested and undigested portions is determined. The value is expressed as a percentage. The results of reassociation of ³H-labeled EBV DNA are shown in Fig. 7 and are plotted as the percentage of renatured DNA versus *Cot*. The *Cot* values are in terms of [³H]EBV DNA. The time of incubation is also shown. It is evident from the *Cot*$_{1/2}$ value of each curve that the reassociation rate increases as the concentration of lymphocyte DNA increases. The increase in reassociation rate is due to the presence of viral DNA sequences in the lymphocyte DNA preparations.

4. *Problems*

To avoid some of the pitfalls which accompany reassociation experiments, several precautions can be taken (see Britten *et al.*, 1974). If it is

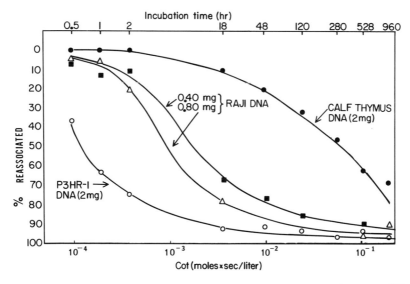

FIG. 7. Reassociation of ³H-labeled EBV DNA. Each reaction contains per milliliter: 0.02 μg [³H]EBV DNA, 2 mg calf thymus DNA (●——●); 0.02 μg [³H]EBV DNA, 0.8 mg Raji DNA, 1.6 mg calf thymus DNA (■——■); 0.02 μg [³H]EBV DNA, 0.8 mg Raji DNA, 1.2 mg calf thymus DNA (△——△); 0.02 μg[³H]EBV DNA, 2 mg P3HR1 DNA (○——○). ³H-labeled EBV DNA is 2 × 10⁶ cpm/μg. The DNA in each reaction was denatured and incubated at 70°C in buffered 1.5 *M* NaCl. At the time indicated 0.1 ml fractions were removed and the amount of double-stranded DNA in each was determined. The percentage of single-stranded DNA which reassociated is plotted versus *Cot* and incubation time. (From Shaw and Pagano, 1976.)

possible to prepare labeled DNA *in vivo* with a high specific activity, then many of the problems associated with *in vitro* labeling of DNA can be avoided. Probes prepared *in vitro* should be characterized before use by hydroxyapatite chromatography and by a single-strand-specific nuclease. Each new batch of hydroxyapatite and each new preparation of single-strand-specific nuclease should be tested with a DNA whose structure has been well established. DNA fragments of different lengths reassociate at different rates. Therefore, in order to obtain controlled rates of reassociation, it is important to control the shearing procedure. This can be accomplished by routine sizing of the sheared DNA fragments. Sizing is also recommended if overdigestion with DNase is suspected.

Although the chelating agents help to protect DNA during reassociation, their presence can interfere with the separation of single- and double-stranded DNA on hydroxyapatite. If S1-nuclease is used instead of hydroxyapatite, the concentration of $ZnCl_2$ should be in excess of the chelator, since Zn^{2+} is necessary for S1 activity. The presence of sodium phosphate in concentrations as low as 10 mM can inhibit S1 activity.

Because of the possibility of concentrating salts with DNA during ethanol precipitation, stock solutions of DNA should be dialyzed against buffer before heat denaturation. This ensures that the temperature used for denaturation is well above the melting point of the DNA.

G. Partial Homology: Detection by DNA–DNA Renaturation Kinetics

In addition to its utility as a sensitive and quantitative assay for amounts of homologous DNA, renaturation kinetics analyses are also of value for the determination of degree of homology between related but not identical genomes. With present technology, this technique is applicable if there is at least 10% homology between the genomes being compared. For lesser degrees of homology, heteroduplex formation provides a more sensitive method; this technique can be used to detect as little as 5% homology.

Examples of reciprocal analyses of radiolabeled EBV and human CMV genomes are shown in Figs. 8 and 9. These results indicate a lack of detectable homology between the AD169 strain of HCMV and the P3HR1 strain of EBV. Had there been some degree of homology between the unlabeled DNA under test and the labeled probe DNA, an initial acceleration of renaturation of the index DNA would have been produced. Since the variation around the control points is not large in these tests, it is clear that in the neighborhood of 5–10% homology in the added unlabeled test DNA would produce a perceptible acceleration of renaturation of the labeled DNA.

Epstein–Barr virus DNA sequences are found in cell lines derived from

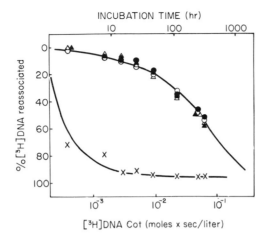

FIG. 8. Reassociation kinetics analysis of tritiated human cytomegalovirus DNA with the DNA of HSV 1, HSV 2, and EBV. Sonically disrupted ³H-labeled CMV DNA (strain AD-169), 0.02 μg (3.4 × 10⁴ cpm), 20 μg of calf thymus DNA, and 4 μg of unlabeled viral DNA were mixed and denatured in the presence of 0.01 M Tris-HCl (pH 7.4) and 0.0025 M EDTA. The salt concentration was then adjusted to 1.2 N NaCl. The hybridization was carried out at 66 °C, and the fraction of reassociated [³H]DNA was analyzed by Sl enzyme differential digestion as described above. Calf thymus control DNA, 4 μg (○); HSV 1 DNA (△); HSV 2 DNA (▲); EBV DNA (●); and AD-169 DNA (×). (From Huang and Pagano, 1974.)

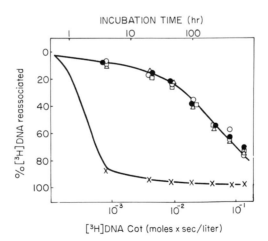

FIG. 9. Reassociation kinetics and analysis of tritiated EBV DNA with the DNA of HSV 1, HSV 2 and CMV DNA. Sonically disrupted ³H-labeled EBV DNA, 0.02 μg (5.4 × 10⁴ cpm), 20 μg of calf thymus DNA and 3 μg of unlabeled viral DNA were mixed. The experiment was carried out as described in Fig. 8. Calf thymus DNA (○); HSV 1 DNA (△); HSV 2 DNA (●); AD-169 DNA (□); and EBV DNA (×). (From Huang and Pagano, 1974.)

patients with infectious mononucleosis (IM). The question of whether the EBV genomic material found in Burkitt's lymphoma and in infectious mono-nucleosis are in fact identical has been a persistent one. Results shown in Fig. 10 indicate clearly that 25–35% of the DNA sequences found in the P3HRl-derived EBV cannot be detected in the four cell lines derived from two patients with infectious mononucleosis. These results do not distinguish between the presence of an extensively deleted EBV genome in the IM cell line as against an "IM strain" of EBV with up to 35% heterologous DNA sequences. In order to discriminate between these possibilities, it would be necessary to prepare radiolabeled IM EBV DNA from purified IM virus and to conduct a reciprocal type of DNA–DNA renaturation kinetics analysis with cold P3HRl EBV DNA. At present sufficient quantities of IM EBV are not available. Another approach would be to determine the molecular weight of the IM virus DNA by sedimentation analysis or contour length measurements by electron microscopy. Ultimately it will be possible to explore homology between the genomes derived from these two sources by heteroduplex formation (see Section IV,B). A similar lack of homology rang-ing up to 40% has been found in analysis of homologous DNA sequences contained in nasopharyngeal carcinomas from Tunisia in comparison with the HRl EBV genome (Pagano *et al.*, 1975; J. S. Pagano and J.-L. Li, un-published data).

IV. Special Methods

A. Specific DNA Fragments and Blot-Transfer Hybridization

Restriction endonucleases have become powerful tools for analysis not only of smaller viral genomes but also of genomes of increasing complexity and molecular size. Cleavage of DNA into specific terminal fragments and construction of a DNA fragment map provide elements needed for detailed characterization of the viral genome, for the determination of the direction of DNA replication, and for the regulation of gene transcription. The sensitivity of detection of viral genetic material in various tumor and virus-transformed cells by DNA–DNA hybridization can be enhanced by the use of specific DNA fragments as probes. Location of viral mRNA on a genetic map can also be approached by hybridization of the mRNA, isolated from lytically infected or transformed cells, to restriction endonuclease-generated fragments.

There are several ways of recovering the specific viral DNA fragments. The DNA is first subjected to restriction enzyme digestion, and the DNA

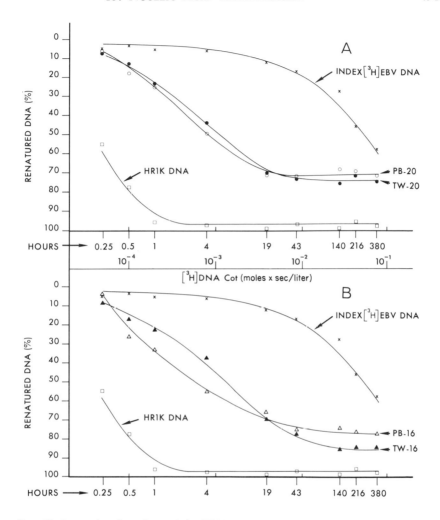

Fig. 10. Incomplete homology of the EBV genome of the viral DNA sequences in lymphocyte lines from infectious mononucleosis. The determinations were by DNA–DNA renaturation kinetics analyses. (A) [³H]EBV DNA (0.02 µg) + calf thymus DNA (2 mg); PB20 DNA (1.1 mg) is from a cell line derived from peripheral blood; TW20 DNA (1.8 mg) is a cord-lymphocyte line transformed by throat washings from the same patient; HRIK DNA (2 mg) is homologous DNA from the EBV-producing cell line that was the source of the index EBV DNA. (B) PB16 DNA (1.5 mg) is from a cell line established from the peripheral blood of another patient with infectious mononucleosis; TW16 DNA (1.5 mg) was from a cell line produced by exposure to throat washings from the same patient.

Radiolabeled EBV DNA (0.02 µg, specific activity of 1.8×10^6 cpm/µg) is present in all renaturation mixtures. The total concentration of DNA in all mixtures was brought to 2 mg by the addition of calf thymus DNA. (From Pagano *et al.*, 1976.)

fragments are then electrophoresed to separate them according to size. The DNA fragments can be located in agarose gels by staining with 0.5 μg/ml of ethidium bromide and observation under UV light or by ^{32}P autoradiography (Fig. 11). The DNA fragments can be recovered from the gel either by electrophoretic elution or by potassium iodide solubilization. The solubilization method of Hayward et al. (1975) is briefly described as follows. The agarose-gel blocks are dissolved by shaking 5 volumes of saturated potassium iodide solution in TE buffer (0.01 M Tris-HCl, pH 8.5; 0.001 M EDTA) at 37 °C for 20 minutes. DNA-grade hydroxyapatite is added with shaking to the mixture to absorb the DNA. The hydroxyapatite granules which absorb the DNA fragments are then collected by centrifugation. The pelleted granules are washed once with 0.1 M potassium phosphate buffer, pH 6.8. Double-stranded DNA is then eluted by 0.4 M potassium phosphate buffer, pH 6.8. The eluant containing the DNA fragments is then phenol-extracted once. The EtBr which is intercalated in the double-stranded DNA is removed by hydrogen Form Dowex 50 (Ag 50W-X8, 200–400 mesh, Bio-Rad) in high salt buffer (1 M NaCl, 0.1 M Tris-HCl, pH 8, and 0.01 M EDTA). The DNA is then dialyzed against H$_2$O and precipitated with alcohol.

Hybridization of viral mRNA to viral DNA fragments has been successfully applied to adenovirus and herpesvirus systems (for reference, see Flint et al., 1976a,b). Assignment of viral mRNA, purified from transformed cells or from different stages of infected cells, to viral DNA fragments, and assignment of isolated specific RNA, e.g., tRNA or 5 S RNA, to a set of restriction endonuclease-generated fragments can be approached by blot transfer and hybridization. This transfer technique is also called Southern's technique (1974). The DNA fragments are denatured in situ in the agarose-slab gel by immersing the gel in 0.5 N NaOH for 20 minutes. After denaturation, the solution is neutralized with 1.1 N HCl in 0.2 M Tris. The gel is then immediately transferred onto 4–5 layers of Whatman #1 filter papers immersed in or in contact with 6 × SSC. A sheet of nitrocellulose membrane paper (Millipore) is laid on the gel. Paper towels or Whatman filter papers are put on top of the nitrocellulose filter sheet to absorb the liquid from the gel and the 6 × SSC in which the filter papers under the gel are immersed. The DNA fragments are then transferred to and immobilized on the nitrocellulose by blotting and by capillary diffusion of 6 × SSC. The membrane filter with the set of restriction endonuclease-generated fragments is then dried at room temperature for 3 hours and baked at 80°C in a vacuum oven for 3 hours. The membrane filter is then cut into long strips for hybridization against radioactive mRNA or cRNA; each strip should contain a whole set of DNA fragments. The hybridization is carried out as with RNA–DNA hybridization on membrane filters (Section III,C). The DNA fragments from which mRNA is derived can be detected by autoradiography.

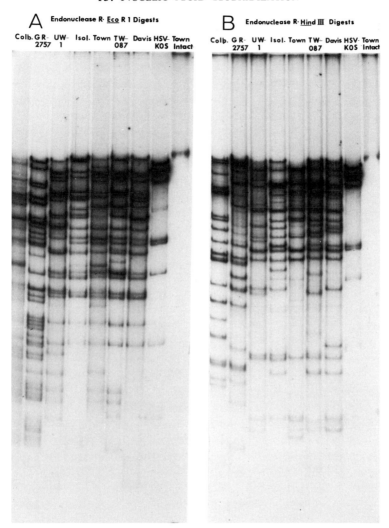

FIG. 11. Comparison and analysis of various CMV strains by cleavage of their DNA with restriction endonucleases EcoR1 and Hind. III. CMV DNA labeled with ^{32}P and purified as described in the text was dissolved in TBS (0.05 M Tris-HCl, pH 7.4, and 0.15 M NaCl) and digested with either enzyme in the presence of 10 mM MgCl$_2$ and 5 mM β-mercaptoethanol for 24 hours at 37°C. All samples were electrophoresed on a 1% agarose-slab gel in 1 E buffer until the tracking dye (bromphenol blue) migrated to the bottom of the gel. The gels were dried and exposed to x-ray film as described (Huang *et al.*, 1976b). KOS, a strain of herpes simplex virus, was used as a marker and a control for digestion conditions. (A) EcoR1 digests. (B) Hind III digests. KOS in both panels was digested with EcoR1 enzyme. (From Huang *et al.*, 1976b.)

B. Electron Microscopy: Heteroduplex Formation and R-Loop Formation

Heteroduplex formation provides a powerful direct method to localize specific gene sequences in genomes of high complexity and to compare the degree of homogeneity between two gene sets. The general principle, considered in detail by Davis and Davidson (1968) and Davis *et al.* (1971), is that after denaturation two partially complementary or partially noncomplementary strands are allowed to renature under optimal conditions, and the resulting products are observed by electron microscopy for sequence homology. By differentiating single- and double-stranded DNA with electron microscopy in suitable preparations, the regions of homology and nonhomology can be located and mapped. The technique has been widely used in gene localization, for example, mapping the position and size of bacteriophage deletion and insertion mutations (Westmoreland *et al.*, 1969) and localizing SV40 integration sites in adenovirus 7–SV40 hybrid DNA molecules.

The procedures and methods for heteroduplex formation and mounting DNA for electron microscopy have been described in publications of Davis *et al.* (1971). In this technique, summarized from Davis *et al.* (1971), the DNA can be either purified or directly prepared from virus or phage particles by heat or chelation shocking of the phage, or by treatment with urea or $NaClO_4$. The phage protein present does not significantly interfere in the reaction and the electron microscopy (Davis *et al.*, 1971). DNA from two different phages or viruses are denatured by alkali (in 0.5 ml of 0.1 M NaOH, 0.02 M EDTA) for 10 minutes at room temperature and then neutralized by adding one-tenth volume of 1.8 M Tris-HCl, 0.2 M Tris-OH. Formamide (99%) in a volume of 0.5 ml is added to the neutralized DNA solution. The final pH of the solutions should be around 7.5–8.5. Under these conditions, 50% of the DNA renatures in 1–2 hours. After reassociation, the reaction mixture is cooled to 0 °C to stop the reassociation and dialyzed against 0.01 M Tris-HCl, pH 8.5, and 1 mM EDTA at 4 °C for mounting. Two mounting techniques—the aqueous technique with ammonium acetate and the formamide technique—were described in the original work of Davis *et al.* (1971).

R-loop formation is a technique that makes use of the principle of heteroduplex formation, but with RNA instead of with DNA as in D-loop. RNA is hybridized to double-stranded DNA in the presence of formamide to displace the segment of identical DNA sequence in a genome of high complexity. The resulting triplex structure of an RNA–DNA duplex with single-stranded DNA is called the R-loop; the techniques have been described in detail recently by Thomas *et al.* (1976) and White and Hogness (1977) in a *Drosophila* rRNA (rDNA) system.

R-loop technology can be used to localize genes for early functions in the intact viral genome by hybridizing early mRNA (for example, SV40 early mRNA) to the linear viral genome (in this case EcoRl-digested SV40 DNA). Integrated viral gene sequences can be localized by hybridizing the virus-specific mRNA isolated from virus-transformed cells to the intact linear viral DNA molecule. DNA fragments generated by restriction enzymes also can be assigned locations on the intact genome with the use of cRNA synthesized from fragment templates and R-loop mapping techniques.

The method and buffer used for R-loop formation by Thomas *et al.* (1976) are as follows: 10 μl of a solution containing approximately 0.3 μg each of RNA and DNA, 0.1 M NaCl, and 0.05 Tris-HCl, pH 7.5, were added to 50 μl of buffered formamide solution. The buffered formamide was prepared by mixing 0.42 ml of formamide, 50 μl of 1 M pipes [piperazine-N,N'-bis(2-ethanesulfonic acid) $N_{1.4}$], pH 7.8, 12 μl of 0.5 M Na$_3$ EDTA, and 18 μl of H$_2$O. The final formamide and cation concentration were 70% and 0.17 M, respectively. The mixture was covered with paraffin oil, sealed in a siliconized glass tube, and heated at 47°C for 20 hours. The sample was then mounted for electron microscopy by the formamide technique of Davis *et al.* (1971).

The rate of R-loop formation, as reported by Thomas *et al.* (1976), reaches its maximum at the temperature at which half of the duplex DNA is irreversibly converted to single-stranded DNA and falls precipitously a few degrees above or below that temperature. Once the R-loop forms, it displays considerable stability; the formamide can be removed, and the DNA can be cleaved with restriction enzyme without loss of the structure. The R-loop is quite sensitive to RNase; incubation in 10 μg/ml of RNase (in 0.1 M NaCl, 0.5 M Tris-HCl, pH 7.5, 0.1 M EDTA) for 5 minutes at 37°C leads to complete loss of R-loops. The loop is also sensitive to alkaline hydrolysis in 0.2 N NaOH at 37°C for 10 minutes.

Acknowledgments

We would like to thank Drs. J. Shaw, J. Estes, and B. Kilpatrick for their valuable discussion, Shu-Mei Huong Chien-Hui Huang, Carolyn Smith, and Mohmed Fatteh for their technical assistance, and Catherine Reinhardt for helping with manuscript preparation. This work was supported by the National Institute of Allergy and Infectious Disease (1-ROl-AI12717), the National Cancer Institute (1-POl-CA19014 and NOl-CP33336), and the National Heart and Lung Institute (NHLI-72-2911-NIH).

References

Adesnik, M., Salditt, M., Thomas, W., and Darnell, J. E. (1972). *J. Mol. Biol.* **71**, 21–30.
Ando, T. (1966). *Biochim. Biophys. Acta* **144**, 158–168.
Aposhian, H. V., and Kornberg, A. (1962). *J. Biol. Chem.* **237**, 519–525.

Aviv, H., and Leder, P. (1972). *Proc. Natl. Acad. Sci. U.S.A.* **69**, 1408–1412.

Borenfreund, E., Rosenkruz, H. S., and Bendich, A. (1959). *J. Mol. Biol.* **1**, 195–203.

Brawerman, G. (1974). *In* "Methods in Enzymology" (L. Grossman and K. Moldave, eds.), Vol. 30, Part F, pp. 605–612. Academic Press, New York.

Brawerman, G., Gold, L., and Eisenstadt, J. (1963). *Proc. Natl. Acad. Sci. U.S.A.* **50**, 630–638.

Brawerman, G., Mendecki, J., and Lee, S. Y. (1972). *Biochemistry* **11**, 637–641.

Britten, R. J., and Kohne, D. E. (1968). *Science* **161**, 529–540.

Britten, R. J., Graham, D. E., and Neufeld, B. R. (1974). *In* "Methods in Enzymology" (L. Grossman and K. Moldave, eds.), Vol. 29, Part E, pp. 363–418. Academic Press, New York.

Burgess, R. R. (1969). *J. Biol. Chem.* **244**, 6160–6170.

Commerford, S. L. (1971). *Biochemistry* **10**, 1993–2000.

Davis, D. B., and Kingsbury, D. T. (1976). *J. Virol.* **17**, 788–793.

Davis, R. W., and Davidson, N. (1968). *Proc. Natl. Acad. Sci. U.S.A.* **60**, 243–250.

Davis, R. W., Simon, M., and Davidson, N. (1971). *In* "Methods in Enzymology" (L. Grossman and K. Moldave, eds.), Vol. 21, Part D, pp. 413–428. Academic Press, New York.

Duesberg, P. H. (1970). *Curr. Top. Microbiol. Immunol.* **51**, 79–104.

Duesberg, P. H., and Canaani, E. (1970). *Virology* **42**, 783–788.

Duesberg, P. H., and Vogt, P. K. (1973). *Virology* **54**, 207–219.

Dunn, A. R., Gallimore, P. H., Jones, K. W., and McDougall, J. K. (1973). *Int. J. Cancer* **11**, 628–636.

Edmonds, M., and Caramela, M. G. (1969). *J. Biol. Chem.* **224**, 1314–1324.

Edmonds, M., and Kopp, D. W. (1971). *Biochem. Biophys. Res. Commun.* **41**, 1531–1537.

Flint, S. J., Gallimore, P. H., and Sharp, P. A. (1975). *J. Mol. Biol.* **94**, 47–68.

Flint, S. J., Berget, S. M., and Sharp, P. A. (1976a). *Virology* **72**, 443–455.

Flint, S. J., Sambrook, J., Williams, J. F., and Sharp, P. H. (1976b). *Virology* **72**, 456–470.

Frenkel, N., Roizman, B., Cassai, E., and Nahmias, A. (1972). *Proc. Natl. Acad. Sci. U.S.A.* **69**, 3784–3789.

Fujinaga, K., and Green, M. (1966). *Proc. Natl. Acad. Sci. U.S.A.* **55**, 1567–1574.

Gall, J. G., and Pardue, M. L. (1971). *In* "Methods in Enzymology" (L. Grossman and K. Moldave, eds.), Vol. 21, Part D, pp. 470–480. Academic Press, New York.

Garapin, A. C., Varmus, H. E., Faras, A. J., Levinson, W. E., and Bishop, J. M. (1973). *Virology* **52**, 264–274.

Gelb, L. D., Aaronson, S. A., and Martin, M. A. (1971). *Science* **172**, 1353–1355.

Georgiev, G. P., and Mantieva, V. L. (1962). *Biochim. Biophys. Acta* **61**, 153–154.

Gillespie, D., and Spiegelman, S. (1965). *J. Mol. Biol.* **12**, 829–842.

Gillespie, D., Saxinger, W. C., and Gallo, R. C. (1975). *Prog. Nucleic Acid Res. Mol. Biol.* **15**, 1–108.

Girard, M. (1967). *In* "Methods in Enzymology" (L. Grossman and K. Moldave, eds.), Vol. 12, Part A, pp. 581–588. Academic Press, New York.

Hadjivassilions, A., and Brawerman, G. (1966). *J. Mol. Biol.* **20**, 1–7.

Hayward, G. S., Jacob, R. J., Wadsworth, S. S., and Roizman, B. (1975). *Proc. Natl. Acad. Sci. U.S.A.* **72**, 4243–4247.

Huang, E.-S., and Pagano, J. S. (1974). *J. Virol.* **13**, 642–645.

Huang, E.-S., Nonoyama, M., and Pagano, J. S. (1972). *J. Virol.* **9**, 930–937.

Huang, E.-S., Chen, S.-T., and Pagano, J. S. (1973). *J. Virol.* **12**, 1473–1481.

Huang, E.-S., Huang, C.-H., Huong, S.-M., and Selgrade, M. J. (1976a). *Yale J. Biol. Med.* **49**, 93–98.

Huang, E.-S., Kilpatrick, B. A., Huang, Y.-T., and Pagano, J. S. (1976b). *Yale J. Biol. Med.* **49**, 29–43.

Jones, K. W. (1970). *Nature (London)* **225**, 912–915.

Jones, K. W., and Corneo, G. (1971). *Nature (London) New Biol.* **233**, 268–271.

Jovin, T. M., Englund, P. T., and Bertsch, L. L. (1969). *J. Biol. Chem.* **244**, 2996–3008.

Junghaus, R. P., Duesberg, P. H., and Knight, C. A. (1975). *Proc. Natl. Acad. Sci. U.S.A.* **72**, 4895–4899.

Kates, J. (1970). *Cold Spring Harbor Symp. Quant. Biol.* **35**, 743–752.

Kohne, D. E., and Britten, R. J. (1971). *Prog. Nucleic Acid Res.* **2**, 500–507.

Lee, S. Y., Mendecki, J., and Brawerman, G. (1971). *Proc. Natl. Acad. Sci. U.S.A.* **68**, 1331–1335.

Levine, A. S., Levin, M. J., Oxman, M. N., and Lewis, A. M. (1973). *J. Virol.* **11**, 672–681.

McDougall, J. K., Dunn, A. R., and Jones, K. W. (1972). *Nature (London)* **236**, 346–348.

Maio, J. J., and Schildkraut, C. L. (1967). *J. Mol. Biol.* **24**, 29–39.

Nonoyama, M., and Pagano, J. S. (1971). *Nature (London) New Biol.* **233**, 103–106.

Nonoyama, M., and Pagano, J. S. (1972). *Nature (London) New Biol.* **238**, 169–171.

Nonoyama, M., Huang, C. H., Pagano, J. S., Klein, G., and Singh, S. (1973). *Proc. Natl. Acad. Sci. U.S.A.* **70**, 3265–3268.

Nygaard, A. P., and Hale, B. D. (1964). *J. Mol. Biol.* **9**, 125–142.

Oxman, M. N., Levine, A. S., Crumpacker, C. S., Levin, M. J., Henry, P. H., and Lewis, A. M. (1971). *J. Virol.* **8**, 215–226.

Pagano, J. S., and Huang, E.-S. (1974). *In* "Viral Immunodiagnosis" (E. Kurstak and R. Morisset, eds.), pp. 279–299. Academic Press, New York.

Pagano, J. S., Huang, C.-H., Klein, G., de-Thé, G., Shanmugaratnam, K., and Yang, C. S. (1975). *In* "Oncogenesis and Herpesviruses II." (G. de-Thé, M. A. Epstein, and II. zur Hausen, eds.), pp. 179–190. IARC Sci. Publ. No. 11. International Agency for Cancer Research, Lyon, France.

Pagano, J. S., Huang, C.-H., and Huang, Y.-T. (1976). *Nature (London)* **263**, 787–789.

Phillipson, L., Wall, R., Glickman, R., and Darnell, J. E. (1971). *Proc. Natl. Acad. Sci. U.S.A.* **68**, 2806–2809.

Prensky, E., Steffensen, D. M., and Hughes, W. L. (1973). *Proc. Natl. Acad. Sci. U.S.A.* **70**, 1860–1864. .

Robinson, W. S., Pitkanen, A., and Rubin, H. (1965). *Proc. Natl. Acad. Sci. U.S.A.* **54**, 137–144.

Rokutanda, M., Rokutanda, H., Green, M., Fujinaga, K., Ray, R. K., and Gurgo, C. (1970). *Nature (London)* **227**, 1026–1028.

Scherberg, N. H. and Refetoff, S. (1974). *J. Biol. Chem.* **249**, 2143–2150.

Sharp, P. A., Gallimore, P. H., and Flint, S. J. (1974). *Cold Spring Harbor Symp. Quant. Biol.* **39**, 457–676.

Shaw, J. E., and Pagano, J. S. (1976). *In* "Manual of Techniques in Immunology and Virology for Cancer Research Workers" (J. P. Lamelin and G. Lenoir, eds.). International Agency for Cancer Research, Lyon, France.

Shaw, J. E., Huang, E.-S., and Pagano, J. S. (1975). *J. Virol.* **16**, 132–140.

Sheldon, R., Jurale, C., and Kates, J. (1972). *Proc. Natl. Acad. Sci. U.S.A.* **69**, 417–421.

Smith, D. K., Armstrong, J. L., and McCarthy, B. J. (1967). *Biochim. Biophys. Acta* **142**, 323–330.

Southern, E. M. (1974). *Anal. Biochem.* **62**, 317–318.

Temin, H. M., and Baltimore, D. (1972). *Adv. Virus Res.* **17**, 129–186.

Tereba, A., and McCarthy, B. J. (1973). *Biochemistry* **12**, 4675–4679.

Thomas, M., White, R. L., and Davis, W. R. (1976). *Proc. Natl. Acad. Sci. U.S.A.* **73**, 2294–2298.

Westmoreland, B. C., Saybalski, W., and Ris, H. (1969). *Science* **163**, 1343–1348.

Wetmur, J. G., and Davidson, N. (1968). *J. Mol. Biol.* **31**, 349–370.

White, R. L., and Hogness, D. (1977). In preparation.

zur Hausen, H., and Schulte-Holthausen, H. (1970). *Nature (London)* **227**, 245–248.

AUTHOR INDEX

Numbers in italics refer to the pages on which the complete references are listed.

SUBJECT INDEX

A

Abelson mouse leukemia virus, transformation by, 84

AC20 cells. *(See also under Agallia constricta)*
  growth of, 408–410
  indexing of *in situ,* 415
  karyotyping of, 411–413
  morphology of, 411, 412

Acarina, cells lines from, 325

Aceratagallia sanguinolenta
  cell lines from, 325, 401
    availability of, 430
    storage of, 427

Acetone, disinfection of, 21

Acholeplasma as cell line contaminant, 350

Acyrthosiphum pisum cell cultures from, 429, 430–431

Adenovirus(es)
  classification of, 296
  cytohybridization of, 477
  DNA extraction from, 305
  DNA fragment studies on, 492
  nucleic acid hybridization studies on, 489
  transfection studies on, 295, 298

Adenovirus-1, transfection of, 298

Adenovirus-2
  SV40 hybrid of, nucleic acid hybridization studies on, 475, 494
  transfection studies on, 295

Adenovirus-5
  DNA extraction from, 305
  transfection of, 312

Adoxophyes orana, cell line from, 431

Aedes albopictus
  cell lines from, 325, 345, 347, 348
    carrier cultures from, 379–381
    contamination of, 411
    CPE in, 373–374
    viral attenuation in, 382

Aedes albopictus (cont.)
    cell lines from: virus assays in, 362–363, 366, 367
    virus culture in, 327–338, 354, 356–360, 368–369, 371–372, 376–377

Aedes aegypti
  cell lines from, 325, 346, 348, 351
    virus culture in, 327–335, 338, 357, 360, 374

Aedes malayensis
  cell lines from, 325
    virus culture in, 328–329, 373

Aedes novo-albopictus, cell lines from, 325

Aedes pseudocutellaris
  cell lines from, 325
    virus culture in, 328–329, 354, 373

Aedes sollicitans cytoplasmic polyhedrosis virus, invertebrate cell culture of, 338

Aedes taeniorhynchus, cell lines from, 325

Aedes vexans
  cell lines from, 325, 351
    virus culture in, 327–331

Aedes vittatus
  cell lines from, 325virus culture in, 327

Aedes w-albus
  cell lines from, 325
    virus culture in, 327–329, 335

African horse sickness virus, invertebrate cell culture of, 333

Agallia constricta
  cell lines from, 325, 347, 401, 405, 407, 408, 411–413
    availability, 430
    indexing *in situ,* 415
    media for, 408–410
    storage of, 427
    virus culture in, 395, 424, 425
  egg collection from, 395–396
  embryos of, media for, 403

Agallia quadripunctata, cell lines from, 325, 401, 430

521